MARINE INTERFACES ECOHYDRODYNAMICS

Elsevier Oceanography Series, 42

MARINE INTERFACES ECOHYDRODYNAMICS

Edited by

J.C.J. NIHOUL

University of Liège, B5 Sart Tilman, B-4000 Liège, Belgium

ELSEVIER

Amsterdam — Oxford — New York — Tokyo 1986

ELSEVIER SCIENCE PUBLISHERS B.V.
Sara Burgerhartstraat 25
P.O. Box 211, 1000 AE Amsterdam, The Netherlands

Distributors for the United States and Canada:

ELSEVIER SCIENCE PUBLISHING COMPANY INC.
52, Vanderbilt Avenue
New York, NY 10017, U.S.A.

ISBN 0-444-42626-4 (Vol. 42)
ISBN 0-444-41623-4 (Series)

Printed in The Netherlands

FOREWORD

The International Liège Colloquia on Ocean Hydrodynamics are organized annually. Their topics differ from one year to another and try to address, as much as possible, recent problems and incentive new subjects in physical oceanography.

Assembling a group of active and eminent scientists from different countries and often different disciplines, they provide a forum for discussion and foster a mutually beneficial exchange of information opening on to a survey of major recent discoveries, essential mechanisms, impelling question-marks and valuable recommendations for future research.

The Scientific Organizing Committee and all the participants wish to express their gratitude to the Belgian Minister of Education, the National Science Foundation of Belgium, the University of Liège, the Intergovernmental Oceanographic Commission and the Division of Marine Sciences (UNESCO) and the Office of Naval Research for their most valuable support.

The editor is indebted to Dr. Jamart for his help in editing the proceedings.

Jacques C.J. NIHOUL

TABLE OF CONTENTS

LIST OF PARTICIPANTS

ALLANSON, B., Prof. Dr., Department of Zoology, Rhodes University, South Africa

BAEYENS, W.F., Dr., ANCH-Wetenschappen, Vrije Universiteit te Brussel, Belgium

BAH, A., Dr., Département de Biologie, Université de Laval, Canada

BILLEN, G., Dr., Laboratoire de Chimie Industrielle, Université Libre de Bruxelles, Belgium

BOUKARY, S., Dr., University of Niamey, Niger

BOUQUEGNEAU, J.M., Dr., Laboratoire d'Océanologie, Université de Liège, Belgium

BRANDT, A., Dr., Applied Physics Laboratory, The Johns Hopkins University, USA

BRUNDRIT, G.B., Prof. Dr., Department of Oceanography, University of Cape Town, South Africa

BUTMAN, C.A., Dr., Ocean Engineering Department, Woods Hole Oceanographic Institution, USA

CAREY, D.A., Dr., Department of Earth and Environmental Sciences, Wesleyan University, USA

CARTER, R.A., Dr., National Research Institute for Oceanology, South Africa

CHABERT D'HIERES, G., Ing., Institut de Mécanique de Grenoble (I.M.G.),France

CLEMENT, F., Mr., GeoHydrodynamics and Environment Research (GHER), University of Liège, Belgium

COACHMAN, L.K., Prof. Dr., School of Oceanography, University of Washington, USA

CZITROM, S., Dr., Instituto de Ciencias del Mar y Limnologia, Ciudad Universitaria, Mexico

DELEERSNIJDER, E., Ing., GeoHydrodynamics and Environment Research (GHER), University of Liège, Belgium

DEMERS, S., Dr., Centre Champlain des Sciences de la Mer, Canada

DENMAN, K.L., Dr., Institute of Ocean Sciences, Canada

DIEBEL-LANGOHR, D., Mrs., Fachbereich 8 - Physik, Universität Oldenburg, Germany

DIETERLE, D., Mr., Department of Marine Sciences, University of South Florida, USA

DISTECHE, A., Prof. Dr., Laboratoire d'Océanologie, Université de Liège, Belgium

DJENIDI, S., Ing., GeoHydrodynamics and Environment Research (GHER), University of Liège, Belgium

DUPOUY, C., Miss, Antenne ORSTOM, Centre de Météorologie Spatiale, France

ESTRADA, M., Dr., Instituto de Investigaciones Pesqueras de Barcelona, Paseo Nacional, Spain

EVERBECQ, E., Ing., GeoHydrodynamics and Environment Research (GHER),University of Liège, Belgium

FLEBUS, C., Mr., GeoHydrodynamics and Environment Research (GHER), University
    of Liège, Belgium

FORTIN, M.J., Miss, Université de Montréal, Canada

FRANKIGNOULLE, M., Mr., Laboratoire d'Océanologie, Université de Liège, Belgium

FRONTIER, S., Prof. Dr., Université des Sciences et Techniques de Lille, France

GALLARDO, Y., Dr., Antenne ORSTOM/IFREMER, France

GLAS, P., Mr., Delft Hydraulics Laboratory, The Netherlands

GODEAUX, J., Prof. Dr., Institut de Zoologie, Université de Liège, Belgium

GUNTHER, K., Dr., Fachbereich 8 - Physik, Universität Oldenburg, Germany

HAPPEL, J.J., Ing., GeoHydrodynamics and Environment Research (GHER), University
    of Liège, Belgium

HARRIS, G.P., Dr., CSIRO Fisheries Research, Marine Laboratories, Tasmania

HECQ, J.H., Dr., GeoHydrodynamics and Environment Research (GHER), University
    of Liège, Belgium

HEIP, C., Dr., Marine Biology Section, Zoology Institute, University of Gent,
    Belgium

HERMAN, P., Dr., Marine Biology Section, Zoology Institute, University of Gent,
    Belgium

HUERTA, M.A., Dr., Consejo Nacional de Ciencia y Tecnologia (CONACYT), Ciudad
    Universitaria, Mexico

HUTCHINGS, L., Dr., Sea Fisheries Research Institute, South Africa

INGRAM, R.G., Prof. Dr., Department of Oceanography, Mc Gill University, Canada

JAMART, B.M., Dr., Unité de Gestion du Modèle Mathématique Mer du Nord et
    Estuaire de l'Escaut (UGMM), Institut de Mathématique, Belgium

JOIRIS, C., Dr., Laboratorium voor Ekologie en Systematiek, Vrije Universiteit
    te Brussels, Belgium

KLEIN, P., Dr., Laboratoire d'Océanographie Physique, Université de Bretagne
    Occidentale, France

KRAUSE, G., Prof. Dr., Institut für Meeresforschung, Germany

LEBON, G., Prof. Dr., Irreversible Thermodynamics, University of Liège, Belgium

LEGENDRE, L., Prof. Dr., GIROQ, Département de Biologie, Université de Laval,
    Canada

LEWIS, M.R., Dr., Department of Oceanography, Dalhousie University, Canada

MASO, M.A., Miss, Instituto de Investigaciones Pesqueras de Barcelona, Paseo
    Nacional, Spain

MONREAL, A., Mrs, Consejo Nacional de Ciencia y Tecnologia (CONACYT), Ciudad
    Universitaria, Mexico

MOUCHET, A., Miss, GeoHydrodynamics and Environment Research, (GHER), University
    of Liège, Belgium

NIHOUL, J.C.J., Prof. Dr., GeoHydrodynamics and Environment Research (GHER),
    University of Liège, Belgium

PARTHENIADES, E., Prof. Dr., Department of Engineering Sciences, University of
    Florida, USA

PHINNEY, D.A., Mr., Bigelow Laboratory for Ocean Sciences, USA

PICHOT, G., Dr., Unité de Gestion du Modèle Mathématique Mer du Nord et Estuaire
    de l'Escaut (UGMM), Belgium

QUADER, Md. O., Mr., Bangladesh Space Research and Remote Sensing Organization
    (SPARRSO), Bangladesh

REES, J.M., Mr., Fisheries Laboratory, Ministry of Agriculture, Fisheries and
    Food, UK

REUTER, R., Dr., Fachbereich 8 - Physik, Universität Oldenburg, Germany

ROBLES, F.L.E., Dr., Intergovernmental Oceanographic Commission, UNESCO, France

RONDAY, F.C., Dr., GeoHydrodynamics and Environment Research (GHER), University
    of Liège, Belgium

RYDBERG, L., Mr., Institute of Oceanography, University of Gothenburg, Sweden

SALAS, D.A., Mr., Consejo Nacional de Ciencia y Tecnologia (CONACYT), Ciudad
    Universitaria, Mexico

SMITZ, J., Ing., GeoHydrodynamics and Environment Research (GHER), University
    of Liège, Belgium

SPIES, A., Dr., Alfred-Wegener-Institute for Polar Research, Germany

SPITZ, Y., Miss, Unité de Gestion du Modèle Mathématique Mer du Nord et Estuaire
    de l'Escaut, Institut de Mathématique, Belgium

SCHLITTENHARDT, P., Commission of the European Communities, Joint Research
    Center, Italy

STIGEBRANDT, A., Dr., Department of Oceanography, University of Gothenburg,
    Sweden

SUNDBERG, J., Dr., Institute of Oceanography, University of Gothenburg, Sweden

TANKE, M., Dr., Elsevier Scientific Publishing Company, The Netherlands

TAUPIER-LETAGE, I., Miss, Centre d'Océanologie de Marseille, Centre
    Universitaire de Luminy, France

TAYLOR, A.H., Mr., Institute for Marine Environmental Research, UK

THERRIAULT, J.Cl., Dr., Centre Champlain des Sciences de la Mer, Canada

TOPLISS, B.J., Dr., Department of Fisheries and Oceans, Bedford Institute of
    Oceanography, Canada

TIJSSEN, S.B., Mr., Netherlands Institute for Sea Research, The Netherlands

VALKE, A., Mr., GeoHydrodynamics and Environment Research (GHER), University
    of Liège, Belgium

VAN HEIJST, G.J.F., Dr., Institute for Meteorology and Oceanography, The
    Netherlands

VETH, C., Dr., Nederlandse Instituut voor Onderzoek der Zee (NIOZ), The Netherlands

WALEFFE, F., Ing., GeoHydrodynamics and Environment Research (GHER), University of Liège, Belgium

WHITLEDGE, T.E., Dr., Oceanographic Sciences Division, Brookhaven National Laboratory, USA

WOODS, J.D., Prof. Dr., Institut für Meereskunde, Universität Kiel, Germany

BIOLOGICAL PRODUCTION AT MARINE ERGOCLINES*

L. LEGENDRE[1], S. DEMERS[2] and D. LEFAIVRE[2]

1 Département de biologie, Université Laval, Québec, Québec G1K 7P4 (Canada)

2 Centre Champlain des sciences de la mer, C.P. 15500, 901 Cap Diamant, Québec, Québec G1K 7Y7 (Canada)

ABSTRACT

Ergoclines are aquatic interfaces which have the common characteristic of involving spatial and/or temporal gradients where physical processes can produce structures associated with enhanced biological production. Biological production is taken here as the storage of primary (solar) energy by autotrophs and its transfer to or among heterotrophs. It is hypothesized that enhanced biological production occurs at ergoclines as the consequence of the matching or resonance of physical scales with biological scales. Physical scales act on biological production through the proximal agency of resources. High biological production most often exhausts the resources available at the lower trophic level; auxiliary energy on proper scales makes possible the replenishment of these limiting resources, and thus biological production at aquatic ergoclines. It is considered that transitions on various time scales (annual, meteorological, tidal and so on) and spatial structures such as the ice-water and bottom-sediment interfaces, the pycnocline, tidal fronts, and others, all belong to the general category of marine ergoclines. Hypotheses as to the mechanisms that govern biological production at ergoclines are developed, and the conditions to test these hypotheses at sea are discussed.

THE ERGOCLINE HYPOTHESIS

Legendre and Demers (1985) have proposed that the input of mechanical energy in the aquatic environment not only improves biological production but is an essential requirement for it. This hypothesis has evolved from energetic considerations. The energy that is stored by photosynthetic organisms

---

* Contribution to the program of GIROQ (Groupe interuniversitaire de recherches océanographiques du Québec)

and subsequently flows through aquatic ecosystems is called by ecologists "primary energy". The primary source of energy for ecosystems is the photosynthetically active radiation of the sun. In addition to primary energy, the productivity of marine ecosystems is influenced by the input of mechanical energy caused by winds, tides, freshwater runoff, air-ocean heat exchanges, and so on, called "auxiliary energy". Auxiliary energy is not directly used by living organisms, but it is efficient in increasing the storage of primary (solar) energy by the photosynthetic organisms (autotrophs), and the transfer of this stored energy to heterotrophs or between the heterotrophic components of the food web. Biological production is taken here as the flow of primary energy through ecosystems. The observed spatio-temporal variations in marine biological production are much more related to the space and time distributions of auxiliary energy than to those of primary energy. For example, Margalef and Estrada (1980) have shown that the overall distribution of phytoplankton in the oceans corresponds to that of air-sea heat exchanges. A similar conclusion has been reached by Odum (1980) for the auxiliary energy of tides in estuaries.

Legendre (1981) has proposed that phytoplankton blooms generally occur at the spatio-temporal transition from unstable to stable conditions. This model is verified in the marine environment, where several spatial and/or temporal interfaces are highly productive. Tidal fronts, nutriclines, ice-water and water-sediment interfaces, temporal transitions in vertical stability of the water column on various scales (annual, meteorological, tidal), and so on, have the common characteristic of involving spatial and/or temporal gradients where physical processes can produce structures associated with enhanced biological production. According to Legendre and Demers (1985), all these structures are spatial and/or temporal gradients in auxiliary energy, which they termed "ergoclines" (εργον: work).

The above definition of ergoclines leads to a general hypothesis:

Enhanced biological production occurs at ergoclines as the consequence of the matching or resonance of physical scales with biological scales.

Before exploring some of its implications, the very concept of ergoclines and their associated high biological production will be confronted to data from a number of marine systems. Various types of ergoclines will be briefly reviewed. Among the temporal ergoclines, the annual, meteorological and semidiurnal scales will be considered. In space, such horizontal ergoclines as the nutricline and the water-sediment interface will be examined. Tidal fronts, as examples of vertical ergoclines, will also be investigated.

ERGOCLINES IN THE MARINE ENVIRONMENT

The spring phytoplankton bloom (Fig. 1a) is probably the best known biological response to the physical scales of an aquatic ergocline. According to the "critical depth" model (Sverdrup, 1953), the spring bloom cannot occur before the mixed layer shallows to such a depth that carbon fixation by the phytoplankton exceeds respiration per unit area. The bloom is therefore determined by the balance between decreased vertical mixing and increased solar radiation (Riley, 1942). However, changes in solar radiation do not play a major role except in very high latitudes, since increased vertical stability of the water column at mid-latitudes results in a phytoplankton bloom even in wintertime (e.g. the Canadian Scotian Shelf: Fournier et al., 1979). Thus, the main factor of the spring phytoplankton bloom is the vertical stability of the water column.

Increased vertical stability of the water column, at the end of the winter in temperate regions, can be caused by either higher surface temperatures or lower salinities. In the oceans, higher temperatures result from air-sea heat exchanges. In coastal waters, reduced surface salinities have been associated with the spring phytoplankton bloom. For instance, Legendre et al. (1981) have proposed that phytoplankton blooms under the sea ice probably result from the deepening of the photic layer (seasonal increase of under-ice irradiance) combined with the increased stratification caused by the low-salinity melting water. Similarly, the phytoplankton bloom near a receding ice edge in the Ross Sea (Antarctica) was attributed by Smith and Nelson (1985) to the enhanced stability brought about by the melting of the ice. Another source of freshwater in coastal areas is river runoff, which can cause increased vertical stability of the water column and has thus been invoked by Sinclair et al. (1981) to explain why several estuaries bloom earlier than adjacent water bodies. In other estuaries, which are blooming later than the adjacent waters, the shortened residence time of the surface layer within the estuary during the freshet would prevent the initiation of an early bloom. Similarly, in such environments as Indian Arm (a fjord of Northwestern Canada: Gilmartin, 1964), the increased river runoff destabilizes the water column, so that a phytoplankton bloom only occurs upon reduction of the freshwater flow. Whatever the actual mechanism, the spring phytoplankton bloom only occurs when there is matching or resonance of physical scales with biological scales.

At a shorter time scale, that is the scale of a few days that corresponds to the passage of frontal disturbances (Heath, 1973; Walsh et al., 1977), physical transient phenomena (wind storms, intermittent upwelling, etc.) are causing aperiodic inputs of auxiliary energy, and thus of nutrients, in the water column. Intermittent phytoplankton blooms (Fig. 1b), that follow

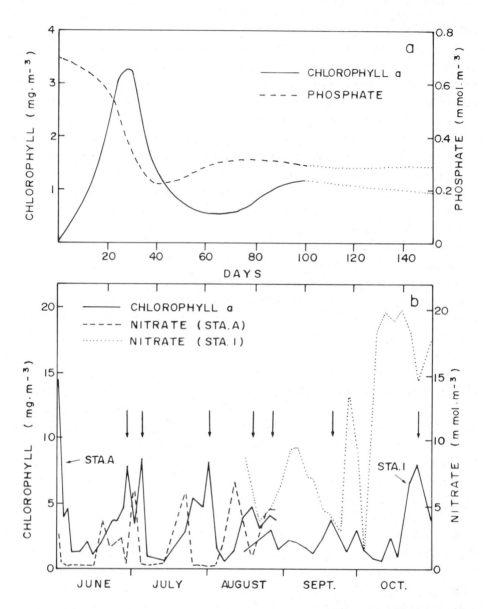

Fig. 1 Changes in biomass (chlorophyll) and in limiting nutrients, associated with phytoplankton blooms on two time scales: (a) The spring bloom (Riley, 1963: model of Steele, 1958, using a C:Chl ratio of 100). (b) Summer blooms (Takahashi et al., 1977; the arrows indicate the blooms); these blooms follow sudden increases of nitrate, that are caused by either winds or fortnightly tides.

stabilization of a water column previously destabilized by strong winds, have been reported in several studies (Iverson et al., 1974; Takahashi et al., 1977; Walsh et al., 1978; Legendre et al., 1982). At an even shorter time scale, Fortier and Legendre (1979) and Fréchette and Legendre (1982) have reported, in the lower St. Lawrence Estuary, bursts of phytoplankton photosynthetic activity and biomass that occurred on the semidiurnal cycle of tidal destratification.

The mechanisms for phytoplankton blooms on these various time scales are the same. As explained by Legendre and Demers (1985), increased vertical mixing may have several effects on phytoplankton: (1) increased loss rate of cells from the photic layer, (2) lowered photosynthetic activity and production due to the deepening of the mixed layer (light limitation of the vertically mixed cells), and (3) nutrient replenishment of the mixed layer. According to the production model of Legendre (1981), a phytoplankton bloom occurs upon stabilization of the water column, on any time scale. This bloom will last as long as nutrients do not become limiting, which depends on the duration of the stable phase. When the stable phase is long enough to result in nutrient limitation (Fig. 2a), nutrients are utilized at the beginning of the stable phase, after which photosynthetic activity is limited by the rate of in situ nutrient regeneration. During the next unstable phase, photosynthesis becomes limited by light. When the duration of stable periods is short, nutrient limitation of the phytoplankton seldom occurs (Fig. 2b). On the contrary, in environments where the input of auxiliary energy is high and persistent (ergoclines absent or very small), hydrodynamics act on phytoplankton through the sole agency of light, hence circadian cycles of phytosynthesis and possible light limitation (Legendre and Demers, 1985). This general mechanism of phytoplankton blooms, that applies to all the time scales (Fig. 1), leads to the following conclusions: (1) Physical scales do not act directly on phytoplankton, but rather through the proximal agency of light and nutrients (Table 1; the concept of "proximal agents" will be further developed below). (2) A phytoplankton bloom generally occurs at temporal ergoclines, whatever their periodicities (annual, meteorological, semidiurnal, or other). This makes phytoplankton blooms a very general biological response to physical processes.

As explained by Legendre and Demers (1985), animals can also respond to temporal changes in physical scales. On an interannual scale, Sutcliffe (1972, 1973) has related the landings of several commercial species in the Gulf of St. Lawrence to variations in St. Lawrence River discharge years before. This may be explained by the influence of river discharge, in the springtime, on the production of plankton. Similarly, the early-survival

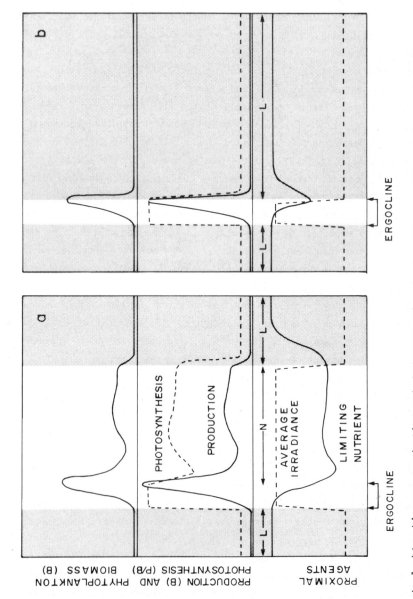

Fig. 2 Phytoplankton biomass, production and photosynthesis at schematic temporal ergoclines, in response to variations of nutrients and average irradiance (proximal agents) in the mixed layer, caused by the input of auxiliary energy (shaded). (a) Stable phase (auxiliary energy low or null) is long enough to result in nutrient limitation. (b) Stable phase is short enough to prevent nutrient limitation. L: light limitation; N: nutrient limitation. Adapted from Riley (1963, Fig. 5) and Legendre and Demers (1985, Fig. 2).

approach in fisheries biology hypothesizes that the recruitment of commercial species depends on the success or failure of the annual colonization of the environment by the larvae. Most of the hypotheses explain the observed variations in recruitment, and thus in early survival, by the availability of suitable food (e.g. Pearcy, 1962; Cushing, 1972; Lasker, 1975). As recognized by Fortier (1982) and Legendre and Demers (1984), food is the proximal agent through which physical scales act on larvae (Table 1). On a much shorter time scale, Legendre and Demers (1985) have suggested that the metabolic activity of such pelagic organisms as zooplankton perhaps responds to semidiurnal tidal variations in auxiliary energy.

The dynamics of vegetal and animal components of aquatic ecosystems can therefore be related to temporal changes in the physical scales. In all cases, one or several "proximal agents" are involved, and increased biological production occurs at ergoclines.

Ergoclines do not only exist along the time axis, and they are encountered as well in space. Spatial ergoclines can be either horizontal or vertical. Among the horizontal ergoclines, the water-sediment interface and the nutricline will now be briefly reviewed. A third type of horizontal ergoclines, the ice-water interface, is discussed by Demers et al. (this book) and by Lewis (this book).

The water-sediment interface, among other interfaces, has been recognized by microbiologists as a preferred biotic habitat (Marshall, 1976). As reported by Legendre et al. (1985b), there is, at the water-sediment interface, "a sharp gradient of kinetic energy which enables organisms to utilize the power of fluid motion for their mechanical work (e.g. for pumping dissolved or particulate substances, Nixon et al. 1971). Sharp gradients in heat or chemical concentrations across these interfaces create potentials which facilitate the transport of food and wastes via convective and diffusive mechanisms, thus stimulating metabolic processes."

A good example of the dynamics at the water-sediment ergocline is provided by a series of studies on the grazing of phytoplankton by mussels in the St. Lawrence Estuary. Fréchette and Bourget (1985) found that vertical depletion of particulate organic matter occurs frequently immediately above the mussel bed, and that the resulting vertical gradients can be destroyed by waves and currents, and even inverted by the inputs of mechanical energy of waves. This supports the idea of Wildish and Kristmanson (1979) that food is often depleted immediately above suspension-feeder populations, and that hydrodynamic processes are critical in determining food availability. Fréchette (ms.a) demonstrated that vertical turbulent diffusion indeed results in downwards transport of phytoplankton. Both current velocity and bottom roughness influence vertical diffusion. Low current velocity obviously results in poor downwards transport of phytoplankton, which leads to depletion of phytoplankton above the mussel bed and ultimately to reduced mussel growth. On the other hand, the very development of the mussel bed increases bottom roughness and thus vertical turbulent diffusion. The scales of physical processes at the water-sediment ergocline therefore control the production of suspension feeders.

Subsurface chlorophyll maxima are often observed in the sea, and their significance has been reviewed by Cullen (1982). He first cautions the reader against the usual interpretation of chlorophyll vertical profiles as indices of phytoplankton biomass: some heterogeneities in the vertical distributions of chlorophyll may well reflect variable chlorophyll content of phytoplankton, and not changes in biomass. In shallow seas and on the continental shelf of temperate regions, subsurface chlorophyll maxima occur at the nutricline, often within short distance of the pycnocline. A possible explanation is that sinking phytoplankton become "trapped" in the pycnocline. Another explanation relies on the fact that the upper part of the water column is a two-layer system, with a nutrient-depleted surface layer and deeper water in which light becomes rapidly limiting (Dugdale, 1967). Maximum phytoplankton biomass is often observed at the boundary between the two layers. This suggests that subsurface chlorophyll maxima might be a biological response to an ergocline, between the upper well-mixed layer and the more stable underlying waters.

Comparing the nutricline to one of the ergoclines discussed above (for instance, a temporal ergocline) leads to an interesting conclusion (Fig. 3). In the nutricline configuration, the two proximal agents (light and nutrients) are inverted relative to the temporal ergocline. Despite this fact, maximum biomass (and probably production) seems to occur at the ergocline, in both systems. This supports the idea that biological production responds primarily to physical scales. The subsurface chlorophyll maxima discussed here thus occur at the depth at which both proximal agents (light and nutrients) are not limiting. This is similar to what has been found above for other ergoclines. Whether this same depth also corresponds to a zone of high stability remains to be demonstrated in most cases. Pingree et al. (1975) have explained the development of high concentrations of dinoflagellates, within sharp pycnoclines near tidal fronts, by a longer characteristic mixing time of the water. Holligan et al. (1984) report that dinoflagellate maxima in the Gulf of Maine occur below the stability maximum (pynocline) but are centered in regions of zero to slightly positive Brunt-Vaïsälä frequencies. Lewis et al. (1983) show that the very presence of a subsurface chlorophyll maximum causes a differential heating of the water column, with the result that the depth stratum

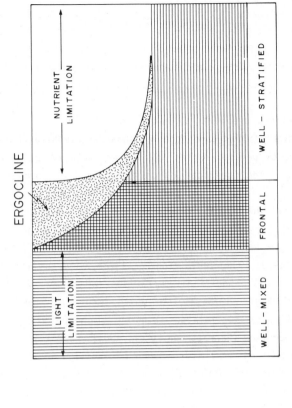

Fig. 4 Surface and subsurface chlorophyll maxima in the ergocline of a tidal front. Nutrients are abundant in the well-mixed waters, in the frontal zone and under the pycnocline of the well-stratified water column. Schematized from Pingree et al. (1975) and Demers et al. (1985).

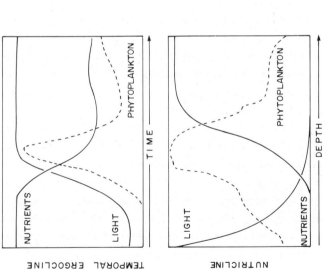

Fig. 3 Light (temporal ergocline: average irradiance in the mixed layer; nutricline: irradiance); concentration of limiting nutrient(s) and phytoplankton biomass (chlorophyll) at temporal and horizontal ergoclines. Temporal ergocline: see Figs. 1a and 2; nutricline: schematized from Holligan et al. (1984, Fig. 5).

containing the deeper portion of the chlorophyll maximum becomes increasingly more stable with time. This increased stability may permit the maintenance and development of the subsurface chlorophyll maximum. It is therefore possible that at least some subsurface chlorophyll maxima be other examples of increased biological production in resonance with physical scales at an horizontal ergocline.

Other spatial ergoclines are vertical rather than horizontal. Among the vertical ergoclines, tidal fronts have received considerable attention in the last decade. Such fronts develop during the summer, between well-mixed and well-stratified waters. Their effects on phytoplankton have been recently reviewed by Demers et al. (1985). In shallow seas, the high phytoplankton biomass associated with the fronts (Fig. 4) have been explained by the fortnightly tidal excursions of these fronts, which contribute to periodically replenish nutrients in the waters on the stable side of the front (Pingree et al., 1975, 1976, 1977, 1978, 1982, 1983; Pingree, 1978a; Simpson and Pingree, 1978; Parsons et al., 1983). Additional mechanisms for cross-frontal mixing are cyclonic eddies (Pingree 1978a,b), to which horizontal phytoplankton distributions were found to be related (Pingree et al., 1979), and the frictionally induced mean flow (Garrett and Loder, 1981). These mechanisms of frontal nutrient enrichment do not apply, however, to shallow areas where both sides of the tidal front are nutrient limited and where the high phytoplankton biomass associated with the front cannot therefore be explained by cross-frontal mixing (Perry et al., 1983).

High phytoplankton biomasses do not necessarily reflect high phytoplankton production, since mechanical aggregation of the cells does also result in high biomasses. Both indirect evidences (Savidge, 1976; Holligan, 1981; Tett, 1981) and direct measurements of primary production (Parsons et al., 1983) indicate that nutrients entering the frontal zone from the well-mixed waters are the driving force for new phytoplankton production. It has been stressed by Demers et al. (1985) that the frontal ergocline coincide with two disconti-nuities in stability: (1) the frontal transition, which is located between well-mixed and well-stratified waters, and (2) the temporal transition, that occurs from spring to neap tides.

Vertical fronts have also been associated with high animal concentrations. For example, Iles and Sinclair (1982) and Sinclair and Iles (1985) have proposed that yearclass variability in Atlantic herring is largely determined by the confinement of larvae in "larval retention areas". These areas lie to a large extent within well-mixed zones bounded by temperature fronts.

Tidal fronts are often identified by contouring the horizontal distribution of s, the Simpson and Hunter's (1974) stratification parameter (e.g. Garrett et al., 1978; Pingree and Griffiths, 1978, 1980; Bowman et al., 1980; Bowman and Esaias, 1981; Griffiths et al., 1981). In order to explain phytoplankton abundances in terms of both stratification (s) and water column illumination, Pingree et al. (1978) have suggested to plot these abundances in the s-kh diagram, where k is the light extinction coefficient and h is the depth of the station. This has been tried by Bowman et al. (1981), for Long Island Sound, and by Bah and Legendre (1985), for the Middle St. Lawrence Estuary. In agreement with the ergocline hypothesis, the latter found that high biomasses were concentrated in the marginally stable part of the transition zone of the s-kh diagram, between well-mixed and well-stratified waters.

The hypothesis that high biological production occurs chiefly at aquatic ergoclines also applies to the coastal zone (Fig. 5), where high autotrophic production can occur as long as nutrients do not become limiting (Demers et al., ms). In the coastal zone, the depth of the well-mixed water column is shallow enough to prevent light limitation of photosynthesis, so that it is an environment providing suitable spatio-temporal scales for primary production. Comparison of the coastal zone and the nutricline shows that, as far as the ergocline is concerned, the two systems are homologous.

The examples discussed above, of temporal and spatial ergoclines, do stress two major concepts: (1) Where energetics is concerned, biological production at a given trophic level is the same as the flow of primary energy into this trophic level. (2) This flow of energy can be limited by the availability of resources at the previous trophic level. These resources are the proximal agents, which have been mentioned above and through which physical scales act on biological production. High biological production most often leads to the local (temporal and/or spatial) exhaustion of the previous trophic level. The input of auxiliary energy on proper scales makes possible the replenishment of the exhausted resources, and thus biological production at the ergocline. Outside the ergocline, the scales of the physical structures are either too large or too small to have any direct influence on biological production, or the physical structures are on such scales that they impede the production.

In the case of autotrophs (phytoplankton, etc.), the limiting resources (proximal agents) can be either light (primary energy) or nutrients (limiting materials). For heterotrophs, primary energy and materials are combined into food resources, so that there is generally a single class of proximal agents (e.g. food) (Table 1).

Making the general ergocline hypothesis (first section, above) operational at sea rests on the ability for biological oceanographers to specifiy the biological scales which are conducive to high production, and for physical oceanographers to identify corresponding scales in the physical environment.

TABLE 1

Proximal agents through which auxiliary energy acts on the various trophic
levels.

Trophic level (examples)	Proximal agents
Primary (phytoplankton, etc.)	Light and nutrients
Secondary (herbivorous zooplankton, fish larvae, molluscs, etc.)	Organic particles (phytoplankton, detritus, etc,)
Carnivory (fishes, etc.)	Preys

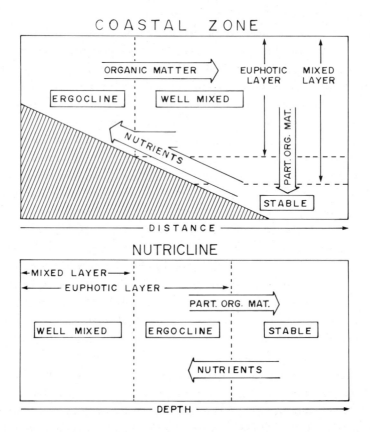

Fig. 5 Schematic representations of a coastal zone where nutrients are
nonlimiting, and of a nutricline with a subsurface chlorophyll maximum.
Fluxes of nutrients and of organic matter; in the coastal zone, the fluxes can
also be along the shore. In both systems, nutrients flow from the stable
reservoir to the ergocline, where autotrophic production is maximum, and the
particulate organic matter ultimately sinks into the stable reservoir where it
is mineralized.

Specifying the spatio-temporal scales of the most significant biological processes in oceans has been attempted by various authors, among which Haury et al. (1978) and Harris (1980). As far as the transfer of primary energy in ecosystems (i.e. biological production) is concerned, the physical structures of interest must (1) in the time domain, be sufficiently active to resupply the resources depleted by biological production and persist long enough to allow some accumulation of biomass, and (2) in the space domain, have such an extension as to sustain biological activity of the trophic level under consideration. As explained just above, physical structures that are outside the specified range of scales either do not have any direct influence on biological production or, on the contrary, do impede it.

As examples, possible characteristic time and space scales are given in Table 2 for phytoplankton, zooplankton, actively growing fishes and mussels. The range of physical scales matching these biological scales obviously extends above and below the characteristic values of Table 2. In the time domain, for instance, Harris (1980) suggests that the characteristic time period for each phytoplankton cell is the generation time, that is the period over which the environment must be integrated while one cell grows and divides in two. Phytoplankton growth requires replenishment of nutrients by vertical processes, which occur in the oceans on time scales between a few hours and a few days; such replenishment must also persist one or two weeks, for significant phytoplankton biomass to develop. The range of physical scales matching phytoplankton time scales therefore encompasses the mean phytoplankton doubling time. Similarly for mussels, vertical gradients in particulate organic matter above the bed must be destroyed by waves and currents several times every hour, for the mussels to grow; at the other end of the time range, it takes several months of renewed food supply before a mussel bed becomes established. The same applies mutatis mutandis to zooplankton and actively growing fishes, with the added complexity that the periodicity of resources uptake can be partially controlled by the animals themselves through vertical and/or horizontal migrations.

In the space domain, on the other hand, only the production of those organisms that cannot vertically migrate (e.g. some phytoplankters, and also benthic organisms) is critically dependent on the vertical extent of the physical structures. Vertical mixing must maintain phytoplankton within the photic layer for a bloom to occur, and mussels cannot grow if food particles are not continuously resupplied in the benthic boundary layer. This is not to say that vertically migrating organisms are not influenced by the vertical scales of such physical structures as the pycnocline, internal waves, and so

TABLE 2

Characteristic time and space scales of various groups of marine organisms.

	Phytoplankton	Zooplankton	Fishes	Mussels
Temporal: Mean doubling time (days) of the biomass	1[a]	10-40[a]	100-900[b]	120-500[c]
Horizontal (km): Characteristic scales of patches, swarms, schools, etc.	0.1-1[d]	0.1-1[e]	1-100[e]	?
Vertical: Extent of the physical control (m)	5-50[f]	---	---	0.5[g]

[a] Parsons (1980).
[b] Banse and Mosher (1980).
[c] Dare (1976).
[d] Harris (1980), Legendre and Demers (1984).
[e] Haury et al. (1978).
[f] Depth of the photic layer.
[g] Thickness of the particle depleted benthic layer: Fréchette (ms.b).

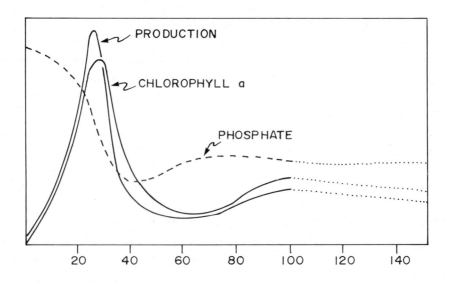

Fig. 6 Relationships between phytoplankton production and biomass, and a limiting nutrient, at an ergocline (see Fig. 1a).

on, but it does not seem that their production is critically dependent on these vertical scales.

On the horizontal, all the organisms show characteristic spatial scales of organization (Table 2). These result from the interplay between biological production and the environment. Phytoplankton patches are often explained by the "KISS" model (Kierstead and Slobodkin, 1953; Skellam, 1951), as the result of both horizontal diffusion and phytoplankton growth. This model has been modified by Platt and Denman (1975) and Wroblewski et al. (1975) to include the effect of zooplankton grazing. Alternatively, Riley (1976) proposed that phytoplankton patchiness is caused by differential grazing, resulting from the interaction of zooplankton vertical migrations with tides and residual drift. It is noteworthy that the characteristic space scale of phytoplankton patches (1 km: Table 2) is equivalent to a time scale of about 1 day (Harris, 1980), which corresponds to the mean doubling time of phytoplankton (Table 2). The explanatory mechanisms of zooplankton heterogeneous distribution generally refer to some physical structures or biological properties or a combination of both physical and biological factors (Legendre and Demers, 1984). This is also true for herring larvae and juveniles (Iles and Sinclair, 1982; Sinclair and Iles, 1985). The horizontal scales of the physical structures are therefore significant as to the production of several trophic levels in the oceans.

The general ergocline hypothesis, which explains enhanced biological production as the consequence of the matching or resonance of physical scales with biological scales can therefore be operationally applied to various temporal and/or spatial ergoclines. This indicates that hypotheses can be developed with respect to the common characteristics of physical structures and biological production at marine ergoclines. These hypotheses could provide oceanographers with an unified biological-physical approach to the hydrodynamic mechanisms that control biological production in the sea.

THE RESOURCES-ERGOCLINE DIAGRAM

In the field, there is no coincidence between maximum resources, maximum biological production and maximum biomass. This is because biological production uses the resources to build up biomass. In the model of Figs. 1a and 6, it is assumed that, as long as the resources remain nonlimiting, production depends only on the accumulated biomass [production = biomass x specific production rate]; when the resources become limiting, production is assumed to be proportional to the concentration of resources [production = biomass x f (resources)]. As a result, maximum production and biomass occur somewhere between the maximum and minimum values observed for the resources. In addition, maximum biomass generally follows maximum production (Fig. 6).

Studying the spatial or temporal distributions of either production or biomass cannot therefore be used to locate ergoclines at sea.

Ergoclines correspond to gradients of physical scales in space and/or time. Sustained biological production, at ergoclines, requires high-rate replenishment of the resources. This occurs where and when the time scales of the physical processes are short. As the time scale lengthen, the increasingly lower rate of supply of the resources progressively limits the biological production. An operational approach to ergoclines, at sea, would be therefore to look at resources. Plotting resources along the observation axis results in the Resources-Ergocline (R-E) diagram, where ergoclines are easily identified (Fig. 7, bottom panels). In the R-E diagram, the observed values of each resource are standardized by the level of the resource that limits biological production. $(R/R_{lim}) = 1$ is then the threshold between productive (> 1) and nonproductive (< 1) waters, which is true for any resource since (R/R_{lim}) is dimensionless. When there are several resources, only the smallest of the standardized values is plotted in the R-E diagram, since a single limiting resource is enough to limit the biological production. Specific cases of spatial ergoclines will now be examined using the R-E diagram.

In the coastal zone (Figs. 5 and 7), the production of autotrophs (kelp beds, benthic microalgal mats, or phytoplankton) is generally limited by light, not by nutrients. Both bottom irradiance (which is critical for benthic algae) and the average irradiance in the mixed layer (for the vertically mixed phytoplankton) depend on water depth. As a result, the coastal ergocline is defined, for autotrophs, by the offshore gradient in irradiance, that is water depth. Nutrients do not become limiting in the coastal zone when the time scale of water movements (advection, etc.) is short enough to meet the nutrient requirements of autotrophs (Fig. 7, top panel). On the other hand, there is no light limitation of primary production as long as the dimensionless space scale (depth of the photic layer / depth of the mixed layer, to which phytoplankton is known to respond: Harris, 1978) remains above a critical value. When the depth of the mixed layer exceeds that of the photic layer, light limitation progressively develops, which sets the offshore limit of the coastal ergocline (Fig. 7, bottom panel).

At tidal fronts, the situation is somewhat more complex, since phytoplankton production can be limited by light and also by nutrients (Figs. 4 and 7). The frontal ergocline corresponds to changes in both the space and time scales (Fig. 7, top panel). On the inshore side of the tidal front, the depth of the mixed layer generally exceeds that of the photic layer (dimensionless space scale < 1), so that phytoplankton cells can be light limited; simultaneously,

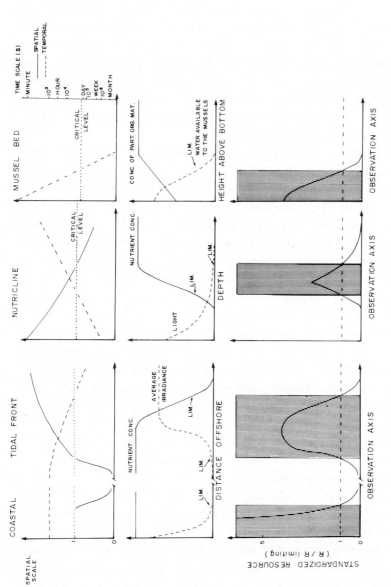

Fig. 7 Top panels: schematic changes in the space and time scales of physical energy inputs at four different spatial ergoclines. In the coastal zone and at tidal fronts, the dimensionless space scale is [depth of the photic layer / depth of the mixed layer], while for the nutricline it is [depth of the photic layer / depth in the water column]. Intermediate panels: schematic changes in limiting resources. Bottom panels: Resources-Ergocline (R-E) diagrams corresponding to the panels above. In these diagrams, standardized resources (observed/limiting) are plotted along the observation axis; values > 1 delimitate the ergocline (shaded). When there are several resources, the smallest of the standardized values is plotted.

the time scale of vertical water movements is short enough to maintain high nutrient concentrations in the mixed layer. On the offshore side of the tidal front, the dimensionless space scale exceeds 1, so that phytoplankton cells are kept in the photic layer and thus do not experience light limitation; hovever, as the time scale of vertical water movements increases up to a few hours, phytoplankton become progressively nutrient limited. The frontal ergocline corresponds to the region where both the physical time and space scales match the physiological scales of the phytoplankton. At the tidal front ergocline, where both resources exceed their limiting values (R-E diagram), phytoplankton production and biomass are often high. Whether this primary production is actually transferred, at the ergocline, to other trophic levels is still a subject of active research.

The nutricline (Figs. 3 and 7) is located between an upper layer where phytoplankton is often nutrient limited, and nutrient-rich deep waters where primary production is light limited. Nutrients are replenished in the upper layer by the action of winds, currents, internal tides, and so forth. These physical processes set the time scale of nutrient replenishment at various depths in the water column. The upper waters, where the time scale is below the critical value for primary production, experience nutrient limitation. On the other hand, the dimensionless space scale (defined here as: depth of the photic layer / depth in the water column) sets the vertical extension of the productive zone. Here again, the ergocline is located in the layer where the time and space scales of the physics match the biological scales. Biological production occurs in the nutricline when light intensity at that depth is high enough for photosynthesis. The zone of the R-E diagram where standardized resources exceed one (both resources are nonlimiting) defines the depth of the ergocline. It has been suggested that the production maximum is located a few metres above the biomass maximum. This could be explained by the sinking of phytoplankton cells once produced. However, Cullen (1982) rightly stresses that most investigations do not have a fine enough vertical resolution to discriminate between two such peaks.

It is noteworthy that, in all the cases above which dealt with primary production, light limitation was related to the physical space scales, while nutrient limitation depended on the physical time scales. This is simply because light cannot be physically mixed downwards in the water column, so that light availability for the autotrophic organisms depends on their position relative to the light gradient, that is on the spatial scales of vertical water movements. By contrast, enhanced primary production increases the rate of nutrient uptake, which results in nutrient limitation if the rate of nutrient supply is too low, hence the role of hydrodynamic time scales.

For organisms that do not depend on light, it is therefore expected that physical time scales will be more critical than the space scales, as will be verified for mussel beds.

Fréchette (ms.b) presents data on the concentrations of phytoplankton above a mussel bed which show (1) that filtration by the mussels can reduce the concentration of the resource, near the bottom, to about half the ambient concentration, and (2) that the effects of filtration can be observed up to at least 0.5 m above the bed. Despite the fact that mussels are attached to the bottom, their filtration activity thus influences the water column up to at least 0.5 m. It has been explained above that replenishment of phytoplankton in the benthic boundary layer is effected by vertical turbulent diffusion (Fig. 7). Due to the fact that mussels are attached to the bottom, the ergocline cannot be defined by simply looking at the vertical distribution of particulate organic matter above the bed. The decreasing availability of the resource with the distance from the bed must also be taken into account. In Fig. 7, the concentration of particulate organic matter is nonlimiting above the mussel bed, since if it was limiting the mussels could not grow there. The critical parameter here is the time required for an horizontal layer of particle-rich water to come in contact with the bed, since phytoplankton must be actively replenished in the benthic layer by downwards water movements. With increasing distance above the bottom, it takes longer for the water to come in contact with the mussel bed. As the time scale of the vertical replenishment of the benthic layer increases with the height above the bottom, it reaches a critical value where the particle supply cannot meet the requirements of the mussel bed. The actual height of this critical value above the bottom is determined by the angle with the axis of ordinates in the top panel of Fig. 7. This angle depends on the dynamics of the processes of vertical diffusion and advection. The vertical extent of the ergocline is thus a function of the volume of water available to mussel filtration. To really quantify this idea, some measurements of the vertical distribution of the particulate organic matter attainable by the mussel bed would be needed. The ergocline for mussels is limited to a thin layer, since they cannot swim up in the water column. The growth of mussels into hummocks might be a way to raise some of them slightly above the bottom. The ultimate solution has however been found by mussel growers, who suspend the mussels from the surface, thus extending the ergocline to the whole water column.

All the above examples show that ergoclines can be identified at sea by monitoring the limiting resources, which is much easier than measuring biological production. R-E diagrams can be used to identify potential ergoclines, when limiting values for the resources are known. The same would be true for temporal ergoclines (Figs. 1 and 2). The explicit assumption of the R-E diagram is that biological production can occur when resources are nonlimiting; the implicit assumption is that such conditions are encountered where and when physical scales match biological scales.

TESTING HYPOTHESES AT AQUATIC ERGOCLINES

The general ergocline hypothesis, that followed the definition of ergoclines in the first section above, was:

Enhanced biological production occurs at ergoclines as the consequence of the matching or resonance of physical scales with biological scales.

One must be aware that this hypothesis does not concern the relationship between ergoclines and biomass, but rather the relationship between ergoclines and production. This is due to the fact that high biomass may either result from high production or from mechanical aggregation of the organisms. Conversely, low biomass may reflect low production, but it can also result from mechanical dispersion of the biomass, grazing or predation. Primary production at ergoclines is termed "new production", that is production resulting from allochtonous nutrient inputs, versus "regenerated production", which results from nutrient regeneration in the surface waters (Eppley and Peterson, 1979). The above general hypothesis is not easily testable, as originally formulated. Testable hypotheses must be so formulated that sampling at sea be aimed at falsifying or rejecting them, since there is no way of accepting a scientific hypothesis as being univocally true. The first step in testing the general ergocline hypothesis would be to falsify the following hypothesis:

The temporal, horizontal or vertical distribution of biological production is independent from gradients in the scales of physical structures.

To test this hypothesis, one must measure at sea (1) buoyancy and velocity fields, from which horizontal and vertical fluxes can be computed and spectra of physical scales be specified, as well as (2) biological production and biomass, in order to characterize the distribution of specific biological production (i.e. production per unit biomass). Biological production has been defined above as the flow of primary energy into a given trophic level. High specific production indicates that the high biological production measured at sea results from enhanced biological transfer of primary energy and not only from high accumulated biomass. If the hypothesis of independence is rejected, the spatio-temporal zone of high biological production most likely corresponds to an ergocline, as all ergoclines have the common characteristic of involving spatial and/or temporal gradients where physical processes can produce structures associated with enhanced biological production (see above). The general ergocline hypothesis can be further investigated by testing this next hypothesis:

At ergoclines, the temporal, horizontal or vertical distribution of physical scales does not necessarily correspond to the characteristic scales of biological production.

In order to try falsifying this second hypothesis, one must (1) determine the characteristic time and space scales of biological production at the ergocline (e.g. Table 2) and (2) compare the distribution of the physical scales associated with the ergocline to these characteristic biological scales. If dominant physical scales at the ergocline (which is the spatio-temporal zone of enhanced biological production) do coincide with the characteristic biological scales, the second hypothesis can be rejected and the implications of the ergocline hypothesis can be further explored.

The first interesting question raised by the ergoclines hypothesis concerns the mechanisms through which biological production increases at ergoclines. According to the well-known papers of Riley (1942) and Sverdrup (1953), the usual hypothesis explaining phytoplankton blooms is that they result from changes in the physical environment, and not from physiological changes of the organisms. However, it has been observed in the lower St. Lawrence Estuary that photosynthetic characteristics of the phytoplankton do change according to the semidiurnal cycle of tidal mixing, thus leading to semidiurnal blooms (Fortier and Legendre, 1979; Fréchette and Legendre, 1982). Some physiological properties of phytoplankton can even show endogenous variations, phased on the semidiurnal cycles of auxiliary energy (Auclair et al., 1982; Legendre et al., 1985a). Zooplankton physiology perhaps also responds to the semidiurnal tidal mixing (data from Maranda and Lacroix, 1983, for the Middle St. Lawrence Estuary, reinterpreted by Legendre and Demers, 1985). The hypothesis of a solely physical effect of ergoclines on production cannot therefore be accepted before rejecting first the hypothesis that:

The high biological production that occurs at ergoclines is (in part) caused by changes in the physiological state of the organisms, as the consequence of the matching or resonance of physical scales with biological scales.

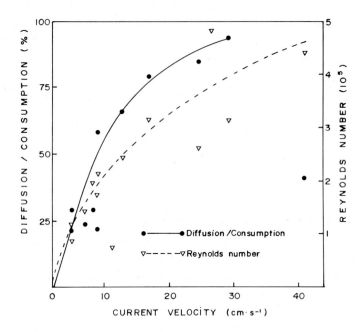

Fig. 8 Ratio of mean vertical diffusion to mean consumption of phytoplankton by a mussel bed, and Reynolds number, as a function of current velocity near the bottom. Curves freehand adjusted. Data selected from Fréchette (ms.a).

   To falsify this hypothesis, one must refine the measurements of biological production at sea, that were already needed to test the previous hypotheses. To do so, one must first identify the proximal agent(s) through which auxiliary energy enhances the biological production. Once the proximal agent is known, it becomes possible to measure its flux across the ergocline, together with the physio- ecological responses of the living organisms. At the water-sediment ergocline, for example, Fréchette and Bourget (1985) have shown that mussels can deplete the phytoplankton (proximal agent) immediately above the bed. Advection and vertical diffusion insured seston replenishment in the benthic boundary layer on physical scales that matched the biological requirements (Fréchette, ms.a). Phytoplankton depletion above the mussel bed was negatively correlated with current velocity (Fréchette and Bourget, 1985). As current velocity (and thus turbulence, i.e. Reynolds number) increased, diffusion did account for a larger proportion of the replenishment (Fig. 8). Filtration rates of the mussels did tend to decrease when depletion became more severe (Fréchette and Bourget, ms). This example shows that the measurements needed to test the above hypothesis are both physical and physio-ecological. Physical measurements concern fluxes of the proximal agent(s) at the ergocline, while the physio-ecological measurements concern such biological characteristics as uptake rates, excretion rates, and so forth.

The enhanced biological activity at ergoclines can modify the physical environment. As mentioned above, Lewis et al. (1983) have shown that, due to differential heating in the water column, the depth stratum containing the deeper portion of the subsurface chlorophyll maximum becomes increasingly more stable in time. This may permit the maintenance and development of the chlorophyll maximum. As another example, coral reefs increase the tidal currents (auxiliary energy available for mechanical work) in overlying waters, by decreasing the water depth above their growing calcareous mounds (Odum and Odum, 1955). Similarly, the growth of mussels, and of groups of mussels that form hummocks in high density situations, increases bottom roughness and thus vertical turbulent diffusion, which might favour phytoplankton replenishment of the boundary layer and ultimately mussel growth (Fréchette, ms.a). Hummocks probably act as density-dependent structures that exert a positive feedback on the carrying capacity of a bed by enhancing phytoplankton supply (Carefoot, 1977). Sea-grasses and other macrophytes can improve their light environments by enhancing sediment deposition and reducing resuspension (Ginsburg and Lowenstam, 1958; Ward et al., 1984). Similar properties are being discussed by Lewis (this book) as to the growth of ice microalgae at the ice-water interface. Such examples of positive feedbacks led Legendre et al. (1985b) to the conclusion that, at some ergoclines, biological activity can modify the physical environment so as to enhance the growth and/or survival of ecosystem components.

Such a general hypothesis cannot be tested without demonstrating first that some environmental characteristics are indeed modified by the biological activity at the ergocline. To do so, the following hypothesis must be rejected:

Environmental characteristics at an ergocline are not influenced by biological production.

Falsifying this hypothesis requires to simultaneously measure environmental properties and biological activity, at ergoclines characterized by different levels of biological production. The environmental properties measured would be those for which a response to biological production is predicted by models (e.g. the thermal structure at the pycnocline, or vertical turbulent diffusion at the water-sediment interface). The hypothesis would be rejected if significant differences in the environmental properties are found as a function of biological production. It is also possible to approach the question experimentally, by introducing artificial mimics of the biological systems in either the natural environment or laboratory conditions, and measuring the resulting changes in environmental variables.

The above hypothesis concerns, however, only the first part of the positive feedback mechanism. Some researchers would choose to test independently whether the environmental changes brought about by biological production do enhance production itself. Such an approach is however questionable, due to the possibility of unknown, and therefore uncontrolled interactions among variables. The only convincing test of the positive feedback hypothesis would be to measure changes in biological production when a single critical environmental characteristic is manipulated. For example, one should measure the assimilation of particulate organic matter by mussels when manipulating the sole bottom roughness or vertical turbulent diffusion, all the other characteristics of the system remaining unchanged. Similarly at the pycnocline, the sole temperature gradient should be manipulated; at the ice-water interface, it should be the rate of ice melt only; and so on. Whether such critical manipulations of environmental conditions can be technically achieved remains debatable.

At ergoclines, by definition, biological production is not limited by the resources (proximal agents). Under such conditions, the biomass can theoretically develop to very high concentrations. As an example, the average concentration of chlorophyll in the surface waters of the oceans does not exceed 1 mg.m^{-3}. At some tidal fronts in the approaches to the English Channel, chlorophyll concentrations can reach 100 mg.m^{-3} (Pingree et al., 1975); at the ice-water interface, these concentrations can exceed 500 mg.m^{-3} (e.g. Apollonio, 1965; Alexander et al., 1974). In such cases, it is not impossible that the very high concentrations locally impede biological production (self-shading of autotrophs, competition for space among animals, and so on). The rate of production would then become a function of the rate of dispersal of the biomass at the ergocline. The dispersal may be effected by physical processes, such as sedimentation or turbulence, or it may be the result of biological processes such as grazing, predation, and so forth. In order to test this idea, it would be necessary to falsify the following hypothesis:

At ergoclines, the rate of biological production is not limited by the accumulated biomass, so that it remains independent of the rate of dispersal of this biomass.

It is easy enough to test the hypothesis, if the rate of biomass dispersal is not the same everywhere along the ergocline. This may occur naturally, in different parts of the ergocline or at different times. Such variations can also be induced experimentally, by selective removal of the biomass. Testing the hypothesis then requires to measure simultaneously both production and biomass dispersal rates, along the ergocline.

The above hypotheses are examples of those that can be derived from the ergocline perspective. The power of the ergocline approach lies in the fact that several biologically productive structures can be studied using common hypotheses, since all these structures belong to the general category of ergoclines. The biological- physical approach to be developed in studying marine ergoclines correspond to the terms of reference described by Legendre and Demers (1984) for "dynamic biological oceanography", and it is expected that the ergocline perspective will shed a new light on the mechanisms that govern biological production in the oceans.

ACKNOWLEDGMENTS

The definition of ergoclines and the general ergocline hypothesis were discussed during a workshop, organized by SCOR Working Group 73 in Liège, from 17 to 19 May 1985, following the Seventeenth International Liège Colloquium on Ocean Hydrodynamics. The authors wish to thank Prof. E. Bourget (Université Laval, Québec), Drs. M. Fréchette, L. Fortier, J.C. Therriault and Mr. M. Levasseur (Pêches et Océans Canada, Québec) and Prof. S. Frontier (Université des sciences et techniques de Lille) for their most useful suggestions. A grant from the Natural Sciences and Engineering Research Council of Canada to the first author was instrumental in the completion of this work.

REFERENCES
Alexander, V., Horner, R. and Clasby, R.C., 1974. Metabolism of Arctic sea ice organisms. Rep. Inst. mar. Sci. Univ. Alaska, R74-4, 120 p.
Apollonio, S., 1965. Chlorophyll in Arctic sea-ice. Arctic, 18: 118-122.
Auclair, J.C., Demers, S., Fréchette, M., Legendre, L. and Trump, C.L., 1982. High frequency endogeous periodicities of chlorophyll synthesis in estuarine phytoplankton. Limnol. Oceanogr., 27: 348-352.
Bah, A. and Legendre, L., 1985. Biomasse phytoplanctonique et mélange de marée dans l'estuaire moyen du Saint-Laurent. Naturaliste can., 112: 39-49.
Banse, K. and Mosher, S., 1980. Adult body mass and annual production/biomass relationships of field populations. Ecol. Monogr., 50: 355-379.
Bowman, M.J. and Esaias, W.E., 1981. Fronts, stratification and mixing in Long Island and Block Island Sounds. J. geophys. Res., 86: 4260-4264.
Bowman, M.J., Esaias, W.E. and Schnitzer, M.B., 1981. Tidal stirring and the distribution of phytoplankton in Long Island and Block Island Sounds. J. mar. Res., 39: 587-603.

Bowman, M.J., Kibblewhite, A.C. and Ash, D.E., 1980. M_2 tidal effects in Greater Cook Strait, New Zeland. J. geophys. Res., 85: 2728-2742.

Carefoot, T., 1977. Pacific seashores. Douglas, Vancouver, Canada, 208 p.

Cullen, J.J., 1982. The deep chlorophyll maximum: comparing vertical profiles of chlorophyll a. Can. J. Fish. aquat. Sci., 39: 791-803.

Cushing, D.H., 1972. The production cycle and the number of marine fish. Symp. Zool. Soc. London, 29: 213-232.

Dare, P.J., 1976. Settlement, growth, and production of the mussel, Mytilus edulis L., in Morecambe Bay, England. Fishery Investigations, Minis. Agriculture, Fisheries and Food, London, Ser. II, 28: 25 pp.

Demers, S., Legendre, L. and Therriault, J.C., 1985. Phytoplankton responses to vertical tidal mixing. In: M.J. Bowman, W.T. Petersen and C.M. Yentsch (Editors), Tidal mixing an plankton dynamics. Springer-Verlag, New York, in press.

Demers, S., Therriault, J.C. and Bourget, E., ms. Phytoplanktonic productivity of the littoral zone: Turbidostat analogy. In preparation.

Dugdale, R.C. 1967. Nutrient limitation in the sea: dynamics, identification, and significance. Limnol. Oceanogr. 12: 685-695.

Eppley, R.N. and Peterson, B.J., 1979. Particulate organic matter flux and planktonic new production in the deep ocean. Nature (London), 282: 677-680.

Fortier, L., 1982. Environmental and behavioral control of large-scale distribution and local abundance of ichthyoplankton in the St. Lawrence Estuary. Ph. D. Thesis, McGill Univ., Montréal, Québec, 162 p.

Fortier, L. and Legendre, L., 1979. Le contrôle de la variabilité à court terme du phytoplancton estuarien: stabilité verticale et profondeur critique. J. Fish. Res. Board Can., 36: 1325-1335.

Fournier, R.O., van Det, M., Wilson, J.S. and Hargreaves, N.B., 1979. Influence of the shelf-break front off Nova Scotia on phytoplankton standing stock in winter. J. Fish. Res. Board Can., 36: 1228-1237.

Fréchette, M., ms.a. The importance of diffusion in supplying phytoplankton to benthic suspension feeders. In preparation.

Fréchette, M., ms.b. Food availability for an intertidal Mytilus edulis L. population: short-term relationships. In preparation.

Fréchette, M. and Bourget, E., 1985. Energy flow between the pelagic and benthic zones: factors controlling particulate organic matter available to an intertidal mussel bed. Can. J. Fish. aquat. Sci., 42: 1158-1165.

Fréchette, M. and Bourget, E., ms. The significance of small-scale spatio-temporal heterogenity in phytoplankton abundance for benthic energy flow. In preparation.

Fréchette, M. and Legendre, L., 1982. Phytoplankton photosynthetic response to light in an internal tide dominated environment. Estuaries, 5: 287-293.

Garrett, C.J.R., Keeley, J.R. and Greenberg, D.A., 1978. Tidal mixing versus thermal stratification in the Bay of Fundy and Gulf of Maine. Atm. Ocean 16: 403-423.

Garrett, C.J.R. and Loder, J.W., 1981. Dynamical aspects of shallow sea fronts. Phil. Trans. R. Soc. Lond., A302: 563-581.

Gilmartin, M., 1964. The primary production of a British Columbia fjord. J. Fish. Res. Board Can., 21: 505-538.

Ginsburg, R.N. and Lowenstam, H.A., 1958. The influence of marine bottom communities on the depositional environment of sediments. J. Geol., 66: 310-318.

Griffiths, D.K., Pingree, R.D. and Sinclair, M., 1981. Summer tidal fronts in the near-Arctic regions of Foxe Basin and Hudson Bay. Deep-Sea Res., 28: 865-873.

Harris, G.P., 1978. Photosynthesis, productivity and growth. The physiological ecology of phytoplankton. Arch. Hydrobiol. Beih. Ergeb. Limnol., 10: 1-171.

Harris, G.P., 1980. Temporal and spatial scales in phytoplankton ecology. Mechanisms, methods, models, and management. Can. J. Fish. aquat. Sci., 37: 877-900.

Haury, L.R., McGowan, J.A. and Wiebe, P.H., 1978. Patterns and processes in the time-scale spaces of plankton distributions. In: J.H. Steele (Editor), Spatial pattern in plankton communities. Plenum Press, New York, pp. 277-327.

Heath, R.A., 1973. Flushing of coastal embayments by changes in atmospheric conditions. Limnol. Oceanogr., 18: 849-862.

Holligan, P.M., 1981. Biological implications of fronts on the northwest European continental shelf. Phil. Trans. R. Soc. Lond., A302: 547-562.

Holligan, P.M., Balch, W.M. and C.M. Yentsch, 1984. The significance of subsurface chlorophyll, nitrite and ammonium maxima in relation to nitrogen for phytoplankton growth in stratified waters of the Gulf of Maine. J. mar. Res., 42: 1051-1073.

Iles, T.D. and M. Sinclair, 1982. Atlantic herring: stock discreteness and abundance. Science, 215: 627-633.

Iverson, R.L., Curl, H.C. Jr., O'Connors, H.B. Jr., Kirk, D. and Zakar, K., 1974. Summer phytoplankton blooms in Auke Bay, Alaska, driven by wind mixing of the water column. Limnol. Oceanogr., 19: 271-278.

Kierstead, H. and Slobodkin, L.B., 1953. The size of water masses containing plankton blooms. J. mar. Res., 12: 141-147.

Lasker, R., 1975. Field criteria for the survival of anchovy larvae: the relation between inshore chlorophyll maximum layers and successful first feeding. Fish. Bull., 73: 453-462.

Legendre, L., 1981. Hydrodynamic control of marine phytoplankton production: the paradox of stability. In: J.C.J. Nihoul (Editor), Ecohydrodynamics. Elsevier, Amsterdam, pp. 191-207.

Legendre, L. and Demers, S., 1984. Towards dynamic biological oceanography and limnology. Can. J. Fish. aquat. Sci., 41: 2-19.

Legendre, L. and Demers, S., 1985. Auxiliary energy, ergoclines and aquatic biological production. Naturaliste can., 112: 5-14.

Legendre, L., Demers, S., Therriault, J.C. and Boudreau, C.A., 1985a. Tidal variations in the photosynthesis of estuarine phytoplankton isolated in a tank. Mar. Biol., in press.

Legendre, L., Ingram, R.G. and Poulin, M. 1981. Physical control of phytoplankton production under sea-ice (Manitounuk Sound, Hudson Bay). Can. J. Fish. aquat. Sci., 38: 1385-1392.

Legendre, L., Ingram, R.G. and Simard, Y., 1982. A periodic changes of water column stability and phytoplankton in an Arctic coastal embayment, Manitounuk Sound, Hudson Bay. Naturaliste can., 109: 775-786.

Legendre, L., Kemp, W.M., Atlan, H., Conrad, M., Fréchette, M., Lane, P., Platt, T., Rodriguez, G., Tundisi, J. and Yentsch C.S. 1985b. Possible holistic approaches to the study of biological-physical interactions in the oceans. Can. Bull. Fish. aquat. Sci., 213: 248-253.

Lewis, M.R., Cullen, J.J. and Platt, T., 1983. Phytoplankton and thermal structure in the upper ocean: consequences of nonuniformity in chlorophyll profile. J. geophys. Res., 88: 2565-2570.

Maranda, Y. and Lacroix, G., 1983. Temporal variability of zooplankton biomass (ATP content and dry weight) in the St. Lawrence Estuary: advective phenomena during neap tide. Mar. Biol., 73: 247-255.

Margalef, R. and Estrada, M., 1980. Las areas océanicas mas productivas. Investigacion y Ciencia (Spanish edition of Scientific American), 49: 8-20.

Marshall, K.C., 1976. Interfaces in microbial ecology. Harvard Univ., Cambridge, MA, 156 p.

Nixon, S.W., Oviatt, C.A., Rogers, C. and Taylor, K., 1971. Mass and metabolism of a mussel bed. Oecologia, 8: 21-30.

Odum, E.P., 1980. The status of three ecosystem-level hypotheses regarding salt-marsh estuary: tidal subsidy, outwelling, and detritus based food chains. In: V.S. Kennedy (Editor), Estuarine perspectives. Academic Press, New York, pp. 485-495.

Odum, H.T. and Odum, E.P., 1955. Trophic structure and productivity of a windward coral reef community on Eniwetok Attoll. Ecol. Monogr., 25: 292-320.

Parsons, T.R., Perry, R.I., Nutbrown, E.D., Hsieh, W. and Lalli, C.M., 1983. Frontal zone analysis at the mouth of Saanich Inlet, British Columbia, Canada. Mar. Biol., 73: 1-5.

Parsons, T.R., 1980. Zooplanktonic production. In: R.S.K. Barnes and K.H. Mann (Editors), Fundamentals of aquatic ecosystems. Blackwell, Oxford, pp. 46-66.

Pearcy, W.G., 1962. Ecology of an estuarine population of winter flounder, Pseudopleuronectes americanus (Walbaum). II. Distribution and dynamics of larvae. Bull. Bingham oceanogr. Collect., 18: 16-38.

Perry, R.I., Dilke, B.R. and Parsons, T.R., 1983. Tidal mixing and summer plankton distribution in Hecate Strait, British Columbia. Can. J. Fish. aquat. Sci., 40: 871-887.

Pingree, R.D., 1978a. Mixing and stabilization of phytoplankton distributions on the Northwest European continental shelf. In: J.H. Steele (Editor), Spatial pattern in plankton communities. Plenum Press, New York, pp. 181-220.

Pingree, R.D., 1978b. Cyclonic eddies and cross-frontal mixing. J. Mar. biol. Ass. U.K., 58: 955-963.

Pingree, R.D. and Griffiths, D.K., 1978. Tidal fronts on the shelf seas around the British Isles. J. geophys. Res., 83: 4615-4622.

Pingree, R.D. and Griffiths, D.K., 1980. A numerical model of the M_2 tide in the Gulf of St. Lawrence. Oceanol. Acta, 3: 221-225.

Pingree, R.D., Holligan, P.M. and Head, R.N., 1977. Survival of dinoflagellate blooms in the western English Channel. Nature (London), 265: 266-269.

Pingree, R.D., Holligan, P.M. and Mardell, G.T., 1978. The effects of vertical stability on phytoplankton distributions in the summer on the northwest European shelf. Deep-Sea Res., 25: 1011-1028.

Pingree, R.D., Holligan, P.M. and Mardell, G.T., 1979. Phytoplankton growth and cyclonic eddies. Nature (London), 278: 245-247.

Pingree, R.D., Holligan, P.M. and Mardell, G.T. and Head, R.N., 1976. The influence of physical stability on spring, summer and autumn phytoplankton blooms in the Celtic Sea. J. mar. biol. Ass. U.K., 56: 845-873.

Pingree, R.D., Mardell, G.T., Holligan, P.M., Griffiths, D.K. and Smithers, J., 1982. Celtic Sea and Armorican current structure and the vertical distributions of temperature and chlorophyll. Cont. Shelf Res., 1: 99-116.

Pingree, R.D., Mardell, G.T. and Maddock, L., 1983. A marginal front in Lyme Bay. J. mar. biol. Ass. U.K., 63: 9-15.

Pingree, R.D., Pugh, P.R., Holligan, P.M. and Forster, G.R., 1975. Summer phytoplankton blooms and red tides along tidal fronts in the approaches to the English Channel. Nature (London), 258: 672-677.

Platt, T. and Denman, K.L., 1975. A general equation for the mesoscale distribution of phytoplankton in the sea. Mem. Soc. R. Sci. Liège (Ser. 6), 7: 31-42.

Riley, G.A., 1942. The relationship of vertical turbulence and spring diatom flowerings. J. mar. Res., 5: 67-87.

Riley, G.A., 1963. Theory of food-chain relations in the ocean. In: M.N. Hill (Editor), The sea, Vol. 2, Interscience, New York, pp. 438-463.

Riley, G.A., 1976. A model of plankton patchiness. Limnol. Oceanogr., 21: 873-880.

Savidge, G., 1976. A preliminary study of the distribution of chlorophyll a in the vicinity of fronts in the Celtic and Western Irish Seas. Estuar. coast. mar. Sci., 4: 617-625.

Simpson, J.H. and Hunter, J.R., 1974. Fronts in the Irish Sea. Nature (London), 250: 404-406.

Simpson, J.H. and Pingree, R.D., 1978. Shallow sea fronts produced by tidal stirring. In: M.J. Bowman and W.E. Esaias (Editors), Oceanic fronts in coastal processes. Springer-Verlag, Berlin Heidelberg, pp. 29-42.

Sinclair, M., Subba Rao, D.V. and Couture, R., 1981. Phytoplankton temporal distribution in estuaries. Oceanol. Acta, 4: 239-246.

Sinclair, M. and Iles, T.D., 1985. Atlantic herring (Clupea harengus) distributions in the Gulf of Maine-Scotian Shelf area in relation to oceanographic features. Can. J. Fish. aquat. Sci., 42: 880-887.

Skellam, J.G., 1951. Random dispersal in theoretical populations. Biometrika, 78: 196-218.

Smith, W.O. and Nelson, D.M., 1985. Phytoplankton bloom produced by a receding ice edge in the Ross Sea: spatial coherence with the density field. Science, 227: 163-166.

Steele, J.H., 1958. Plant production in the northern North Sea. Scot. Home Dep. Mar. Res., 1958, (7): 1-36.

Sutcliffe, W.H. Jr., 1972. Some relations of land drainage, nutrients, particulate material, and fish catch in two eastern Canadian Bays. J. Fish. Res. Board Can., 29: 357-362.

Sutcliffe, W.H. Jr., 1973. Correlations between seasonal river discharge and local landings of American lobster (Homarus americanus) and Atlantic halibut (Hippoglossus hippoglossus) in the Gulf of St. Lawrence. J. Fish. Res. Board Can., 30: 856-859.

Sverdrup, H.U., 1953. On conditions for the vernal blooming of phytoplankton. J. Cons. perm. int. Explor. Mer, 18: 287-295.

Takahashi, M., Siebert, D.L. and Thomas, W.H., 1977. Occasional blooms of phytoplankton during summer in Saanich Inlet, B.C., Canada. Deep-Sea Res., 24: 775-780.

Tett, P., 1981. Modelling phytoplankton production at shelf-sea fronts. Phil. Trans. R. Soc. Lond., A302: 605-615.

Walsh, J.J., Whitledge, T.E., Barvenik, F.W., Wirick, C.D. and Howe, S.O., 1978. Wind events and food-chain dynamics within the New York bight. Limnol. Oceanogr., 23: 659-683.

Walsh, J.J., Whitledge, T.E., Kelley, J.C., Huntsman, S.A. and Pillsbury, R.D., 1977. Further transition states of the Baja California upwelling ecosystem. Limnol. Oceanogr., 22: 264-280.

Ward, L.G., Kemp, W.M. and Boynton, W.R., 1984. The influence of waves and seagrass communities on suspended particulates in an estuarine embayment. Mar. Geol., 59: 85-103.

Wildish, D.J. and Kristmanson, D.D., 1979. Tidal energy and sublittoral macrobenthic animals in estuaries. J. Fish. Res. Board Can., 36: 1197-1206.

Wroblewski, J.S., O'Brien, J.J. and Platt, T., 1975. On the physical and biological scales of phytoplankton patchiness in the ocean. Mem. Soc. R. Sci. Liège (Ser. 6), 7: 43-57.

BIOLOGICAL PRODUCTION AT THE ICE-WATER ERGOCLINE *

S. DEMERS[1], L. LEGENDRE[2], J.C. THERRIAULT[1] and R.G. INGRAM[3]

[1] Centre Champlain des sciences de la mer, Ministère des Pêches et des Océans, C.P. 15500, 901 Cap Diamant, Québec, Québec G1K 7Y7, Canada.

[2] Département de biologie, Université Laval, Québec, Québec G1K 7P4, Canada.

[3] Institute of Oceanography, McGill University, 3620 University, Montréal, Québec H3A 2B2, Canada.

ABSTRACT

The ice-water interface is the site of high microalgal productivity. These microalgae constitute an important part of the productivity of polar seas. The growth of ice microalgae during the spring and perhaps during the autumn extends the short growing season in the water column. Herbivores have been observed to actively feed on the ice microalgae. Sea-ice microalgae respond to variations in salinity (which controls biomass and taxonomic composition in coastal areas influenced by freshwater runoff), temperature (the survival of microalgae depends on their ability to develop a protection mechanism against freezing), light (the photosynthetic activity of ice microalgae is a function of both light intensity and quality) and nutrients (nutrient limitation has been demonstrated even when ambient nutrient concentrations were high). The biological production at this energetic interface is examined in the specific context of the ice-water ergocline.

According to the hypothesis of Legendre and Demers (1985) and Legendre et al. (this book), energetic interfaces (ergoclines) are preferential sites for biological production in the oceans. The ice-water interface is one example of such ergoclines. In this paper we will review the major characteristics of biological production at the ice-water interface.

SEA-ICE MICROALGAE

The first mention of colonization of the Arctic sea-ice by microalgae date from

---

*Contribution to the program of GIROQ (Groupe interuniversitaire de recherches océanographiques du Québec).

the mid 1800's, when Ehrenberg (1841, 1853) listed the diatoms collected during the research trips of Sir John Franklin in the Canadian Arctic Archipelago. The first observations in the Antarctic are attributed to Hoocker (1847). Since that time, an impressive body of literature has accumulated on sea-ice microalgae. The papers published before 1960 were however mainly descriptive, and considered the sea-ice microalgae merely as a curiosity. After 1960, the interest in the physiological aspects of microalgal growth in the sea-ice rose considerably and researchers began to investigate the physiological adaptations of those algae to such an extreme environment (Apollonio, 1961).

Three types of ice microalgal communities are usually recognized, depending on the level at which the maximum biomass is observed in the ice column (Ackley et al., 1979). These communities are defined as (1) the snow community, (2) the ice interior community and (3) the "epontic" or ice bottom community.

The snow community was first described by Meguro (1962) and is observed when the weight of snow depresses the ice surface below the water level, causing a flood of water containing microalgae through the ice. Few algae are seen in this layer. The ice interior community (Fig. 1) results from the trapping of ice bottom cells in the ice matrix as the thickness of the ice increases over time (Hoshiai, 1969; Ackley et al., 1979). These microalgae constitute vestigial populations since the cells are probably not growing after being trapped in the ice matrix (Ackley et al., 1979). The rate of freezing of the ice determines the cell density inside the ice (Grainger, 1977; Demers et al., 1984). The highest chlorophyll biomasses are observed at the bottom of the ice (Apollonio, 1965; Bunt, 1963, 1968; Bunt and Wood, 1963; Meguro et al., 1967; Poulin et al., 1983; Demers et al., 1984) (Table 1). The ice bottom community is the most metabolically active (Palmisano et al., 1985). Microalgal growth occurs at the ice-water interface, in the unconsolidated ice layer (Meguro et al., 1967; Alexander et al., 1974) and/or in the interstitial water of the ice matrix (Bunt, 1963). In the Arctic, the microalgae form a coloured layer 1 to 4 cm thick at the bottom of the ice, while in the Antarctic, microalgal growth occurs in the under-ice slush formed by a layer (0.2 to 1.0 m) of unconsolidated ice crystals (Bunt, 1963; Bunt and Wood, 1963; Andriashev, 1968; Hoshiai, 1972; Gruzov, 1977). At the beginning of the spring growth period, the epontic community is usually composed of a mixed population of pelagic and benthic species (Horner and Schrader, 1982; Gosselin et al., 1985; Rochet et al., 1985) but, with time, the benthic diatoms become largely dominant and may represent as much as 99% of the epontic community (Bunt and Wood, 1963; Meguro, 1962; Bunt, 1963, 1964, 1968; Burkholder and Mandelli 1965; Poulin and Cardinal, 1982a, b, 1983; Hsiao, 1980). This suggests that ice algae are present in the water column in low abundance, perhaps as resting spores. Once trapped in the ice with other pelagic species, natural selection favours species adapted to the ice habitat (Horner and Schrader, 1982). Other micro-organisms such as bacteria (Sullivan and Palmisano, 1981; Kaneko et al., 1978;

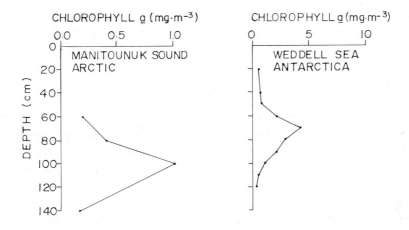

Fig. 1   Vertical distribution of microalgae (chlorophyll a) in the sea-ice of two polar environments.   Data from Poulin et al. (1983: Arctic) and Ackley et al. (1979: Antarctica).

TABLE 1

Chlorophyll concentrations in the upper and bottom layers of the ice, in various locations.

Location	Chlorophyll concentration (mg.m^{-3})		References
	Upper layer (30-120 cm)	Bottom layer (20 cm)	
Robeson Channel, Canada	0.33	31.69	Dunbar and Acreman (1980)
Barrow Strait, Canada	0.24	60.31	Dunbar and Acreman (1980)
Austin Channel, Canada	0.08	60.31	Dunbar and Acreman (1980)
Hudson Bay, Canada	0.75	109.08	Dunbar and Acreman (1980)
Gulf of St.Lawrence, Canada	1.87	8.08	Dunbar and Acreman (1980)
St.Lawrence Estuary, Canada	0.87	4.55	Demers et al. (1984)
Frobisher Bay, Canada	0.99	52.1 to 300.5	Hsiao (1980)
McMurdo Sound, Antartica	0.50	656.00	Palmisano and Sullivan (1983)

Griffiths et al., 1978), protozoa (Lipps and Krebs, 1984) or even copepods are also present at the bottom of the ice, but they never constitute an important part of the total carbon biomass. The biomass at the bottom of the ice thus results from the growth and dispersion of cells and from the mechanical accumulation of microalgae (Horner and Schrader, 1982).

ECOLOGICAL SIGNIFICANCE OF SEA-ICE MICROALGAE

The growth of microalgae in the sea ice constitutes an important part of the primary productivity of polar seas (Meguro et al., 1966; Horner and Alexander, 1972; McConville and Wetherbee, 1983; Alexander, 1974; McRoy and Goering, 1974; Clasby et al., 1976). Microalgal growth starts during the initial period of ice formation in autumn, and often reaches abundances comparable to those observed in the spring (in the Arctic: Horner and Schrader, 1982; in the Antarctic: Hoshiai, 1977). Then, microalgal growth at the ice bottom slows down gradually and stops completely when very low winter light intensities are reached. Heterotrophy has been invoked as a mechanism by which the algae would survive the prolonged darkness of the Arctic winter (Rodhe, 1955; Wilce, 1967; Allen 1971), but experiments using a variety of labelled organic substrates failed to demonstrate any heterotrophic growth for the Arctic (Horner and Alexander, 1972) or the Antarctic (Bunt and Lee, 1972) ice algae. It has been suggested that these autumn algae could be at the origin of the ice interior vestigial community, after being trapped in the ice matrix following ice formation (Poulin et al., 1983). However, it is difficult to generalize this conclusion, since autumn data are scarce. At the end of the winter period and in early spring, an intense peak of microalgal growth generally occurs at the ice-water interface when the under-ice light intensity reaches a minimum critical light level (I_{cr}) (Table 2). The active growth of ice algae during spring and perhaps during autumn extends the short growing season in the water column (Andriashev, 1968; Bunt, 1968; Allen, 1971; McRoy and Goering, 1974; Horner, 1976, 1977; Bradford,

TABLE 2

Critical light levels below which the photosynthetic activity of ice microalgae was undetectable, at various locations.

Location	Critical light level $\mu Einst.m^{-2}.s^{-1}$	References
Manitounuk Sound, (Hudson Bay)	7.6	Gosselin et al. (1985)
Barrow Strait, (Alaska)	2.3 to 9.3	Alexander et al. (1974) Clasby et al. (1973)
Narwhal Island, (Alaska)	7.7	Horner and Schrader (1982)
McMurdo Sound, (Antarctica)	0.3 to 0.9	Bunt (1964) Palmisano and Sullivan (1983)

1978). As a matter of fact, the high concentration of ice microalgae at the ice-water interface (4 to 656 mg Chl a.m^{-3}: Table 1) represents a significant fraction of the primary biomass of polar ecosystems, where chlorophyll concentrations in the water column are generally low (< 1 mg.m^{-3}: Horner and Schrader, 1982; Bunt and Lee, 1970; Holm-Hansen et al., 1977). The biomass of ice algae is often comparable to water column values integrated over the photic layer in many productive planktonic systems (Table 3). The annual carbon production per unit area of ice microalgae is comparable to the pelagic production of the Arctic Ocean (Table 4). However, productivity of the ice algae is only ∼ 3% of the estimated total annual primary production in the water column, since young ice only covers a small area of the Arctic Ocean (Table 5). Some isolated studies in the Southern Bering Sea have, however, reported very high production values at the ice edge (600 to 725 mg C.m^{-2}.h^{-1}: Niebauer et al., 1981, as reported by Subba Rao and Platt, 1984), compared to values (0.3 to 57 mg C.m^{-2}.h^{-1}) reported previously in the literature (Clasby et al., 1973; Matheke and Horner, 1974).

Although no comprehensive studies have been carried out in situ to evaluate the utilization of the ice microalgal biomass by different grazers, some observations suggest that herbivores can feed on ice algae during springtime when water column productivity is low. Andriashev (1968) reports that a dozen animal species use the lower layer of the Antarctic fast ice, both as refuge and feeding ground. Richardson and Whitaker (1979) also report grazing of ice microalgae by the amphipod Pontaginia antarctica at the ice-water interface. Cross (1982) and Grainger et al. (1985) found high densities of nematodes and harpacticoid and cyclopoid copepods associated with the under-ice microalgae. These under-ice organisms are believed to represent a link in the transfer of energy from ice algal production to amphipods to sea-birds and other mammals living in the Arctic (Horner, 1976; Bradstreet and Cross, 1982; Gulliksen, 1984).

RESPONSES OF SEA-ICE MICROALGAE TO ENVIRONMENTAL VARIATIONS

Salinity, temperature, light and nutrients have been identified as the main factors regulating the growth of sea-ice microalgae. However, unlike phytoplankton, ice algae usually grow under relatively stable conditions of temperature and salinity, and they are not subjected to high frequency light fluctuations. Nutrients are generally believed to be the factors which are the most influenced by hydrodynamical variations. Nutrient concentrations at the ice-water interface are however generally high and most researchers have not considered them as limiting.

Salinity

A number of studies have suggested that salinity was not important in regulating the growth of the epontic community. For example, Grant and Horner (1976) reported that ice diatoms were able to grow in a wide range of salinities (5 to 60). Meguro

TABLE 3

Chlorophyll concentrations in some productive planktonic systems (values integrated over the photic zone) and at the ice-water interface.

Location	Chlorophyll concentration mg.m^{-2}	References
Planktonic systems		
St.Lawrence Estuary	2 to 230	Therriault and Levasseur (1985)
Puget Sound	10 to 160	Winter et al. (1975)
Hudson River Estuary	150	Malone (1977)
Ice-water interface		
Manitounuk Sound, Canada	0.85	Gosselin et al. (1985)
Manitounuk Sound, Canada	0.76	Poulin et al. (1983)
Robeson Channel, Canada	12.1	Dunbar and Acreman (1980)
Barrow Strait, Alaska	12.5	Dunbar and Acreman (1980)
Jones Sound, Canada	23.0	Apollonio (1965)
Narwhal Island, Alaska	26.5	Horner and Schrader (1982)
Frobisher Bay, Canada	30.1	Hsiao (1980)
Frobisher Bay, Canada	19.6	Grainger and Hsiao (1982)
Barrow Strait, Alaska	30.5	Clasby et al. (1973)
Hudson Bay, Canada	39.6	Gosselin et al. (ms)
McMurdo Sound, Antarctica	16 to 34	Bunt and Lee (1970)
McMurdo Sound, Antarctica	131	Sullivan and Palmisano (1983)
Factory Cove, Antarctica	309	Whitaker (1977)

TABLE 4

Annual microalgal production in the water column and at the ice-water interface of the Arctic Ocean.

Location	Annual production[a] gC.m^{-2}.a^{-1}	References
Pelagic production		
Shelf (< 200 m)	27	Moiseev (1971) as reported by Subba Rao and Platt (1984)
Offshore (> 200 m)	9	Moiseev (1971) as reported by Subba Rao and Platt (1984)
Ice-algal production		
Barrow, Alaska	0.9 to 22.0	Clasby et al. (1973)
Chukchi Sea	1.4 to 164.0	Matheke and Horner (1974)
Beaufort Sea	5.0 to 10.0	Alexander (1974)
Narwhal Island	0.7	Horner and Schrader (1982)
Barrow, Alaska	5.0	Horner and Schrader (1982)
Antarctica	10.0	Fogg (1977)
Syowa Station, Antarctica	1.5 to 3.3	Hoshiai (1981)
McMurdo Sound, Antarctica	4.1[b]	Palmisano and Sullivan (1983)

[a] Values from Subba Rao and Platt (1984)
[b] In Subba Rao and Platt (1984), this value is incorrectly reported as gC.m^{-2}.d^{-1} and qualified as being substantially high.

TABLE 5

Total annual production of the pelagic system and the ice-water interface of the Arctic Ocean, and their respective areas.

Location	Area[a] 10^6 km^2	Total production 10^6 tC.a^{-1}
Arctic Ocean	13.1	205.8
Arctic sea-ice	0.6	6.0[b]

[a] Values from Subba Rao and Platt (1984)
[b] This value is calculated from an average value for the annual ice production of 10 gC.m^{-2}.a^{-1}

et al. (1967) found ice algae that were exposed to salinities up to 45, in brine cells near the bottom of the ice in Barrow Strait. Grainger (1977) found ice algae at very low salinity (~ 0.5 to 7.0) at the time of ice break-up in Frobisher Bay. These studies, however, did not present any actual data on the growth response of ice microalgae to salinity variations. The fact that ice algae do not develop in freshwater ice suggests that salinity probably has a strong effect on the growth of ice algae in polar seas. Suggestions to that effect were made by Bunt (1964) and Sullivan and Palmisano (1981), the latter proposing a synergetic effect of temperature and salinity on the light response of the algae. Poulin et al. (1983) observed a decreasing gradient of cell abundance with decreasing salinity, along transects of stations in Manitounuk Sound and in Hudson Bay off the mouth of the Great Whale River (Fig. 2). They explained the observed effect of salinity on microalgae by suggesting (1) that the number of brine cells, and consequently the surface available for colonization by the ice algae, decreases with decreasing salinity and (2) that a taxonomic gradient parallels the salinity gradient at the interface, as fewer taxa were tolerant to lower salinities. This was supported by the observation of a lower number of taxa in the less saline waters of Manitounuk Sound (151 taxa) than in those of the Canadian Arctic (> 200 species). Therefore, it might be suggested that seasonal variations in salinity would not seriously affect the growth of ice algae offshore, but that salinity certainly plays an important role in the distribution of ice algae in coastal areas influenced by freshwater runoff.

During the period of ice melting, the drastic decrease in surface salinity may have important consequences for phytoplankton and the entire epontic community. At the time of ice melting and breakup, the algal biomass produced at the ice-water interface is released in the water column. There is conflicting evidence as to the relative importance of these ice microalgae for the initiation and formation of the spring bloom in the water column (Horner, 1976; Grainger, 1977). Some papers suggest

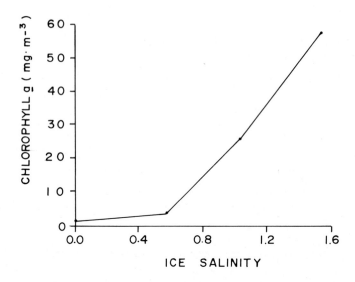

Fig. 2 Chlorophyll a concentration (sea-ice microalgae) in Manitounuk Sound, Hudson Bay, as a function of ice salinity. Data from Poulin et al. (1983).

that the phytoplankton bloom in the water column is directly initiated by the release of ice microalgae from the melting ice, whereas others claim that the release of cells into the underlying water is not a major source for the spring phytoplankton bloom. However, according to Allen (1971), low salinity kills or seriously affects the diatoms growing under the ice, and the population does not fully recover until vertical mixing restores the salinity of the surface waters. Using data from Horner (1972), Legendre et al. (1981) have shown that low salinities do not preclude high photosynthetic activity, which is contrary to the contention of Allen (1971). Saito and Taniguchi (1978) consider the melt water as a positive trigger for the under-ice bloom because, at the same time as ice algae are released from the ice, a bloom occurs in the shallow surface layer which is stabilized by the presence of low salinity waters. A similar positive effect of reduced salinity on phytoplankton blooms was also observed at ice edges (Alexander and Niebauer, 1981; El-Sayed and Taguchi, 1981; Smith and Nelson, 1985). In Manitounuk Sound, Legendre et al. (1981) suggested that the phytoplankton bloom under the ice probably resulted from the simultaneous deepening of both the photic layer (seasonal light increase) and the stratified layer (low-salinity melting water). The co-occurrence of ice microalgae and phytoplankton in the blooming water column would considerably increase the algal biomass, under conditions of ice melt. Similar results were found in the Chukchi Sea (Hameedi, 1978), in Stefansson Sound and at Narwhal Island (Horner and Schrader, 1982). Other studies have suggested that the increased algal biomass (chlorophyll a) during the ice melt does not necessarily result in an increase of the productivity of the water column (Clasby et al., 1973; Grainger, 1977). Microscopic examinations of cells collected during ice melting in the Arctic have indicated that the ice microalgae in the water column were not healthy (Horner, 1977; Horner and Schrader, 1982). In contrast, Smith and Nelson (1985) suggested that the epontic algae could play an important role as a bloom inoculum. They found that Nistchia curta, a member of the ice community, was present in significant numbers and photosynthetically active (microautoradiographic analysis) in the bloom. We believe that the discrepancies between those studies may have resulted from the different methodologies used for assessing the activity of the cells. Also, many of the contradictions reported above may have been caused by the time of sampling, relative to the sequence of melting (release of ice algae) - stratification - phytoplankton bloom (other species than the ice microalgae), or because no real under-ice bloom of phytoplankton was present (Legendre et al., 1981).

Physical processes that modify the under-ice salinity include brine drainage during the period of ice growth, melt water input during ice melt and freshwater runoff. The actual salinity values depend on ambient water characteristics and the degree of mixing. Since the water temperature is usually close to freezing, the salinity field determines the temperature at the interface. Across the ice cover, downward heat flux values vary from negative to positive over the ice season.

Temperature

The ability to grow under conditions of low in situ irradiance and temperature is determinant for the survival of sea ice microalgae and is an important factor in the control of species succession and community structure. A number of studies have reported physiological adaptations of algae growing under extreme light and temperature conditions (Neori and Holm-Hansen, 1982; Platt et al., 1982; Li et al., 1984). For example, Bunt (1968) showed in laboratory that sea ice algae from the Antarctic were extremely shade-adapted and obligate psychrophilic in their temperature response. Seaburg et al. (1981) found different growth rates in the temperature range from 2 to 34°C, for different species among 35 taxa isolated from the Antarctic sea ice. It is not known whether or to what extent net photosynthesis is limited by low temperature at the low light intensities prevailing under the sea ice, but net carbon fixation by sea ice algae is definitively temperature dependent, particularly in the range -1.5° C to 5° C (Bunt, 1964; Sullivan and Palmisano, 1981). A distinctive psychrophilic acclimation caused by the seasonally increasing light intensity was observed by Rochet et al. (1985) for sea ice algae in Hudson Bay (Fig. 3). They suggested that this acclimation was occurring because of the increasing sensitivity of photosynthesis to low temperature as light intensity increases. Rochet et al. (1985) also found a seasonal decrease in species diversity, which might suggest that the survival of sea ice algae in the ice habitat is strongly dependent on their ability to develop a protection mechanism against freezing (Horner and

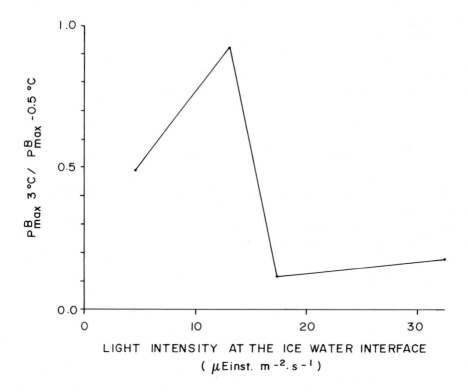

Fig. 3   Ratio of maximum photosynthetic rate per unit chlorophyll a of sea-ice microalgae (P^B_{max}), measured at 3°C and -0.5°C (ambient temperature), plotted as a function of irradiance (I_z) at the ice-water interface in Hudson Bay.   Data from Rochet et al. (1985).

Schrader, 1982).   Therefore, natural selection probably favours species capable of growing at low temperature in the ice habitat, as the ambient light intensity seasonally increases.   A parallel can be made with the microalgae that live in extremely hot waters (hot springs), where optimal growth also occurs at the environmental temperature (Brock, 1967a,b).   As there is no form of adaptability entirely without biological cost (Conrad, 1983), continued photosynthesis at freezing temperature, when the light intensity increases, tends to narrow the range of temperatures within which optimum photosynthesis can occur.   Even if the mechanism of this physiological adaptation is not yet known, some studies have suggested a relationship with intrinsic cell characteristics (Ben-Amotz and Gilboa, 1980) or with the synthesis of specific enzymes (e.g. RuBPC) at low temperatures (Li and Morris, 1982).   The acclimation of ice microalgae to low temperature conditions opens a new avenue for research, which could be determinant in understanding the ecology of the ice-water interface.

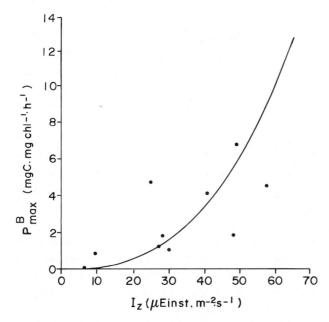

Fig. 4   Maximum photosynthetic rate per unit chlorophyll a (P^B_{max}) of sea-ice microalgae in Manitounuk Sound, Hudson Bay, as a function of the irradiance (I_z) at the ice-water interface.   Freehand adjusted curve.   Data from Gosselin et al. (1985).

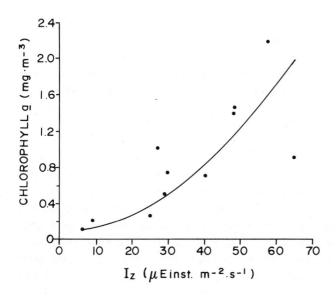

Fig. 5 Chlorophyll a concentration (sea-ice microalgae) in Manitounuk Sound, Hudson Bay, as a function of the irradiance (I_z) at the ice-water interface.   Freehand adjusted curve.   Data from Gosselin et al. (1985).

Light

The intensity of light at the ice-water interface is low, typically ~ 0.02%, of that incident to the surface of the ice. For this reason, light is considered as a critical factor for regulating the growth of ice algae (Meguro et al., 1967; Bunt and Lee, 1970; Clasby et al., 1976; Grainger, 1979; Hsiao, 1980; Horner and Schrader, 1982; Gosselin et al., 1985). The under-ice illumination is mainly controlled by air temperature and snow depth (Gosselin et al., 1985), since the optical properties of sea ice and snow depend on air temperature (Grenfell and Maykut, 1977; Grenfell, 1983). The ice albedo decreases with increasing air temperature, while snow albedo decreases with an increase in the proportion of its liquid phase. The seasonal changes in light intensity are reflected in the photosynthetic response of the ice algae (Fig. 4), in the chlorophyll biomass (Fig. 5) and in species diversity (Gosselin et al., 1985) which could result from a synergetic effect of light and temperature (Sullivan and Palmisano, 1981; Rochet et al., 1985). Algal growth in the ice does not start before some critical light level (I_{cr}) is reached (Table 2). The difference between Arctic and Antarctic critical values may be related to differences in species composition or in the physical characteristics of the sea ice (McConville and Wetherbee, 1983; Clarke and Ackley, 1984). Ice algae were found to be markedly shade adapted (Bunt 1964; Bunt, 1968; Burkholder and Mandelli, 1965). Maximal growth of microalgae at the ice-water interface was generally observed at ~ 1000 lux and photoinhibition occured at ~ 10000 lux (Bunt, 1964; Burkholder and Mandelli, 1965). Interestingly, Gosselin et al. (1985) observed that the photosynthetic parameter I_k was approximately equal to the under-ice light (I_z) (Fig. 6), suggesting a strong adaptation of epontic cells to the ambient light conditions.

As the snow cover presents great variability in time and in space, it is expected that the horizontal light distribution under the ice will be quite variable. This is the cause of an important small-scale spatial heterogeneity in the horizontal distribution of ice microalgae (Alexander et al., 1974; Clasby et al., 1976; Horner and Schrader, 1982; Bunt and Lee, 1970; Sasaki and Watanabe, 1984). Gosselin et al. (ms) were able to demonstrate a relationship between the spatial distribution of ice algae and snow depth in Hudson Bay. This relationship however changed during the season. At the beginning of the growing season (when $I_z > I_{cr}$) maximum algal biomass was observed under areas covered by the smallest snow depths. Towards the end of the season, when the solar input was much higher, maximum algal biomass was observed under areas covered by the deepest snow. This suggests that ice algae have two critical light levels: a minimum level and a maximum one. The minimum level is the critical light intensity below which photosynthetic activity is not detectable (I_{cr} ~ 7.6 μEin.m^{-2}.s^{-1} for Hudson Bay), and the maximum level at a given time corresponds to the light intensity above which photosynthetic activity is inhibited. The snow cover thus extends the length of the growing season for the epontic community, by offering areas where light intensity in springtime is compatible with the physiological limits of the cells. The observed variability in the distribution of ice algae might correspond to different stages of development of the epontic community, as was suggested for phytoplankton by Steele (1978). The snow cover, by its optical and insulating properties, thus protects the ice microflora against photoinhibition, and delays the dispersion of the cells by desaltation.

In addition to changing the light intensity, the snow cover also changes the spectral quality of the under-ice irradiance (Thomas, 1963). Maykut and Grenfell (1975) showed that the spectral distribution of light ranged between 430 and 550 nm at the ice-water interface, so that the red part of the spectrum was highly attenuated by the ice cover. This could have an important impact on the pigment composition and the photosynthetic activity of the ice algae, since it has long been recognized that light quality plays an important role for phytoplankton (Atlas and Bannister, 1980; Vesk and Jeffrey, 1977; Haxo, 1960). For example, Wallen and Geen (1971a,b) found that the photosynthetic activity of the diatom Cyclotella nana and

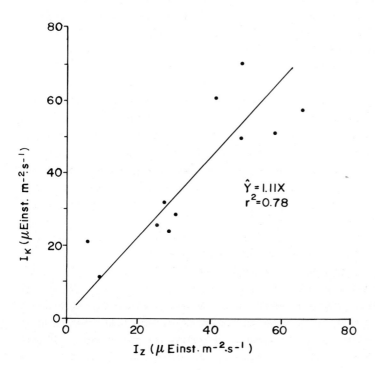

Fig. 6   Photosynthetic parameter I_k of sea-ice microalgae in Manitounuk Sound, Hudson Bay, as a function of the irradiance (I_z) at the ice-water interface.   Data from Gosselin et al. (1985).

the green alga <u>Dunaliella</u> <u>tertiolecta</u> was higher under blue light than under green or white light.   Faust et al. (1982) also found higher growth of <u>Prorocentrum</u> <u>mariae-lebouriae</u> under blue and red light than under white light. Changes in pigment composition (Vesk and Jeffrey, 1977; Jones and Myers, 1965; Brody and Emerson, 1959; Fujita and Hattori, 1959), growth rate and respiration were also shown to depend on light quality. The effect of light quality on the growth of ice algae has never been studied before, but Rochet et al. (ms) recently showed changes in the pigment composition of microalgae in Hudson Bay, in response to light quality.

Nutrients

Several researchers have concluded that nutrients probably did not limit the growth of the epontic community (Meguro et al., 1967; Bunt and Lee, 1970; Alexander et al., 1974; Sullivan and Palmisano, 1981; Horner and Schrader, 1982; Poulin et al., 1983; Clarke and Ackley, 1984; Holm-Hansen et al., 1977). The concentration of dissolved nutrients in the ice was found to be about one order of magnitude higher than in the underlying water (Horner, 1976; Demers et al., 1984). Three processes are usually invoked to explain nutrient abundance in the ice habitat: (1) replenishement of nutrients by exchange between the lower part of the ice and the underlying water, (2) desaltation and (3) <u>in situ</u> bacterial regeneration (Meguro et al., 1967; Alexander et al., 1974; Sullivan and Palmisano, 1981). Only Grainger (1977, 1979) has ever mentioned the possibility of nutrient limitation of ice algae for the Arctic. Recently, however, bioassays carried out by Maestrini et al. (ms), in southeastern Hudson Bay, gave direct evidence of nutrient limitation of the ice microalgae. Nitrogen was identified as the limiting nutrient, but surprisingly, the limiting threshold was quite high (~ 15 μmol N.l^{-1}). Such a high limiting threshold for nitrogen had also been observed by Maestrini et al. (1982) for benthic diatoms in an oyster pond. In Hudson Bay, the observed low N:P ratios (~ 5.2) suggest a large excess of phosphate over nitrogen. However, when nitrogen values reach or exceed the threshold level, phosphate and silicate might become the limiting nutrients (Maestrini et al., ms). In Manitounuk Sound, Gosselin et al. (1985) found evidence for a relationship between the photosynthetic efficiency (α^{B}) of the ice algae and phosphate replenishment of the ice-water interface by fortnightly tidal mixing, thus suggesting a possible phosphate limitation of microalgal growth.

Cota et al. (ms), by indirect calculations, showed that a substantial flux of nutrients from the underlying water is necessary to sustain the observed rate of increase in plant biomass. Gosselin et al. (ms) observed, in Hudson Bay, chlorophyll concentrations over 30 mg m^{-2}. Assuming no losses, we can conclude from the N:Chl ratios that at least 150-300 mg N were fixed by microalgae per square metre. If only regeneration and desaltation had been involved, the water forming the ice ($\sim$ 2m thick) should have contained at least 75-150 mg N m^{-3}, values which are much higher than those reported for this area (Legendre et al., 1981). Thus, active upward nutrient transport from the water column to the ice-water interface is required for the growth of the ice algae. This is supported by the findings of Gosselin et al. (1985), mentioned above, which indicated nutrient enrichment of the ice-water interface by fortnightly tidal mixing. In Barrow Strait, low frequency fluctuations of the growth of ice algae were also related to nutrient pulses, driven by the tides and atmospheric events (Cota et al., ms). The biological dynamics of the ice-water interface is therefore coupled to the hydrodynamics of the underlying waters.

The rate of nutrient transport at the ice-water boundary depends both on the downward flux by salt rejection and the brine cell drainage from the ice and the upward flux from the underlying waters. During the ice growth period, brine rejection and/or brine drainage can cause local destabilisation of the water layer adjoining the ice (Lake and Lewis, 1970; Lewis, 1972). Whereas, during the northern spring, ice melting both at the lower and upper ice boundaries can occur. Increased energy absorption on the lower boundary can occur in response to algal presence (Lewis, this book). The addition of melt water at the ice-water boundary leads to a stabilization of the adjacent thin layer. The degree to which melting occurs depends on the distribution of snow cover, solar radiation and temperature at the upper and lower interfaces. Nutrient input to the interface relies on destabilization occurring frequently. However, it is more likely that in tranquil areas, entrainment processes associated with internal wave generation on the melt-ambient water interface leads to nutrient replenishment. Undulations in a thin layer immediately under the ice can be seen visually during the melt period in Hudson Bay.

The physical processes occuring at the ice-water interface which determine the rate of upward nutrient transport are found to vary both in space and time. The boundary layer characteristics depend on the smoothness of the local and surrounding ice, the regularity of the current regime and other factors related to the gravitational stability in the layer. The degree of smoothness and its effect on the velocity structure in the boundary layer has been investigated by Langleben (1982) for first year sea ice and by Chriss and Caldwell (1984) for a smooth ocean floor. Obviously, in areas adjacent to pressure ridging, the turbulence characteristics are greatly modified. The regularity of the ambient motion depends both on low frequency and tidal forcing. From detailed observations of the velocity, density and nutrient fields, it is possible to estimate the vertical nutrient flux. A vertical eddy diffusion coefficient can be estimated from an equation of the form $K_z = K_0 (1+mRi)^{-n}$ where K_0 is the coefficient for neutral stability, Ri is the Richardson number and m and n are positive numbers (Jones, 1973). Richardson number is defined as $Ri = g.\rho^{-1} .d\rho/dz. (dU/dz)^{-2}$ where ρ is density, g is gravity, dz is the vertical length scale and dU is horizontal velocity difference over dz (Turner, 1973). Using a Fick's law formulation, the diffusive nutrient flux (Q) can be estimated from $Q = K_z dN/dz$ where dN is the difference in nutrient concentration

over dz.   The extended time series required to calculate actual fluxes are in
practice difficult to obtain in the field.

Stability of the boundary layer varies from convectively unstable
during brine rejection to very stable near fresh water sources.   In areas sufficiently far from coastal fresh water sources such that only weak stratification is
present or during periods of meltwater input, conditions suitable for the retention
of algae in the melt layer and input of nutrients from deeper waters can coexist, the
rate of nutrient input varying with the local stability.   In a Richardson number
sense, values of order one indicate onset of turbulence.   Values larger than one
indicate a reduction of vertical diffusion, as described in the preceding paragraph.
For example, a pycnocline of 10 cm in thickness separating melt and ambient waters
requires a velocity difference of greater than 15 cm.s^{-1} to become unstable.
Richardson numbers calculated by Ingram (1981) in the interior of the under-ice plume
of the Great Whale River (Hudson Bay) were sufficiently large (30+) that little or no
mixing occurred between the 5 m thick fresh water layer and the underlying sea water.
Salinity, nutrient and algal concentration levels were negligible, the latter in
response to the low salinity.   In contrast, Gosselin et al. (1985) showed that
nutrient transport to the interface and algal growth occurred in areas where the
Richardson number varied between 1 and 10, the variability resulting from neap-spring
tidal forcing and low frequency current changes.   The actual structure of the
boundary layer subject to tidal forcing shows a departure from a logarithmic profile
(Soulsby and Dyer, 1981).

Thus, the dynamics of the interface not only determine the nutrient supply, but
also the stability of the adjoining fluid layer.   The alternation of stable and
unstable conditions in the melt water layer may affect algal concentrations if
positive buoyancy is essential to their retention at the interface.   Figure 7 shows
in sketch form the various factors influencing the light, temperature, salinity as
well as nutrient and algal levels at the interface.   The similarity of algal growth
at the periphery of river-derived fresh water sources and in melt water regions may
be through the presence of a lighter layer of intermediate stability relative to the
ambient fluid.   Destabilisation occurs in response to tidal or other current
variability.   These conditions can be parameterized by the Richardson number values
in the under-ice layer.

THE ICE-WATER INTERFACE AS AN ERGOCLINE

The model implicit in most of the literature on the ice-water interface is that of
a temporal ergocline, located between the autumn vertical mixing of the water column
and the spring/summer melt of the ice followed by summer stratification.   In a sense,
the growth of microalgae at the ice-water interface was often considered as an
extended spring bloom, frozen in the ice sheet.   In this model, nutrients used by
microalgal growth were assumed to be replenished by the autumn vertical mixing,

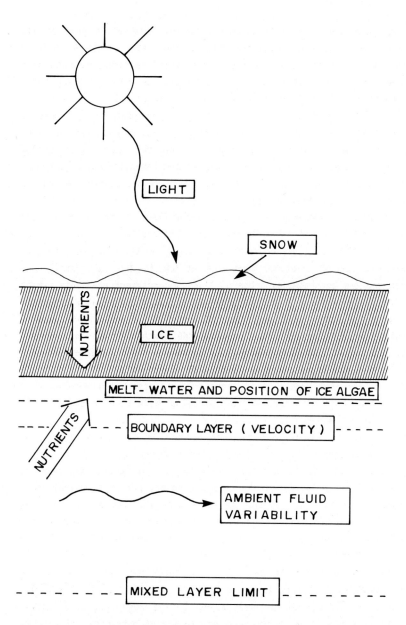

Fig. 7   Schematic representation of various factors that influence temperature, salinity, irradiance, nutrients and microalgal concentration at the ice-water interface.

frozen in the ice, and made available to the cells by desaltation and in situ regeneration. It has been explained above that such an hypothesis cannot account for the observed high concentrations of epontic microalgae, which leads to the conclusion that the ice-water interface must be dynamically linked to the underlying water column (e.g. Demers et al., 1984, Gosselin et al., 1985; Cota et al., ms).

In order to explain the high productivity of the ice-water interface, Demers et al. (1984) have compared it with frontal areas where microalgal productivity is often very high. In more general terms, it is proposed that the ice-water interface belongs to the general category of ergoclines (Legendre and Demers, 1985; Legendre et al., this book). The ice-water ergocline is located between the stable ice and the highly energetic nutrient-rich water column.

The ice-water interface is often described as a mirror image of the water-sediment ergocline. Both ergoclines offer a stable physical substrate on which or in which organisms can grow; as a consequence, the microalgae which colonize the two ergoclines are mainly pennate diatoms, that can attach to the substrate. One difference between the two environments concerns light and nutrient gradients, which are parallel at the water-sediment ergocline and inverted at the ice-water interface (Fig. 8). A more significant difference lies in the flux of particulate organic matter. Particulate organic matter flows towards the benthic interface, where a complex animal and bacterial community has evolved to manage this abundant resource. On the contrary, particulate organic matter falls off the ice-water interface, where a complex animal and bacterial community is therefore unnecessary to take charge of the primary production. This might explain the striking difference in the relative complexities of the benthic and the ice fauna.

As explained by Gosselin et al. (1985), the production of microalgae at the ice-water interface is controlled by both climatic and hydrodynamic phenomena. Light available for photosynthesis depends on the seasonal increase in solar irradiance and also on the meteorological influence on the snow-ice cover, nutrient replenishment relies on hydrodynamic events. In addition to those controls from above and below the ergocline, the existence of the interface itself depends on the seasonal heat flux into the ice and also perhaps on the biological production at the interface (Lewis, this book).

The high microalgal production at the ice-water interface, which can become nutrient limited, indicates that this environment is an ergocline (Legendre et al., this book). At this ergocline, the auxiliary energy of the underlying water column acts on microalgal production through the proximal agency of nutrients. Inputs of auxilary energy in the water column do not affect the light regime at the interface. The fact that the ergocline does not generally extend downwards in the water column is probably related to the lack of vertical stability (inadequate physical scale) rather than to light limitation. This is supported by two observations: (1) if artificial substrates are placed a few metres below the ice-water interface, at times

ICE – WATER ERGOCLINE   WATER–SEDIMENT ERGOCLINE

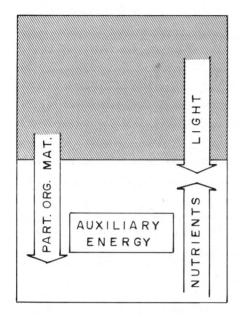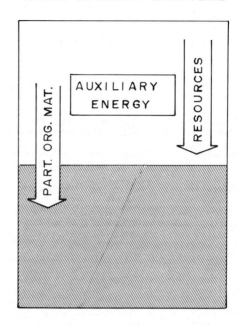

Fig. 8   Schematic representation of the fluxes of light, nutrients and particulate organic matter at the ice-water and water-sediment ergoclines.

when no phytoplankton are found in the water column, these substrates are rapidly colonized by microalgae (M. Gosselin, unpublished data); (2) when the water column stabilizes, as a consequence of ice melt, a phytoplankton bloom can occur under the ice (Legendre et al., 1981). By comparison with other ergoclines that support high microalgal production (temporal transitions, nutriclines, tidal fronts, an so on), the ice-water interface thus offers the advantage of less complexity, since the hydrodynamics do not influence the light regime at the same time as it acts on nutrients.   The ice-water ergocline can therefore be used to test the hypotheses concerning ergoclines proposed by Legendre et al. (this book).

    There are two major advantages in using the ergocline approach to study biological production at the ice-water interface. First, the comparison with other ergoclines can evidence general mechanisms that govern biological production in the oceans. Second, the ergocline perspective can suggest new avenues for research as well as new sampling and/or experimental designs concerning specific mechanisms that govern biological production at the ice-water interface.

50

REFERENCES

Ackley, S.F., Buck, H.R. and Taguchi, s., 1979. Standing crop of algae in the ice of the Weddel Sea region. Deep-Sea Res., 26: 269-281.
Alexander, V., 1974. Primary productivity regimes of the nearshore Beaufort Sea with reference to potential rates of ice biota. In: J.C. Reed, J.E. Later (Editors), The coast and shelf of the Beaufort Sea. The Arctic institute of North America, Calgary, Alberta, pp. 609-632.
Alexander, V., Horner, R.A. and Clasby, R.C., 1974. Metabolism of Arctic Sea-ice organisms. Rep. Inst. mar. Sci. Univ. Alaska R74-4, 120 pp.
Alexander, V. and Niebauer, H.J., 1981. Oceanography of the eastern Bering Sea ice-edge zone in spring. Limnol Oceanogr., 26: 1111-1125.
Allen, M.B., 1971. High latitude phytoplankton. Annu. Rev. Ecol. Syst. 2: 261-276.
Andriashev, A.P., 1968. The problem of the life community associated with Antarctic fast ice. In: SCAR-SCOR-IAPO-IUBS Symposium on Antarctic Oceanography; Santiago, Chile, pp. 147-155.
Apollonio, S., 1961. The chlorophyll content of Arctic sea-ice. Arctic, 14: 197-200.
Apollonio, S., 1965. Chlorophyll in Arctic sea-ice. Arctic, 18: 118-122.
Atlas, D. and Bannister, T.T., 1980. Dependence of mean spectral extinction coefficient of phytoplankton on depth, water color, and species. Limnol. Oceanogr., 25: 157-159.
Ben-Amotz, A. and Gilboa, A., 1980. Cryopreservation of marine unicellular algae with regard to size, culture age, photosynthetic activity and chlorophyll-to-cell ratio. Mar. Ecol. Prog. Ser., 2: 157-161.
Bradford, J.M., 1978. Sea-ice organisms and their importance in the Antarctic ecosystem. N.Z. Antarct. Res., 1: 43-50.
Bradstreet, M.S.W. and Cross, W.E., 1982. Trophic relationships at high Arctic ice-edges. Arctic, 35: 1-12.
Brody, M. and Emerson, R., 1959. The effect of wavelength and intensity of light on the proportions of pigments in Pophyridium cruentum. Am. J. Bot., 43: 433-440.
Brock, T.D., 1967a. Microorganisms adapted to high temperatures. Nature, 214: 882-885.
Brock, T.B., 1967b. Life at high temperatures. Science, 158: 1012-1019.
Bunt, J.S., 1963. Diatoms of Antarctic sea-ice as agents of primary production. Nature, Lond., 199: 1255-1257.
Bunt, J.S., 1964. Primary productivity under sea-ice in Antarctic waters. 2. Influences of light and other factors on photosynthetic activities of Antarctic marine microalgae. In: M.O. Lee (Editor), Biology of Antarctic Seas, Antarct. Res. Ser., 1: 27-31.
Bunt, J.S., 1968. Microalgae of the Antarctic pack ice zone. In: R.I. Currie (Editor), Proceedings of the symposium on Antarctic Oceanography; Santiago, Chile, 13-16 September, 1966, pp. 198-218.
Bunt, J.S. and Lee, C.C., 1970. Seasonal primary production in Antarctic sea-ice at McMurdo Sound in 1967. J. Mar. Res.. 28: 304-320.
Bunt, J.S. and Lee, C.C., 1972. Data on the composition and dark survival of four sea-ice microalgae. Limnol. Oceanogr., 17: 458-461.
Bunt, J.S. and Wood, E.J.F., 1963. Microalgae and Antartic sea-ice. Nature, Lond., 199: 1254-5.
Burkholder, D.R. and Mandelli, E.F., 1965. Productivity of microalgae in Antarctic sea-ice. Science (Wash D.C.), 149: 872-874.
Chriss, T.M. and Caldwell, D.R., 1984. Universal similarity and the thickness of the viscous sublayer at the ocean floor. J. Geophys. Res., 89: 6403-6414.
Clarke, D.B. and Ackley, S.F., 1984. Sea ice structure and biological activity in the Antarctic marginal ice zone. J. Geophys. Res., 89: 2087-2095.
Clasby, R.C., Alexander, V. and Horner, R.A., 1976. Primary productivity of sea-ice algae. In: D.W. Hood and D.C. Burrel (Editors), Assessment of the Arctic marine environment. Selected topics. Institute of Marine Science, University of Alaska, pp. 289-304.

Clasby, R.C., Horner, R.A. and Alexander, V., 1973. An in situ method for measuring primary productivity of sea-ice algae. J. Fish. Res. Board Can., 30: 835-838.

Conrad, N., 1983. Adaptability. Pelnum Press, New-York and London. 408 p.

Cota, G.F., Lewis, M.R., Loder, J.W., Bennett, E.B., Prinsenberg, S.J., Anning, J.L., Watson, N.H.F. and Harris, L.R. ms. Mechanisms and dynamics of nutrient supply during extended blooms of Arctic ice algae.

Cross, W.E., 1982. Under-ice biota at the Pond Inlet ice edge and in adjacent fast ice areas during spring. Arctic, 35: 13-27.

Demers, S., Therriault, J.C. and Descolas-Gros, C., 1984. Biomasse et composition spécifique de la microflore des glaces saisonnières: influences de la lumière et de la vitesse de congélation. Mar. Biol., 78: 185-191.

Dunbar, M.J. and Acreman, J.C., 1980. Standing crops and species composition of diatoms in sea-ice from Robeson Channel to the gulf of St.Lawrence. Ophelia, 19: 61-72.

Ehrenberg, C.G., 1841. Ein Nachtrag zu dem Vortrage über Verbrutung und Einfluss des mickoskopeschen lebens in Süd und Nord-America. Monatsber. d. Berl. Akad., 1841: 202-207.

Ehrenberg, C.G., 1853. Über neue Anschauungen des Kleinsten nördlichen Polarlebens, Monastsber. d. Berl. Akad., 1853: 522-529.

El-Sayed, S.Z. and Taguchi, S., 1981. Primary production and standing crop of phytoplankton along the ice-edge in the Weddell Sea. Deep-Sea Res., 28: 1017-1032.

Faust, M.A., Sager, J.C. and Meeson, B.W., 1982. Response of Prorocentrum mariae-Lebouriae (dinophyceae) to light of different spectral qualities and irradiances: Growth and pigmentation. J. Phycol., 18: 348-356.

Fogg, G.E., 1977. Aquatic primary production in the Arctic. Philos Trans. R. Soc. London Ser. B., 279: 27-38.

Fujita, Y. and Hattori, A., 1959. Formation of phycobilin pigments in a blue-green alga, Tolypothrix tenuis, as induced by illumination with colored lights. J. Biochem. (Tokyo), 46: 521-524.

Gosselin, M., Legendre, L., Demers, S. and Ingram, R.G., 1985. Responses of sea-ice microalgae to climatic and fortnightly tidal energy inputs (Manitounuk Sound, Hudson Bay). Can. J. Fish. Aquat. Sci., 42: 999-1006.

Gosselin, M., Legendre, L., Therriault, J.C. and Demers, S., ms. Physical control of the horizontal patchiness of sea-ice microalgae.

Grainger, E.H., 1977. The annual nutrient cycle in sea-ice. In M.J. Dunbar (Editor), Polar oceans, Proceedings of SCOR/SCAR Polar Oceans Conference, Montreal, May 1974. Arctic Institute of North America, Calgary, pp. 285-299.

Grainger, E.H., 1979. Primary production in Frobisher Bay, Arctic Canada. In: M.J. Dunbar (Editor), Primary production mechanisms, International biological programme 20. Cambridge University Press, London, pp. 9-30.

Grainger, E.H. and Hsiao, S.I.C., 1982. A study of the ice biota of Frobisher Bay, Baffin Island 1979-1981. Can. MS Rep. Fish. Aquat. Sci. 1647. 128 pp.

Grainger, E.H., Mohammed, A.A. and Lovrity, J.E., 1985. The Sea Ice Fauna of Frobisher Bay, Arctic Canada. Arctic, 38: 23-30.

Grant, W.S. and Horner, R.A., 1976. Growth responses to salinity variation in four arctic ice diatoms. J. Phycol., 12: 180-185.

Grenfell, T.C., 1983. A theoretical model of the optical properties of sea-ice in the visible and near infrared. J. Geophys. Res., 88: 9723-9735.

Grenfell, T.C. and Maykut, G.A., 1977. The optical properties of ice and snow in the Arctic basin. J. Glaciol., 18: 445-463.

Griffiths, R.P., Hayasaka, S.S., McNamara, T.M. and Morita, R.Y., 1978. Relative microbial activity and bacterial concentrations in water and sediments samples taken in the Beaufort Sea. Can. J. Microbiol., 24: 1217-1226.

Gruzov, E.N., 1977. Seasonal alternations in coastal communities in the Davis Sea. In: G.A. Llano (Editor), Adaptations within Antarctic ecosystems, Proc. Third SCAR Symp. Antarc. Biol. Smithsonian Institution Washington, D.C., pp. 263-268.

Gulliksen, B., 1984. Under-ice fauna from Svalbard waters. Sarsia, 69: 17-23.

Hameedi, M.J., 1978.   Aspects of water column primary productivity in Chukchi Sea during summer. Mar. Biol. 48: 37-46.

Haxo, F.T., 1960.   The wavelenght dependence of photosynthesis and the role of accessory pigments.   In: M.B. Allen (Editor), Comparative biochemistry of photoreactive pigments. Academic Press, New-York, pp. 339-360.

Holm-Hansen, O., El-Sayed, S.Z., Franceschini, G.A. and Gubel, R.L., 1977. Primary production and the factors controlling phytoplankton growth in the Southern Ocean. In: G.A. Llano (Editor), Adaptations within the Antarctic Ecosystems, Proc. Third SCAR Symp. Antarct. Biol. Smithsonian Institution, Washington, D.C., pp. 11-50.

Hoocker, J.D., 1847.   The botany of the Antarctic voyage of H.M. Discovery ships Erebus and Terror in the years 1839-1843. I. Flora Antarctica. Reprinted 1963, J. Cramer, Weinhein, pp. 503-519.

Horner, R.A., 1972. Ecological studies on arctic sea-ice organisms. Progress report to the Office of Naval Research. Contract N00014-67-A-0317-003, 79 pp.

Horner, R.A., 1976.   Sea-ice organisms. Oceanogr. Mar. Biol. Annu. Rev., 14: 167-182.

Horner, R.A., 1977.   History and recent advances in the study of ice biota. In: M.J. Dunbar (Editor), Polar oceans Conference, Montreal, May 1974.   Arctic Institute of North America, Calgary, pp. 269-283.

Horner, R.A. and Alexander, V., 1972.   Algal populations in arctic sea-ice: an investigation of heterotrophy. Limnol. Oceanogr., 17: 454-458.

Horner, R.A. and Schrader, G.C., 1982.   Relative contributions of ice algae, phytoplankton and benthic microalgae to primary production in nearshore regions of the Beaufort Sea. Arctic, 35: 485-503.

Hoshiai, T., 1969.   Ecological observations of the colored layer of the sea-ice at Syowa Station. Antarct. Res., 34: 60-72.

Hoshiai, T., 1972.   Diatom distribution in sea-ice near McMurdo and Syowa Station. Antarct. J. of U.S., 7: 84-85.

Hoshiai, T., 1977. Seasonal changes of ice communities in the sea ice near Syowa Station, Antarctica.   In M.J. Dunbar (Editor), Polar oceans, Proceeding of SCOR/SCAR Polar Oceans Conference, Montreal, May 1974. Arctic Institute of North America, Calgary, pp. 285-299.

Hoshiai, T., 1981. Proliferation of ice algae in the Syowa Station area, Antarctic. Mem. Natl. Inst. Polar Res., 34: 1-12.

Hsiao, S.I.C., 1980. Quantitative composition, distribution, community structure and standing stock of sea-ice microalgae in the Canadian Arctic.   Arctic, 33: 768-793.

Ingram, R.G., 1981. Characteristics of the Great Whale River Plume. J. Geophys. Res., 86: 2017-2023.

Jones, J., 1973.   Vertical mixing in the Equatorial undercurrent.   J. Phys. Oceanogr., 3: 286-296.

Jones, L.W. and Meyers, J., 1965. Pigment variations in Anacystis nidulans induced by light of selected wavelengths. J. Phycol., 1: 7-14.

Kaneko, T., Roubal, G. and Atlas, R.M., 1978.   Bacterial populations in the Beaufort Sea. Arctic, 31: 97-107.

Lake, R.A. and Lewis, E.L., 1970.   Salt rejection by sea-ice during growth. J. Geophys. Res., 75: 583-597.

Langleben, M.P., 1982. Water drag coefficient of first-year sea-ice. J. Geophys. Res., 87: 573-578.

Legendre, L. and Demers, S., 1985. Auxilary energy, ergoclines and aquatic biological production, Naturaliste can. 112: 5-14.

Legendre L., Demers, S. and Lefaivre, D., (1986). Biological production at marine Ergoclines. (this book).

Legendre, L., Ingram, R.G. and Poulin, M., 1981. Physical control of phytoplankton production under sea-ice (Manitounuk Sound, Hudson Bay). Can. J. Fish. Aquat. Sci., 38: 1385-1392.

Lewis, E.L., 1972. Convection beneath sea ice. Coll. Internat. du CNRS, no. 215: 49-55.

Lewis, M.L., (1986).   Radiation, absorption by ice-algae: influence on the heat budget of springtime sea-ice. (this book).

Li, W.K.W. and Morris, I., 1982. Temperature adaptation in Phaeodactylum tricornutum bohlin: photosynthetic rate compensation and capacity.   J.   Exp. Mar. Biol. Ecol., 58: 135-150.

Li, W.K.W., Smith, J.C. and Platt, T., 1984. Temperature response of photosynthetic capacity and carboxilase activity in Arctic marine phytoplankton.   Mar. Ecol. Prog. Ser., 17: 237-243.

Lipps, J.H. and Krebs, W.N., 1974. Planktonic foraminefera associated with Antarctic sea-ice. J. Foraminiferal Res., 4: 80-85.

Maestrini, S.Y., Rochet, M., Legendre, L., and Demers, S., (ms). Nutrient limitation of the ice-microalgal biomass (southeastern Hudson Bay, Canadian Arctic).

Malone, T.C., 1977. Environmental regulation of phytoplankton productivity in the lower Hudson Estuary. Estuar. Coastal mar. Sci., 5: 157-171.

Matheke, G.E.M. and Horner, R.A., 1974.   Primary productivity of the benthic microalgae in the Chukcki Sea near Barrow, Alaska. J. Fish Res. Board Can., 31: 1779-1786.

Maykut, G.A. and Grenfell, T.C., 1975. The spectral distribution of light beneath first-year sea ice in the Arctic Ocean. Limnol. Oceanogr. 20: 554-563.

McConville, M.J. and Wetherbee, R., 1983. The bottom-ice microalgal community from natural ice in the inshore waters at east Antarctica. J. Phycol., 19: 431-439

McRoy, C.P. and Goering, J.J., 1974. The influence of ice the primary productivity of the Bering Sea. In: D. Hood and E. Kelly (Editors), The Oceanography of the Bering Sea.   University of Alaska Institute of Marine Sciences, Fairbanks., pp. 403-421.

Meguro, H., 1962.   Plankton ice in the Antarctic Ocean, Antarct. Record., 14: 1192-1199.

Meguro, H., Ito, K., and Fukushima, H., 1966. Diatoms and the ecological conditions of their growth sea-ice in the Arctic ocean. Science (Washington, D.C.), 152: 1089-1090.

Meguro, H., Kuniyuki, I. and Fukushima, H., 1967.   Ice flora (bottom type):   a mechanism of primary production in the polar seas and growth of diatoms in sea ice. Arctic, 20: 114-133.

Moiseev, P.A., 1971.   The living resources of the world Oceans (translated from Russian).   Published for National Marine Fish Service, National Oceanic and Atmospheric Administration, U.S.   Dept. of Commerce and the National Science foundation, Washington D.C., by Israël Program for Scientific Translation, Jerusalem.

Neori, A., and Holm-Hansen, O., 1982.   Effect of temperature on rate of photosynthesis in Antarctic phytoplankton. Polar Biol., 1: 33-38.

Niebauer, H.J., Alexander, V. and Cooney, R.T., 1981. Primary production at the Eastern Bering Sea-ice edge. The physical and biological regimes. In: D.W. Hood and T.A.   Colder (Editors), the Eastern Bering Sea Shelf: Oceanography and Resources. U.S. Deept. commerce, pp. 763-772.

Palmisano, A.C., Kottmeier, S.T., Moe, R.L. and Sullivan, C.W., 1985.   Sea-ice microbial communities. IV. The effect of light perturbation on microalgae at the ice-seawater interface in McMurdo Sound, Antartica. Mar. Ecol. Prog. Ser., 21: 37-45.

Palmisano, A.C. and Sullivan, C.W., 1983. Sea-ice microbial communities (SIMLO). 1. Distribution, abundance and primary production of ice microalgae in McMurdo Sound, Antarctic in 1980. Polar Biol., 2: 171-177.

Platt, T., Harrison, W.G., Irwin, B., Horne, E.P. and Gallegos, C.L., 1982. Photosynthesis and photoadaptation of marine phytoplankton in the Arctic. Deep-Sea Res., 29: 1159-1170.

Poulin, M. and Cardinal, A., 1982a.   Sea-ice diatoms from Manitounuk Sound, southeastern, Hudson Bay (Quebec, Canada). I. Family Naviculaceae. Can. J. Bot., 60: 1263-1278.

Poulin, M. and Cardinal, A., 1982b.   Sea-ice diatoms from Manitounuk Sound, Southeastern Hudson Bay (Quebec, Canada). II.   Naviculaceae, genus Navicula. Can. J. Bot., 60: 2825-2845.

Poulin, M. and Cardinal, A., 1983. Sea-ice diatoms from Manitounuk Sound, Southeastern Hudson Bay (Quebec, Canada). III. Cymbellaceae, Entomoneidaceae, Gomphonemataceae, and Nitzschiaceae. Can. J. Bot., 61: 107-118

Poulin, M., Cardinal, A. et Legendre, L., 1983. Réponse d'une communauté de diatomées de glace à un gradient de salinité (Baie d'Hudson). Mar. Biol., 76: 191-202.

Richardson, M.G. and Whitaker, T.M., 1979. An Antarctic fast-ice food chain: observations on the interaction of the amphipod Pontogeneia antarctica chevreux with ice-associated microalgae. Br. Antarct. Surv. Bull., 47: 107-115.

Rochet, M., Legendre, L. and Demers, S., 1985. Acclimation of sea-ice microalgae to freezing temperature. Mar. Ecol. progr. Ser. 24: 187-191.

Rochet, M., Legendre, L. and Demers, S., ms. Réponses des microalgues des glaces aux variations quantitatives et qualitatives de la lumière à l'interface glace-eau: Photosynthèse et pigments.

Rodhe, W., 1955. Can plankton production proceed during winter darkness in subarctic lakes? Verh. int. ver. theo. Ang. Limnol., 12: 117-122.

Saito K and Taniguchi, A., 1978. Phytoplankton communities in the Bering Sea and adjacent seas. II. Spring and summer communities in seasonally ice covered areas. Astarte 11: 27-35.

Sasaki, H., and Watanabe, K., 1984. Underwater observations of ice algae Lützow-Holm Bay, Antarctica. Antarct. Res., 81: 1-8.

Seaburg, K.G., Parker, B.C., Nharton, R.A., Jr and Simmons, G.M., Jr., 1981. Temperature growth responses of algal isolated from Antarctic oases. J. Phycol., 17: 353-360.

Smith, W.O. and Nelson, D.M., 1985. Phytoplankton bloom produced by a receding ice edge in the Ross Sea: Spatial coherence with the density field. Science, 227: 163-166.

Soulsby, R.L. and Dyer, K.R., 1981. The form of the near-bed velocity profile in a tidally accelerating flow. J. Geophys. Res., 86: 8067-8074.

Subba Rao, D.V. and Platt, T., 1984. Primary production of Arctic waters. Polar Biol., 3: 191-201.

Sullivan, C.W. and Palmisano, A.C., 1981. Sea-ice microbial communities in McMurdo Sound. Antarct. J. U.S., 16: 126-127.

Steele, J.H., 1978. Some comments on plankton patches. In: J.H. Steele (Editor), Spatial pattern in plankton communities. Plenum Press, New-York and London, pp. 1-17.

Therriault, J.C. and Levasseur, M., 1985. Control of phytoplankton production in the Lower St.Lawrence Estuary: Light and Freshwater runoff. Naturaliste Can., 112: 77-96.

Thomas, C.W.,1963. On the transfer of visible radiation through sea-ice and snow. J. Glaciol., 4: 481-484.

Turner, J.S., 1973. Buoyancy effects in fluids. Cambridge Univ. Press, New-York, 367 pp.

Vesk, M. and Jeffrey, S.W., 1977. The effect of blue-green light on photosynthetic pigments and chloroplast structure in unicellular marine algae from six classes. J. Phycol., 13: 280-288.

Wallen, D.G., and Geen, G.M., 1971a. Light quality in relation to growth, photosynthetic rates and carbon metabolism in two species of marine plankton algae. Mar. Biol., 10: 34-43.

Wallen, D.G. and Geen, G.H., 1971b. Light quality and concentration of proteins, RNA, DNA and photosynthetic pigments in two species of marine plankton algae. Mar. Biol., 10: 44-51.

Whitaker, T.M., 1977. Sea-ice habitats of Signy Island (South Orkneys) and their productivity. In: G.A. Llano (Editor), Adaptations within Antarctic ecosystems. Proceedings of the third SCAR Symposium on Antarctic Biology, Smithsonian Institution, Washington, 75-82.

Wilce, R.T., 1967. Heterotrophy in arctic sublittoral seaweeds Bot. Mar., 10: 185-197.

Winter, D.F., Banse, K. and Anderson, G.C., 1975. The dynamics of phytoplankton blooms in Puget Sound, a Fjord in the northwestern United States. Mar. Biol., 29: 139-176.

STUDYING FRONTS AS CONTACT ECOSYSTEMS

Serge FRONTIER

Station Marine, BP 41, F62930 Wimereux (France)

ABSTRACT

At the level of an oceanographic front an enhanced primary productivity and/or accumulation of biomass can be observed, generating a particular ecosystem. Two main types of frontal ecosystems are to be distinguished :

(1) When the new or accumulated biomass is exploited by a trophic chain involving vagile (i.e., migrating) organisms, the biomass is exported to remote oligotrophic areas. Overaccumulation (eutrophication) is avoided and the exploited community remains a juvenile and opportunistic one.

(2) When no active food chain arises, the new or accumulated biomass is degraded through bacterial activity. A mature, complex and very diversified microheterotrophic community arises, with Protozoans, and a local regenerated primary production can occur.

Intermediate cases exist, for example filtering macroorganisms feeding on microheterotrophs, and possibly eaten by fish.

Enhanced vegetal production or biomass necessitates a tuning of the physical regime (nutrient advection or diffusion; light supply) and the biological phenomena. An "auxiliary energy", coming from the degradation of the kinetic energy of water masses until reaching a time-and-space scale compatible with biological phenomena, is then used. Tuning between various trophic fluxes and migrations along the trophic chain system also use an auxiliary energy able to bring into contact, at the right rythm, the various elements of the trophic network. That "secondary auxiliary energy" arises from the energy assimilated by consumer organisms which utilize a part of this energy for movements and migrations.

The interaction of two parts of the system, a productive one and a consumer and vagile one, may be considered as an exploitation (= biomass exportation) of an ecosystem by another one. The interaction needs an interpenetration between them. Auxiliary energy at various levels is used to establish the fractal spatio-temporal structure of the entire system.

INTRODUCTION

This paper gives the conclusion of the Working Group "Thème 5"[*] having participated in the elaboration of the FRONTAL Program, a French large-scale oceanographic experiment (PIROCEAN-INSU, 1985).

---

[*] M. Bianchi, F. Blanc, S. Frontier, F. Ibanez, G. Jacques, P. Klein, J. Le Fevre, M. Leveau, P. Mayzaud, L. Prieur, F. Rassoulzadegan, A. Sournia, D. Viale

Fronts are contact zones between two different water masses. At this level, particular biological and ecological phenomena occur, which perhaps determine more the ecological properties of the area, than the phenomena occurring inside the two water masses do. I give here an overview of the frontal phenomena from the point of view of the General Ecosystem Theory in its present (and provisional !) state, and particularly of the Theory of contact or interface between ecosystems, recently emphasized as *Ergocline Theory*
I also refer to the *Theory of exploitation of an ecosystem by another one.*

Remember that an ecosystem is not "a biomass" but is better defined as the interaction between biomass and the physical medium. This interaction has to be observed at various observation levels.

MAIN FEATURES IN FRONTAL ECOSYSTEMS

The main characteristics of oceanographic fronts are :
- often mesoscale phenomenons, in fact phenomena of very varying range of scales;
- rather well localized in some permanent areas, and
- frequently impulsional or quasi-periodic, that is, not permanent but showing important temporal variations.

The problem of frontal ecosystems can be formulated as follows : at the level of any front a particular ecosystem seems to appear which is distinct from those of the two adjacent water masses. An enhanced primary production is observed or postulated here but in some cases only a passive accumulation of biomass occurs. Sometimes, that locally increased biomass is exploited by various trophic chains which, through the migration of the organisms involved, redistribute the assimilated energy into more oligotrophic areas. Sometimes, on the contrary, the trophic chains do not develop well, and the accumulated biomass turns out to be degraded by microheterotrophic organisms. In all cases, a modification of the community composition is observed in the vicinity of fronts, but in various ways according to the type of front. It denotes a particular organization of the trophic network following the enhanced production of biomass, and adapted to it. According to the kind of biological organization, nutrients are regenerated either immediately within the area (allowing a regenerated production) or far away after a migration over a long period of time.

The ecological phenomena associated with fronts are indeed very diverse and, moreover, are to a certain extent unpredictable. This unpredictability is brought about by physical phenomena involved and instabilities, resulting in the unstable localization of the front and its often sporadic or quasi-periodic existence. It follows that a variety of descriptions of fronts have been proposed in the literature.

TUNING OF THE PHYSICAL REGIME AND THE BIOLOGICAL PHENOMENA

An analysis of the frontal phenomenon has already been outlined following
a number of recent works in various frontal zones (see for ex. Grall & al.,
1980; Le Fevre & al., 1970, 1981, 1983 a and b; Le Fevre, 1985; Holligan,
1979, 1981; Holligan & al., 1983, 1984, 1985; Pingree, 1978; Pingree & al.,
1974, 1975, 1978 a and b, 1979, 1981; Legendre, 1981; Legendre & Demers, 1984;
etc...). A thorough list of references and a complete review of the subject
are given in Le Fevre, 1985.

These analyses gave rise to a number of general ideas also encountered in
the analysis of non-pelagic ecosystems, particularly of continental ecosystems.
The main common feature seems to be the fact that, at the spatio-temporal
scales involving the physical phenomena associated with various interfaces,
a great deal of non-trophic energy is used by the ecosystem as "auxiliary
energy" (Margalef, 1974, 1985), allowing the "covariance" (Margalef, 1978),
that is, a spatio-temporal tuning between the proximal agents of the productivity namely light, nutrients and photosynthetic pigments. The kinetic
energy of water masses, when degraded until being able to act at a specific
scale compatible with the biological phenomena, has the role of realizing
that spatio-temporal "covariance". More precisely, the upwelling of nutrient
enriched water into the trophic zone, the turbulent mixing, eddies and
shearings, bring into contact the complementary elements, hence allow the
primary production.

Auxiliary energy is issued from a compartment of the solar radiation other
than the photosynthetic energy, namely the infrared, which is principally
used in moving fluids. This feature has an exact equivalent in the terrestrial
ecosystems For example, it can be shown that in order to produce any vegetal
biomass, a grassland needs 30 to 100 times more energy for its evapotranspi-
ration than for its photosynthesis in the same time. Obviously, photosynthesis
only occurs when a sufficient flow of sap goes upwards through the plants,
carrying the saline solution of soil up to the living and lightened leaves.
The evaporation energy driving this ascendent flow can be calculated
(Frontier, unpublished). We also may consider that the energy - derived from
the primary production - involved in the edification of anatomical structure
allowing the flow of sap, such as vessels and wood, is also an auxiliary
anergy (Margalef, pers. com.), which may be called "secondary auxiliary
energy".

In aquatic ecosystems, the "primary auxiliary energy" is much greater
than in terrestrial vegetations, because it involves the whole energy which
moves the concerned water masses, mixes them and provokes alternances of
stabilized and destabilized situations (Legendre, 1981; Legendre & Demers,1984).

The more or less favourable *periodicity* of the latter is more important, for
the ecosystem, that the amount of energy involved. *Turbulence* is a progressive
degradation of the kinetic energy of water masses into smaller and smaller
eddies down to the viscous range, and *at a given scale* of space and time, the
energy brings about the coincidence at any point of light, nutrients and living
cells. The synchronism, or tuning, of physical and biological rythms explains,
in that hypothesis, the particular dynamism observed in the vicinity of fronts.
A sustained productivity demands a *covariance* accomplished *at the right rythm*,
as suggested by Le Fevre & al. (1983). The generalization of these notions to
other sharp gradients in the environment gave rise to the definition of
ergoclines.

Summarizing, the proximal features of frontal ecosystems depend on the
interaction between spatio-temporal characteristics of the physical regime
of the front and the response time of biological phenomena resulting in
primary production.

But not only the primary production is concerned in the tuning of complementary phenomena. Secondary productivity depends on a tuning or resonance
between the response time of successive fluxes along the trophic chain. It
follows that a great diversity of problems of organization inside the frontal
biomass influence the ecological properties of the latter. Different cases can
be distinguished :
a) A divergence or a mixing leads to an enhanced "new" production linked
with the advection of nutrients. A "juvenile" community then arises, with
high P/B, low diversity, predominance of *r* strategies, predominance of vegetal
biomass (principally Chlorophyceas and Diatomeas) and of small size, fastly
multiplying planktonic herbivores, etc..., that is typically an opportunistic
community which is colonizing a new resource - the "resource" being here
generally defined as a covariance between the various conditions for production.

If that production is exploited, the regenerated production is low and the
flux of matter through the biomass is open. The microheterotrophic organisms
are scarce because they are unfavoured in the competition against vagile macroorganisms. Indeed, the latter export biomass outside the area of enhanced
primary productivity, hence the regeneration of nutrients occurs elsewhere,
sometimes far away, and is dispersed in a large oligotrophic area.

A tuning of the biomass consumption with respect to the speed of primary
production is necessary in order for phytoplankton to be efficiently grazed
by herbivorous zooplankton. Counter-examples exist : along the Ushant tidal
front, off French Brittany, Le Fevre & al. (1983 a) and Le Fevre (1985) suggested that the existing zooplankton is not adapted to the periodic variations
of the frontal primary production. The exporting trophic network hence does not

exist, and phytoplankton is rapidly degraded by bacterias, as seen in many other circumstances (Fenchel, 1964; Holligan & al., 1984; Jordan & Joint, 1984; Newell & Linley, 1984; Le Fevre, 1985).

The bacterial community is very important, and still largely disregarded in the frontal zones. This community seems to adapt very rapidly to environmental conditions and to the nature of the organic material, due to their rapid response (a few hours) - an opportunistic community -, and their high potential biochemical diversity. It results in a very precise matching between anabolism and catabolism (M. Bianchi, Personal communication).

b) A convergence, on the contrary, induces a passive accumulation leading to a degradation of the organic matter in the same area. The responsible, often complex and very diversified community is mainly constituted of bacterias, microzooflagellates and ciliates. It can be observed in aquatic ecosystems that a large concentration of organic matter never occurs without being immediately, or almost immediately, degraded[*]. The living matter has to be permanently renewed, and the more intensive the accumulation process, the faster the destructive process is. Consequently if a "juvenile" biomass issued from an enhanced primary production gets exported (either by trophic chains associated with horizontal and vertical migrations of the successive consumers, or by physical phenomena such as horizontal dispersion by currents or vertical sinking), then a stationnarity of the whole process can be obtained. When, on the contrary, the produced or passively accumulated biomass is neither biologically nor physically exploited, it rapidly breaks down due to the activity of bacterial and protozoan communities.

c) Such a description is of course oversimplified. Intermediate or composite phenomena can occur. In some cases, the microheterotrophic community can, after a time, be exploited by filtering macroorganisms, which may in turn be consumed and migrate, or do not. For example, Salp swarms can appear inside the phytoplankton blooms and exploit them avoiding the accumulation of vegetal biomass. It can be shown that these swarms, in the frontal zones, are in no way trophic deadlocks, for they contain a great quantity of semi-parasitic Hyperiid Amphipods (Laval, 1980) which are, in turn, exploited by tunas or slipjacks (Frontier, unpublished). Exploitation of phytoplankton accumulations by Doliolids, Appendicularians (which are eaten by fish larvae, specially flatfishes) and Pteropod *Limacina* has also been described (Southward & Barrett, 1983; Le Fevre, 1985; Deibel, 1985).

---

[*] In continental waters, limnologists call "eutrophication" such a huge accumulation of biomass, which after a short time gets degraded; it never occurs with the same importance in marine waters, except sometimes in the red tides (Wyatt & Horwood, 1973; Wyatt, 1975).

In other cases, there is little grazing by microorganisms, and the local degradation of living matter produces nutrients in the same area, inducing a regenerated production. In this way, the area can be shared into a mosaic of heterotrophic and autotrophic communities. In the Ushant front, a permanent convergence inducing biomass accumulation and regenerated production is associated with the periodic new production linked to tidal rythms, but is insufficiently exploited by zooplankton. In such a composite systems, some features of a mature community are to be found (hich diversity, predominance of dinoflagellates, low P/B, high rate of nutrient recycling, etc...) near other features characteristic of opportunistic ecosystem (Diatomeas)(Legendre & al., this volume).

Summarizing, front ecosystems are of two main types :
(i) those in which a macrotrophic network develops, up to large fishes and cetaceans; these ecosystems possess a system of displacements of the consumer organisms, with a precise matching of these displacements along the trophic chains, so that an overall migration of biomass occurs;
(ii) those in which such a trophic network does not exist, and the biomass decays in the field through the action of bacterias.

This contrast is nothing but a realization of the ancestral competition between micro and macroheterotrophic organisms, as it occurs in all kinds of ecosystems, including terrestrial ones. Intermediate types of fronts occur when particular macroorganisms are developing and multiplying by exploiting the microheterotrophic organisms. The first type of system demands a well adjusted matching of the various populations of consumers and their behaviour. But the conditions in which this matching does appear and maintains itself, as a well regulated structure, remains to be searched. It is a real problem of ecological evolution.

BIOLOGICAL EXPLOITATION OF THE FRONTAL PRODUCTIVITY

Coming back to the organization of the consumer networks which starts from any frontal area, we may prefer to consider two adjacent ecosystems, the one producing a biomass, and the other consuming it. We rejoin here the *Theory of exploitation of an ecosystem by another one*. "Exploitation" is here considered in a general view, as an *exportation of biomass* whatever the cause (Frontier, 1978). We have a picture of a highly productive, juvenile ecosystem exploited by a mature one whose diversity is higher, and which particularly contains some high trophic levels, able to come and graze the juvenile biomass at the interface - unless it enters it following a geometry of interpenetration which has something to do with the Fractal theory (Mandelbrot, 1977 a, 1982, 1985).

That exploitation of a juvenile ecosystem by a mature one has two
consequences :

1.- The juvenile ecosystem is maintained in this state, or even drops in
organization, with lowered diversity, being forced into a rapid production role.
2.- The mature ecosystem develops more (and faster) than it would if limited
by its own resources in energy and matter. By the way, there is a transfer of
energy and matter (and also of information) from the former to the latter, and
the dissymmetry is maintained (or even enhanced) between the two parts of the
coupling system, instead of disappearing by mixing. That dissymmetry, or
organization, is strictly depending on the permanence of the fluxes of matter,
energy and information from one system to the other. It can be considered as
another ergocline, between the two different ecosystems, and problems of tuning
are to be investigated here too.

Once again a comparison can be made with terrestrial ecosystems. Various
interfaces have been described and analyzed in continental systems. A contact
between a forest and a savanna is a front. A savanna is an ecosystem with low
diversity, a rapid turnover, and is actively exploited by herbivorous animals
coming by night from the forest, and feeding at the edge where primary productivity is high; during the day, these herbivores are enriching the forest with
their dejections and carcasses. As a consequence, there is a flux of energy
and matter from the savanna into the forest. In other words, an open flux of
matter is continuously travelling through the juvenile, or less organized,
ecosystem and through the interface, allowing a carnivorous biomass to be more
important and more diversified than it can be at some distance from the contact.
Hunters indeed have known for a long time that game is more abundant and more
diversified in the contact zone and in the mosaic forest-savanna, as well as
tuna and whale fishers know that their preys are often located near the
oceanographic fronts.

An active exportation of biomass by active consumers, resulting in a dispersion of the assimilated energy throughout a large, generally oligotrophic area,
is consuming energy. That energy is obviously coming from the metabolism of
the consumer organisms themselves and, after complete examination, comes from
the primary photosynthetic energy. It follows that *at each trophic level* of an
ecosystem involving vagile organisms, *a part of the assimilated energy is
withdrawn from the growth or maintenance of biomass, and devoted to the
edification and functionning of the exporting structures,* which are able to
make resilient the whole system (Frontier, 1978). I called "*secondary auxiliary
energy*" that part of assimilated energy which allows the matching (at a favourable speed and periodicity) between the various elements of the ecosystem,
and which is consequently, used as a "covariance energy" in the same sense as

62

Figure 1.

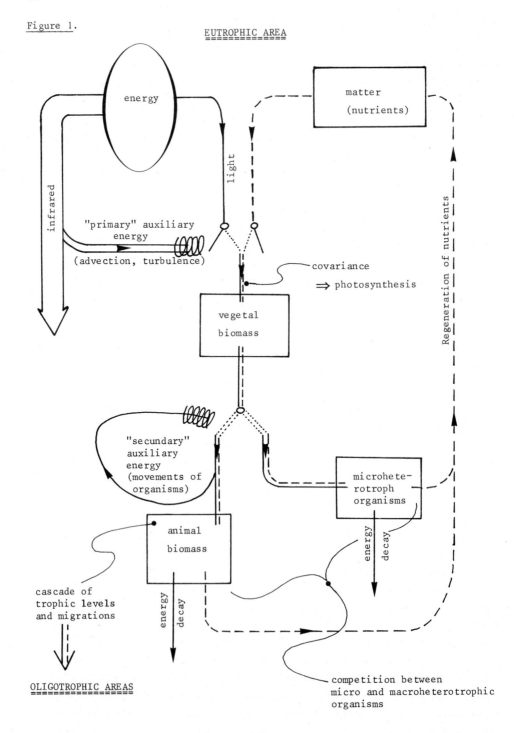

the "primary" auxiliary energy is.

Figure 1 represents a conceptual model summarizing the respective roles of the open energy flows throughout the ecosystem, the matter cycling, and the necessary spatio-temporal matching between the various fluxes. Primary auxiliary energy guarantees the matching of the vegetal biomass and the physical environment as secondary auxiliary energy does between the various partners of the interaction system in the community. The compartment called "Animal Biomass" symbolizes, in fact, the whole cascade of trophic levels and associated migrating cycles, resulting in a dispersion of the trophic energy into some large, remote or deep, oligotrophic areas.

AUXILIARY ENERGY, FRONTS AND FRACTALS

We can compare further the two kinds of auxiliary energy, the one directly issued from the kinetic energy of fluids, the other from the photosynthetic energy transmitted through the food chains. Both act much more through their fitness, and particularly through the tuning of the temporal and spatial scales involved, than by the total amount of energy. Primary auxiliary energy is useful by advecting nutrients, by dispersing biomass when the maturation of eocsystem occurs along a drift, and above all by giving rise to *turbulence* which mixes water masses. Turbulence has a fractal geometry (Mandelbrot, 1974, 1975, 1976, 1977 a and b, 1982), that is, the smaller and smaller eddies indefinitely increase the contact surface, and multiply the contact points between the parts of the water masses, increasing hugely the "covariance". In the terrestrial ecosystems, such an enhanced contact between plants and the medium (soil and atmosphere) is accomplished by a ramified geometry, which is also a fractal (Mandelbrot, 1977 a, 1978, 1982). We saw above that these ramified structures are built up using an amount of energy coming from the photosynthetic assimilation, and that I called "secondary auxiliary energy" too.

With regard to motile organisms, as well in terrestrial as in aquatic environments, secondary auxiliary energy is used in order to permit displacements which realize an interpenetration between the two subsystems - the productive one and the consumer one. This reciprocal penetration through more or less "random" movements, is also a fractal. Finally, the very sense of the (primary and secondary) auxiliary energy seems to be to allow the edification of a fractal spatio-temporal geometry of biomass and surrounding medium, which is necessary for all living processes because it conditions the covariance and tuning between all the partners of the ecosystem (Frontier & Legendre, 1985).

In conclusion, I have not presented here new results, but rather a new general vision based on previous observations, and demanding new ones. The analysis of the frontal systems following the framework of the Ecosystem Theory provides some specific working hypotheses. We have now to confirm (or negate, or complete), the model, and to fit it by measuring fluxes, by investigating the relations between fluxes and structures (particularly in the field of exportation of biomass from the frontal into the surrounding areas), and finally by attempting to estimate the importance of the fluxes and the structures in the production balance, first in the frontal zone, then in the entire ocean.

ACKNOWLEDGEMENTS

I thank Dr. Jacques Le Fevre for useful suggestions about this paper.

REFERENCES

Deibel, D., 1985. Blooms of the pelagic Tunicate Dolioletta gegenbauri : are they associated with Gulf Stream frontal eddies ? J. Mar. Res., 43: 211-236.
Fenchel, T., 1968. On "red waters" in the Isefjord (inner Danish waters) caused by the Ciliate Mesodinium rubrum. Ophelia, 5: 245-253.
Frontier, S., 1978. Interfaces entre deux écosystèmes : exemples dans le domaine pélagique. Ann. Inst. Océanogr., Paris, 54: 95-106.
Frontier, S., 1977. Réflexions pour une théorie des écosystèmes. Bull. Ecol., 8: 445-464.
Frontier, S. and Legendre, P., 1985. Théorie des fractals : applications à l'écologie, implications dans l'échantillonnage. Rapport CNRS, ATP 9.82.65. Evaluation et optimisation des plans d'échantillonnage en écologie littorale. 32 pp.
Frontier, S. and Legendre, P.. Submitted, Fractals in Ecology.
Grall, J.R., Le Corre, P., Le Fevre, J., Marty, Y., and Tournier, B., 1980. Caractéristiques de la couche d'eau superficielle dans la zone des fronts thermiques Ouest-Bretagne. Oceanis, 6: 235-249.
Holligan, P.M., 1979. Dinoflagellate blooms associated with tidal fronts around the British Iles. In D.L. Taylor and H.H. Seliger, eds., Toxic Dinoflagellate Blooms, Elsevier/north Holland, 249-256.
Holligan, P.M., 1981. Biological implications of fronts on the northwestern European continental shelf. Phil. Trans. Roy. Soc. London, ser. A, 302: 547-562.
Holligan, P.M., Viollier, M., Dupouy, C. and Aiken, J., 1983. Satellite studies on the distribution of chlorophyll and dinoflagellate blooms in the Western English Channel. Continental Shelf Res., 2: 81-96.
Holligan, P.M., Williams, P.J., Purdie, D. and Harris, R. P., 1984. Photosynthesis, respiration and nitrogen supply in stratified frontal and tidally mixed shelf waters. Mar. Ecol. Progr. Ser., 17: 201-213.
Holligan, P.M., Pingree, R.D. and Mardell, G.T., 1985. Oceanic solitons, nutrient pulses and phytoplankton growth. Nature, London, 314: 348-350.
Jordan, M.B. and Joint, I.R., 1984. Studies on phytoplankton distribution and primary production in the western English Channel in 1980 and 1981. Continental Shelf Res., 3: 25-34.
Laval, Ph., 1980.Hyperiid Amphipods as Crustaceans parasotoids associated with gelatinous zooplankton. Oceanogr. Mar. Biol. Ann. Rev., 18: 11-56.
Le Fevre, J. and Grall, J.R., 1970. On the relationship of Noctiluca swarming off the western coast of Brittany with hydrological features and plankton characteristics of the environment. J. Exp. Mar. Biol. Ecol., 4: 287-306.
Le Fevre, J., Cochard, J.C. and Grall, J.R., 1981. Physical characteristics of an inshore area on the Atlantic coast of Brittany and their influence on the pelagic ecosystem : the case of the "Rivière d'Etel". Estuarine, coastal and Shelf Science, 13: 131-144.

Le Fevre, J., Le Corre, P., Morin, P. and Birrien, J.L., 1983 a. The pelagic ecosystem in frontal zones and other environments off the western coast of Brittany. Oceanologica Acta, Proceed., 17th European Marine Biology Symposium, 125-129.

Le Fevre, J., Viollier, M., Le Corre, P., Dupouy, C. and Grall, J.R., 1983 b. Remote sensing observations of biological material by Landsat along a tidal thermal front and their relevancy to the available field data. Estuarine, Coastal and Shelf Science, 16: 37-50.

Le Fevre, J., 1985. Aspects of the biology of frontal systems. Adv. Mar. Biol., in press.

Legendre, L., 1981. Hydrodynamic control of marine phytoplankton production : the paradox of stability. In J. Nihoul ED., Ecohydrodynamics, Elsevier Scientific Publish., Amsterdam, 191-207.

Legendre, L. and Demers, S., 1984. Towards dynamical biological oceanography and limnology. Can. J. Fish. Aquat. Sci., 41; 2-19.

Legendre, L., Demers, S. and Lefaivre, D., 1985. Biological production at marine ergoclines. Proceed. 17th Internat. Liège Colloquium on Ocean Hydrodynamics, this volume, J. Nihoul ed.

Mandelbrot, B., 1974. Intermittent turbulence in self-similar cascades : divergence of high moments and dimension of the carrier. J. Fluid Mech., 62: 331-358.

Mandelbrot, B., 1975. On the geometry of homogeneous turbulence, with stress on the fractal dimension of the iso-surface of scalars. J. Fluid Mech., 72: 401-416.

Mandelbrot, B., 1976. Intermittent turbulence and fractal dimension : kurtosis and the spectral exponent $5/3 + B$. Turbulence and Navier-Stokes equations. Lecture notes in Mathematics, 565; 121-145.

Mandelbrot, B., 1977 a. Fractals, Form, Chance and Dimension. W.H. Freeman and Co., San Francisco, 365 pp.

Mandelbrot, B., 1977 b. Fractals and turbulence : attractors and dispersion. Lecture Notes in Mathematics, New York, 615: 83-93.

Mandelbrot, B., 1978. The fractal geometry of trees and other natural phenomena. Lecture Notes in Biomathemactics, New York, 23: 235-249.

Mandelbrot, B., 1982. The fractal geometry of Nature. W.H. Freeman & Co., San Francisco, 468 pp.

Mandelbrot, B., 1985. Les objets fractals. Flammarion, Paris, 204 pp.

Margalef, R., 1974. Ecologia. Ediciones Omega, Barcelona, 951 pp.

Margalef, R., 1978. Life forms of phytoplankton as survival determinatives in an unstable environment. Oceanol. Acta, 1: 493-510.

Margalef, R., 1985. From hydrodynamic processes to structure (information) and from information to process. In R.E. Ulanowicz and T. Platt eds. Proceed. Sympos. on Ecological Theory in Relation to Biological Oceanography (Québec, March 1984), Can. J. Aquat. Fish. Sci., 213: 200-220.

Newell, R.C. and Linley, E.A.S., 1984. Significance of microheterotrophs in the decomposition of phytoplankton : estimates of carbon and nitrogen flow based on the biomass of plankton communities. Mar. Ecol. Progr. Ser., 16: 105-119.

Pingree, R.D., Forster, G.R. and Morrison, G.K., 1974. Turbulent convergent tidal fronts. J. Mar. Biol. Assoc. U.K., 54: 469-479.

Pingree, R.D. and Pennycuik, L., 1975. Transfer of heat, fresh water and nutrients through the seasonal thermocline. J. Mar. Biol. Assoc. U.K., 55: 261-274.

Pingree, R.D., 1978. Cyclonic eddies and cross-frontal mixing. J. Mar. Biol. Assoc. U.K., 58: 955-963.

Pingree, R.D. and Griffiths, D.K., 1978a. Tidal fronts on the shelf seas around the British Isles. J. Geophys. Res., 83: 4615-4622.

Pingree, R.D., Holligan, P.M. and Mardell, G.T., 1978 b. The effect of vertical stability on phytoplankton growth and cyclonic eddies. Nature, London, 278: 245-247.

Pingree, R.D. and Mardell, G.T., 1981. Slope turbulence, internal waves and phytoplankton growth at the Celtic Sea shelf break. Phil. Trans. Roy. Soc. London, ser. A, 302: 663-682.

PIROCEAN-I.N.S.U., 1985. Projet de faisabilité du Programme "FRONTAL". Doc. C.N.R.S. (PIROCEAN), multigr. 90 pp.

Southward, A.J. and Barrett, R.L., 1983. Observations on the vertical distribution of zooplankton, including postlarval Teleosts, off Plymouth in the presence of a Chlorophyll-dense layer. J. Plankton Res., 5: 599-618.

Wyatt, T. and Horwood, J., 1973. Model which generates red tides. Nature, London, 244: 238-240.

Wyatt, T., 1975. Further remarks on red tide models. Environmental Letters, 9: 217-224.

THE FRONTAL ZONE IN THE SOUTHERN BENGUELA CURRENT

L. HUTCHINGS[1], D.A. ARMSTRONG[1] and B.A. MITCHELL-INNES[2]

[1]Sea Fisheries Research Institute, Private Bag X2, Rogge Bay 8012,
Cape Town, South Africa.

[2]Marine Biology Research Institute, University of Cape Town,
Rondebosch 7700, Cape Town, South Africa.

INTRODUCTION

The southern region of the Benguela Current is an area of complex mixing, where Agulhas Current and South Atlantic Surface Water meet and mingle with upwelled South Atlantic Central Water. The relationships between the water masses (see Fig. 1) are presented in the conceptual image of

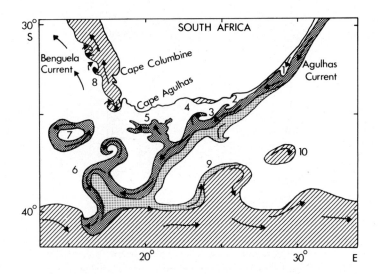

Fig. 1. A conceptual image of the Agulhas Current System. Heavy dots indicate warm water, lighter dots colder water and hatching cold, subtropical surface water and upwelled water. The line demarcating the subtropical water at about 40°S latitude is the Subtropical Convergence. 1. The embryonic stages of a Natal pulse. 2. Small waves or disturbances on the Agulhas Current border. 3. Shear edge effects, warm-water plumes or shear edge eddies. 4. Cold upwelled water. 5. The dispersion of a shear edge feature attached to an Agulhas Current meander. 6. The Agulhas Current retroflection. 7. An Agulhas Current Ring being advected northward. 8. The Benguela upwelling regime with frontal eddies in evidence. 9. A planetary wave on the Agulhas Return Current-Subtropical Convergence. 10. An independent cold-wave eddy spawned by an unstable planetary wave. (From Lutjeharms, 1981a).

Lutjeharms (1981a), showing advection of filaments of the Agulhas Current onto the Agulhas Bank, Agulhas Rings breaking off and moving northwards into the Atlantic, and entrainment of Agulhas Bank Water along the west coast offshore of the Benguela upwelling zone by seasonal southeasterly winds.   These factors combine and peak during summer months to create an intense front from Cape Agulhas (35°S) in the extreme south of the continent up to Cape Columbine (33°S).

Beyond this point gradients generally weaken (Shannon, 1985).   Scientific attention has been focussed mainly on the inshore area, and numerous transects across the shelf have been made (see Fig. 2).   The most frequently sampled line was the Upwelling Monitoring (UM) line, sampled 36 times between March 1971 and March 1983 (Andrews and Hutchings, 1980).   The Bathythermograph (BT) line at 255° from Cape Town (Bang and Andrews, 1974), and the December 1984 Frontal Zone (FZ) transect at 32° 28 ' due west of Cape Columbine, were perpendicular to the usual orientation of the front.

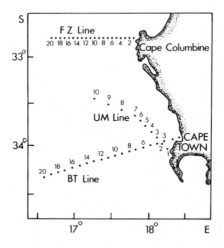

Fig. 2.   The transect lines off the Cape Peninsula.   Hydrological and plankton data were collected along the Upwelling Monitoring (UM) line and Frontal Zone (FZ) transect, and temperature only along the Bathythermograph (BT) line.

SURFACE FEATURES

An appreciation of the variability of the front has grown in direct proportion to the frequency with which images were obtainable from shipboard studies, from aircraft and lately, from satellites.   Colour images from the NIMBUS-7 satellite (Shannon et al., 1984a) show that the thermal front roughly follows the shelf-break between Cape Columbine and Cape Point, but further south it cuts eastwards across the bottom contours.   Temperature gradients are often 1°C per nautical mile but can be several degrees over a few hundred meters (Bang, 1973). Maximum gradients occur in midsummer to autumn when southeasterly winds predominate. Imbedded within the coastal waters are active upwelling sites (see Fig. 3) off the Cape Peninsula, Cape Columbine and Hondeklip Bay (Nelson and Hutchings, 1983). Surface features are extremely sensitive to changes in wind stress (Bang and Andrews, 1974; Andrews and Hutchings, 1980; Lutjeharms, 1981b; Taunton-Clark, 1985), varying extensively within a few hours of a change in wind direction. During winter the area south of 32°S is subject to westerly winds as cyclones

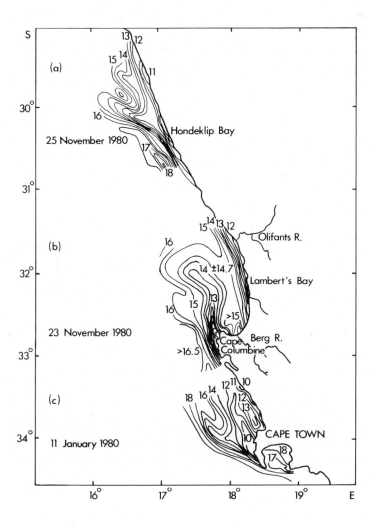

Fig. 3. A montage of three upwelling sites in the southern Benguela Current (a, b and c) defined by aerial radiation thermometry. (From Nelson and Hutchings, 1983).

associated with depressions move eastward past the southern tip of Africa. Upwelling activity declines and cross-shelf surface temperature gradients decrease from 11°C (10-21°C) down to 1-2°C (13-15°C) (Andrews and Hutchings, 1980).

VERTICAL STRUCTURE

Bang (1971, 1973, 1974) did pioneering work on the frontal structure using detailed bathythermography between 31°S and 34°S. He distinguished an upwelling front associated with the growing upwelling centre at 10-15 n.miles offshore, and a shelf-break front at 50-60 n.miles offshore. After prolonged upwelling these two fronts might coalesce; this in fact occurs further south where the shelf is very steep, resulting in the steepest temperature gradients observed during active upwelling. The subsurface structure of the shelf-break front appears to persist perennially (Hutchings et al., 1984), weakening in midwinter but intensifying in August to September with the intrusion of cold water (8°C) onto the shelf, before the local southeasterly wind stress begins, indicating that at least part of the formation of the front is due to larger scale processes. The narrow shelf and the convolutions of the shelf-break with two major submerged canyons, the Cape Point valley and the Cape Canyon associated with the Cape Peninsula and Cape Columbine respectively, facilitate the movement of cold, low salinity water onto the shelf.

Associated with the steeply sloping isotherms of the frontal region is a strong equatorward jet current (Bang and Andrews, 1974) with some evidence of compensating flows on the bottom inshore and also offshore (see Fig. 4). In a recent description of the currents off the Cape Peninsula, Nelson (1985) indicates that the jet current is not in geostrophic balance; he postulates that vorticity changes, which occur when the South Atlantic gyre is deflected by the shelf, combine with other large scale forces to play an important role in maintaining the shelf-edge jet. Near-surface frontal features alter rapidly with changes in wind stress and southward flowing currents can occur after prolonged northwesterly winds (Shelton and Hutchings, 1982).

North of Cape Columbine the shelf widens and the shelf profile alters, with the shelf-break varying between 200-500m (Shannon, 1985). Upwelling activity occurs both in a narrow coastal strip and on the shelf-break (Hart and Currie, 1960; Shannon, 1966; Bang, 1971; Nelson and Hutchings, 1983), and frontal features become less definite. Therefore we have focussed our attention on the front between Cape Point and Cape Columbine (33°50' -34°30'S).

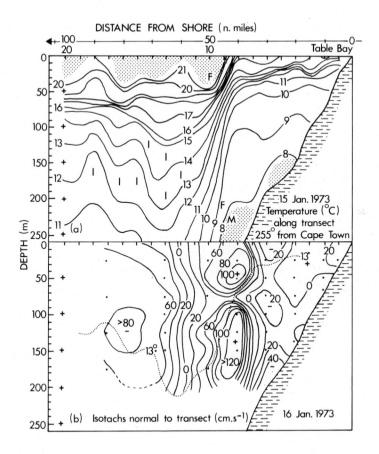

Fig. 4. Subsurface frontal features along the BT line and the shelf-edge jet west of the Cape Peninsula during January 1973. (From Bang and Andrews, 1974).

BIOLOGICAL CHANGES ACROSS THE FRONT

Marked changes in species composition and abundance of phytoplankton, zooplankton and ichthyoplankton occur across the front (De Decker, 1973, 1984; Thiriot, 1978; Hutchings, 1979; Shelton, in prep.), in keeping with similar findings in other frontal systems (Holligan, 1981; Boucher, 1984; Vinogradov and Shushkina, 1978; Shushkina et al., 1978). A low diversity, cool water fauna extends along the coast inshore, from 31°S to 19°S (Kollmer, 1963; Unterüberbacher, 1964; Thiriot, 1978; De Decker, 1984). The offshore extensions of the distribution of Centropages brachiatus (see Fig. 5a), a typical surface member of this inshore community, may indicate occasional advection into the South Atlantic. Immediately beyond the front is warm water, often of Agulhas Bank or Current origin, with a high diversity of species (De Decker, 1984; Hutchings, 1979). Centropages furcatus, a characteristic Agulhas Current copepod, disappears as this water moves northwards and mixes (see Fig. 5b), its penetration into the Atlantic being proportional to its tolerance to changing conditions. Further offshore, species from the South Atlantic gyre and occasionally from Subantarctic waters (e.g. Calanus tonsus and Calanoides macrocarinatus) are apparent (De Decker, 1984).

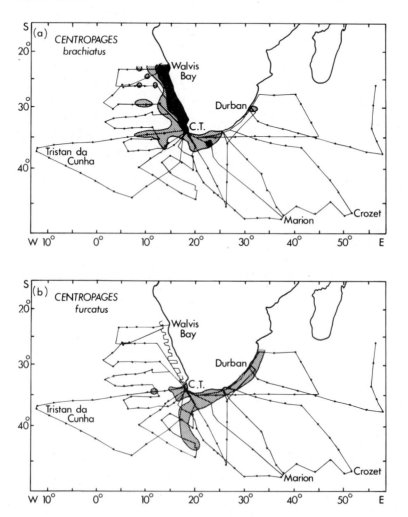

Fig. 5. The distribution of (a) Centropages brachiatus, a cold water copepod species, and (b) Centropages furcatus, a characteristic Agulhas Current copepod, around the coast of South Africa. (From De Decker, 1984).

Nutrients and phytoplankton

Vigorous growth of phytoplankton in newly upwelled waters occurs close
inshore. Nutrient concentrations are often low in aged upwelled waters
at the surface immediately inshore of the shelf-break front, resulting
in the absence of a pronounced nutricline across the front in the upper
layers (see Fig. 6). Nutrients increase rapidly below the shallow pycnoclines.
Strong gradients in chlorophyll 'a' across the front, however, do exist
(see Fig. 6), with very low concentrations in the upper layers on the
warm water side, while higher concentrations inshore sink below the front
and extend offshore at subsurface depths for some distance. Raised chlorophyll 'a' levels can extend to a depth of 75-100m at the front, indicating
strong sinking motions, as illustrated by the November 1972 transect of
the UM line (see Fig. 7).

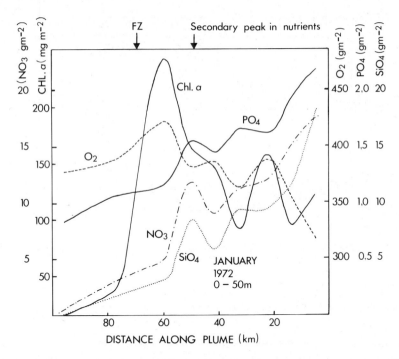

Fig. 6. Distributions of nitrate, silicate, phosphate, oxygen and chlorophyll in the photic layer in a section along the UM line in January 1972.
Note the position of the frontal zone (FZ) and the secondary nutrient
peak inshore of the front. (From Andrews and Hutchings, 1980).

During quiescent periods or onshore winds, complicated motions occur
in the frontal region as a surface layer of warm water floods shorewards.
High subsurface chlorophyll 'a' maxima occur, as illustrated by the January
1973 transect of the UM line (see Fig. 8). The coarse spacing of stations

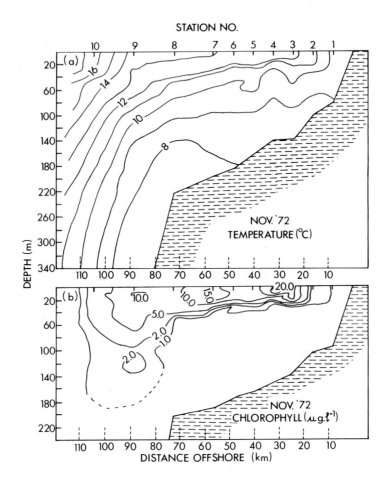

Fig. 7. Temperature (a) and chlorophyll 'a' (b) sections along the UM line in November 1972, showing a pronounced front with chlorophyll sinking at the front.

in time (monthly) and space (5–10 n.miles apart) preclude detailed analyses of these changes.

Results from the December 1984 Frontal Zone cruise (stations were 3 n.miles apart) showed a weakly developed front (see Fig. 9b) after prolonged northwesterly winds prior to sampling. Currents measured with a Niel-Brown acoustic current meter showed strong northward-flowing currents on either side of the front (see Fig. 9a), up to $1m.sec^{-1}$ immediately above the Cape Canyon. Surface temperatures ranged from 15.2°C inshore, to 18.9°C offshore and nitrates were low (<1 mmol.m^{-3}) on both sides of the front, rising slightly close inshore. However, there were pronounced differences

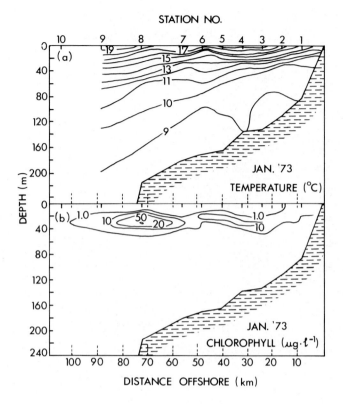

Fig. 8. Temperature (a) and chlorophyll 'a' (b) sections along the UM line in January 1973, showing conditions during quiescent periods, with a relaxed front and a pronounced subsurface chlorophyll 'a' maximum.

in the phytoplankton standing stock (see Fig. 10) along the transect. Low concentrations occurred offshore of the frontal zone, with approximately equal concentrations of nano- and net-phytoplankton, whereas, net-phytoplankton predominated at the biomass peak close inshore, as well as at the secondary peak in the frontal zone (stations 4 and 5).

Microzooplankton

During an extensive survey of the distribution of pelagic ichthyoplankton off the south western Cape in 1977-1978 (Cape Egg and Larval Programme, CELP), microplankton were collected at five depths near interfaces (surface, thermoclines etc.) in the upper 75m by sieving 2 litres of water through 37μm nylon mesh. The main concentration of microzooplankton was located inshore of the front and comprised mostly egg and naupliar copepod stages in the upwelled waters. Scattered dinoflagellate blooms occurred on the

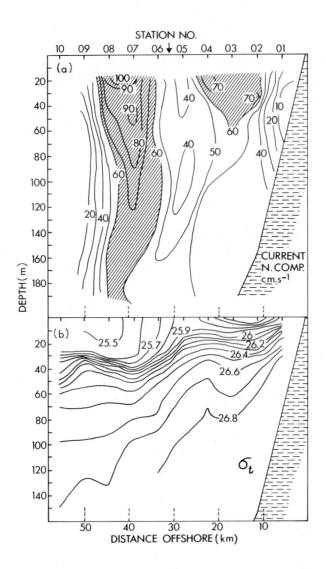

Fig. 9. Current (a) and density (b) sections along the FZ line in December 1984, showing strong equatorward jets flanking the frontal zone (marked by an arrow).

Agulhas Bank, with low concentrations beyond the front. No noticeable accumulation occurred within the frontal zone, which effectively formed a barrier to the offshore dispersion of the abundant young copepod stages.

Zooplankton

Shelton and Hutchings (in prep.) show that high zooplankton stocks,

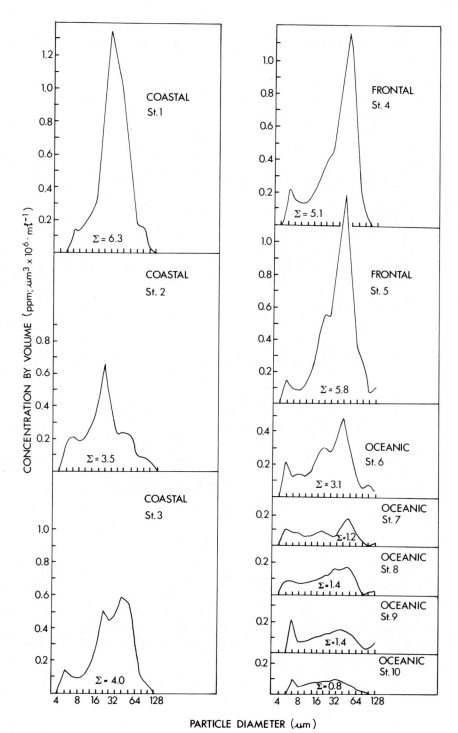

Fig. 10. Particle spectra of samples from the chlorophyll maximum layer along the FZ line in December 1984. Stations are characterized as coastal, frontal or oceanic based on the water characteristics.

expressed as displaced volume, are limited to the nearshore region by strong frontal features (see Fig. 11). If zooplankton standing stock were plotted in dry weight or carbon units, the inshore, offshore differences would probably be enhanced, as the crustacean-dominated, cool-water plankton inshore is replaced by more gelatinous species offshore. Pillar (in prep.) has shown significantly higher biomasses of euphausiids and copepods, expressed as dry weight, in coastal waters than in oceanic waters.

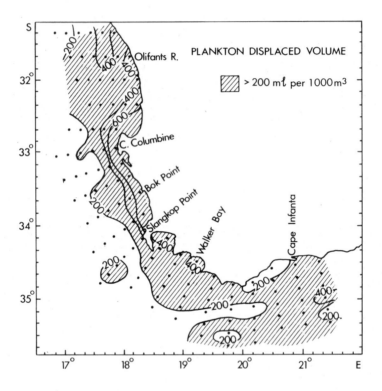

Fig. 11. The distribution of zooplankton (dispaced volume, ml. 1000m^{-3}) around the south western Cape coast measured during the CELP surveys, 1977-1978 . (From Shelton and Hutchings, in prep.)

Hutchings (1981) examined 13 of 36 monthly transects of the southern Benguela shelf region off the Cape Peninsula, in terms of mesozooplankton dry weight (WP-2 net) in the whole water column, chlorophyll 'a' in the upper 100m and surface temperature. Highest zooplankton concentrations were observed close inshore, and there was no particular association of higher zooplankton stocks with the frontal zone. Andrews and Hutchings (1980) showed a strong shoreward movement of high zooplankton concentrations

as oceanic water flooded shorewards during winter and onshore wind periods
in summer. A more detailed analysis of the same data by Armstrong (in
prep.) showed that when a strong front was present above the shelf-break
after prolonged upwelling, there was a pronounced peak of mesozooplankton
in the upper mixed layer, immediately inshore of the front (see Figs.
12a, b). In February 1973, a double front was present, with high zooplankton
biomass at the inshore upwelling front and only a slight increase at the
shelf-edge front (see Fig. 12d). When the front was weakly developed,
with an intrusion of warm water onshore (e.g. January 1973), relatively
low zooplankton concentrations were observed along the entire line (see
Fig. 12c).

Hutchings (1979) collected an intensive data set of zooplankton samples
over 10 days in the Cape Peninsula upwelling system. Multivariate analysis
of these data showed that different zooplankton communities occurred across
the frontal zone (see Fig. 13), along the front. Vertical distribution
of species and biomass at a frontal station showed a clear separation
in the vertical plane with distinct warm water, thermocline and cool water
communities in separate layers. How these communities are maintained
in the presence of the strong water motions in the region (Bang and Andrews,
1974; Shannon et al., 1981; Nelson, 1985), still remains uncertain. Armstrong
(in prep.) also found sharp changes in communities across a well defined
frontal feature (see Fig. 14) with a low diversity, high biomass community
inshore and a high diversity, low biomass commmunity in the warm water
offshore.

Pelagic fish

Egg and larval species assemblages in the southern Benguela are clearly
influenced by the strong front. Very few fish species spawn inshore of
the front in cool upwelled waters; a larger number (including the commercially important anchovy and pilchard) spawn in warm shelf waters on the Agulhas
Bank, and many more species (largely mesopelagic forms) spawn in deep
water offshore of the front (Shelton, in prep). In December 1984, pelagic
fish eggs and larvae were observed along most of the transect, with peak
larval abundance beyond the front in the warm surface waters and eggs
closer to the front.

The cumulative annual distribution of anchovy eggs and larvae during
1977-1978 (see Fig. 15) clearly show the centre of spawning on the Agulhas
Bank, with an advected tail of eggs and larvae entrained in the frontal
jet current extending up the west coast. The subsequent movement of older
larvae and juveniles is the subject of much conjecture (Badenhorst and

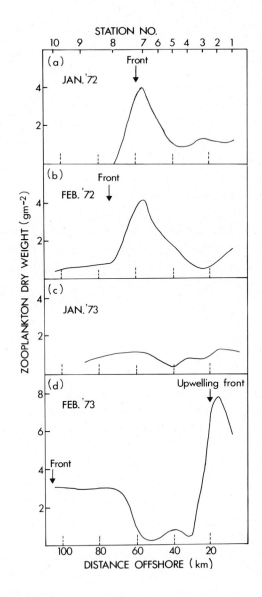

Fig. 12. The distribution of mesozooplankton biomass (g.dry wt.m^{-2}) above the thermocline along the UM line under various frontal conditions. In January 1972 (a) and February 1972 (b) there was a pronounced front; in January 1973 (c) the front was weakly developed and in February 1973 (d) a double front was present.

Boyd, 1980; Boyd and Hewitson, 1983). Other species such as round herring Etrumeus whiteheadi) and mackerel (Scomber japonicus) are associated with warmer waters outside the front. A pole-and-line fishery for longfin

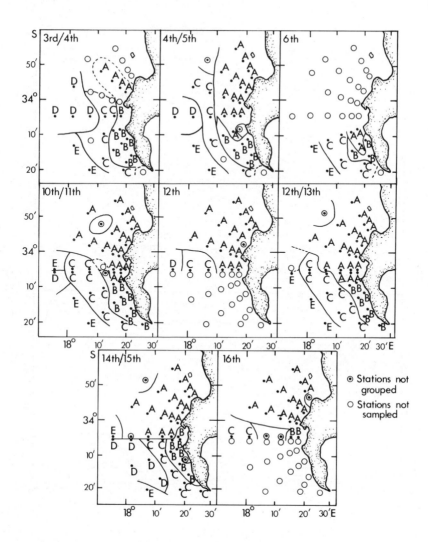

Fig. 13. Geographic distribution of sample-groups based on zooplankton species abundance derived using the Bray-Curtis measure of similarity. Group A samples represent the zooplankton community of recently upwelled water; Group B samples are inshore neritic species, Group C samples are frontal zone groups, and Groups D and E samples are warm water groups. (From Hutchings, 1979).

and yellowfin tuna exists along the frontal zone in the region of the shelf-break from 28°S to 35°S. Fishing boats characteristically cross the front, steam for a few more miles and then begin trolling parallel with the bottom isobaths until shoals are encountered. Recently satellite imagery has been used to concentrate fishing activities in frontal regions.

82

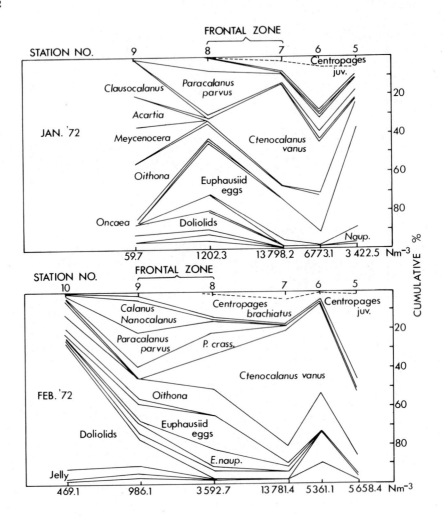

Fig. 14. Changes in species composition and concentration of mesozooplankton, collected above the thermocline (upper 30-60m), across sharp frontal gradients along the UM line during January 1972 and February 1972.

DYNAMIC BIOLOGICAL PROCESSES

Phytoplankton development in newly upwelled waters

Drogues, placed at 10m depth in large patches of newly upwelled water off the Cape Peninsula, were tracked for 5-9 day periods during six Plankton Dynamics Cruises. Barlow (1982a, b), Olivieri et al. (1985), and Brown and Hutchings (1985) found that nitrate is rapidly removed from the stabilizing upper mixed layer by a developing phytoplankton bloom, usually but not always dominated by diatoms (Olivieri et al., 1985; Hutchings et al., 1984; Mitchell-Innes, in prep.). Preliminary estimates of grazing in

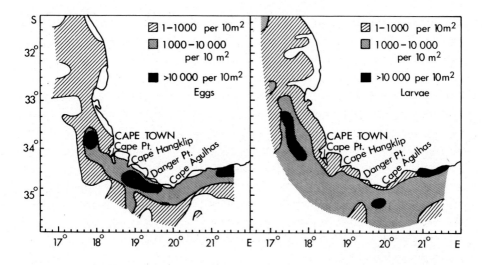

Fig. 15. Cumulative distribution of anchovy (<u>Engraulis capensis</u>) eggs and larvae off the south western Cape during 1977-1978; (a) eggs (August 1977-August 1978), (b) larvae (August 1977-February 1978, April and June 1978). (From Shannon et al., 1984b).

newly upwelled waters (Olivieri and Hutchings, in prep.) show that only a small proportion of phytoplankton production is grazed by mesozooplankton, with major losses probably due to dispersion over a wide plume area (Boyd, 1982) or to sinking.

During these studies none of the initial blooms approached the frontal region, yet Andrews and Hutchings (1980), Shannon et al. (1984a) and Hutchings et al. (1984) showed high chlorophyll 'a' levels at the front. This may be due to rapid offshore transport of developing blooms from the extreme south of the upwelling centre, or the slower development of patches due to poor seeding, leading to increased offshore transport prior to a bloom developing. An alternative explanation of the peak at the front is the presence of a mixing cell where the aged upwelled water sinks, with some uplift of water immediately inshore of the front. This is supported by the presence of a secondary nutrient peak occasionally detected just inshore of the frontal peak in chlorophyll 'a' (see Fig. 6). Bang (1973) suggests mixing processes in the frontal region may be caused by internal waves propagating up the steep isopycnals and "shaking themselves out", creating a series of eddies where phytoplankton may be retained.

Phytoplankton production, nutrient uptake and regeneration

During the December 1984 Frontal Zone survey daily integrated

productivity values at the frontal zone stations were 2096 and 2916 mgC.m^{-2}.
day^{-1}. The production rates in the frontal area, although higher than in
adjoining waters, were nevertheless comparatively low when compared to the
production measured in upwelled waters of the Benguela upwelling system.
Brown (1984) reports integrated daily production rates in summer ranging
from barely detectable levels in newly upwelled water to 11056 mgC.m^{-2}.day^{-1}.
Since P:B ratios did not show any enhanced productivity at the frontal
stations, the higher production may be due to the accumulation of phytoplankton in an area of lower current speed flanked by stronger currents to
either side (see Fig. 9a). Size fractionation of samples showed that
at oceanic stations the nanoplankton fraction (3–10μm) formed 33–70% of
the total carbon production. At stations in frontal or inshore waters
the nanoplankton fraction only contributed 6–31% of the total carbon fixed.

Recent work by Probyn (1985) on nitrogen uptake by different size-fractions
of phytoplankton at inshore, mid-shelf (100–200m) and offshore (1000m)
locations in the Southern Benguela show that, in common with other upwelling
zones, net plankton (> 10μm) dominated inshore in terms of both biomass
and nitrogen uptake (mainly nitrate), while offshore, nano- and pico-plankton
(1–10μm and < 1μm respectively) were relatively more important, with ammonia
and urea the major sources of nitrogen. A lower total rate of nitrogen
uptake occurred offshore, beyond the front. The presence of aged, low–nitrate
water and a large quantity of phytoplankton well below the euphotic zone
at the front suggests that microheterotrophic processes may be enhanced
in the frontal region.

Water column stability and phytoplankton growth

Although sunwarming and mixing create shallow discs of moderately stable
aged upwelled water overlying cooler water inshore, the pycnocline is
weaker and shallower than that present offshore of the front, a situation
similar to that off Oregon (Mooers et al., 1978). Advected filaments
of Agulhas water (Bang, 1971, 1973; Lutjeharms, 1981b) increase the vertical
gradients considerably beyond the front. Strong winds combined with wave
action can periodically cause some mixing of nutrient rich water into
the euphotic zone across the weak pycnocline inshore of the front, maintaining
some growth potential compared with offshore waters.

During southeasterly wind relaxations or reversals in wind stress as
occurred in January 1973 (see Fig. 8), a layer of warm water can mix across
the front (Bang and Andrews, 1974) rapidly stabilizing the water to a
degree unusual in most coastal waters. Legendre (1981) has pointed out
the role of alternation of stable and mixed water columns in stimulating

phytoplankton growth. With a 3 to 6 day wind cycle characteristic of summer months (Andrews and Hutchings, 1980; Nelson and Hutchings, 1983; Nelson, 1985), waters in close proximity to the front could be highly productive for short but frequent periods between wind events during summer.

Sinking processes

Sinking processes in frontal regions have been proposed by both biologists (Packard et al., 1978; Andrews and Hutchings, 1980) and physical oceanographers (Mooers, Collins and Smith, 1976; Bang, 1973, 1976; Simpson, 1981; Brink, 1983; Nelson, 1985). Many models of upwelling circulation containing one, two and three cells have been proposed, some with sinking at the front at the end of a wind cycle (Stevenson et al., 1974), and during active upwelling (Andrews and Hutchings, 1980; Nelson, 1985). Andrews and Hutchings (1980) show strong evidence for large-scale frontal sinking during active upwelling as well as sinking at the coast during onshore winds. The November 1972 transect off the Cape Peninsula (see Fig. 7) illustrates clearly the deepening chlorophyll-rich layer in the vicinity of the front, to depths well below the euphotic zone. Much of the phytoplankton would be entrained in the jet current (Bang and Andrews, 1974; Shannon et al., 1981) and transported rapidly northwards.

The December 1984 Frontal Zone transect, sampled during quiescent conditions, showed chlorophyll 'a' and particle distributions indicative of coastal downwelling, but even the close (3 n. miles) spacing of stations failed to reveal any frontal sinking motions during this phase. Holligan (1981), discussing tidal fronts, shows maximum phytoplankton stocks on the stratified side close to the thermocline region, where lateral and vertical diffusion of nutrients may be sufficient to maintain production, combined with some migration of motile species. Clearly there are very big differences between upwelling and tidal fronts, although the physical features may look similar.

Transport of pelagic fish eggs and larvae

The drogue study of Shelton and Hutchings (1982) and the distribution of anchovy eggs and early larvae (see Fig. 15) show the entrainment and rapid alongshore transport of spawning products to the recruitment grounds on the west coast. Vertical distribution studies (Shelton, 1984) show the eggs to be closely associated with the frontal zone (see Fig. 16). Successful feeding of early stage larvae in this turbulent environment, the processes of cross-frontal mixing necessary to move young fish to the inshore nursery areas, and the possibility of offshore advection and consequent starvation of a large proportion of the early life history stages of anchovy are of current interest to local scientists, stimulating the focus on frontal processes.

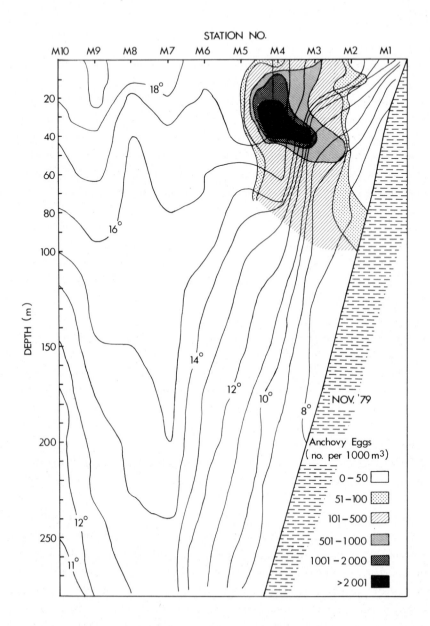

Fig. 16. Vertical temperature structure and distribution of anchovy eggs along a line which transects the front close to Cape Town in November, 1979. Note the maximum egg distribution coinciding with a sharp front. (Modified from Shelton, 1984).

CONCEPTUAL IMAGES OF THE FRONTAL SYSTEMS IN THE SOUTHERN BENGUELA

All the models discussed are, of necessity, descriptive ones as both Brink (1983) and Nelson (1981, 1985) point out the difficulties of modelling frontal systems in regions of rapid change with strong topographic features. Brink (1983) emphasizes the complex nature of cross-shelf and alongshore flow in upwelling regions. Secondary cross-shelf flows and frontal downwelling may only occur during particular phases of the upwelling cycle. The situation becomes even more complex as topographic effects need to be considered, as well as the shelf-edge jet which appears to be responding to larger oceanic phenomena rather than upwelling processes alone. For simplicity we will list the ideas chronologically.

Hart and Currie (1960) proposed a two-cell structure over a wide continental shelf with inshore and shelf-break upwelling, and sinking inshore of the shelf-break front (see Fig. 17).

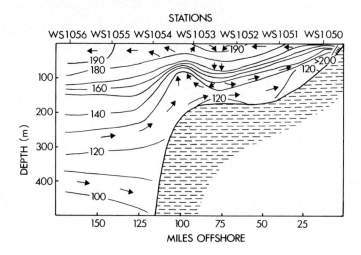

Fig. 17. Distribution of specific volume anomaly along the Orange River line, September 1950, showing cellular structure. (From Hart and Currie, 1960).

Bang (1973, 1976) proposed a complex two front, three cell type of circulation pattern (see Fig. 18) with: (i) the coastal divergence slightly displaced from the coast by an inner closed circulation pattern; (ii) the "upwelling" front at 20-30 n.miles offshore with sinking on the inshore side; (iii) the shelf-break front, separated from the upwelling front by a lens of Agulhas water some 30-50m thick. There is a subsurface uplift of cool water at the shelf-break with both shoreward and offshore sinking

88

to either side, in a similar manner to that proposed by Hart and Currie (1960).

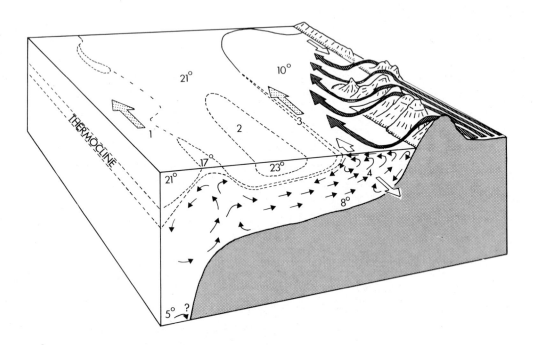

Fig. 18. An impression of the three-dimensional structure created when southerly winds force warm, predominantly tropical water away from the coast, thus causing cold water of predominantly Antarctic origin to well up in replacement. Key elements include 1) the Jet, with sporadic eddies and partial upwelling along its inner-side; 2) shallow patches of Agulhas Current water; 3) the Benguela Front, possibly associated with a shallow northward jet; 4) the De Decker undercurrent. Note: scales are not consistent but the block is approximately 700m thick and 100 km wide. Note also that N/S speeds are very much greater than E/W speeds. (From Bang, 1976).

Lutjeharms (1981b), on the basis of infrared satellite imagery, showed a series of frontal eddies (see Fig. 1) which he considered to be a result of instabilities in the jet current. Shannon et al. (1983), using satellite imagery, showed a similar set of eddies based on water colour, although Nelson (1985) cautions on distinguishing between actual physical processes in the region and biological consequences which may persist long after all physical manifestations have disappeared.

Nelson (1985), in a preliminary analysis of the hydrography and currents off the Cape Peninsula, based on drogues, current profiles and moored meters, proposes no less than six different types of fronts (see Fig.

19). Some of these are more pronounced than others, particularly the surface front over the shelf-edge between the 230-250m contours. A strong jet flowing northwards is associated with this front.

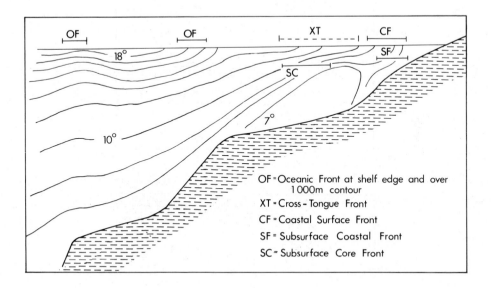

Fig. 19. Conceptual image of isotherm formation typically found during the Cape Upwelling Experiment (CUEX), 1978-1980 . (From Nelson, 1985).

Andrews and Hutchings (1980) proposed a one-celled circulation with sinking at the front during upwelling conditions. They suggest that much of the high phytoplankton biomass is entrained into the powerful jet current associated with the front. However, re-examination of some of the chlorophyll 'a' and nutrient transects along the UM line made in 1971-1973 revealed a peak of chlorophyll 'a' at the front, and a minor peak of nutrients just inshore of the chlorophyll 'a' peak on some transects (see Fig. 6). This is at variance with the downward trend in nutrients along the transect and suggests that sinking at the front may be accompanied by a slight uplift in an anti-cyclonic cell (see Fig. 20a). There is little evidence, however, to indicate that any uplift of water is taking place at the outer edge of the front. In agreement with Andrews and Hutchings (1980), sinking is shown to occur at the coast during relaxations of the upwelling favourable wind stress (see Fig. 20b).

CONCLUSIONS

In comparison with newly upwelled water at the inshore sites of active

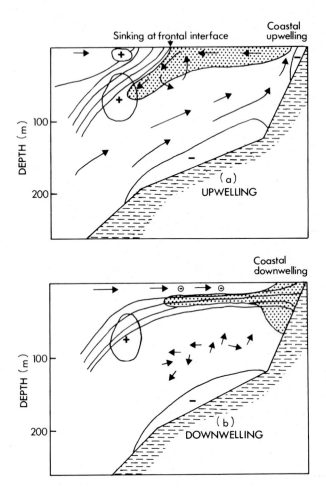

Fig. 20. Schematic presentation of the main direction of water movement proposed to occur during (a) upwelling and (b) downwelling.

upwelling, the frontal region is, despite strong gradients, not a good example of a productive ergocline in terms of the description by Legendre et al. (1985). Rapid changes in the position of surface features of the front and the strong alongshore transport within the front, make it a difficult region to study on suitable time and distance scales. The front is nevertheless of considerable ecological importance in the southern Benguela current, as summarised below:-

1) The frontal zone in the southern Benguela current displays very pronounced gradients compared with many other frontal regions, suggesting strong convergent flow.

2) Very complicated water motions which fluctuate rapidly with tidal,

diurnal, wind and seasonal cycles occur, to the despair of the local modelling fraternity.

    3) The front may affect biological processes in a number of ways:

        a) as a convergent boundary region between cool eutrophic and warm oligotrophic water masses, restricting energy and material fluxes with little input to the offshore environment.

        b) where the strong sinking motions and subsequent re-entrainment of organisms could be important in providing a mechanism for population maintenance within the system but, paradoxically, could also limit their growth potential in upwelled waters.

        c) as a region where a large stock of aged phytoplankton exists well below the euphotic zone, which may stimulate increased microheterotrophic activity and nutrient regeneration leading to enhanced primary productivity within the frontal zone.

        d) as an area of enhanced productivity induced by intermittent surface stabilization and/or upwelling of nutrients at the shelf-break or from an anticyclonic uplift of nutrients inshore of the front. Vertical migration by zooplankton and mesopelagic fish from the shelf-break region may also increase grazing pressures in the region of the front.

        e) as the location of a jet current which transports pelagic fish spawning products to the recruitment area, with the attendant problems of: alongshore transport, offshore dispersion and onshore displacement, and feeding of and predation on the larvae, which are closely associated with the physical dynamic processes in the frontal zone. Of particular interest in this regard are the eddies, common features in satellite images of the frontal zone, which may provide important mechanisms for cross frontal mixing or transport of organisms out of the system.

ACKNOWLEDGEMENTS

The authors would like to express their appreciation to their colleagues at Sea Fisheries and the University of Cape Town for their useful comments on the original manuscript, and for access to unpublished data. Thanks are also due to Mr T. van Dalsen and staff for the artwork and to Ms M. van Niekerk for typing the final document.

REFERENCES

Andrews, W.R.H. and Hutchings, L., 1980. Upwelling in the southern Benguela current. Prog. Oceanog., 9: 1-81.
Badenhorst, A. and Boyd, A.J., 1980. Distributional ecology of the larvae and juveniles of the anchovy Engraulis capensis Gilchrist in relation to the hydrological environment off South West Africa, 1978/79. Fish. Bull. S. Afr., 13: 83-106.

Barlow, R.G., 1982a. Phytoplankton ecology in the southern Benguela current. I. Biochemical composition. J. Exp. Mar. Biol. Ecol., 63: 209-227.

Barlow, R.G., 1982b. Phytoplankton ecology in the southern Benguela current. III. Dynamics of a bloom. J. Exp. Mar. Biol. Ecol., 63: 239-248.

Bang, N.D., 1971. The southern Benguela current region in February, 1966: Part II. Bathythermography and air-sea interactions. Deep-Sea Res., 18: 209-224.

Bang, N.D., 1973. Characteristics of an intense ocean frontal system in the upwell region west of Cape Town. Tellus, 25: 256-265.

Bang, N.D., 1974. The southern Benguela system: finer oceanic structure and atmospheric determinants. Ph.D. thesis, University of Cape Town, South Africa, 181 pp.

Bang, N., 1976. On estimating the oceanic mass flux budget of lateral and cross circulations of the southern Benguela upwelling system. National Research Institute for Oceanology, Internal General Report, SEA IR 7616, 14 pp.

Bang, N.D. and Andrews, W.R.H., 1974. Direct current measurements of a shelf-edge frontal jet in the southern Benguela system. J. mar. Res., 32: 405-417.

Boucher, J., 1984. Localization of zooplankton populations in the Ligurian marine front: role of ontogenic migration. Deep-Sea Res., 31: 469-484.

Boyd, A.J., 1982. Small-scale measurements of vertical shear and rates of horizontal diffusion in the southern Benguela current. Fish. Bull. S. Afr., 16: 1-9.

Boyd, A.J. and Hewitson, J.D., 1983. Distribution of anchovy larvae off the west coast of southern Africa between 32°30' and 26°30'S, 1979-1982. S. Afr. J. mar. Sci., 1: 71-75.

Brink, K.H., 1983. The near-surface dynamics of coastal upwelling. Prog. Oceanog., 12: 223-257.

Brown, P.C., 1984. Primary production at two contrasting nearshore sites in the southern Benguela upwelling region, 1977-1979. S. Afr. J. mar. Sci., 2: 205-215.

Brown, P. and Hutchings, L., 1985. Phytoplankton distribution and dynamics in the southern Benguela current. In: C. Bas, R. Margalef and P. Rubies (Editors), International Symposium on the Most Important Upwelling Areas off Western Africa (Cape Blanco and Benguela), Barcelona, November 1983. Instituto de Investigaciones Pesqueras, Barcelona, 1: 319-344.

De Decker, A.H.B., 1973. Agulhas Bank plankton. In: B. Zeitschel, (Editor), Ecological studies. Analysis and synthesis, Vol. 3. Springer-Verlag, Berlin, pp. 189-219.

De Decker, A.H.B., 1984. Near-surface copepod distribution in the south-western Indian and south-eastern Atlantic ocean. Ann. S. Afr. Mus., 93: 303-370.

Hart, T.J. and Currie, R.I., 1960. The Benguela Current. Discovery Reports, 31: 123-298.

Hutchings, L., 1979. Zooplankton of the Cape Peninsula Upwelling Region. Ph.D. thesis, University of Cape Town, South Africa, 223 pp.

Hutchings, L., 1981. The formation of plankton patches in the southern Benguela Current. In: F.A. Richards (Editor), Coastal Upwelling. Coastal and Estuarine Sciences, American Geophysical Union, Washington D.C., 1: 496-506.

Hutchings, L., Holden, C. and Mitchell-Innes, B., 1984. Hydrological and biological shipboard monitoring of upwelling off the Cape Peninsula. S. Afr. J. Sci., 80: 83-89.

Holligan, P.M., 1981. Biological implications of fronts on the northwest European continental shelf. Phil. Trans. R. Soc. Lond., A302: 547-562.

Kollmer, W.E., 1963. Notes on zooplankton and phytoplankton collections made off Walvis Bay. Investl Rep. mar. Res. Lab. S.W. Afr., 8: 78pp.

Legendre, L., 1981. Hydrodynamic control of marine phytoplankton production: the paradox of stability. In: J.C.J. Nihoul (Editor), Ecohydrodynamics. Elsevier, Amsterdam, pp. 191-207.

Legendre, L., Demers, S. and Lefaivre, D., 1985. Biological production at marine ergoclines. In: J.C.J. Nihoul (Editor), Proceedings of the 17th International Liège Colloquium on Ocean Hydrodynamics. Elsevier, Amsterdam. This volume.

Lutjeharms, J.R.E., 1981a. Features of the southern Agulhas current circulation from satellite remote sensing. S. Afr. J. Sci., 77: 231-236.

Lutjeharms, J.R.E., 1981b. Satellite studies of the South Atlantic upwelling system. In: J.F.R. Gower (Editor), Oceanography from Space. Plenum Press, New York. Mar. Sci., 13: 195-199.

Mooers, C.N.K., Collins, C.A. and Smith, R.L., 1976. The dynamic structure of the frontal zone in the coastal upwelling region off Oregon. J. Phys. Oceanogr., 6: 3-21.

Nelson, G. and Hutchings, L., 1983. The Benguela upwelling area. Prog. Oceanog., 12: 333-356.

Nelson, G., 1985. Notes on the physical oceanography of the Cape Peninsula upwelling system. In: L.V. Shannon (Editor), South African Ocean Colour and Upwelling Experiment. Sea Fisheries Research Institute, Cape Town, pp. 63-95.

Olivieri, E.T., Hutchings, L., Brown, P.C. and Barlow, R.G., 1985. The development of phytoplankton communities in terms of their particle size frequency distribution, in newly upwelled waters of the southern Benguela current. In: C. Bas, R. Margalef and P. Rubies (Editors), International Symposium on the Most Important Upwelling Areas off Western Africa (Cape Blanco and Benguela), Barcelona, November 1983. Instituto de Investigaciones Pesqueras, Barcelona, 1: 345-371.

Packard, T.T., Blasco, D. and Barber, R.T., 1978. Mesodinium rubrum in the Baja California upwelling system. In: R. Boje and M. Tomczak (Editors), Upwelling Ecosystems. Springer-Verlag, Berlin, Heidelberg, New York, pp. 73-89.

Probyn, T.A., 1985. Nitrogen uptake by size-fractionated phytoplankton populations in the southern Benguela upwelling system. Mar. Ecol. Prog. Ser., 22: 249-258.

Shannon, L.V., 1966. Hydrology of the south and west coasts of South Africa. Investl Rep. Div. Sea Fish. S. Afr., 58: 62 pp.

Shannon, L.V., 1985. The Benguela Ecosystem. Part 1. Evolution of the Benguela, physical features and processes. Oceanogr. Mar. Biol. Ann. Rev., 23: 105-182.

Shannon, L.V., Nelson, G. and Jury, M.R., 1981. Hydrological and meteorological aspects of upwelling in the southern Benguela current. In: F.A. Richards (Editor), Coastal Upwelling. Coastal and Estuarine Sciences, American Geophysical Union, Washington D.C., 1: 146-159.

Shannon, L.V., Mostert, S.A., Walters, N.M. and Anderson, F.P., 1983. Chlorophyll concentrations in the southern Benguela current region as determined by satellite (Nimbus-7 coastal zone colour scanner). J. Plankton Res., 5: 565-583.

Shannon, L.V., Schlittenhardt, P. and Mostert, S.A., 1984a. The Nimbus-7 CZCS experiment in the Benguela current region off Southern Africa, February 1980. 2. Interpretation of imagery and oceanographic implications. J. Geophys. Res., 89: 4968-4976.

Shannon, L.V., Hutchings, L., Bailey, G.W. and Shelton, P.A., 1984b. Spatial and temporal distribution of chlorophyll in southern African waters as deduced from ship and satellite measurements and their implications for pelagic fisheries. S. Afr. J. mar. Sci., 2: 109-130.

Shelton, P., 1984. Notes on the spawning of anchovy during the summer of 1982-3. S. Afr. J. Sci., 80: 69-71.

Shelton, P.A. and Hutchings, L., 1982. Transport of anchovy, Engraulis capensis Gilchrist, eggs and early larvae by a frontal jet current. J. Cons. int. Explor. Mer., 40: 185-198.

Shushkina, E.A., Vinogradov, M.Ye., Sorokin, Yu.I., Lebedeva, L.P. and Mikheyev, V.N., 1978. Functional characteristics of planktonic communities in the Peruvian upwelling region. Oceanology, 18: 579-589.

Simpson, J.H., 1981. The shelf-sea fronts: implications of their existence
    and behaviour. Phil. Trans. R. Soc. Lond., A302: 531-543.
Stevenson, M.R., Garvine, R.W. and Wyatt, B., 1974. Lagrangian measurements
    in a coastal upwelling zone off Oregon. J. Phys. Oceanogr., 4: 321-336.
Taunton-Clark, J., 1985. The formation, growth and decay of upwelling
    tongues in response to the mesoscale wind field during summer. In:
    L.V. Shannon (Editor), South African Ocean Colour and Upwelling Experiment.
    Sea Fisheries Research Institute, Cape Town, pp. 47-61.
Thiriot, A., 1978. Zooplankton communities in the West African upwelling
    area. In: R. Boje and M. Tomczak (Editors), Upwelling Ecosystems.
    Springer-Verlag, Berlin, Heidelberg, New York, pp. 32-61.
Unterüberbacher, H.K., 1964. Zooplankton studies in the waters off Walvis
    Bay with special reference to the Copepoda. Investl Rep. mar. Res.
    Lab. S.W. Afr., 11: 1-42.
Vinogradov, M.E. and Shushkina, E.A., 1978. Some development patterns
    of plankton communities in the upwelling areas of the Pacific Ocean.
    Mar. Biol., 48: 357-366.

THE DYNAMIC CONTROL OF BIOLOGICAL ACTIVITY IN THE SOUTHERN BENGUELA UPWELLING
REGION

G.B. BRUNDRIT

Department of Oceanography, University of Cape Town, Private Bag, 7700 Rondebosch (South Africa)

ABSTRACT

   Dynamic processes have been recognised to be responsible for the control of
spatial and temporal variability in biological activity. This is indicated in
the results from a field study in the Southern Benguela upwelling region in
which water type structures can be consistently identified on physical, chemical
and biological criteria. Various candidates for the dynamic processes which
control interactions both within and between these structures are considered and
assessed.

INTRODUCTION

   Dynamical processes in the ocean have been recognised to be fundamental to
an understanding of the spatial and temporal variability in biological production. Indeed, the inter-relations between the two have led to the emergence
of a new discipline, which has been termed "dynamic biological oceanography"
(Legendre and Demers, 1984). In common with other coastal upwelling regions
(Brink, 1983), the Southern Benguela (Shannon, 1985) provides a dramatic contrast to the open ocean surface norm with the high levels of biological activity
which are to be found intermittently within its boundaries. It is thus an ideal
field laboratory for the study of dynamic biological oceanography.

   During active wind forcing of coastal upwelling, a pronounced surface front
is established some 50 km offshore, which forms the ocean boundary of the
region in which the biological production is initiated. Within a matter of
days, the wind forcing abates and the upwelling relaxes, whilst the biological
production develops and matures. In this passive phase, new dynamical processes leading to the eventual breakdown of the surface front become important,
and distinctive spatial structures become apparent in the frontal region.

   Recent field studies have provided detailed information on the physical,
chemical and biological variability of the water masses within the Southern
Benguela region (Hutchings et al., 1985). A "snapshot" from one such field
study is used to examine the potential role of dynamic processes in character-
ising the scope of the biological productivity.

   It is shown that surface advection of upwelled water, progressively warmed

through insolation, is the dominant process in fueling the biological production during the active phase of the upwelling cycle. Thus the production is
confined to the surface mixed layer being driven longshore and offshore from
the topographically determined centre of upwelling. Thereafter, during the
passive phase, isopycnal interleaving of water masses leads to a smearing of
the definition of the front and to "sinking" of productive surface waters
below the photic zone. These conclusions are aided by the detailed profiles
of salinity which is distinguished as a conservative variable throughout the
upwelling processes.

FIELD DATA

The field data was collected during a 40 km transect, offshore from Cape
Columbine (see Fig. 1), as part of the Frontal Zone Cruise of the South
African Sea Fisheries Research Institute during December 1984 (Hutchings et
al., 1985). This transect had been chosen so as to cross the Cape Columbine
upwelling plume, which is shown in its active phase in the sea surface
temperatures also marked in Fig. 1 (after Nelson and Hutchings, 1983). The
transect consisted of nine stations at a spacing of 5 km, which crossed an
ill-defined temperature front between stations FZ06 and FZ07. Station FZ05
was repeated as station FZ05A.

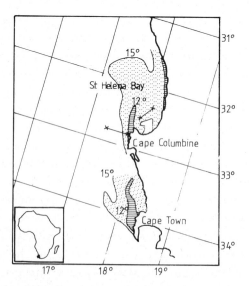

Fig. 1. Location map of the study area in the Southern Benguela upwelling
region, showing the line of the Cape Columbine transect and the typical
sea surface temperatures found north of Cape Columbine and north of
Cape Town in the active phase of the upwelling cycle (after Nelson and
Hutchings, 1983).

At each station, CTD profiles were taken, currents measured, a rosette
sampler collected bottle samples of water from various depths, and net samples
were taken. From among the results of the various physical, chemical and biological analyses undertaken later, the particular quantities used in this study
were temperature, salinity, nitrate and chlorophyll. Other studies are being
undertaken and will appear in the literature (Hutchings et al., 1985).

CTD profiles of temperature and salinity with depth at the innermost station
FZ01, and at stations FZ05A and FZ09 on either side of the front, are taken as
typical and are shown in Fig. 2. There is a separation between surface and
subsurface waters at about 30 m depth on each profile. At station FZ09, the
surface water appears to be a wind mixed, sun warmed extension of the subsurface water. This is not the case at station FZ05A, where the surface water
bears no direct relation to the water beneath it. Taking insolation into
account, the surface water is too cool and not sufficiently saline to have been
formed by wind mixing of subsurface water at the station. There is thus a contrast in both temperature and salinity between the surface waters at these two
stations. This contrast is not reflected in subsurface waters, though these
appear to rise towards the coast. At the innermost station FZ01, the subsurface
water consists of a bottom mixed layer, while the surface water shows active
sun warming in its temperature profile.

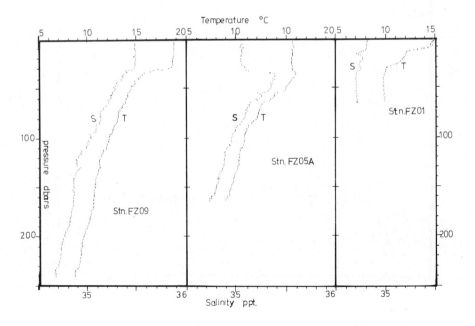

Fig. 2. CTD profiles of temperature and salinity with pressure (depth) at
        stations FZ01, FZ05A and FZ09.

98

A complete set of temperature and salinity values from rosette sampler depths at each station on the transect are displayed on a T-S diagram (see Fig. 3). These results extend the comments already made concerning the surface and subsurface waters. In particular, it should be noted that it is the increase or decrease in salinity between the surface and subsurface water at each station which identifies the surface water on either side of the front.

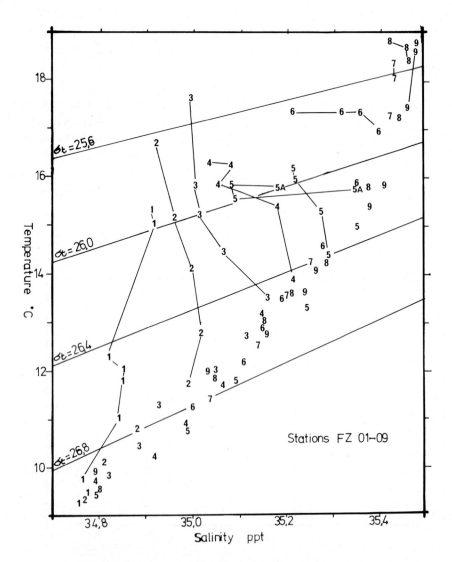

Fig. 3. T-S diagram, with density as sigma-t, for all stations of the Cape Columbine transect. The thin lines join values at each station, between 0 and 30 m depth.

More importantly, it can be seen that the subsurface water, below 30 m depth, exhibits a consistent correlation between temperature and salinity. This axis of temperature and salinity is South Atlantic Central Water, which is the source of all the water which eventually forms the surface waters. At each station, the T-S values in the surface layer are joined together (see Fig. 3). It can be seen that there are cross-shore gradients in the values of temperature, salinity and sigma-t at the top of the subsurface water. This is useful in following the subsurface water into the surface layer above 30 m.

Despite the modification of the temperature of surface water by insolation, the origin of the surface water can still be traced through its salinity. Outside the front, the surface water is a sun-warmed extension of its subsurface counterpart. Inside the front, it is clear that upwelling, in the sense of subsurface water entering the surface layer, is only taking place at the innermost stations. The surface water at other stations inside the front is advected from inshore locations upstream where upwelling has previously occurred. Current metering confirms this view, and it may be conjected that some of the surface water had its origin in the centre of upwelling at Cape Town (see Fig. 1).

In order to examine the variability in nutrients and biological productivity, the T-S diagrams are repeated in Figs. 4 and 5, but show spot values of nitrate and chlorophyll rather than station number. In Fig. 4, the nitrate values along the Central Water axis confirm previous observations (Andrews and Hutchings, 1980) that nutrients are correlated with temperature and salinity within Central Water. The exception to this is to be found in the bottom mixed layer at the innermost stations FZ01 and FZ02, where temperature and salinity are correlated in the manner of Central Water, but nutrient values are enhanced.

In the surface waters, the nitrate level is reduced below the Central Water values. It remains relatively high at the innermost station, but decreases to low levels offshore, indicating how the surface waters have been stripped of nutrients by biological production. It should be noted that there are no gradients in surface nutrient values across the front.

The T-S diagram with spot values of chlorophyll (see Fig. 5) confirms that Central Water can be considered to be conservative in nutrients as there is no biological production present. In the surface water, the chlorophyll levels provide contrasts between the innermost station, the stations inshore of the front, and the stations outside the front. This confirms the distinctions made on the basis of salinity despite the lack of contrast in nutrient levels across the front. This agreement between salinity and chlorophyll can be highlighted in an offshore transect for the upper 50 m depth, which shows salinity contours and spot values of chlorophyll (see Fig. 6). Except at the innermost station FZ01, the production is confined to water which is above the salinity maximum.

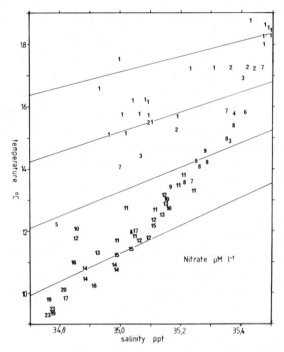

Fig. 4. T-S diagram with spot values of nitrate for all stations.

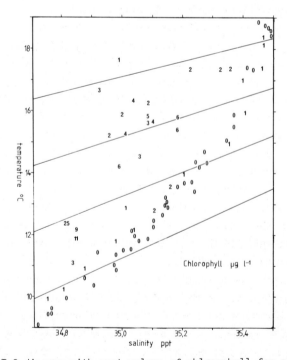

Fig. 5. T-S diagram with spot values of chlorophyll for all stations.

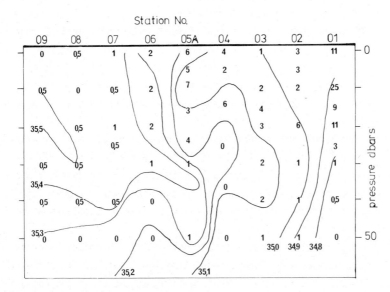

Fig. 6. Contours of salinity and spot values of chlorophyll for all stations of the Cape Columbine transect, above 50 m depth.

WATER TYPE STRUCTURES AND DYNAMIC CONTROL

The various water type structures within the surface and subsurface waters of the Southern Benguela region can be identified on physical, chemical and biological grounds, and there is a consistency between the results of the identifications which confirms the basic hypothesis of dynamic biological oceanography.

South Atlantic Central Water can be identified on the basis of the demonstrated correlations between its physical and chemical properties. All the other water type structures can then be considered to be modifications of Central Water in which outside fluxes render the various properties non-conservative. Salinity is distinctive in retaining its conservative character.

The subsurface water in the bottom mixed layer at the innermost stations is enhanced in nutrients. The surface water there is sun-warmed, and is losing its nutrients due to rapidly developing production. The surface waters inside the front are inshore surface waters further modified in temperature and progressively stripped of nutrients as they are advected offshore. Outside the front, the surface waters are a sun warmed modification of the subsurface waters beneath, with nutrients lost in production, long since completed.

Having established this set of water type structures which are consistently identified on the basis of their physical, chemical and biological properties, it is possible to examine candidates for the dynamic processes which may be responsible for the control of these structures.

South Atlantic Central Water is itself a mixture of South Atlantic Surface Water and Antarctic Intermediate Water (Shannon, 1985). Over a long period of time, various mixing processes will have been responsible for the uniform trend of properties which characterise the water. Residual evidence of these properties can be seen in the familiar step features present in the CTD profile of station FZO9 (see Fig. 2).

For the remaining water type structures the dynamic controls may act within the structure or between adjacent structures. Clearly the wind will mix the surface waters, but the evidence is that only outside the front does the wind mixed layer extend completely through to the subsurface waters. Indeed, it is this contact which has partially determined the character of the South Atlantic Central Water.

Inside the front, the wind is not actively entraining subsurface water but the onset of high winds may well change this situation. Rather, it is the indirect action of the wind in driving the upwelling at the coast and the off-shore advection of the surface water which provides the important dynamic control of the surface water inshore of the front. Insolation must also play an important role in this process, and investigations are needed which take both insolation and wind driven surface transport into account. Only then will the detail of the dynamic control of the surface water inshore of the front be-come apparent.

In the bottom mixed layer at the innermost stations, it should be noted that the time scale needed for the nutrient enhancement is long compared to the up-welling cycle time scale. Thus this layer must be considered to be relatively permanent and the upwelling must proceed over it. Candidates for the control of this structure should be sought within these constraints.

It is now possible to turn to the interaction between the various water type structures. Mention has already been made of the possibility of a deepening wind mixed surface layer entraining subsurface water. The advection of the surface water across the subsurface water beneath can also give rise to a shear flow entrainment of subsurface water. These two mechanisms driving the entrainment and subsequent upward vertical fluxes are interconnected and the relative importance of each at different phases of the upwelling cycle should be assessed.

In the horizontal, advection is important in bringing subsurface waters in-shore and upward towards the coast and moving surface waters away from the coast inshore of the front. Across the front, the evidence suggests that in the passive stages of upwelling, the warming of the surface waters inshore of the front will lead to weakened density (sigma-t) gradients. The possibility of lateral interleaving of water type structures across the front then occurs.

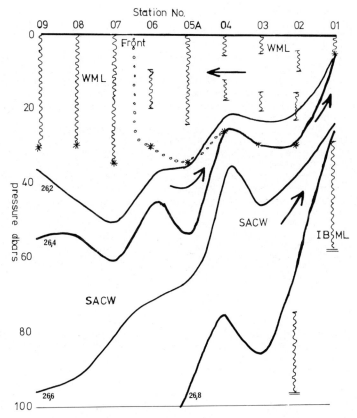

Fig. 7. A conceptual diagram of the stations of the Cape Columbine transect, above 100 m depth, showing contours of sigma-t for South Atlantic Central Water (SACW), the mixed layers ( ) at the surface (WML) and on the bottom at the innermost stations (IBML), and the interleaving in the region of the front. The depth of the salinity maximum is denoted by an asterisk at each station.

The strength of this process may be enhanced by the vertical salinity gradients which will tend to destabilise the water column through double diffusion (Ruddick and Turner, 1979). The interleaving will lead to the very important cross frontal fluxes which accompany the breakdown of the front. Evidence of interleaving can be seen in the CTD profile at station FZ05A (see Fig. 2) and in the salinity section (see Fig. 6).

The conclusions reached concerning the identification of the water type structures in the Southern Benguela region and the dynamic controls of these structures are summarised in a conceptual diagram (see Fig. 7). This study forms part of the Benguela Ecology Programme and acknowledgement should be made to the work of members of the Sea Fisheries Research Institute during and following the Frontal Zone Cruise.

REFERENCES

Andrews, W.R.H. and Hutchings, L., 1980. Upwelling in the Southern Benguela
    current. Prog. Oceanogr., 9: 1-81.
Brink, K.H., 1983. The near surface dynamics of coastal upwelling. Prog.
    Oceanogr., 12: 223-257.
Hutchings, L., Armstrong, D.A. and Mitchell-Innes, B.A., 1985. The frontal
    zone in the Southern Benguela current. In: J.C.J. Nihoul (Editor),
    Proceedings of the 17th International Liege Colloquium on Ocean Hydro-
    dynamics. Elsevier, Amsterdam.
Legendre, L. and Demers, S., 1984. Towards dynamical biological oceanography
    and limnology. Can. J. Fish. Aquat. Sci., 41: 2-19.
Nelson, G. and Hutchings, L., 1983. The Benguela upwelling area. Prog.
    Oceanogr., 12: 333-356.
Ruddick, B.R. and Turner, J.S. 1979. The vertical length scale of double-
    diffusive intrusions. Deep-Sea Res., 26A: 903-913.
Shannon, L.V., 1985. The Benguela ecosystem, I Evolution of the Benguela,
    physical features and processes. Oceanogr. Mar. Biol. Ann. Rev., 23: 105-
    182.

FRONTAL ZONES, CHLOROPHYLL AND PRIMARY PRODUCTION PATTERNS IN THE
SURFACE WATERS OF THE SOUTHERN OCEAN SOUTH OF CAPE TOWN

J.R.E. LUTJEHARMS[1], B.R. ALLANSON[2] and L. PARKER[2]

[1]National Research Institute for Oceanology, C.S.I.R.,
  P.O.Box 320, Stellenbosch 7600, South Africa
[2]Department of Zoology and Entomology, Rhodes University,
  P.O.Box 94, Grahamstown 6140, South Africa

ABSTRACT

   The enhancing effects of certain frontal regions in the world
ocean to biological activity is becoming recognised to an increasing degree. A number of underlying mechanisms for such enhancing
effects are being hypothesised. In the Southern Ocean a series of
coherent fronts are formed due to a variety of physical factors,
making it an ideal area to study the proposed hypotheses. Fronts
in the Southern Ocean sector south of Africa are particularly good
candidates for such investigations since they exhibit some of the
most extreme horizontal gradients in physico-chemical variables.
It has, furthermore, been noted that the primary production of the
Southern Ocean as a whole is considerably lower than would be
suggested by the usual controlling factors. This process is not
well understood. Results of recent research cruises in the area
between Africa and Antarctica show that many fronts in this area
exhibit relatively high chlorophyll concentrations at the sea
surface, increases in potential primary production as well as
increases in photosynthetic efficiency. These fronts, because of
their unique and special characteristics, may thus offer superior
areas for studying not only the interaction between physico-
chemical factors and open ocean biological activity in general,
but also Southern Ocean primary productivity in particular.

INTRODUCTION

   Increased biological activity at ocean fronts has been observed
with increasing frequency in the last decade. Pingree et al.
reported summer blooms at tidal fronts in the English channel in
1975 and showed that this was related to greater stratification in
those areas. In most continental shelf seas, primary production

blooms are restricted to a short period of spring when an increase in stratification of the water column traps phytoplankton in a nutrient rich euphotic zone. Primary production decreases once the nutrients are depleted. Nutrient recycling is then restricted by the increased stability of the water column. Breakdown of this stratification at shelf-edge fronts (Pingree, 1982; Simpson, 1979; Marra et al., 1982) or due to tidally induced turbulent mixing (Dooley, 1981) makes nutrients more readily available and may enhance levels of primary production.

Increased primary production has also been observed at the edges of deep-sea eddies or current rings. Yentsch and Phinney (1985) show that phytoplankton populations in these peripheral regions experience a near steady state growth related to the rotational velocity of the rings. This supports the findings of Lutjeharms and Walters (1985) who show increased concentrations of chlorophyll at the borders of the Agulhas Current and of filaments of Agulhas Current water. The affinity for fronts by organisms is not restricted to the lower trophic levels. Olson and Backus (1985) have described the concentration of fish at the border of a warm-core Gulf Stream ring, while Maul et al. (1984) have shown that the Japanese bluefin tuna fisheries in the Gulf of Mexico is most efficient when concentrated at the boundary of the Gulf Loop Current. Similar, anecdotal, results on tuna fishing have been reported by Lutjeharms et al. (1985). The concentration of birds at some fronts has been described by Abrams (1985) and by Ainley and Jacobs (1981).

Various mechanisms have been proposed for the increased biological activity at fronts. Most rely on increased stratification (Pingree et al., 1975; Simpson, 1979; Smith and Nelson, 1985) while some assume a measure of horizontal convergence which is believed to concentrate buoyant, non-motile organisms at fronts (Olson and Backus, 1985; Ainley and Jacobs, 1981). Such mechanisms would lead to a decaying population at fronts, which is certainly not true of all instances where increased phytoplankton concentrations were observed at the sea surface. Simpson et al. (1979) in fact report a healthier phytoplankton standing crop at a shelf sea front than in the mixed water inshore. Additional mechanisms based on current shear have been proposed by Lutjeharms et al. (1985).

SOUTHERN OCEAN FRONTS

The Southern Ocean is divided into zonal, circum-global bands
by a series of well-defined, distinctive fronts (Deacon, 1937).
More closely spaced measurement in the last decade have resolved
detail of these fronts and of some surface fronts not recognised
before (Sievers and Emery, 1978; Lutjeharms and Emery, 1983).
This meridional zonation is also reflected in the distribution of
organisms as has, for instance, been shown by Hedgpeth (1969) in
regards to a selected group of marine invertebrates. Although
Southern Ocean fronts may act as biogeographical barriers
(Tranter, 1982) they are by no means totally impervious to the
movement of species (Voronina, 1962). Certain fronts in the
Southern Ocean have also been shown to be areas in which a greater
concentration of phytoplankton standing stock (Plancke, 1977),
primary productivity (Allanson et al., 1981) and birds (Ainley and
Jacobs, 1981; Abrams, 1985) are to be found. Fronts also seem to
be preferred areas of spawning for krill, the main Antarctic
macro-zooplankton (Tranter, 1982).

Deacon (1982) has discussed each individual frontal area and
its biogeographical impact in detail. He shows that the Antarctic
Divergence, the line between average easterly and westerly winds,
may be an area of local, short-lived upwelling due to Ekman divergence. Tranter (1982) believes that enhanced primary productivity
at this front is less likely to be due to nutrient enrichment from
deeper layers than to light-associated factors related to the wind
regime, that is, vertical movement will inhibit loss of organisms
from the euphotic zone. The Antarctic Polar Front is a potent
biogeographical barrier. Tranter (1982) considers that this may
be partially due to the vertical migration, in the case of krill
larvae, between surface waters which have a northerly drift component and deeper waters which have a southerly component in the
Antarctic zone. Apart from being a biogeographical boundary, the
Antarctic Polar Front also has some species specific enhancing and
degrading properties (Deacon, 1982). Some species are least
abundant at the front, with higher catches north or south of it,
while some have their highest concentrations at the front. Deacon
(1982) states that the Sub-Antarctic Front, about which little is
as yet known, has also been observed to act as a biogeographical
front. The Sub-Tropical Convergence, which is the most intense
front at the sea surface in a number of Southern Ocean locations,

marks the limit of many warm-water species and forms the conventional northern boundary of the Southern Ocean.

From the above it seems clear that although all major fronts in the Southern Ocean have been shown to probably possess significant biogeographical effects, data about the exact causes for these effect and the underlying mechanisms are sparse. Enhanced primary productivity, as observed at some fronts, holds special import since the low primary productivity of the Southern Ocean as a whole is not well understood.

PRIMARY PRODUCTION IN THE SOUTHERN OCEAN

In most parts of the world ocean, primary productivity is nutrient limited. In the Southern Ocean nutrient concentrations, with the exception of silica, are high south of the Sub-Tropical Convergence. South of the Antarctic Polar Front all nutrients, including silica, are present in high concentrations. One would therefore expect high primary production, but this has not been observed (Holm-Hansen et al., 1977). It was thought that low concentrations of silica might be an inhibiting factor in the Sub-Antarctic zone, between the Sub-Tropical Convergence and the Antarctic Polar Front, while low temperatures of antarctic waters might limit algal growth rates elsewhere. Witek et al. (1982), however, considered nutrients as being totally non-limiting in the western Antarctic and in fact observed net-phytoplankton only in water exhibiting a well-developed thermal stratification. Fogg and Hayes (1982) also consider hydrographic stratification to be of paramount importance. Tranter (1982) was convinced that primary production in the Southern Ocean is limited primarily by light and recent results seem to support his conclusion (Tilzer et al., 1985). During summer, when sufficient light is available, wind stress is frequently so great that phytoplankton is mixed well below the compensation depth where gains by photosynthesis are lost in respiration. Such a mechanism is not unique to the Southern Ocean. In the area south of 28°S in the southwestern tropical Pacific a combination of available light and mixed-layer depth is the limiting factor to sea surface chlorophyll (Dandonneau and Gohin, 1984). Marra and Boardman (1984), working on the ice edge zone in the Weddell Sea, have also come to the conclusion that phytoplankton distributions are regulated by the availability of light. Jennings et al. (1984) have recently used a new productivity estimate for the Weddell Sea, based on the seasonal depletion of nitrate, phosphate and silicic acid in the surface layer. Their estimate is far in excess of most reported measurements of productivity in the open ocean areas of the Southern Ocean casting doubt on previous assessments.

In a review on nutrient cycles in marine antarctic ecosystems, Holm-Hansen (1985), considering many of the conflicting conjectures on primary productivity in the Southern Ocean mentioned above, comes to the conclusion that our understanding of the interaction between biological-physical-geochemical processes in the Antarctic is meagre. The frontal systems south of Africa, because of their special characteristics, may be particularly amenable to studies that could resolve some of these problems.

FRONTS AND BIOLOGICAL ACTIVITY SOUTH OF AFRICA

The frontal systems between Africa and Antarctica consist of the Sub-Tropical Convergence, sometimes enhanced by the Agulhas Front, the Sub-Antarctic Front, the Antarctic Polar Front, an ephemeral Antarctic Divergence and the Continental Water Boundary at the edge of the Antarctic continental shelf. The detail of these fronts has been studied extensively in the past few years (Lutjeharms and Emery, 1983; Lutjeharms and Foldvik, 1985; Lutjeharms and Rickett, 1985; Lutjeharms, 1985b). Their average locations, widths and characteristics have been established (Lutjeharms, 1985a) both from measurements by expendable bathythermograph probes and by conductivity-temperature-depth units. The results are in good agreement with those of a statistical analysis of all available sea surface temperature measurements (Lutjeharms and Valentine, 1984).

The results show that the Sub-Tropical Convergence south of Africa exhibits some of the strongest sea surface temperature gradients observed. It is an area of extreme mesoscale variability (Lutjeharms and Baker, 1980) which stretches from the Agulhas Retroflection region to the longitude of the Crozet Islands (Lutjeharms and van Ballegooyen, 1984). This variability is due to meanders in the front and to active eddy shedding to both sides of the front (Lutjeharms and Valentine, 1985). The Sub-Antarctic Front was observed on almost all cruises, in the area, lies at about 46°S (Lutjeharms and Valentine, 1984) and exhibits an increasing step-like structure in the vertical with the onset of summer (Lutjeharms and Foldvik, 1985). This step-like morphology is also found to develop in the Antarctic Polar Front which furthermore shows little meridional wandering. The characteristic vertical thinning of the subsurface temperature minimum in the Antarctic sector, which is assumed to correspond to the Antarctic Divergence, is seldom found. Few direct measurements of the Continental Water Boundary are available, but when observed it is a very explicit feature.

Some measurements of phytoplankton in the Sub-Tropical Convergence area in 1973 and 1974 showed increases in biomass at the northern edge of the convergence (Plancke, 1977). Ichimura and Fukushima (1963) had found very similar features, but have also reported peaks in chlorophyll a content at the Antarctic Polar Front. The influence of these fronts was also observed in the distribution of pelagic birds (Abrams, 1985).

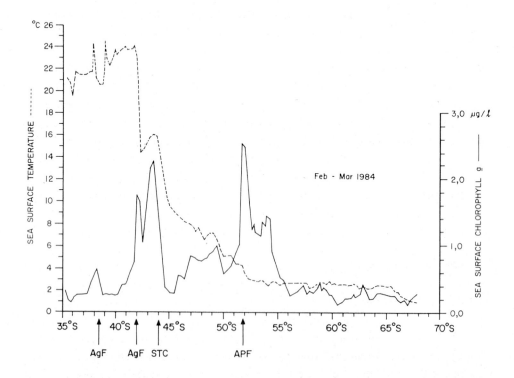

Fig. 1. The meridional distribution of sea surface temperature and sea surface concentration of chlorophyll a along a transect between Antarctica and Cape Town during the months February to March 1984. These data were collected from the icebreaker Shirase and reported by Hamada et al. (1985). The significant peaks in surface chlorophyll at the Antarctic Polar Front (APF), the Sub-Tropical Convergence (STC) and the Agulhas Front (AgF) are evident.

Further simultaneous measurements of chlorophyll and hydrographic variables have shown some very distinctive patterns (Allanson et al., 1981; Lutjeharms et al., 1985). An example is given in Fig. 1. A line of continuous chlorophyll a measurements from a depth of 8 m was undertaken between Syowa Station on the Antarctic continent and Cape Town. These readings formed part of the Japanese contribution to the international SIBEX programme, were undertaken from the icebreaker Shirase and subsequently reported by Hamada et al. (1985). On proceeding northward the surface temperatures rose slowly from about $1^{o}C$ to $3,1^{o}C$ at 52° where there was an notable increase of more than $2^{o}C$ over a short meridional distance. This corresponds exactly to the latitudinal and thermal ranges given for the Antarctic Polar Front south of Africa by Lutjeharms and Valentine (1984). The sea surface chlorophyll readings exhibit a very significant peak at this surface expression of the Antarctic Polar Front with values ten times that found in normal Antarctic Surface Water. Two very strong horizontal thermal gradients were also found at 44° and $42^{o}S$. The former's thermal signal is that of the Sub-Tropical Convergence, but it is located further south than usual. The latter is the Agulhas Front. Both frontal features have corresponding peaks in the sea surface chlorophyll distribution (Fig. 1). A small peak at $38^{o}S$ lies at the location of the presumed northern border of the Agulhas Current. Although it might not be a statistically significant feature in this instance, it is interesting to note that Lutjeharms and Walters (1985) observed enhanced chlorophyll concentrations at the same landward edge of the Agulhas Current on a previous occasion. The Sub-Antarctic Front may have been located at $49^{o}S$ according to surface temperature gradients. This corresponds to an increased chlorophyll a value, but without subsurface readings to determine accurately the location of the Sub-Antarctic Front this apparent correlation cannot be ascertained unambiguously.

These high values found at the abovementioned fronts are extreme by Southern Ocean standards. Their strong geographical correlations with the fronts indicate that these areas may be important features of primary production in the Southern Ocean as a whole. However, since only surface chlorophyll a was determined in this particular case, it is quite possible that these concentrations were due to advective processes only which could have accumulated surface organisms, and not due to enhanced primary productivity.

Allanson et al. (1981) have also undertaken such readings between Africa and Antarctica and have presented very similar results. Some of these are portrayed in the two upper panels of Fig. 2. Concurrent subsurface temperature readings were taken to a depth of 500 m making it possible to locate accurately all fronts, but in particular the Sub-Antarctic Front and the Antarctic Divergence. A full suite of readings was also taken right up to the ice edge of Antarctica, thus resolving the Continental Water Boundary (Fig. 2). Once again significant peaks in the chlorophyll a concentrations were latitudinally correlated with the Sub-Tropical Convergence region, with the Antarctic Polar Front and particularly with the Continental Water Boundary. Smaller increases were observed at the Sub-Antarctic Front and in the stratified surface layer of Antarctic Surface Water. Potential primary production values were also determined, but not with the latitudinal resolution as the surface chlorophyll measurements (Fig. 2A). A statistically significant correlation between chlorophyll a concentrations at the sea surface and potential primary production was established (Allanson et al., 1981). Although it is a tentative conclusion, based on only one set of data, these results agree with those of Plancke (1977) and do point to underlying mechanisms enhancing the productivity at these fronts and not to mechanical concentration processes.

Accumulating all the potential primary production estimates of four cruise tracks between Cape Town and SANAE, the South African Antarctic base, a distribution is obtained which is shown in Fig. 2C. These values have been related to the statistical average locations for these fronts according to Lutjeharms and Valentine (1984). A noticeable overlap between peaks in the combined potential primary productivity from 1980 to 1982 and the meridional range for the various fronts is found. These data are for different periods and for overlapping, but not identical, longitudinal bands. A perfect correspondence can thus not be expected. It may also be noted that the attenuation coefficient measured at the fronts, and given in Fig. 2C, is an order of magnitude greater than the average value for most of the Antarctic Surface Water, which is about 0,05.

The maximum potential primary production at each station was determined by taking in situ samples of the phytoplankton, at various depths and incubating them on board at the light intensity measured at those depths. The depth at which maximum photosynthe-

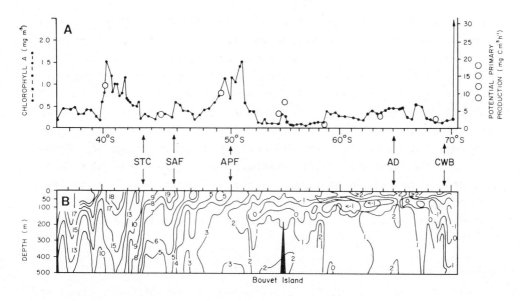

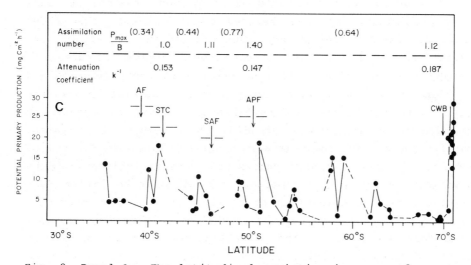

Fig. 2. Panel A: The latitudinal variation in sea surface concentration of chlorophyll a along a cruise track between Antarctica
and Cape Town during January 1981. Data were collected from the
research and supply vessel S.A.Agulhas and results reported by
Allanson et al. (1981). Peaks in the surface chlorophyll
correspond to the latitudinal locations of the Antarctic Polar
Front (APF), the Sub-Tropical Convergence (STC), to a lesser
extent the Sub-Antarctic Front (SAF) but strongly so at the Continental Water Boundary (CWB), all shown in panel B. Panel B: The
thermal structure of the upper 500 m of the water column between
Antarctica and Cape Town during January 1981. Panel C: The accumulated potential primary production estimates for tracks of the
vessel S.A.Agulhas between Cape Town and the South African Antarctic base SANAE from December 1980 to January 1982. Assimilation
numbers, in brackets for interfront areas, and high attenuation
coefficients reinforce the biological significance of the fronts.

tic assimilation took place was thus established. This value was then divided by the biomass, as expressed in terms of chlorophyll at that depth, to derive the assimilation numbers for each station. These express the maximum efficiency of the chlorophyll present at the station, or by implication, the health of the phytoplankton population. The values established for frontal and inter-frontal regions are given in Fig. 2C. In all but one instance, the assimilation numbers for frontal areas are double that of the adjacent water masses. This shows that the phytoplankton populations at these fronts are most probably thriving there for specific environmental reasons and have not been accumulated there as part of the average advection patterns at the sea surface.

CONCLUSIONS

Detailed work on the frontal systems of the Southern Ocean and their interaction with the biota has only just started, but some of the preliminary results, sketched above, indicate that these frontal systems may play an important role in the overall primary productivity of this ocean. These fronts consist of areas of convergence or divergence, of high degrees of stratification and of borders between areas with highly disparate nutrient concentrations. A more intensive and detailed study of these fronts may thus lead to a better understanding of a number of different frontal processes and their impact on surface layer ecology all of which are at present poorly understood. These include problems concerning the production limiting role of silicate in the Sub-Antarctic zone, the role of stratification in enhancing productivity, the dynamics of the various frontal zones themselves as well as the role of fronts in the overall primary production of the Southern Ocean as a whole.

REFERENCES

Abrams, R.W., 1985. Energy and food requirements of plegic aerial seabirds in different regions of the African sector of the Southern Ocean. In: W.R. Siegfried, P.R. Condy and R.M. Laws (Editors), Antarctic Nutrient Cycles and Food Webs, Springer-Verlag, Heidelberg, pp. 466-472.

Ainley, D.G. and Jacobs, S.S., 1981. Sea-bird affinities for ocean and ice boundaries in the Antarctic. Deep-Sea Res., 28A: 1173-1185.

Allanson, B.R., Hart, R.C. and Lutjeharms, J.R.E., 1981. Observations on the nutrients, chlorophyll and primary production of the Southern Ocean south of Africa. S. Afr. J. Antarc. Res., 10/11: 3-14.

Dandonneau, Y. and Gohin, F., 1984. Meridional and seasonal variation of the sea surface chlorophyll concentration in the southwestern tropical Pacific (14 to 32°S, 160 to 175°E). Deep-Sea Res., 31: 1377-1393.

Deacon, G.E.R., 1937. The hydrology of the Southern Ocean. Discovery Reports, 15: 1-124

Deacon, G.E.R., 1982. Physical and biological zonation in the Southern Ocean. Deep-Sea Res., 29A: 1-15.

Dooley, H.D., 1981. The role of axially varying vertical mixing along the path of a current in generating phytoplankton production. Phil. Trans. R. Soc. Lond., A302: 649-660.

Fogg, G.E. and Hays, P.K., 1982. The relative importance of nutrients and hydrographic features for the growth of Antarctic plankton. Joint Oceanographic Assembly, Halifax, Abstracts, A6.4: 60-61.

Hamada, E., Taniguchi, A, Okazaki, M. and Naito, Y., 1985. Report on the phytoplankton pigments measured during the JARE-25 cruise to Syowa Station, Antarctica, November 1983 to April 1984. National Institute of Polar Research, Japanese Antarctic Research Expedition, JARE Data Reports No. 103 (Marine Biology 7), 89 pp.

Hedgpeth, J.M., 1969. Introduction to Antarctic Zoogeography. In: V.D. Bushnell (Editor), Distribution of selected groups of marine invertebrates in waters south of 35°S latitude. American Geographical Society, Antarctic Map Folio Series, Folio 11, 44 pp. + 29 plates.

Holm-Hansen, O., 1985. Nutrient cycles in Antarctic marine ecosystems. In: W.R. Siegfried, P.R. Condy and R.M. Laws (Editors), Antarctic Nutrient Cycles and Food Webs, Springer-Verlag, Heidelberg, pp. 6-10.

Holm-Hansen, O., El-Sayed, S.Z., Franceschini, G.A. and Cuhel, R.L., 1977. Primary production and the factors controlling phytoplankton growth in the Southern Ocean. In: G.A. Llano (Editor), Adaptions within Antarctic Ecosystems, Smithsonian Institution, Washington, D.C., pp. 11-50.

Ichimura, S. and Fukushima, H., 1963. On the chlorophyll content in the surface water of the Indian and the Antarctic Oceans. Bot. Mag., Tokyo, 76: 395-399.

Jennings, J.C., Gordon, L.I. and D.M. Nelson, 1984. Nutrient depletion indicates high primary productivity in the Weddell Sea. Nature, 308: 51-54.

Lutjeharms, J.R.E., 1985a. Location of oceanic frontal systems between Africa and Antarctica. Deep-Sea Res., in press.

Lutjeharms, J.R.E., 1985b. Detail of the upper thermal structure of the Southern Ocean between South Africa and Prydz Bay during March-May 1984. S. Afr. J. Antarc. Res., in press.

Lutjeharms, J.R.E. and Baker, D.J., 1980. A statistical analysis of the meso-scale dynamics of the Southern Ocean. Deep-Sea Res., 27A: 145-159.

Lutjeharms, J.R.E. and Emery, W.J., 1983. The detailed thermal structure of the upper ocean layers between Cape Town and Antarctica during the period Jan-Feb 1978. S. Afr. J. Antarc. Res., 13: 4-14.

Lutjeharms, J.R.E. and Foldvik, A., 1985. The thermal structure of the upper ocean layers between Africa and Antarctica during the period December 1978 to March 1979. S. Afr. J. Antarc. Res., in press.

Lutjeharms, J.R.E. and Rickett, L., 1985. Changes in the structure of thermal ocean fronts south of Africa over a three month period. S. Afr. J. Sci., in preparation.

Lutjeharms, J.R.E. and Valentine, H.R., 1984. Southern Ocean thermal fronts south of Africa. Deep-Sea Res., 31A: 1461-1476.

Lutjeharms, J.R.E. and Valentine, H.R., 1985. The formation of eddies at the Sub-Tropical Convergence south of Africa. J. phys. Oceanogr., in preparation.

Lutjeharms, J.R.E. and van Ballegooyen, R.C., 1984. Topographic control in the Agulhas Current system. Deep-Sea Res., 31A: 1321-1337.

Lutjeharms, J.R.E. and Walters, N.M., 1985. Ocean colour and thermal fronts south of Africa. In: L.V. Shannon (Editor), The South African Ocean Colour and Upwelling Experiment, Sea Fisheries Research Institute, Cape Town, in press.

Lutjeharms, J.R.E., Walters, N.M. and Allanson, B.R., 1985. Oceanic frontal systems and biological enhancement. In: W.R. Siegfried, P.R. Condy and R.M. Laws (Editors), Antarctic Nutrient Cycles and Food Webs, Springer-Verlag, Heidelberg, pp. 11-21.

Marra, J. and Boardman, D.C., 1984. Late winter chlorophyll a distributions in the Weddell Sea. Mar. Ecol. Prog. Ser., 19: 197-205.

Marra, J., Houghton, R.W., Boardman, D.C. and Neale, P.J., 1982. Variability in surface chlorophyll a at a shelf-break front. J. mar. Res., 40: 575-591.

Maul, G.A., Williams, F., Roffer, M. and Sousa, F.M., 1984. Remotely sensed oceanographic patterns and variability of blue-fin tuna catch in the Gulf of Mexico. Oceanol. Acta, 7: 469-479.

Olson, D.B. and Backus, R.H., 1985. The concentrating of organisms at fronts: a cold-water fish and a warm-core Gulf Stream ring. J. mar. Res., 43: 113-137.

Pingree, R.D., Mardell, G.T., Holligan, P.M., Griffiths, D.K. and Smithers, J., 1982. Celtic Sea and Armorican current structure and the vertical distributions of temperature and chlorophyll. Cont. shelf Res., 1: 99-116.

Pingree, R.D., Pugh, P.R., Halligan, P.M. and Forster, G.R., 1975. Summer phytoplankton blooms and red tides along tidal fronts in the approaches to the English channel. Nature, 258: 672-677.

Plancke, J., 1977. Phytoplankton biomass and productivity in the Subtropical Convergence area and the shelves of the western Indian subantarctic islands. In: G.A. Llano (Editor), Adaptions within Antarctic Ecosystems, Smithsonian Institution, Washington, D.C., pp. 51-73.

Sievers, H.A. and Emergy, W.J., 1978. Variability of the Antarctic Polar Frontal Zone in the Drake Passage - Summer 1976-1977. J. geophys. Res., 83: 3010-3022.

Simpson, J.H., Edelsten, D.J., Edwards, A., Morris, N.C.G. and Tett, P.B., 1979. The Islay Front: physical structure and phytoplankton distribution. Est. coast. mar. Sci., 9: 713-726.

Smith, W.O. and Nelson, D.M., 1985. Phytoplankton biomass near a receding ice-edge in the Ross Sea. In: W.R. Siegfried, P.R. Condy and R.M. Laws (Editors), Antarctic Nutrient Cycles and Food Webs, Springer-Verlag, Heidelberg, pp. 70-77.

Tilzer, M.M., von Bodungen, B. and Smetacek, V., 1985. Light-dependance of phytoplankton photosynthesis in the Antarctic Ocean: implications for regulating productivity. In: W.R. Siegfried, P.R. Condy and R.M. Laws (Editors), Antarctic Nutrient Cycles and Food Webs, Springer-Verlag, Heidelberg, pp. 60-69.

Tranter, D.J., 1982.   Interlinking of physical and biological
    processes in the Antarctic Ocean.   Oceanogr. Mar. Biol. Ann.
    Rev., 20: 11-35.
Voronina, N.M., 1962.   On the dependance of the character of the
    boundary between Antarctic and Sub-Antarctic pelagic zones on
    the meteorological conditions.   American Geophysical Union
    Monographs, 7: 160-162.
Witek, Z., Pastuszak, M. and Grelowski, A., 1982. Net-phytoplank-
    ton abundance in western Antarctic and its relation to environ-
    mental conditions. Meeresforshung, 29: 166-181.
Yentsch, C.S. and Phinney, D.A., 1985. Rotary motions and convec-
    tion as a means of regulating primary production in ware core
    rings. J. geophys. Res., 90: 3237-3248.

FRONTAL SYSTEMS IN THE GERMAN BIGHT AND THEIR PHYSICAL AND
BIOLOGICAL EFFECTS

G. KRAUSE, G. BUDEUS, D. GERDES, K. SCHAUMANN and K. HESSE
Institut für Meeresforschung, Am Handelshafen 12, 2850 Bremerhaven
(Federal Republic of Germany)

ABSTRACT

Results of an interdisciplinary study on fronts in the German
Bight are presented. During normal summer conditions a frontal
system is a permanent feature north of the East Frisian Islands.
It separates well mixed coastal water from stratified North Sea
water. In the transition zone one observes a cold salty belt with
a front towards the well mixed and a separate front towards the
stratified regime. Fronts of the river plume type occur in the
eastern part of the bight. Upwelling water in the region of the
old Elbe Valley near Helgoland is separated from surrounding water
by well developed fronts. Physical observations are discussed together with biological implications as accumulation of organisms
and the role of fronts for the distribution of zoo- and phytoplankton communities. It is shown that strong fronts are not necessarily associated with high biomass accumulation. In a case
study an intensive phytoplankton bloom is traced back to small
horizontal inhomogeneities in the surface layer.

INTRODUCTION

Existence and relevance of a frontal zone in the German Bight
are well known from early hydrographic observations. In the literature this transition area between coastal water influenced by
river discharge, and the open North Sea is described as the "convergence of the German Bight" (Goedecke, 1941; Dietrich, 1950).

More recent measurements by the use of modern profiling equipment, towed instruments and remote sensing techniques have revealed a rather complicated structure within this convergence area
(Becker and Prahm-Rodewald, 1980). Typical space scales of mesoscale fronts, meanders and eddies range between 5 and 20 km. Some
of the fronts have life times in the order of half a tidal cycle,
others may persist throughout summer time. The dynamics of these
structures is superimposed on large tidal variability. The compli-

cated hydrographic situation is reflected by a large heterogeneity of phytoplankton, zooplankton and benthic communities.

In view of the great importance of shelf sea fronts for spreading and mixing processes and as interfaces between different associations of organisms, an interdisciplinary research group of physical and biological oceanographers has been formed to assess the role of fronts for the transport and spreading of substances and organisms in the German Bight.

This paper presents results obtained from case studies at different types of fronts found in the area.

STATE OF KNOWLEDGE AND OBJECTIVES OF THE STUDY

The German Bight is a shallow shelf sea with water depths between 20 and 40 m. Its remarkable topographic features are the Old River Elbe Valley cutting through the flat bottom towards the Northwest and the existence of large tidal flats.

4 basic processes are responsible for the occurrence of fronts in this area.

1. The competition between tidal stirring and heat input at the surface during summer results in a stratification of the water column in the deeper part and a well-mixed region near the coast. At places where the stratification parameter h/u^3 (h water depth, u amplitude of tidal current) assumes a critical value, thermal fronts may be observed. This process has been studied extensively in the Irish Sea (Simpson and Hunter, 1974) and in the waters around the British Isles. The process has been described by potential energy models and verified by satellite IR imagery (Simpson and Bowers, 1979).

2. The fresh water entering the German Bight by the rivers Elbe and Weser gives rise to river plume fronts which have been studied in many other parts of coastal waters (Bowman and Esaias, 1978).

3. Upwelling phenomena during easterly winds are frequently observed in the region of the Old Elbe Valley. The upwelling bottom water with rich nutrient content is separated from surface water by a well developed front. This is a particular regional phenomenon of the German Bight.

4. Due to the shallow depth of the area the stirring action by

strong winds is an important factor in formation and decay of
fronts.

Fronts formed under the combined action of tidal stirring, wind
mixing and fresh water input were studied in Liverpool Bay (Czitrom, 1982).

In the German Bight the relative importance of the different and
combined types of fronts, their influence on the mixing process
and on the marine ecosystem were rather unknown at the beginning
of this study.

The working group agreed to concentrate on the following questions:

Physical Oceanography

The most urgent task was to improve the regional knowledge on
fronts. More specifically, where does what type of front occur in
the German Bight and what is its persistence? Besides the evaluation of satellite images this work relies on in-situ measurements
to obtain the vertical structure and tidal displacement of fronts.

Such regional knowledge is not only required to develop or to
apply adequate mathematical models but also to assess the biological relevance of fronts. For both purposes, special experiments
are required to investigate circulation and cross-frontal mixing.
The use of rhodamin as a tracer is very well suited for such investigations, and respective experiments were planned and carried
out in the meantime (Franz et al., 1982).

The larger scale description of spreading and mixing of river
water into the German Bight requires sequences of synoptic water
mass distribution. As yellow substances are indicators of river
water, a third sub-project investigates the use of these substances as a tracer for remote sensing by an air-borne LIDAR system. This system also provides chlorophyll and transparency distributions. It is described by Diebel-Langohr et al. (1985 b, c),
and results are presented in this volume (Diebel-Langohr et al.,
1985 a).

Chemical and Biological Oceanography

Inorganic nutrients constitute one of the major growth factors
for the development of phytoplankton. Studies on distribution and
dynamics of nutrients is a necessary prerequisite for the understanding of phytoplankton activity, development and distribution
in relation to fronts.

The oxygen regime in the German Bight is of particular interest because of the repeated observation of low oxygen concentrations in the bottom water (Rachor and Albrecht, 1983). As this phenomenon is closely associated with the upwelling type of front a special project is devoted to the generating mechanisms of such water masses.

The scientific goals of the biological projects concentrate on the process of accumulation and crossfrontal exchange of phyto-, myco- and zooplankton in close cooperation with the exchange of physical quantities.

The bottom fauna group is particularly interested in the question whether the possible downward transport of larvae in a front is important for the recruitment of benthic communities. The vertical transport of particles and organisms would be important for quality and composition of the bottom substrate and the benthic associations. E.g. we suppose that the occurrence of large mud deposits in the areas where fronts are observed owe their existence to this mechanism.

RESULTS

Many ship surveys throughout the year have not only confirmed the existence of the 3 basic types of fronts to be expected, but they have also shown some new phenomena. The eastern part of the Bight (fig. 1) is characterized by a predominant occurrence of fronts of the river plume type, upwelling fronts were found near Helgoland, and fronts in the transition zone between well-mixed and stratified water occur north of the East-Frisian Islands.

This subdivision of the Bight into at least 3 different hydrographic regimes is reflected by the bottom sediments (fig. 2) and the associations of the macrozoobenthos (fig. 3).

In the water column 4 different zooplankton communities can be identified which are characteristic for
- water of the inner German Bight (stations P2 - P4, fig. 3) which is directly influenced by the rivers Elbe and Weser
- North Sea water in the western part of the Bight (stations P12 - P16)
- near-shore water in the south-westerly part (stations P0, P1, P10, P11), a mixing product of coastal water and water of the open North Sea

- near-shore water in the north-easterly part (stations P17 -
  P19), where mixing of coastal water, water of the inner German
  Bight and North Sea water takes place.

Details on the zooplankton communities can be found in Gerdes
(1985) and later in the text where their interrelations and accumulation at fronts are discussed.

In the following we present results from case studies for the
different types of fronts which occur in the area.

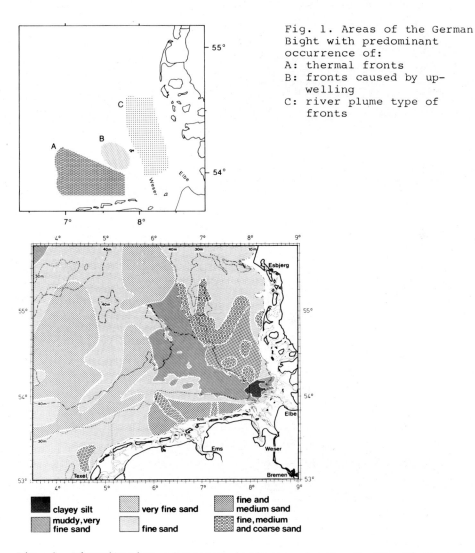

Fig. 1. Areas of the German
Bight with predominant
occurrence of:
A: thermal fronts
B: fronts caused by up-
   welling
C: river plume type of
   fronts

Fig. 2. Distribution of bottom sediments in the German Bight and
adjacent areas (after Salzwedel et al., 1985)

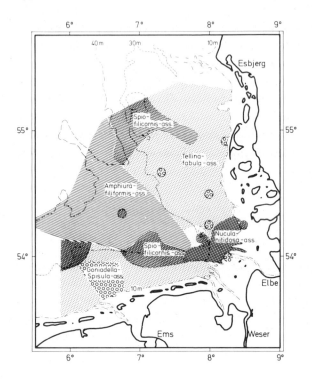

Fig. 3.
Associations of
the macrobenthos
in the deeper
sublittoral of the
German Bight and
adjacent areas
in October 1975
(after Salzwedel
et al. 1985)

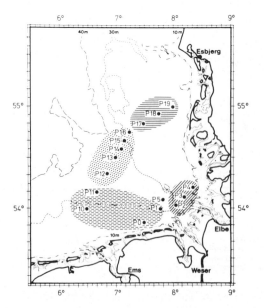

Fig. 4
4 water masses as
identified by 4
different zooplankton
communities.
(after Gerdes, 1985)

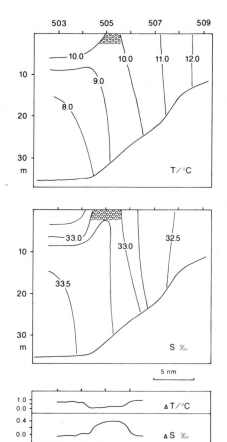

Fig. 5.
Temperature and salinity
distribution on a section
perpendicular to the coast
along 7°40'E
in June 1984.
The colder and saltier
water in the vicinity of
station 505 suggests
upwelling. The bottom
picture is a record
of the ship's thermo-
salinograph. This
record shows the
double front system
much better than the
hydrographic sections.

The "cold belt", region A

A summer section through the transition zone perpendicular to
the coast in North-South direction is presented in fig. 5. Temperature, salinity and density show distributions with the following characteristics:

- The transition between well-mixed and stratified water is not
  only true for temperature, as expected, but also for salinity.
  Salinity has even a greater effect on density than temperature.
- Most of the isotherms strike the bottom rather than the surface.
- An area of cold water at the surface separates the stratified
  and the well-mixed region.
- The area of low surface temperature is also characterized by
  higher salinties.

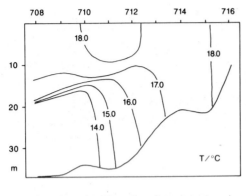

Fig. 6.
Temperature and salinity distribution on the same section as in fig. 5 but at a different time (August 1983).
The very clear situation of fig. 5 is masked by advection of less salty water in the top layer.

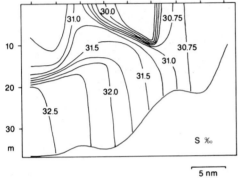

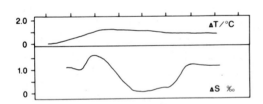

- The isotherms and isohalines tend to approach the surface before bending to the bottom. There is much evidence of upwelling in the transition zone.
- The surface temperature and salinity was additionally measured by a thermosalinograph from the ship under way. The recordings give more information on the real gradients, and it is seen that there are two well pronounced fronts, one towards the stratified and one towards the well mixed water.
- From time to time the clear situation in fig. 5 is disturbed by advection of less haline water in the top layer, but the general structure is maintained.

The distance between the fronts is 5-10 km. Therefore the cold water in between can easily be detected by satellite IR images.

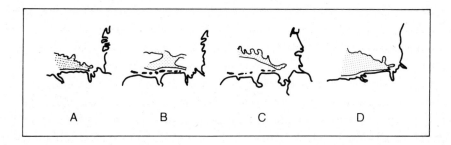

Fig. 7. Areas of colder surface water in the south western
German Bight for 4 cloud-free situations redrawn from IR
satellite images of the AVHRR.
A: 13.5.1980, 14:20 h; B: 18.6.1983, 13:28 h; C: 1.9.1983,
03:24 h; D: 13.5.1984, 14.52 h

Fig. 7 displays the horizontal characteristics of the cold,
salty transition zone. It extends parallel to the coastline, and
due to its usual shape we call the structure the "cold belt". Towards the stratified North Sea the cold belt exhibits very complex
meso-scale structures like meanders and eddies with space scales
in the order of 10 - 20 km.

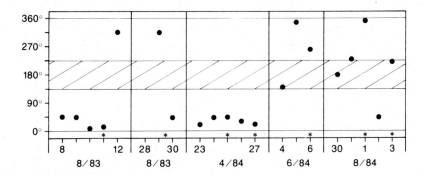

Fig. 8. Wind directions (dots) during periods of in-situ
measurements of the cold belt. The actual time of observation is marked by an asterisk. The hatched area indicates
winds in off-shore direction.

By evaluation of weather maps together with in-situ measurements and IR-images we have found that the cold belt is not caused by the meteorological forcing by offshore winds (fig. 8) but that it is a permanent feature during normal summer conditions.

As fig. 9 shows, our observations coincide with the transition zone where thermal fronts are predicted to occur between stratified and tidally mixed water (Pingree and Griffiths, 1978). Our observed fronts are found closer inshore than those most commonly observed in the Irish Sea. This is to be expected because the stratification parameter log (h/u^3) is derived from a potential energy model which equates heat input and tidal stirring only, whereas the additional presence of river influenced stratification is not taken into account.

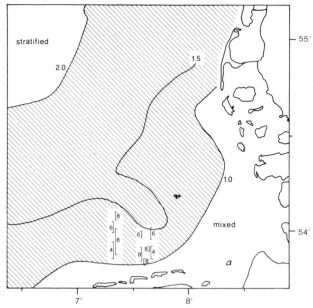

Fig. 9.
The stratification parameter log $(h/c_D u^3)$ in cgs-units and the transition area between stratified and mixed water as predicted by Pingree and Griffith (1978) together with observed fronts in several months indicated by numbers.
h : water depth
c_D: bottom friction coefficient
u : vertically averaged horizontal velocity

The existence of a cold region is easily understood qualitatively. In the deep part the water is stratified. Heat input and wind mixing determine the temperature of the surface mixed layer. In the shallow part the water is well mixed predominantly by tidal stirring. The incoming heat is used for warming of a shallow water column, comparable to the thickness of the well-mixed offshore surface layer.

In the transition zone cold bottom water is mixed with warm surface water by tidal stirring resulting in intermediate temperatures, i.e. colder surface water.

This argument is supported by the experiments of Hachey (1934)
and the numerical model studies of James (1984). In an experimental tank Hachey observed water flowing towards the mixing area
in the top and bottom layer, and the mixed water left the area in
the intermediate layer.

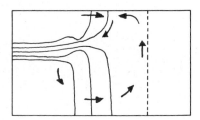

Fig. 10.
Residual circulation and
density distribution in the
transition area between
well-mixed and stratified
water. Simplified picture on
the basis of fig. 4 e in
James (1984)

A much more detailed information on the residual circulation results from the numerical model. A simplified circulation pattern
is presented in fig. 10. It shows upwelling as well as a convergence zone at the surface. It is interesting that James (1984) has
calculated this picture as typical for the Celtic Sea. We can add
that the region A (fig. 1) of the German Bight shows very much the
same characteristics.

On the basis of these results from the physical environment we
can expect the cold area to be richer in nutrients, and a considerable influence on the distribution of organisms seems obvious.

We have not yet confirmed the nutrient situation, but the
following case study on the zooplankton distribution demonstrates
the important role of this front for the near-shore ecosystem.
During the biological sampling the stratification did unfortunately not resemble the simple picture as in fig. 5 but the situation
was that of fig. 6. The additional seaward front (station 2) was
caused by the presence of a water mass of higher temperature and
lower salinity.

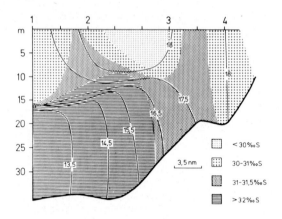

Fig. 11.
The hydrographic
situation on a sec-
tion perpendicular
to the coast along
7°40'E. The bio-
logical samples
were taken at 4
stations in August
1983.

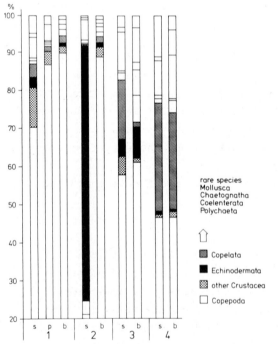

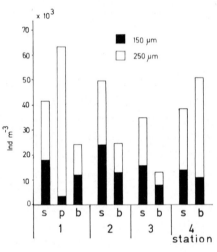

Fig. 12.
Abundance and composi-
tion of the zooplank-
ton populations in dif-
ferent depths at 4 sta-
tions across the cold
belt region. Stations 1
to 4 as in fig. 11. s =
surface, p= pycnocline,
b = close to bottom
sample

As the abundance distribution of zooplankton organisms (fig. 11) shows, the highest abundance in the surface water is at the front (station 2). Even higher values only occur in the pycnocline at station 1.

By means of the composition of the zooplankton populations we can distinguish 3 water masses

- North Sea water near the bottom of stations 1 and 2, dominated by copepods (90 %) and characterized by indicator organisms (Sagitta elegans).
- Near-shore water (station 4) with high values for copelata and lower values for copepods.
- Water of the inner German Bight (surface water at station 3) with medium dominance of copepods and typical organisms like the cladocera Podon intermedius and Evadne nordmanni (compare Gerdes, 1985). Also the chlorophyll-a content was more than two times higher as in any other samples.

At the front station the composition of the zooplankton population is remarkably different with a very high concentration of echinoderm larvae (33,000 Ind. m^{-3}), which have no locomotive power.

Thus, this front appears as a barrier and a region where certain species are accumulated.

Fronts induced by upwelling, region B

In the region of the Old Elbe valley upwelling situations occur mainly during easterly winds. Cold water of higher salinity, occasionally with lower oxygen content (Rachor and Albrecht, 1983), is separated from the surrounding water by a clear front. One such occasion was investigated, and the hydrographical situation is depicted in fig. 13.

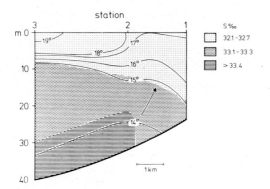

Fig. 13.
A front at station 2 caused by upwelling of bottom water in region B. The section is 3 nautical miles NW of Helgoland at the slope of the Old Elbe valley (August 1982).

As indicated in fig. 14 the highest concentration of organisms
occurs in the front (station 2). In contrast to the cold belt
fronts, some cross-frontal exchange of organisms takes place as it
is obvious from the distribution of echinoderm larvae and the
crustaceans (fig. 15).

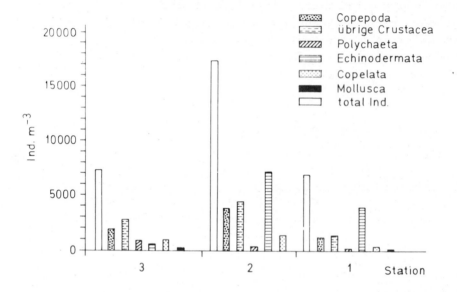

Fig. 14. Abundances of zooplankton organisms in the surface water
of the upwelling front (see fig. 13) and adjacent water masses.

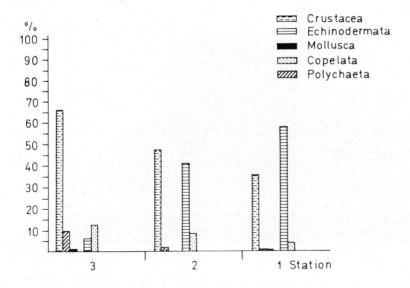

Fig. 15. Corresponding composition of the zooplankton populations on the transect crossing the front.

River plume type of fronts, region C

For a river plume type of front we present a biological phenomenon in the distribution of phytoplankton. In October 1982 this
front was investigated NE of Helgoland. After following the front
over a tidal cycle the ship was anchored, and measurements of
water temperature and salinity were performed at 15 minute intervals. The physical data were processed immediately so that controlled biological sampling could be achieved on the basis of the
physical situation.

The isopleth diagram of salinity (fig. 16) shows that the front
approached the ship, passed it during flood time and left it after
current reversal.

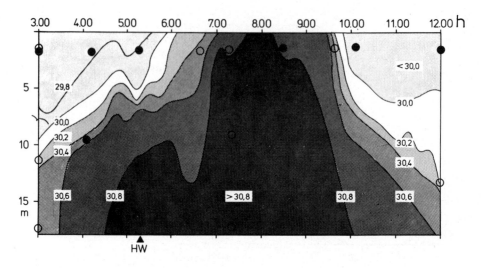

Fig. 16. Isopleths of salinity and biological samples during two passages of a front measured from the anchored ship 10 nm NE of Helgoland island in October 1982. HW = time of tidal high water; circles indicate times and positions of phyto- and mycoplankton samples.

A qualitative and quantitative plankton analysis revealed two distinct phyto- and mycoplankton associations inhabiting the two water masses on either side of the front. This is in agreement with earlier findings. However, it was recognized that there was a third association occuring just within the narrow band of frontal convergence between the two water masses. This specific association was dominated neither by one nor by a mixture of the two adjacent associations of plankton organisms, but it exhibited its very own characteristics (see fig. 17). On the other hand the total biomass of all phytoplankton individuals was not significantly different in the frontal samples as compared with the samples from the two adjacent water masses.

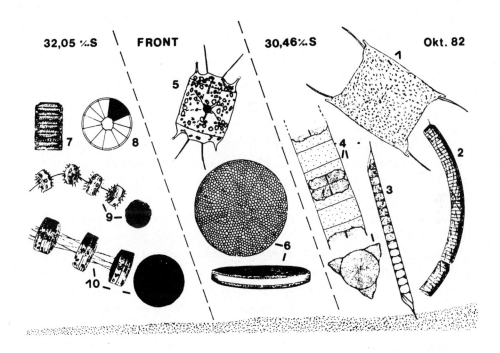

Fig. 17. Sketch of morphology and distribution of marine phytoplankton species (Bacillariophyceae) in a river plume front 10 nm
NE of Helgoland island in October 1982. Illustrated are only dominating and specific organisms of the three different associations resp. water masses. Of course there have been more species,
however, these were indifferent in distribution or too rare for
proper assignment. Please take into account that the species il-
lustrations have been taken from different text books and therefore the scales are different. The true natural dimensions of the
species are given in brackets as follows:

     1 = Biddulphia sinensis (210 x 140 μm)
     2 = Rhizosolenia stolterfothii (119 x 27 μm)
     3 = Rhizosolenia shrubsolei (362 x 10 μm)
     4 = Lithodesmium undulatum (74 x 61 μm)
     5 = Biddulphia regia (163 x 59 μm)
     6 = Coscinodiscus radiatus (∅ 50 μm)
     7 = Paralia sulcata (7 x 20 μm)
     8 = Podosira stelliger (∅ 50 μm)
     9 = Thalassiosira eccentrica (∅ 63 μm)
    10 = Thalassiosira cf. anguste-lineata (∅ 21 μm)

These findings offer new and interesting aspects as to the origin and transport of water masses in such a front. One possible explanation would be the presence of a frontal jet carrying water with different plankton organisms just along the frontal inter-face, but unfortunately at that time this was not investigated by physical measurements, e.g. by current measurements or by dye distribution studies.

Last but not least, the existence of very weak local inhomo-geneities with frontal characteristics and their significance to the distribution of plankton organisms will be presented in an in-teresting case study.

During one of our cruises an extensive red tide patch was ob-served, about 25 nm west of Helgoland. A detailed ad hoc study of the physical, chemical and biological characteristics of this patch and the surrounding waters showed, that the orange-red dis-coloration of the water was caused by an intensive accumulation of Noctiluca miliaris cells, a dinoflagellate, being responsible for the visible phosphorescence of the sea during night time.

The Noctiluca bloom was restricted to the upper 3 to 4 meters (see table 1). On the windward side the patch was homogeneous, and it was bordered by a very sharp line. Towards the lee-side the patch was broken up into long parallel rows, most probably due to Langmuir circulations.

Table 1. Vertical and horizontal biomass distribution of phyto-
and protozooplankton in a section across a red tide patch
in the German Bight, August 2nd 1984.

Station number/ depth	Biomass (mg C/l)		
	Phytoplankton total	Protozooplankton total	Noctiluca (excl.)
672/ 2m	1.006	0.201	0.180
5m	1.300	0.239	0.220
bottom	0.056	0.026	0.020
673/ 2m	1.234	3.064	3.000
5m	–	–	–
bottom	–	–	–
674/ 2m	1.060	0.119	0.100
5m	2.270	0.032	0.003
bottom	0.043	0.023	0.020

There had been a stable meteorological situation with weak winds
before the investigation for at least a week.

It is difficult to present the results in a properly scaled
cross section. The decisive changes occur on a horizontal distance
of some tenth of meters, whereas the total patch phenomenon should
be drawn on a km-scale (fig. 18).

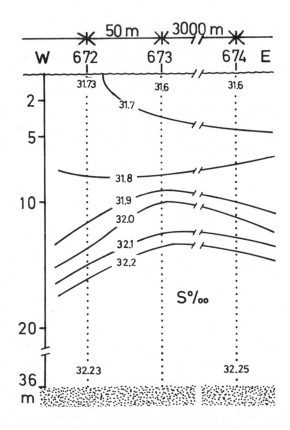

Fig. 18. Isopleths of salinity in a section across a red tide patch (stations
no. 672-674) in the German Bight 25 nm
west of Helgoland island at August 2nd
1984, illustrating a small inhomogeneity
in the surface layer with which an intensive accumulation of <u>Noctiluca miliaris</u>
cells was associated.

In surface water we recognize a very slight local inhomogeneity ("mini-front") with lower salinities to the east. The temperature of this lower salinity water is only 1/10 of a degree higher than in the western waters. Nevertheless, the biological effects are intense as could be traced from photographs and from table 1. This table also shows that east of the Noctiluca bloom there is a secondary biomass maximum (Station 674/5m) being due to the mass development of Ceratium fusus in subsurface waters.

A more detailed presentation of the biological and chemical results of this investigation will be given in a separate paper by Schaumann et al. (1985). In connection with the study of different fronts it is important to note that even very weak fronts may cause surprisingly intense biological phenomena.

DISCUSSION, CONCLUSIONS AND OPEN QUESTIONS

We have presented a variety of case studies on the physical and biological phenomena which are associated with fronts in the German Bight. The first phase of this interdisciplinary study has especially increased the regional knowledge on fronts in this area.

The double-front system which includes a cold belt of water north of the East-Frisian Islands is perhaps the most important new finding. We have not yet modeled the real situation, but it seems that the physics underlying the model of James (1984) for the Celtic Sea (spatial variation in tidal mixing) would also adequately describe the cold belt fronts.

However, spatial variation of tidal mixing on the slope towards the open sea is perhaps not the only important agent for the formation of this front. It might also be possible that deformations of the tidal flow field could cause the front on the sloping bottom. Observations during winter could contribute to this question because under favourable conditions a "warm belt" should exist in the latter case.

Inspite of the considerable influx of fresh water the seaward position of the thermal front follows closely the prediction by a theory which only takes heat input and tidal stirring into account. The reason for this rather puzzling result is still under investigation.

Whereas the stratification parameter log (h/u³) provides an operational guideline for position finding of the thermal fronts, our regional knowledge on fronts of the river plume type is still very poor. We can only name the general area of occurence, but we do not understand the reasons why and when these fronts form at a particular place.

With regard to the biological implications we have shown that all types of fronts are locations with qualitatively differentiated increased phytoplankton, mycoplankton and zooplankton densities.

Of special importance is the existence of front specific plankton communities. Not in all cases it is understood whether selection or advection is the decisive factor for their formation.

It is also interesting to note that certain organism groups can be regarded as indicators for fronts like the echinoderm larvae and certain diatoms.

The study has clearly shown that strong fronts in physical quantities can but do not necessarily produce likewise high accumulations of biomass. It has been demonstrated that high accumulation of biomass can occur in very weak fronts or local inhomogeneities of the surface layer. This fact will make it very difficult to assess quantitatively the relative importance of the different fronts for the overall productivity of the German Bight.

We expect the cold belt area to be more productive than the adjacent regions of the German Bight due to upwelling of bottom water. Besides further studies on the causes of the cold belt and river plume fronts this is the main question currently investigated.

ACKNOWLEDGEMENTS

Many cruises were necessary to gather the material presented in this article. The authors would like to thank to the crew of RV "Victor Hensen" for their permanent assistance and cooperation. We are also grateful to the Deutsche Forschungsgemeinschaft who is funding this work.

REFERENCES

Becker, G.A. and Prahm-Rodewald, G., 1980. Fronten im Meer - Salz-
    gehaltsfronten der Deutschen Bucht. Der Seewart 41: Nr. 1.
Bowman, M.J. and Esaias, W.E. (Editors), 1978. Oceanic Fronts in
    Coastal Processes. Springer Verlag, 114 pp.
Czitrom, S.P.R., 1982. Density stratification and an associated
    front in Liverpool Bay. Ph. D. Thesis, University of College of
    North Wales, Bangor, U.K.
Diebel-Langohr, D., Günther, K.P. and Reuter, R., 1983. Lidar ap-
    plications in remote sensing of ocean properties. Int. Coll. on
    Spectral Signatures of Objects in Remote Sensing, Conf. Proc.,
    Bordeaux.
Diebel-Langohr, D., Hengstermann, T. and Reuter, R., 1985a. Iden-
    tification of hydrographic fronts by air-borne- Lidar-measure-
    ments of Gelbstoff distributions. Proc. 17th International
    Liège Colloquium on Ocean Hydrodynamics.
Diebel-Langohr, D., Günther, K.P., Hengstermann, T., Loquay, K.,
    Reuter, R. and Zimmermann, R., 1985b. An air-borne Lidar system
    for Oceanographic measurements in: Optoelektronik in der Tech-
    nik, Tagungsberichte LASER 85 - Optoelektronic, München 1. - 5.
    Juli 1985, Springer Verlag (in press).
Diebel-Langohr, D., Hengstermann, T. and Reuter, R., 1985c. Depth
    profiles of hydrographic parameters-measurement and interpreta-
    tion of Lidar signals in: Optoelektronik in der Technik, Ta-
    gungsberichte LASER 85 - Optoelektronic, München 1. - 5.
    Juli 1985, Springer Verlag (in press).
Franz, H., Gehlhaar, U., Günther, K.P., Klein, A., Luther, J.,
    Reuter, R. and Weidmann, H., 1982. Airborne fluorescence lidar
    monitoring of tracer dye patches - a comparison with shipboard
    measurements. Deep-Sea Res. 29: 893.
Gerdes, D., 1985. Zusammensetzung und Verteilung von Zooplankton
    sowie Chlorophyll- und Sestongehalte in verschiedenen Wasser-
    massen der Deutschen Bucht in der Jahren 1982/83. Veröff. Inst.
    Meeresforsch. Bremerh. 20: 119-139.
Hachey, H.B., 1934. Movements resulting from mixing of stratified
    water. J. Biol. Bd. Can 1(2): 133-143.
James, I.D., 1984. A three-dimensional numerical model with
    variable eddy viscosity and diffusivity. Cont. Shelf Res. 3:
    69-98.
Pingree, R.D. and Griffiths, D.K., 1978. Tidal fronts on the Shelf
    Seas around the British Isles. J. Geophys. Res. 83: 4615-4622.
Rachor, E. and Albrecht, H., 1983. Sauerstoffmangel im Bodenwasser
    der Deutschen Bucht. Veröff. Inst. Meeresforsch. Bremerh. 19:
    209-227.
Salzwedel, H., Rachor, E. and Gerdes, D., 1985. Benthic macrofauna
    communities in the German Bight. Veröff. Inst. Meeresforsch.
    Bremerh. 20: 199-267.
Schaumann, K., Gerdes, D. and Hesse, K.J., 1985. Biological and
    chemical characteristics of a Noctiluca miliaris red tide patch
    in the western German Bight 1984. Botanica Marina (submitted).
Simpson, J.H. and Hunter, J.R., 1974. Fronts in the Irish Sea.
    Nature 250: 404-406.
Simpson, J.H. and Bowers, D., 1979. Shelf sea fronts' adjustments
    revealed by satellite IR-imagery. Nature 280: 648-651.

ROLE OF THERMAL FRONTS ON GEORGES BANK PRIMARY PRODUCTION

P. KLEIN

Laboratoire d'Océanographie Physique, Faculté des Sciences

29287 Brest Cédex (France)

ABSTRACT

A very simple physical-biological model has been used to simulate the seasonal cycle of Georges Bank plankton ecosystem. Numerical results, which agree
on the whole with available field data, reveal that Georges Bank works as a
chemostat in winter but that recycling processes are dominant in summer. The
transition between the winter and the summer regimes is triggered by the
appearance (in spring) and disappearance (in fall) of the thermal fronts around
the bank.

INTRODUCTION

Georges Bank is a shallow bank, 300 km long and 150 km wide, located along
the seaward edge of the Gulf of Maine (Fig. 1).

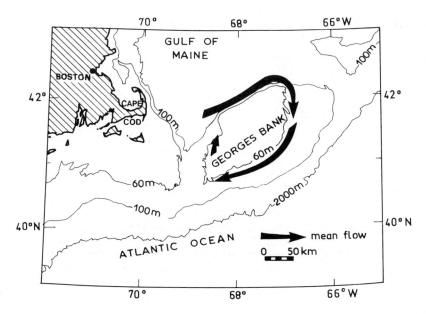

Fig. 1. Schematic map of Georges Bank topography and direction of mean flow
field.

142

Fig. 2 displays the seasonal cycle of inorganic nitrogen (which includes both nitrate and ammonia), $\overline{N}$, phytoplankton, $\overline{P}$, and zooplankton, $\overline{Z}$, on Georges Bank within the 60 m isobath. This cycle, which recurs annually, displays two main distinct periods : the winter and the summer periods separated by transition periods. More precisely, inorganic nitrogen concentration is high in winter but the region appears almost nutrient depleted in summer. Phytoplankton biomass is relatively high year around ; during the summer phytoplankton biomass is about half of the winter value. Note that primary production values range from 0.8 gC/m²/d to 2.5 gC/m²/d (O'Reilly et al., 1985) ; higher values appear between August and December. Zooplankton biomass attains a maximum in July. But zooplankton nitrogen compared with phytoplankton and inorganic nitrogen is very low throughout the year and may not affect phytoplankton-nutrient dynamics. Because of the high phytoplankton biomass, zooplankton may not be food-limited. Moreover total nitrogen $\overline{S}$ defined as $\overline{S} = \overline{Z} + \overline{P} + \overline{N}$ displays the same pattern as inorganic nitrogen, that is, a strong decline in spring and summer. These characteristics raise some important questions. Why are high phytoplankton biomass and production rates not reflected at higher trophic levels ? What are the source of nutrients ? Which process (physical or biological) is responsible of the strong seasonal decline of total nitrogen ? As a preliminary to an examination of these questions, it is useful to briefly examine the seasonal evolution of the physical processes involved in this area.

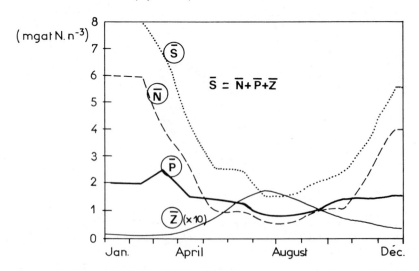

Fig. 2. Seasonal cycle of nitrogen budget estimated from available data : Pasturzack et al. (1982) and O'Reilly et al. (1980) for the inorganic nitrogen ($\overline{N}$), O'Reilly and Evans-Zetlin (1982) and O'Reilly et al. (1980) for the phytoplankton ($\overline{P}$), and Davis (1982) for the zooplankton ($\overline{Z}$). Conversions made use of ratios : mgC/mgat N = 100, mgC/mgChl = 50 (Steele, 1982).

During the winter, Georges Bank area within the 60 m isobath as well as outside areas are vertically well-mixed (at least to a depth of 200 m). There is a
mean (or low frequency) clockwise circulation around the bank (Fig. 1). Corresponding current is about 5-10 cm/s on the south flank, but there is a narrow
jet of 20 cm/s on the north flank (Butman et al., 1982). Because all is well-
mixed, vertical diffusion is high ; consequently, low frequency horizontal dispersion, which results from the combination of the vertical shear of tidal
currents with the vertical turbulent diffusion (Kullenberg, 1978), is large. So,
at this time, exchanges with areas outside the Georges Bank are large. During
the summer, vertical stratification outside the 60 m isobath is established,
but within this isobath, the waters remain well-mixed because of the strong
tidal currents. So the Georges Bank well-mixed area is isolated from Gulf of
Maine and Slope waters by strong thermal fronts around the bank. The resulting
cross-bank density field accelerates the mean clockwise circulation around the
bank. However, vertical turbulent diffusion near the 60 m isobath is small
because of the presence of the thermal fronts and the associated vertical stratification. Then the low frequency horizontal dispersion is low. So, during this
period, exchanges with the outside are small.

This short description of the plankton ecosystem and physical processes observed in the Georges Bank area reveals and puts into evidence a strong correlation between biological and physical processes at the seasonal time scale. In
particular the transition between the winter and the summer regimes seems to be
triggered by the appearance and disappearance of the thermal fronts around the
bank. So some other questions are : how does the seasonal evolution of physical
processes affect the phytoplankton-nutrient cycle ; what is the importance of
physical-biological interactions ? These questions have been adressed through
a very simple numerical model. Simulations have been performed to quantify the
respective effects of physical and biological processes for the different periods of the seasonal cycle of Georges Bank plankton ecosystem. The model used
and its numerical implementation have been described in detail in Klein (1985 a,
b). A short description is given here. Then the numerical results are discussed ;
discussion is focused on the different biological and physical budget components involved during the winter and summer periods.

THE MODEL

Georges Bank within the 60 m isobath can be represented by an ellipse whose
major and minor axes are 170 and 130 km long (Fig. 1). For sake of simplicity,
appropriate transformations (see Klein, 1985a) have been used so that the
transform domain is circular 130 km across, 60 m deep at the periphery and 30 m
deep at the center. The model developed is two-dimensional (x,y), homogeneous-

ly mixed vertically but with varying depth (h). Model equations are written in cylindar coordinates (r,θ). Physical processes are defined by the low frequency velocity field (U_r, U_θ) and the low frequency horizontal dispersion (k_r, k_θ). Biological variables are inorganic nitrogen, N, phytoplankton, P, and zooplankton, Z. An additional variable, the "phytodetritus", M, is included to complete the budget. Units are in mgatN/m^3. Variables are functions of time (t) and space coordinates (r,θ). Equations are :

$$\mathcal{L}(Z) = \overbrace{b_2 \ P \ Z}^{\text{growth}} - \overbrace{d \ Z^2}^{\text{predation}} \tag{1}$$

$$\mathcal{L}(P) = \overbrace{a \ N/(N + K_n) \ P}^{\text{growth}} - \overbrace{b \ P \ Z}^{\text{grazing}} - \overbrace{\alpha \ P}^{\text{"mortality"}} \tag{2}$$

$$\mathcal{L}(N) = \overbrace{-a \ N/(N + K_n) \ P}^{\text{uptake}} + \overbrace{b_1 \ P \ Z + d \ Z^2 + \beta \ P}^{\text{regeneration}} \tag{3}$$

$$\mathcal{L}(M) = \overbrace{\alpha \ P}^{\text{phytodetritus source}} - \overbrace{\beta \ P}^{\text{decomposition}} \tag{4}$$

$\mathcal{L}$ is an operator for the time and space evolution defined as :

$$\mathcal{L} \cdot = \frac{\partial \cdot}{\partial t} + \overbrace{U_r \ \frac{\partial \cdot}{\partial r} + \frac{U_\theta}{r} \ \frac{\partial \cdot}{\partial \theta}}^{\text{advection}} - \overbrace{\frac{1}{rh} \ \frac{\partial}{\partial r} \ rhK_r \ \frac{\partial \cdot}{\partial r} - \frac{1}{r^2} \ \frac{\partial}{\partial \theta} \ K_\theta \ \frac{\partial \cdot}{\partial \theta}}^{\text{diffusion}} \tag{5}$$

Boundary conditions concern the values of the biological variables on the periphery of the area (Klein, 1985a).

The right hand side of eqs. 1 to 4 display the biological parameters involved. Seasonal evolution of the phytoplankton growth rate, a, given on Fig. 3-a contains effects of seasonal varying light integrated over the water column. K_n is the half saturation constant (= 0.3 mgatN/m^3). Estimation of the seasonal cycle of zooplankton growth rate, b_2, shown on Fig. 3-b, is based on the works of Davis (1982) and Steele and Henderson (1981). Zooplankton growth is assumed to be 20 % efficient (Steele, 1982). So : b = 5b_2. The difference b_1 = b - b_2, which represents excretion fecal pellets and mortality, is considered to be recycled in the system at the seasonal time scale. Predation rate on zooplankton is expressed as dZ, which means that the predator population changes in proportion to herbivores. Zooplankton consumed as prey is supposed to be recycled in the system. The phytoplankton "mortality rate", α, and the "phytodetritus" recycling rate, β, have been parameterised and estimated (Fig. 3-c) in relation to the net loss of $\overline{P} + \overline{N}$ ($\sim \overline{S}$ since $\overline{Z}$ is very low). M has been introduced and we have focused on α and β for the following reasons. The strong deficit $\overline{P} + \overline{N}$

during the summer cannot occur through the physical exchanges : this would require a strong seasonal change of the boundary conditions, which is not the
case (Klein, 1985a). The only possible explanation is that the deficit must go
into another biological compartment which is not Z ; hence the introduction of
M. Then we must assume that there is a sink for P which prevents total recycling into nutrients. Parameters α and β are the only ones which have been
fitted. The chosen values are : $\alpha = 0.08$ P ($P > 1.75$), $\alpha = 0.14$ ($P < 1.75$) and
$\beta = 0.08$.

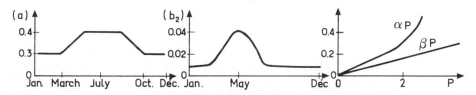

Fig. 3. Seasonal evolution of phytoplankton growth rate (a), zooplankton growth
rate (b) and variations of αP and βP (c).

The low frequency flow field considered has been estimated directly from the
available data (Butman et al., 1982 ; Butman and Beardsley, 1984) and satisfies
the continuity equation. In the domain considered, the flow extends near the
edges over a band with a width between 10 km and 25 km (Klein, 1985a). Velocity
ranges between 1.4 cm/s to 16 cm/s. Horizontal dispersion coefficients K_r, K_Θ
have been estimated by means of classical formulas (Csanady, 1974 ; Kullenberg,
1978) and using tidal current data (Moody and Butman, 1980) and estimations of
vertical turbulent diffusion (James, 1977). Value for the horizontal dispersion
in the well-mixed area is about 350 m^2/s year around ; this value is close to
those found by Loder et al. (1983). On the edges (60 m isobath) values of the
horizontal dispersion across the streamlines (i.e. K_r) range from 350 m^2/s during the winter to 30 m^2/s during the summer.

RESULTS : PHYTOPLANKTON AND NUTRIENT BUDGETS
    The model has been used to simulate the annual cycle of the Georges Bank
plankton ecosystem. Values at the boundary, year around, are $Z_o = 0.05$,
$P_o = 0.05$, $N_o = 8$, $M_o = 0$ (in mgatN/m^3). Fig. 4 displays the spatial distribution of phytoplankton in winter and summer. Phytoplankton biomass is high. In
winter there is a "hot spot" of phytoplankton in the center of the bank ; the
value is 2.65 mgatN/m^3. In summer, there is a high spatial heterogeneity in the
northeast part, due to advection of nutrient. These simulated patterns are similar to observed spatial distributions (O'Reilly et al., 1980). A more precise
comparison with field data, over the seasonal cycle, has been done by averaging

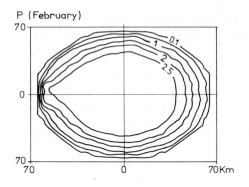

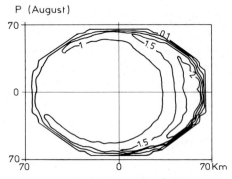

Fig. 4. Phytoplankton distribution (in mgatN/m³) in winter and summer.

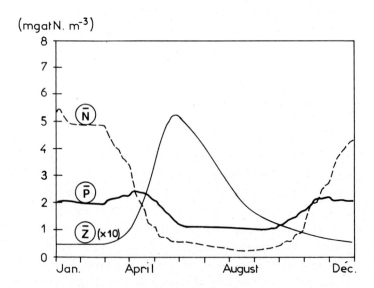

Fig. 5. Seasonal cycle of nitrogen budget from numerical results.

the numerical results over the Georges Bank area (Fig. 5). The modeled time evolution of each mean biological component agrees well with field data (Fig. 2) except for the calculated zooplankton biomass, which is higher than the observed one, but closer to more recent values (Sherman et al., 1985). More detailed results, presented in Klein (1985a), confirm the relative good agreement, on the whole, of the results from this very simple numerical model with available data at the seasonal time scale. In the present study, the model has been used in order to quantify the respective effects of biological and physical processes on the Georges Bank plankton ecosystem. More precisely, attention has been focused on the different phytoplankton and nutrient budget components.

Equations which drive the time evolution of the mean variables can be derived from eqs. 1 to 4 by using a mean operator defined as :

$$\bar{\phi} = \frac{1}{V} \iint_{\Sigma} rh\phi(r,\theta,h)drd\theta \qquad (6)$$

where ϕ designates any biological variable (Z, P, N or M). V and Σ are respectively the volume of water and the area of Georges Bank within the 60 m isobath. The resulting equations are :

$$d\bar{Z}/dt = b_2 \ \overline{P\ Z} - d\ \overline{Z^2} - k_z \ (\bar{Z} - \bar{Z}_0) \qquad (7)$$

$$d\bar{P}/dt = a\ \overline{N/(N + K_n)\ P} - b\ \overline{P\ Z} - \alpha\ \bar{P} - k_p\ (\bar{P} - \bar{P}_0) \qquad (8)$$

$$d\bar{N}/dt = -a\ \overline{N/(N+K_n)\ P} + b_1\ \overline{P\ Z} + d\ \overline{Z^2} + \beta\ \bar{P} - k_n\ (\bar{N} - \bar{N}_0) \qquad (9)$$

$$d\bar{M}/dt = \overline{(\alpha - \beta)\ P} - k_m\ (\bar{M} - \bar{M}_0) \qquad (10)$$

where coefficients k_z, k_p, k_n, k_m are defined by :

$$- k_\phi\ (\bar{\phi} - \bar{\phi}_0) = \frac{1}{V} \int_0^{2\pi} rh \left[\underbrace{U_r\ \phi\big|_{r = R_{max}}}_{\text{advection}} + \underbrace{K_r\ \frac{\partial\phi}{\partial r}\bigg|_{r = R_{max}}}_{\text{diffusion}} \right] d\theta \qquad (11)$$

k_ϕ is in fact an exchange rate between Georges Bank water and waters outside the bank. In this study, explicit contributions of the advective and dispersive processes have been taken into account. They are defined respectively as :

$$\bar{\phi}_U = \frac{1}{V} \int_0^{2\pi} rhU_r\ \phi\big|_{r = R_{max}} d\theta \quad , \quad \bar{\phi}_K = \frac{1}{V} \int_0^{2\pi} rhK_r\ \frac{\partial\phi}{\partial r}\bigg|_{r = R_{max}} d\theta \qquad (12)$$

So we have : $- k_\phi\ (\bar{\phi} - \bar{\phi}_0) = \bar{\phi}_U + \bar{\phi}_K$.

Budget terms for phytoplankton and nutrient have been estimated for the winter and summer periods. Using a ratio of 100 for mgC/mgatN (Steele, 1982) and

assuming a mean depth of 40 m, calculated values for mean primary production
are 2 gC/m^2/d in winter and 0.8 gC/m^2/d in summer. These orders of magnitude
appear realistic compared with field data (Cohen et al., 1982 ; O'Reilly et al.,
1985). Phytoplankton and nutrient budget terms, estimated as percentages of primary production, are displayed on Fig. 6. During the winter, zooplankton biomass
is negligible and has no effect on the phytoplankton-nutrient dynamics. The
phytoplankton flushing-out represents 35 % of the primary production. Phytoplankton "mortality" is high ($\sim$ 65 %) and more than half of the "phytodetritus"
are flushed out ; the other part, 28 %, is recycled in the system. The "new"
production is high : contribution of physical processes to nutrient supply is
72 %. Therefore Georges Bank in winter seems to work as a chemostat with nutrient input and an outflushing of phytoplankton and "phytodetritus". Moreover
90 % of the physical exchanges are due to the cross-bank horizontal dispersion ;
advective processes represent only 10 % of the exchanges. During the summer,
zooplankton biomass is low but not negligible ; grazing stress represents 20 %
of the primary production ; this is in agreement with the order of magnitude
given by Walsh (1983). Again, phytoplankton "mortality" is high : 75 % of the
primary production, but phytoplankton flushing-out is only 5 %. "New" production
represents only 40 % of the primary production ; recycling processes are dominant ($\sim$ 60 %) and concern mainly the "phytodetritus". Reduction of the physical
exchanges in summer concerns mainly the cross-bank horizontal dispersion : they
represent only 25 % of the total physical exchanges.

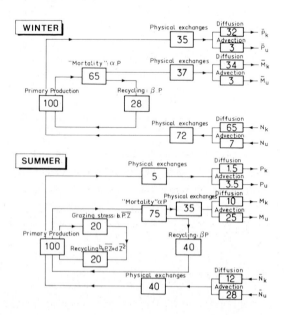

Fig. 6. Flow chart of nitrogen budgets in winter and summer.

From these results it appears that this is the seasonal evolution of the dispersive processes at the edges of the bank, produced by the appearance and disappearance of the thermal fronts, which drives the seasonal cycle of Georges Bank plankton ecosystem. Physical exchanges due to advection are very low year around. In summer, the only explicit effect of advection (which is dominant at this period) could be a lower primary productivity resulting from the spatial heterogeneity (Klein and Steele, 1985). Appearance of the thermal fronts leads to the domination of recycling processes whereas their disappearance leads to the chemostat regime. The strong seasonal change in physical exchanges at the boundary explains the strong deficit of $\overline{P} + \overline{N}$ (Fig. 2) during the summer. Numerical results (Klein, 1985a) show that this deficit is compensated by a strong increase of the "phytodetritus" concentration.

CONCLUSION

A very simple numerical model has been used to illustrate and quantify the respective effects of biological and physical processes on the seasonal cycle of the Georges Bank plankton ecosystem within the 60 m isobath. Numerical results agree, on the whole, with available data. During the winter physical exchanges, which are strongly dominated by the cross-bank dispersive processes, are very important and are responsible of the chemostat regime. However the resulting phytoplankton flushing-out cannot balance the primary production. Phytoplankton "mortality" is high. During the summer, cross-bank dispersive exchanges are strongly reduced because of the presence of the thermal fronts around the bank. Then, physical exchanges (mainly due to advection at this period) are low and as a consequence, recycling processes are dominant. During this period phytoplankton "mortality" is very high.

The high phytoplankton "mortality" and flushing-out of "phytodetritus" in winter and summer are surprising. However this phenomenon is not far from a recent conclusion of Walsh (1983) which shows the significant export of photosynthetic carbon off Georges Bank. Anyway the "mortality" question seems a fundamental one to investigate before going further in modelling work. This study has shown the importance of the low frequency dispersive processes at the edges of the bank and then of the presence and absence of the thermal fronts around the bank. The estimated values used for these dispersive processes seem reasonable, at least at the seasonal time scale. However at a shorter time scale, thermal fronts are non-stationnary. Their evolution in the spring and early in the summer depends on the atmospheric forcings and on the spring-neap cycle (Simpson, 1981). At the end of the summer and during the fall, they are significantly affected by the developpment of barotropic and baroclinic instabilities (Pingree, 1978 ; Garrett and Loder, 1981). These short time-scale events should involve a significant variability of the cross-bank exchanges and then a short time variability of the Georges Bank plankton ecosystem. Investigation of this variability could represent another offshoot of this study.

ACKNOWLEDGEMENTS

I am very grateful to Dr. John Steele for suggesting this problem and for his guidance and encouragement throughout. This work was done while the author was visiting the Woods Hole Oceanographic Institution. It was supported by a French-US exchange award from the National Science Foundation and by the Center for Analysis of Marine Systems of the Woods Hole Oceanographic Institution. I thank C. Mazé and P. Doaré for preparation of the manuscript.

REFERENCES

Butman, B., Beardsley, R., Magnell, B., Frye, D., Vermersch, J., Schlitz, R., Limeburner, R., Wright, W. and Noble, M., 1982. Recent observations of the mean circulation on Georges Bank. J. Phys. Oceanogr., 12 : 569-591.
Butman, B. and Beardsley, R., 1984. Long term observations on the Southern flank of Georges Bank : Seasonal cycle of currents. Submitted to the J. Phys. Oceanogr.
Cohen, E., Grosslein, M., Sissenwine, M., Steimle, F. and Wright, W., 1982. An energy budget of Georges Bank. In : M. Mercer (Editor), Multispecies Approaches to Fisheries Management Advice, Canadian Special Publication in Fisheries Aquatic Sciences, 59 : 95-107.
Csanady, G., 1974. Turbulent diffusion in the environment. D. Reidel Pusblishing Co., Boston, Mass.
Davis, C., 1982. Processes controlling zooplankton abundance on Georges Bank. PhD thesis, Boston University Marine Program, MBL, Woods Hole.
Garrett, C. and Loder, J., 1981. Dynamical aspects of shallow sea fronts. Philosophical Transactions of the Royal Society of London, A302, pp. 563-581.
James, I., 1977. A model of the annual cycle of temperature in a frontal region of the Celtic Sea. Estuarine and Coastal Marine Science, 5 : 339-363.
Klein, P., 1985a. A simulation of some physical and biological interactions. In Georges Bank, MIT Press, Cambridge, MA (in press).
Klein, P., 1985b. A simple numerical method for ecosystem models. J. Computational Phys. (submitted).
Klein, P. and Steele, J., 1985. Some physical factors affecting ecosystems. J. Mar. Res., 43 : 337-350.
Kullenberg, G., 1978. Vertical processes and the vertical horizontal coupling. In : J.H. Steele (Editor), Spatial Patterns in Plankton Communities, Plenum New York, pp. 43-72.
Loder, J., Wright, D., Garrett, C. and Juszko, B., 1982. Horizontal exchanges on Georges Bank. Canadian Journal of Fisheries and Aquatic Sciences, 39 : 1130-1137.
Moody, J. and Butman, B., 1980. Semi-diurnal bottom pressure and tidal currents on Georges Bank and in the Mid-Atlantic Bight. US Geological Survey Report, pp. 80-1137.
O'Reilly, J., Evans, C., Zdanowicz, V., Draxler A., Waldhauer, R. and Matte, A., 1980. Baselines studies on the distribution of phytoplankton biomass, organic production, seawater nutrients and trace metal in coastal waters between Cap Hatteras and Nova Scotia. First Annual Report Northeast monitoring program. NMFS, NE Fisheries Center.

O'Reilly, J. and Evans-Zetlin, C., 1982. A comparison of the abundance (Chla) and size composition of the phytoplankton communities in 20 subareas of Georges Bank and surrounding waters. ICES CM.L : 49, pp. 9.

O'Reilly, J., Evans-Zetlin, C. and Bush, C., 1985. Primary production : Georges Bank, Gulf of Maine and the Mid-Atlantic Shelf. In Georges Bank, MIT Press, Cambridge, MA (in press).

Pastuszak, M., Wright, W. and Patango, D., 1982. One year of nutrient distribution in the Georges Bank region in relation to hydrography, 1975-1976. J. Mar. Res., 40 : 525-542.

Pingree, R., 1978. Cyclonic eddies and cross-frontal mixing. Journal of the Marine Biological Association of the UK, 58 : 955-963.

Sherman et al., 1985. Zooplankton production and the fishery of the northeast shelf. In Georges Bank, MIT Press, Cambridge, MA (in press).

Simpson, J., 1981. The shelf-sea fronts : implications of their existence and behavior. Philosophical Transactions of the Royal Society of London, A302 : 531-546.

Steele, J. and Henderson, E., 1981. A simple plankton model. The American Naturalist, 117 : 676-691.

Steele, J., 1982. The production of Georges Bank. Unpublished Manuscript.

Walsh, J., 1983. Death in the sea : Enigmatic phytoplankton losses. Progress in Oceanography, 12 : 1-86.

THE ROLE OF STREAMERS ASSOCIATED WITH MESOSCALE EDDIES IN THE
TRANSPORT OF BIOLOGICAL SUBSTANCES BETWEEN SLOPE AND OCEAN WATERS

C.S. YENTSCH and D.A. PHINNEY

Bigelow Laboratory for Ocean Sciences, W. Boothbay Harbor, Maine
04575 (U.S.A.)

INTRODUCTION

With the discovery of mesoscale eddies (Parker, 1971; Fuglister, 1972) associated with western boundary currents, it became
evident that a major mechanism for the interchange of slope and
ocean waters had been found. Prior to recognizing the frequency
and magnitude of these eddies, slope and oceanic water interchange
was largely believed to be due to the ageostrophic relationships
associated with the flow of a boundary current (Rossby, 1936; Fig.
1). Although the exchanges due to these motions are slow, they
are extensive, essentially occurring along the entire path of the
high velocity current, hence one can account for considerable
interchange via this mechanism. Mesoscale eddies, both of the
warm and cold core type, offer another exchange system that we
will call entrainment (Fig. 2). This mechanism is initiated when
meanders of the Gulf Stream pinch off to form an eddy that
entrains either slope or Sargasso Sea water. The relative importance of the two processes is not known, but a distinction can be
made between the large scale inshore boundary of a flow such as
the Gulf Stream and intermittent mesoscale eddies that cause more
rapid exchange.

With the advent of satellites that measure ocean color and
temperature, a third mechanism for slope/ocean interchange has
appeared which we will call the streamer mechanism. It is distinct
from the other two transport processes in that it is very small in
scale, occurring at time scales of days to weeks, yet energetic.
The action of mesoscale eddies give rise to these high velocity
ribbons of water when the rotary motions of a ring wrap outlying
waters around its edge producing a streamer. The first real
evidence we have of the existence of streamers comes from satellite observations of ocean color and temperature. This report
describes a persistent streamer observed in the region of Georges

154

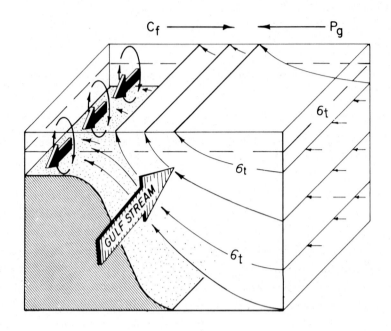

Fig. 1. The Rossby (1936) model of slope/ocean interchange in the
Gulf Stream system. C_f = Coriolis force; Pg = pressure gradient;
σ_t = lines of equal density.

Bank. We offer reasons as to why the streamers appear and
describe the effects they have on the exchange of phytoplankton
between slope and open-ocean waters.

Observations of a Streamer Using Satellite Imagery and Oceano-graphic Measurements

In 1984 we observed a warm core ring and streamer south of
Georges Bank. Figure 2 is an AVHRR (Advanced Very High Resolution
Radiometer) thermal image taken 13 July 1984. Cold, tidally mixed
Georges Bank water appears light colored and warm water masses are
darker. The warm core ring located south of Georges Bank is a
large ring with a strongly delineated warm central core. A
streamer, starting in the region of the 200 m isobath, occupies
the northeastern quandrant of the ring, extending nearly one-half
the entire circumference of the ring, almost to the northern edge
of the Gulf Stream. The streamer is wider nearshore, tapering to
about one-half of the maximum width over the total length. The
streamer appears to be connected to a southerly movement of cold
Georges Bank water. We do not have ocean color (phytoplankton

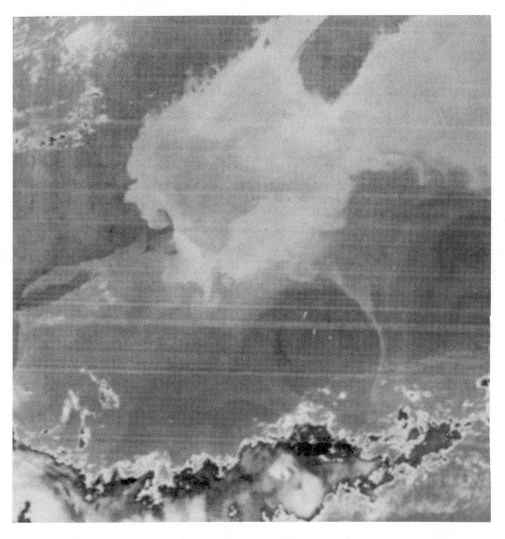

Fig. 2. AVHRR thermal image of 13 July 1985. Cool water appears light colored; warm water is darker. Ring/streamer features can be seen south of Georges Bank.

chlorophyll) imagery for this day, however, on another date Coastal Zone Color Scanner (CZCS) imagery clearly shows that the streamer is associated with marked changes in color as well as temperature. Presumably the color change is derived in part from more productive waters adjacent to Georges Bank.

During a cruise aboard the R/V Cape Hatteras (20 July to 2 August, 1984), the streamer was sampled to establish its size, depth and water column characteristics. Figure 3 shows the surface temperature and chlorophyll signature of the feature in an east to west crossing. The continuous temperature signal showed the streamer to be about 25 km wide and 4–5°C colder than surrounding waters. Surface chlorophyll fluorescence increased by a factor of two; actual surface concentrations measured were ca. 0.3 μg/ℓ in the streamer.

Three CTD/pump stations were occupied, one immediately east (Station 5), one in the streamer (Station 6) and one immediately west in the high velocity region of the ring (Station 7), to create a section of physical and biological parameters.

In Figure 4, the interior of the streamer is seen at 20 m (Station 6), water of <17°C and 33.4 °/oo. Sharp gradients of temperature and salinity are present on either side of the feature. Density (as sigma-t surfaces in Figure 5) shows little horizontal structure. Subsurface chlorophyll concentrations (Fig. 5) form a distinctive pattern. High productivity water carried by the streamer at 20 m forms an additional maximum over the main subsurface chlorophyll maximum (SCM). At Stations 5 and 6, the depth of the SCM was 40–45 m, yet in the high velocity region of

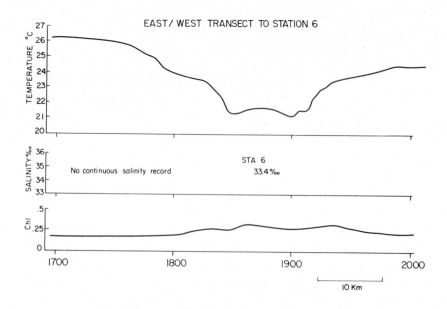

Fig. 3. Surface temperature and chlorophyll signatures, east/west transect of streamer to Station 6.

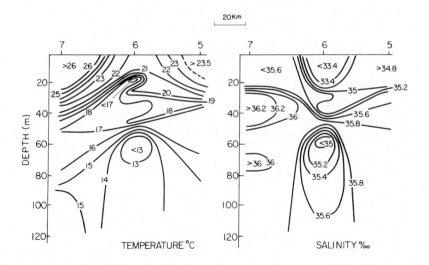

Fig. 4. Temperature and salinity sections from CTD/pump stations. Ring edge is to the left (Station 7).

the ring (Station 7), the SCM was considerably shallower occurring at 35 m. Yentsch and Phinney (1985) found differences in productivity at the center of a ring and the peripheral high velocity region; the upward bowing density surfaces at the edge of a ring cause the SCM to occur at shallower depths.

The presence of a warm core ring adjacent to Georges Bank yields a cross-section of density which can be described as markedly baroclinic. The isotherms at the periphery of the ring sweep upward toward Georges Bank, thus the characteristic of the cross-section extending from Georges Bank through the ring is one of a typical tidal front center coupled with the baroclinic features associated with a warm core eddy. We will now consider the case of such a ring adjacent to Georges Bank and the conditions for streamer formation.

Conceptual Model for Streamer Formation

The general circulation around Georges Bank is clockwise and is believed to be associated with basin geostrophic flow in the Gulf of Maine and tidal processes on Georges Bank (Beardsley and Smith, 1981; Butman et al., 1982). A conspicuous feature of this circulation is the high velocity jet current on the northern flank that follows the overall contours of Georges Bank, with a lessen-

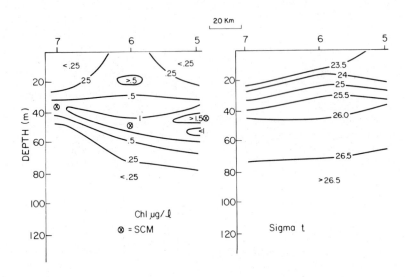

Fig. 5. Chlorophyll and density sections from CTD/pump stations. Ring edge is to the left (Station 7). Depth of the subsurface chlorophyll maximum is denoted by crossed circles.

ing of velocity near the northeast peak and a continued decrease to the south and eventually to the west.

During the summer months, satellite images of ocean color and sea surface temperature clearly outline the Bank, generally in the region of the 60 m isobath (Yentsch and Garfield, 1981). These outlines of color and temperature are believed to arise from the composite activity of both the clockwise flow and tidal stirring. We believe that the general clockwise flow around Georges Bank associated with the frontal regions is interrupted by the presence of a warm core ring (Fig. 6). Specifically, the high velocity region of a ring moves in a northeasterly direction and interrupts the southwesterly flow on the southern flank of Georges Bank. Prior to the discovery of streamers by satellite observation, it was our belief that the high primary productivity suggested by phytoplankton concentrations on Georges Bank was the result of tidal erosion of thermal structure which in turn released nutrients from deep waters to the overlying photic zone (Fig. 7). We further believed that this augmented growth on the Bank was transported laterally off the Bank by a flow arising from the differences in density between the Bank and the surrounding waters. In other words, mixing over the Bank produced a water mass with σ_t greater than surrounding surface waters and this

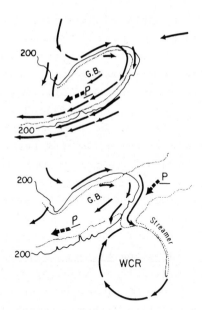

TRANSPORT OF PRODUCTION $\underline{P}$
WITH (above) and WITHOUT (below)
A WARM CORE RING (WCR)

Fig. 6. Transport of production off Georges Bank with and without
the presence of a warm core ring. Normal southwesterly flow off
the Bank is diverted offshore due to counter flow of the ring high
velocity region.

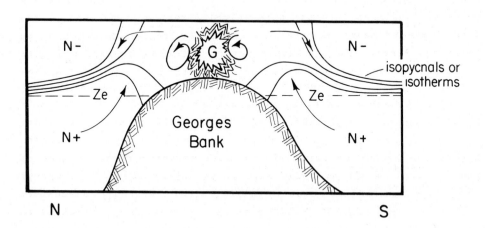

Fig. 7. Conceptual model of density driven nutrient enrichment on
Georges Bank (from Yentsch, 1984).

water mass sinks to a depth of identical density off the Bank. Any excess production that is not consumed locally (and it is estimated that this is considerable) is transported toward the south and west by the mean circulation. Thus the presence of a warm core ring diverts some of this flow seaward as indicated by satellite colorimetry and our sea surface measurements of chlorophyll. Although at this time we have no direct measurements of volume transport of the streamer, we did estimate from ship position that the high velocity region of the ring was in excess of two knots. This velocity is sufficient to divert the slower velocities associated with the frontal regions on the south flank of Georges Bank.

Benefits to Primary Production by Ring Coupling to Tidal Regime

We believe that a ring lying adjacent to the tidal fronts on the southern flank of Georges Bank provides a means for transporting relatively deep nutrient rich waters to the photic zone. The close proximity of the ring to the tidal front argues that there is some coalescence of the frontal flow with the high velocity region of the warm core ring such that a partnership exists between the ring and the tidal front (Fig. 8). We have designated these two regions as tidal regime and geostrophic regime based on the principal forces of each system. Due to the rotary motion of the warm core ring, nutrient rich water is transported along isopycnal surfaces into the high velocity region of the eddy. The water mass adjacent to the high velocity region is stirred by tidal and frontal flow. Nutrient rich water is transported into the tidally stirred euphotic zone and enhanced growth of phytoplankton ensues. As long as the warm core ring stays in this position, we assume that the tidal regime is supplied by nutrients in this manner. A streamer forms as a result of the interaction of the high productivity water carried by the mean circulation around the Bank with water of the high velocity region moving in the opposite direction. The velocities appear to be high enough to offset any of the density differences previously discussed. One could argue that if indeed this is a partnership of processes, further augmentation may occur when productivity of the high velocity region and tidally driven regions are coupled. Our experience at sea in crossing these streamers is that they are not extremely rich in phytoplankton and yet fish and sea mammals are abundant in their presence.

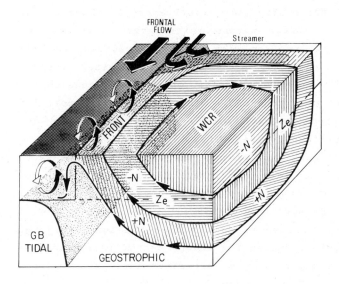

Fig. 8. Conceptual model of ring coupling to tidal regime on the
south flank of Georges Bank. Stippled areas are water of equal
density; arrows indicate direction of flow; Ze = euphotic zone; N+
and N- = nutrient rich and poor water respectively. Deep nutrient
rich water from the geostrophic regime flows along isopycnal
surfaces into the tidal regime of the Georges Bank front.
Opposing directions of flow along the front/ ring high velocity
region interface divert streamer flow.

Need for Study

The statement -- estimates of slope to ocean water interchange cannot be made effectively without the use of satellite
imagery -- is not new, however it needs repeating. The real value
of satellite imagery comes forth in the examination of a streamer;
it does not seem possible that these could be clearly elucidated
without a great deal of shiptime. For those of us who consider
estimates of slope/ocean interchange, it seems futile to rely on
ship crossings alone to gather data in the appropriate time and
space for such estimates. We have used satellite imagery in
concert with shipboard techniques to make measurements of a small,
highly dynamic oceanographic feature. Satellite imagery provides
information concerning frequency, duration and location of
streamers. We sorely need data concerning their physical, biological and chemical characteristics, volume transport and amounts
of materials to make the proper estimates of flux.

It is clear that the observation of streamers is changing our ideas of how slope and ocean waters interchange. Perhaps our original ideas about export from rich areas such as Georges Bank are simplistic. The same may be true for all other regions adjacent to western boundary currents which are occasionally perturbed as eddies move through their area. Transport of nutrients and phytoplankton rich waters may be more rapid and localized than we anticipated heretofore.

The presence of a warm core ring in the vicinity of the south flank of Georges Bank explains our observations of strong baro-clinicity associated with the tidal front regions along the south flank. It remains to be seen how persistent ring features are in this region. We are now tempted to say that the sustained high production on Georges Bank during the summer months is enhanced as a result of the implantation of warm core rings into this region.

ACKNOWLEDGEMENTS

The authors wish to acknowledge the Duke University Marine Consortium for vessel support, the officers and crew of the R/V Cape Hatteras without whom this research would not have been possible, and J. Rollins and P. Boisvert for assistance in preparing the manuscript. This work was supported by funds from the National Science Foundation, the National Aeronautics and Space Administration, the Office of Naval Research and the State of Maine.

This is Bigelow Laboratory for Ocean Sciences contribution no. 85019.

REFERENCES

Beardsley, R.C. and Smith, P.C. 1981. The mean, seasonal and subtidal circulations in the Georges Bank and Gulf of Maine region. In: Third Informal Workshop on the Oceanography of the Gulf of Maine and Adjacent Seas. University of New Hampshire, Durham, NH, pp. 9-16.
Butman, B., Beardsley, R.C., Magnell, B., Frye, D., Vermersch, J.A., Schlitz, R., Limeburner, R., Wright, W.R. and Noble, N.A. 1982. Recent observations of the mean circulation of Georges Bank. J. Phys. Oceanogr., 12: 569-591.
Fuglister, F. 1972. Cyclonic rings formed by the Gulf Stream 1965-66. In: A.L. Gordon (Editor), Studies in Physical Oceanography -- A Tribute to George Wust on his 80th Birth-day, Vol. I. Gordon and Breach, New York, pp. 137-168.
Parker, C.E. 1971. Gulf Stream rings in the Sargasso Sea. Deep-Sea Res., 18: 981-993.

Rossby, C.G.   1936.   Dynamics of steady ocean currents in light of experimental fluid mechanics.   Papers in Phys. Oceanogr. and Meteorol., 5: 3.

Yentsch, C.S. and Garfield, N.   1981.   Principal areas of vertical mixing in the waters of the Gulf of Maine, with reference to the total productivity of the area.   In: J.F.R.   Gower (Editor), Oceanography from Space.   Plenum Publ. Corp., pp. 303-312.

Yentsch,   C.S.   and   Phinney,   D.A..   1985.   Rotary   motions   and convection as a means of regulating primary production in warm core rings. J. Geophys. Res., 90: 3237-3248.

ON THE DYNAMICS OF A TIDAL MIXING FRONT

G.J.F VAN HEIJST

University of Utrecht, Institute of Meteorology and Oceanography,

Princetonplein 5, Utrecht, The Netherlands.

ABSTRACT

This paper addresses some aspects of the dynamics of tidal mixing fronts, as commonly occurring in the continental shelf seas. Attention will be focussed on the along-frontal flow structure, the cross-frontal circulation, and the instability behaviour. The results to be discussed were obtained by analytical and numerical models, by oceanographic observations, and by laboratory experiments.

1. INTRODUCTION

The occurrence of seasonal stratification is a well-known feature of shallow shelf seas: in areas where the tidal stirring is too weak to keep the water column mixed, thermal stratification can be observed during the spring and summer seasons when the enhanced solar heat flux warms the upper layer. On the other hand, in tidally energetic areas the water column will be kept well-mixed throughout the year. The narrow transition zone between well-mixed and stratified areas is commonly called 'tidal mixing front' and forms the subject of this study. The positions of these fronts are well-defined and can be calculated from an energy criterion derived by Simpson & Hunter (1974).

The general structure of a tidal mixing front is depicted schematically in Fig. 1. Note the characteristic 'forking' of the isolines (being isotherms or isopycnals) in the transition zone between the two-layer stratified region and the well-mixed water column. The sea-surface temperature is usually highest in the stratified area, whereas the temperature in the bottom layer is lower than the temperature of the mixed water column. It has been observed in many shelf-sea situations that the surface front is less pronounced than the bottom front – an effect caused mainly by atmospheric forcing.

In general terms, a tidal mixing front acts as a physical boundary between water masses of different thermal, biological and chemical composition. As the exchange of properties between these water masses is locally intense, frontal

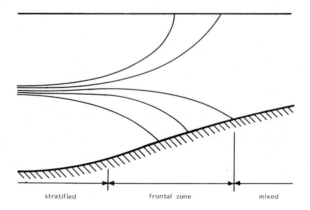

Fig. 1   A schematic picture of a tidal mixing front (the solid curves
represent isotherms or isopynals).

zones usually create particularly favourable conditions for various biological
processes. Fronts are therefore of significant biological importance, and
enhanced primary productivity is often associated with their occurrence (see
e.g. Bowman & Esaias, 1978).

Oceanographic observations have revealed that the basic flow in along-
frontal direction is in first approximation geostrophically controlled, the
Coriolis force being balanced by the density gradient force. In addition to
the along-frontal flow, there is a weak secondary circulation in cross-frontal
direction caused by internal frictional processes. It is generally believed
that this cross-frontal circulation plays an important role with respect to
the enhanced biological activity.

Infrared satellite imagery has provided a wealth of information on fronts,
both on their position and their dynamical behaviour. A general feature is
that most fronts seem to be unstable, as can be observed from the abundant
occurrence of meanders and eddies in the frontal zones. These eddies
contribute significantly to the exchange and mixing of properties across the
front, and are therefore of crucial importance to the biological processes
occurring near the frontal boundary.

This paper concentrates on a few aspects of the frontal dynamics. To start
with, the basic geostrophic along-frontal flow will be considered in the next
section. The secondary circulation in the cross-frontal plane is the subject
of section 3, while section 4 addresses the frontal instability. Special
attention is given to laboratory experiments that have been performed in order
to study the behaviour of unstable bottom and surface fronts.

2. THE ALONG-FRONTAL FLOW

In view of their large along-frontal scales, it seems reasonable to assume that most fronts are, at least in first approximation, geostrophically balanced, with the vertical shear of the along-frontal flow related to the cross-frontal density gradient. In order to study the basic dynamics of the along-frontal flow structure, a simple three-layer model has been designed, which will be described first.

Detailed information about the frontal flow structure was obtained fairly recently by observations in the North Sea. These revealed a number of interesting features, and some of the preliminary results will be discussed at the end of this section.

2.1 A geostrophic adjustment model

The geostrophically controlled basic state of a tidal mixing front has recently been studied by use of a geostrophic adjustment model (Van Heijst, 1985), as schematically shown in Fig. 2. This model considers the geostrophically balanced equilibrium state that results after instantaneous withdrawal of the barrier separating a stably stratified two-layer fluid from a homogeneous fluid of intermediate density. When the barrier is instantaneously lifted, a density-driven flow in cross-frontal direction will arise immediately. As the entire fluid system rotates steadily at angular velocity $f/2$, with f the Coriolis parameter, this cross-frontal flow is subjected to the Coriolis force, and will be deflected accordingly. After some time the flow will be geostrophically adjusted and be directed parallel to the front, such that the pressure gradient force balances the Coriolis force. Referring to a Cartesian coordinate system x,y,z (with x directed perpendicular to the barrier into the stratified fluid, y parallel to the barrier, and z vertically upwards), the equilibrium along-frontal flow in the noses of the layers 1 and 3 (the indices refer to the upper and lower-layer fluids, as shown in Fig. 2a) will be in positive y-direction, while the flow in the nose of the intruding layer 2 will be oppositely directed, i.e. in negative y-direction. The resulting interfacial shapes in the adjusted state are depicted in Fig. 2b.

It is useful to nondimensionalize the problem by scaling the vertical lengths by the total water depth, H, and the horizontal lengths by the

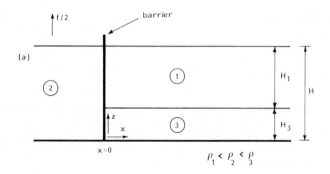

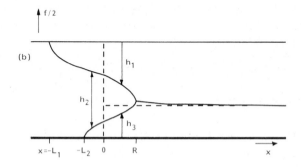

Fig. 2   The geometry of the geostrophic adjustment model: (a) The initial
state with the barrier separating the two-layer stratified fluid
(layers 1 and 3) from the well-mixed fluid (layer 2); the densities
satisfy $\rho_1 < \rho_2 < \rho_3$. (b) The adjusted state, with the along-frontal
flows geostrophically balanced.

internal Rossby radius

$$R = \left[gH(\rho_3 - \rho_1)/\rho_1 \right]^{\frac{1}{2}}/f \ ,$$

with g the gravitational acceleration; ρ_1 and ρ_3 are the upper and lower-
layer fluid densities. The horizontal velocities are then scaled by fR.

The final, geostrophically adjusted state can be calculated by assuming
that mixing and frictional effects are negligible during the adjustment
process, so that potential vorticity is conserved. When the layer depths are
denoted by $h_j(x)$, (j=1,2,3), and along-frontal velocities by $v_j(x)$,
conservation of potential vorticity can be written in nondimensional terms for
each individual layer as

$$\frac{dv_1}{dx} + 1 = \frac{h_1}{\delta} \qquad\qquad (1a)$$

$$\frac{dv_2}{dx} + 1 = h_2 \qquad\qquad (1b)$$

$$\frac{dv_3}{dx} + 1 = \frac{h_3}{1+\delta} \; , \qquad\qquad (1c)$$

with $\delta = H_1/H$. The velocity jump $(v_i - v_j)$ across an interface is directly related to the interfacial slope between the layers i and j according to the Margules equations

$$v_2 - v_1 = - \Delta \frac{dh_1}{dx} \quad \text{for} - L_1 \leqslant x \leqslant - L_2, \; - L_2 \leqslant x \leqslant R \; , \qquad (2a)$$

$$v_3 - v_2 = \overline{\Delta} \frac{dh_3}{dx} \quad \text{for} - L_2 \leqslant x \leqslant R, \qquad\qquad (2b)$$

$$v_3 - v_1 = - \frac{dh_1}{dx} \quad \text{for} \; x \geqslant R, \qquad\qquad (2c)$$

with

$$\Delta = \frac{\rho_2 - \rho_1}{\rho_3 - \rho_1} \; , \quad \overline{\Delta} = 1 - \Delta$$

and L_1, L_2, R the positions of the surface front, the bottom front, and the intrusion nose (see Fig. 2b). In addition, it is useful to assume that the upper surface is covered by a rigid lid, so that

$$h_1 + h_2 = 1 \qquad\quad \text{for} - L_1 \leqslant x \leqslant - L_2 \qquad\qquad (3a)$$

$$h_1 + h_2 + h_3 = 1 \quad \text{for} - L_2 \leqslant x \leqslant R \qquad\qquad (3b)$$

$$h_1 + h_3 = 1 \qquad\quad \text{for} \quad x \geqslant R \; . \qquad\qquad (3c)$$

It is now possible to derive from (1) - (3) general solutions for h_j and v_j in the regions indicated in (3). The integration constants involved can be determined by matching the solutions across the domain boundaries $x = - L_2$ and $x = R$, and by applying some additional conditions associated with conservation of mass and conservation of along-frontal momentum. Details of the analysis are given in Van Heijst (1985).

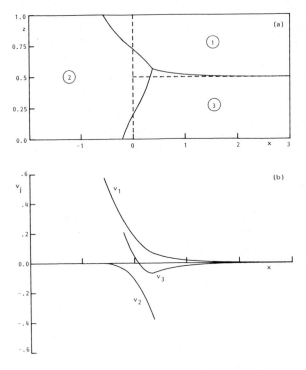

Fig. 3 Model results for the frontal shape (a) and the along-frontal velocity distributions (b) in the geostrophically adjusted equilibrium state. The parameter values are: $\delta = 0.5$, $\rho_1 = 0.95\rho_3$, $\rho_2 = 0.99\rho_3$.

A typical result is shown in Fig. 3 for the parameter values $\delta=0.5$, $\rho_1 = 0.95\rho_3$, $\rho_2 = 0.99\rho_3$. In this case the nondimensional front positions are $L_1 = 0.5737$, $L_2 = 0.2071$, and the position of the intrusion is given by $R = 0.3771$. The shape of the fluid interfaces and the corresponding along-frontal velocities $v_j(x)$ are shown in the upper and lower graphs of Fig. 3, respectively.

In the analysis it is assumed that the flow during the adjustment process is merely a horizontal motion of vertical vortex tubes which are gradually being stretched or shrinked. The resulting velocity distributions $v_j(x)$ can then be understood from conservation of along-frontal momentum and conservation of potential vorticity, which imply that:

. horizontal displacement of a fluid column in positive (negative) x-direction results in a negative (positive) along-frontal velocity v_j , and

. stretching (shrinking) of a fluid column results in a positive (negative) velocity gradient dv_j/dx .

According to these principles it is easily verified why the lower-layer velocity $v_3(x)$ takes both positive and negative values on its solution domain.

Although this model is highly idealized, it provides some insight into the structure of geostrophically adjusted flows. In reality, however, the situation is more complicated (due to the occurrence of turbulent mixing, internal friction, continuous stratification, frontal instabilities), and one should be cautious when drawing detailed comparisons.

2.2. Some observations

It is well-known that seasonal stratification occurs in the North Sea (see e.g. Tomczak & Goedecke, 1964). During the spring, summer and autumn seasons of 1981 and 1982 observations were carried out near the southern edge of the seasonally stratified area south-east of the Dogger Bank. A description of this observational work is presented by Van Aken et al. (1986). Apart from CTD measurements at a number of evenly spaced stations along systematic tracks in the research area, a couple of thermistor chains and current meters were moored for some months near the central position (54°30'N, 4°30'E), close to the boundary of the stratified area. In this region the sea depth measures typically 50 m, and the sea bottom, which is approximately flat in the stratified area, slopes gently upwards into the mixed area, the slope being roughly 10^{-4}.

During the 1982 campaign a strong gale developed soon after deployment of the measuring equipment, causing the front to advect slowly into the stratified area, such that the thermistors and the current meters originally located on the stratified side of the front were able to measure the temperature distribution and the flow structure in a frontal cross-section. Large-scale and small-scale hydrographic surveys revealed that the advection velocity was directed roughly perpendicular to the front at a rate of approximately 20 km in 10 days. The magnitude of the advection speed was very small, and could not be measured directly.

The time-dependent thermistor and current meter signals were averaged by taking running means, and typical graphs of the data obtained in two different stations are shown in Figs. 4 and 5. Note that in both figures the parameters along the abscissa and the ordinate are time (in Julian days) and depth (in meters). In Fig. 5 isotherms are shown in (a), whereas the corresponding isotachs of the along-frontal velocity component are depicted in (b). As the number of thermistors and current meters in the water column is limited (11 thermistors at equidistant depths ranging from 11 m to 31 m, and 6 current

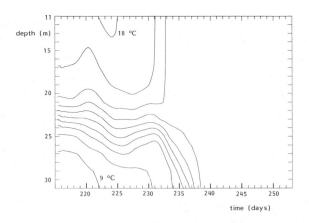

Fig. 4 Graph of time dependent thermistor signals, showing the thermal structure of a slowly advecting front (taken from Van Aken et al., 1986).

meters as well as thermistors at depths of 12, 18, 24, 30, 37 and 44 m), some interpolation had to be made. The isotherms are plotted for differential temperatures of 1°C, ranging from 9°C near the bottom to 18°C in the upper parts of the water column. It appears that there is a well-defined thermocline on the stratified side of the front at roughly 20–25 m depth, with a large temperature jump from 16°C to 10°C over a few meters depth. Although the 'forking' of isotherms, characteristic for tidal mixing fronts, is clearly visible, it appears that the bottom front is much more pronounced than the surface front. This phenomenon was also met at the other stations, and it is believed to be caused by the intense atmospheric forcing.

The isotachs shown in Fig. 5b reveal the existence of a jet-like flow feature at the front, with a maximum velocity of almost 15 cm s^{-1} occurring roughly at the steep slope of the 15°C and 16°C isotherms at day 233. This flow structure is consistent with the thermal-wind balance, assuming that the density distribution has roughly the same structure as the isotherm pattern: the along-frontal velocity is largest in regions with the steepest cross-frontal (hydrostatic) pressure gradients. The observed currents also show agreement with the velocity distribution found semi-analytically by Garrett & Loder (1981, their figure 5).

Because the stratification of the water column is continuous rather than discrete, it is hard to draw a comparison with the results obtained by the geostrophic adjustment model described in section 2.1. Nevertheless, the

velocity distribution $v_2(x)$ on the unstratified side of the front as plotted
in Fig. 3b shows some agreement with the along-frontal jet shown in Fig. 5. As
a matter of fact, along-frontal flows were also calculated by James (1984) in
a numerical model of a tidal mixing front: figure 4f of his paper shows a jet
structure similar, both in magnitude and position, to the observed one.
Besides, James' calculations indicate the presence of a limited region of
small, oppositely directed, along-frontal velocities on the stratified side of
the bottom front – a feature which also shows up in the present observations
(although not visible in the graphical presentation of Fig. 5b).

For a more extensive account of the North Sea observations the reader is
referred to Van Aken et al. (1986).

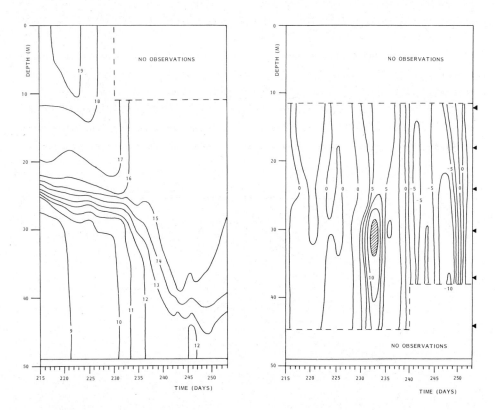

Fig. 5   The temperature distribution and the flow structure of a front as
observed in 1982 at the edge of a seasonally stratified area in the
North Sea at 54°30'N, 4°30'E (see Van Aken et al., 1986). The curves
in (a) represent isotherms, whereas the corresponding isotachs of the
along-frontal velocity component are shown in (b). The hatched area
in (b) represents velocity magnitudes exceeding 12.5 cm s^{-1}.

3. THE CROSS-FRONTAL CIRCULATION

Another important aspect of the mixing-front dynamics is the occurrence of a weak flow in the cross-frontal plane. This secondary circulation, arising from the slow frictional spreading of the front under gravity, has been studied numerically by James (1978, 1984) and semi-analytically by Garrett & Loder (1981). In these studies it was found that the secondary flow pattern of a tidal mixing front has a two-cell structure as schematically drawn in Fig.6. The circulation in the upper and lower cell are in opposite sense, with the flow along the sea bottom directed towards the well-mixed side of the front. The circulation in the upper cell is usually weaker, as the largest horizontal density differences occur in the deeper parts of the water column. This causes an asymmetry in the circulation pattern, resulting in a *convergence* at the surface where the cross-frontal velocity is zero, and an *upwelling flow* on the mixed side of the front.

Surface convergence near fronts is a well-known feature, and can be easily observed from the presence of oil slicks and the accumulation of foam, seaweed, jelly fish and other floating material along the frontal zone.

The vertical velocities associated with the upwelling flow induced by the lower circulation cell are too small to be measured directly (the magnitudes are less than 1 mm/s). However, uplifting of the near-surface isotherms as

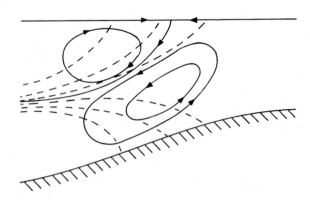

Fig. 6 A schematic representation of the circulation in the cross-frontal plane. Solid curves represent streamlines, whereas isopycnals are denoted by broken lines.

well as a narrow band of minimum sea-surface temperature on the mixed side of the front is frequently observed (see e.g. Simpson et al., 1978; Krause et al., 1986), and this is consistent with the tendency of the circulation pattern to bring dense (cold) bottom water to the sea surface.

An interesting feature visible in most of the isotherm plots obtained from the North Sea observations mentioned before, is a characteristic downward 'dip' in the thermocline at some distance from the bottom front (see Fig. 4). It is believed that this 'dip' is another manifestation of the secondary circulation: it is likely to be caused by the downwelling flow in the lower circulation cell on the stratified side of the front. Such a characteristic feature seems not to have been reported thusfar - probably because it is easily lost when the temperature measurements are performed in too widely-spaced stations. More details of this feature can be found in Van Aken et al. (1986).

4. FRONTAL INSTABILITY

Infrared satellite images of sea-surface temperature have revealed that virtually all surface fronts are unstable, as can be concluded from the abundancy of eddy-like structures in frontal zones. A nice collection of unstable fronts can be seen on Fig. 7, showing the North Sea as viewed by the NOAA-7 satellite on 2 November 1984 at 14h24 G.M.T. Higher sea-surface temperatures show up by darker tones. Apparently, a large band of relatively warm water lies across the North Sea, extending from the English channel to the Skagerrak, bounded by cooler areas in the central North Sea, along the English and Dutch coasts, and in the German Bight. The satellite picture clearly shows that the frontal boundaries are unstable, and that most of them are distorted in a more or less wavelike fashion. An interesting feature is the cold-water tongue extending from the Norfolk Banks in ENE direction, which shows a remarkably regular wave pattern at its southern edge.

Frontal instabilities can be barotropic (when the horizontal shear across the frontal interface is the - kinetic - energy source to feed small perturbations) or baroclinic (when perturbations are fed by the potential energy associated with the horizontal density gradients). In most practical situations, however, both energy sources are present, and the instabilities will be of mixed barotropic-baroclinic type.

The dynamics of unstable surface and bottom fronts has been simulated in laboratory experiments by Griffiths & Linden (1981, 1982) and by Linden & Van Heijst (1984), respectively. This experimental work has provided important

176

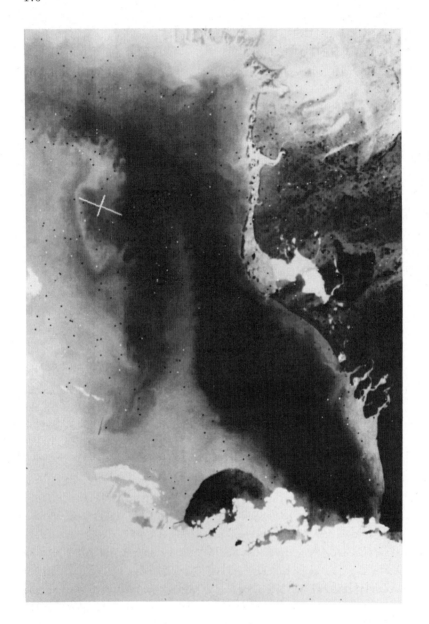

Fig. 7 Infrared satellite picture of the North Sea on 2 November 1984 at 14h24 GMT. (By courtesy of KNMI, De Bilt, The Netherlands).

information about the evolution of the instabilities and the formation of eddies. A numerical simulation of unstable fronts has been performed by James (1984); his calculations show good agreement with available sea-surface observations as well as with the experimental results obtained by Griffiths & Linden (1981, 1982).

4.1. Laboratory experiments

Fronts can be created in the laboratory by allowing a fluid column to collapse in a rotating, environmental fluid of different density (see Saunders, 1973; Griffiths & Linden, 1982). The basic set-up for this experiment is shown schematically in Fig. 8: a central, removable cylinder is filled with a fluid (either homogeneous, or two-layer stratified) and is placed at the centre of a rotating tank containing fluid of a contrasting density. Ideally, the fluids are assumed to be immiscible. When the inner cylinder is withdrawn, a density-driven radial flow will arise in the fluids, until a state of geostrophic balance is reached (similar to the adjustment problem discussed in section 2.1). The resultant flows in this ultimate, geostrophically controlled state are purely azimuthal (ignoring viscous effects, which are confined to the Ekman layers at the bottom and at the interfaces). Provided that the appropriate Rossby radius of deformation is sufficiently smaller than the tank radius, the interfaces between the fluids will in the adjusted state intersect one of the horizontal flow boundaries (or both), thus producing a circular bottom and/or surface front. A surface front is obtained by taking $\rho_1 < \rho_2 = \rho_3$ (as did Griffiths & Linden, 1982; see Fig. 8a), while a bottom front arises when $\rho_3 > \rho_2 = \rho_1$ (see Fig. 8b). When the ambient fluid density ρ_2 is chosen in the range $\rho_1 < \rho_2 < \rho_3$, both a bottom front an a surface front will arise (see Fig. 8c). This particular configuration resembles the tidal mixing front, and is therefore useful to study its dynamics.

In order to visualize the frontal shape, dye may be added to the fluid(s) in the inner cylinder. In addition, small pieces of paper floating on the free surface allow streak photography, which provides essential information about the upper-layer flow. Motions at greater depth can be visualized by dye-producing particles of potassium permanganate to be dropped into the tank during the course of a particular experiment. Observations are most conveniently done by using a co-rotating photocamera mounted at some distance above the fluids.

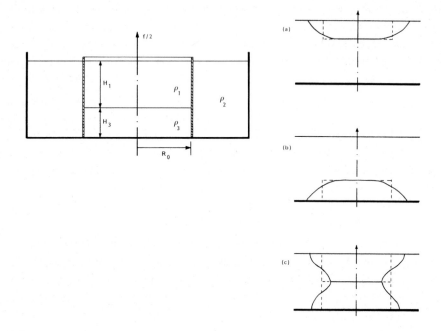

Fig. 8 A schematic representation of the laboratory system used for simulation of unstable fronts. When the inner cylinder is withdrawn the fluids will 'collapse' until a geostrophically adjusted state is reached. Depending on the choice of fluid densities, three different types of fronts can thus be generated: (a) a surface front, for $\rho_1 < \rho_2 = \rho_3$; (b) a bottom front, for $\rho_1 = \rho_2 < \rho_3$; and (c) a 'tidal mixing front', for $\rho_1 < \rho_2 < \rho_3$.
The tank diameter and tank height measure 92.5 cm and 30.0 cm, respectively. The radius of the inner cylinder is $R_0 = 14.4$ cm, and the initial layer depths are denoted by H_i. Another parameter is g'_{ij}, being the reduced gravity based on the density difference between layers i and j.

4.2 Surface fronts

Surface fronts produced by the 'collapse technique' as well as by injection of lighter fluid at the surface have been studied in considerable detail by Griffiths & Linden (1981, 1982) and also by the present author (unpublished). It was observed that these fronts are unstable, and that the circular frontal shape is always distorted by a more or less regular, wavelike pattern. By counting the number of waves (n) the wavelength λ of the frontal instabilities is easily calculated from $\lambda = 2\pi R_0/n$, with R_0 the radius of the inner cylinder. In the early stages of an experiment the (small-amplitude) instabilities have a short wavelength, but this length is usually observed to increase until some ultimate, constant value is reached. The experiments of Griffiths & Linden (1982) have demonstrated that the 'final' wavelength λ of the most unstable instability mode is $\lambda/2\pi\hat{R} = 1.1 \pm 0.3$, with $\hat{R}$ the geometric mean of the Rossby radii of the upper and lower layer (to be defined later). This result agrees well with the value 1.15 derived theoretically by Killworth et al. (1984).

Careful observations have provided a lot of information about the evolution of the frontal instability, which is illustrated in Fig. 9. Initially, the disturbance has a sinusoidal shape (Fig. 9a); the waves move along the front at a speed intermediate to the along-frontal velocities V_1 and V_2 of the fluids on either side of the front. When their amplitudes increase, the waves become slightly asymmetric and show a tendency to break backwards (Fig. 9b).

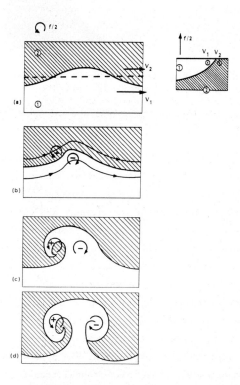

Fig. 9   A schematic representation of the evolution of an unstable surface front. The hatched areas represent the ambient, denser fluid. The drawings, representing subsequent stages of evolution, show (a) the initial sinusoidal perturbation, (b) the asymmetry of the wave, (c) the backwards breaking wave, and (d) the formation of a vortex dipole. Regions where vorticity is being concentrated are indicated by arrows, with the plus and minus signs denoting cyclonic and anticyclonic vorticity, respectively.

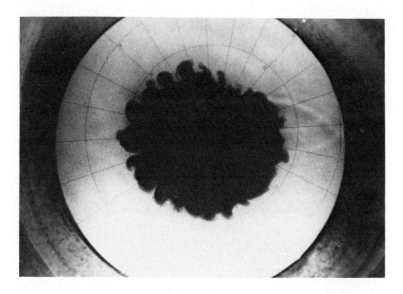

Fig. 10 A photograph showing 'hammer-head'-shaped instabilities at a surface front produced in the laboratory. The picture was taken roughly 6 rotation periods after withdrawal of the inner cylinder (which lower edge is visible). The parameter values are $H_1 = H = 10.0$ cm, $f = 2.30$ s^{-1} and $g'_{12} = 0.5$ cm s^{-2}.

Note that the flow in the wave crest (fluid 1) has anticyclonic vorticity, whereas cyclonic vorticity is being concentrated in the 'wake' (fluid 2). As the amplitude growth continues, the waves are observed to break backwards (Fig. 9c), thereby causing entrainment of denser ambient fluid into the buoyant upper-layer fluid. At this stage the formation of a cyclonic-anticyclonic vortex pair becomes visible.

When the velocity difference $\Delta V = V_1 - V_2$ is not too large the dipole structure becomes more pronounced and the waves take on their characteristic 'hammer-head' appearance (Fig. 9d). The vortex pair may then even be 'pinched off', and propel itself – under the influence of its own vorticity distribution – away from the front, into the ambient fluid. In this way some upper-layer fluid is carried into the environment – a process which is thought to be of considerable importance to the cross-frontal transfer of properties.

Some 'hammer-head'-shaped protrusions can be seen clearly on the satellite picture of the North Sea (Fig. 7), in particular in the frontal zone north of the Netherlands, just at the right-hand side of the white reference mark. Similarly shaped instabilities are visible on Fig. 10, which shows an unstable surface front produced by the collapse of a central column of (dark-coloured) fluid in a rotating fluid of larger density. The observed instability pattern is rather irregular, showing both backwards breaking waves and hammer heads.

When the velocity jump ΔV across the front is too large to allow large protrusions, the backwards-breaking effect is much more pronounced and the waves are curled up cyclonically into a spiral shape. This curling effect is demonstrated in Fig. 11, showing a pair of photographs of an unstable surface front similar to the one shown in Fig. 10, but now under different experimental conditions. (The difference lies in the value of the Froude number $F = (R_0/R)^2$, with R the upper-layer Rossby radius of deformation: for the experiments shown in Figs. 10 and 11 it measures 220 and 91, respectively). The first picture (Fig. 11a), taken approximately 9 rotation periods after the collapse of the central fluid column (the inner cylinder is still visible – it hangs just above the water surface), nicely shows the entrainment of ambient fluid at the back of the waves into the dark-coloured buoyant fluid. Because of the intense shearing motion the mixing in the curling waves is significant, and the spirals soon get a 'blurred' appearance. This can be observed in the second picture (Fig. 11b), taken 2 rotation periods later: details of the curling motion are no longer visible – only a light spot in the centre of the cyclonic vortices indicates the presence of ambient fluid. The presence of denser fluid in the eddy centre is not exclusively caused by entrainment: it is also a result of upwelling driven by the cyclonic upper-layer eddy itself. Similar cyclonic eddies have been frequently observed in the frontal zones as occurring in the shallow seas around the British Isles. Nice examples of 'curling' structures at the Ushant and Flamborough Head fronts as well as some thermal fronts in the Irish Sea and the Celtic Sea have been collected by Pingree (1978, 1979). Detailed hydrographic surveys in the Ushant frontal region revealed that the sea surface density has a maximum value in the central region of such an eddy (see Pingree et al., 1979). This corresponds to the entrainment/upwelling feature observed in the laboratory experiment as shown in Fig. 11.

4.3 Bottom fronts

As described at the beginning of this section, bottom fronts can also be produced in the laboratory by applying the 'collapse technique', now with the centre fluid denser than the environmental fluid. Such experiments have been carried out recently and have revealed a number of striking differences as compared to surface fronts.

For example, it appeared that bottom fronts are not necessarily unstable: when the lower-layer Rossby radius R is sufficiently larger than the radius R_0

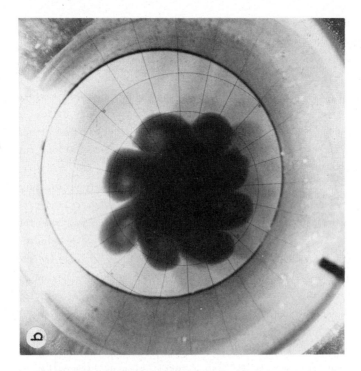

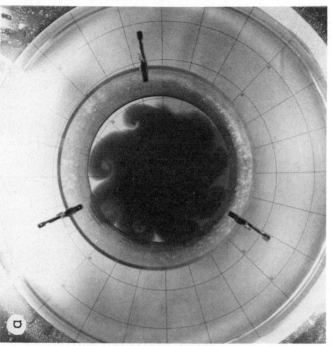

Fig. 11 A sequence of two photographs showing the 'curling' of backwards breaking waves on a surface front. The pictures are taken (a) 9 and (b) 11 rotation periods after the inner cylinder was withdrawn. Parameter values: $H_1 = H = 10.0$ cm, $g'_{12} = .5$ cm s^{-2} and $f = 1.48$ s^{-1}.

of the inner cylinder, the central fluid column was observed to collapse into a smoothly curved 'dome shape' which remained stable throughout the experiment. In fact, one thus produces a single, stable baroclinic vortex (cf. Saunders, 1973). By streak photography and by dropping dye releasing crystals into the fluid information was obtained about the velocity structure in the vortex. It turned out that the upper-layer azimuthal velocity as measured at the free surface was *cyclonic* throughout the experiment. This can be understood from conservation of angular momentum, which requires acceleration in the swirl velocity of a contracting ring of upper-layer fluid during the collapse process. According to the same principle one would expect to measure an *anticyclonic* velocity in the lower layer, and this is indeed what has been observed in the initial stages of the experiment. After a few rotation periods, however, the lower-layer swirl velocity is observed to change sign, and the flow becomes *cyclonic* as in the upper layer. It is believed that this 'spin-up' is caused mainly by the bottom Ekman layer and - to a lesser extent - by the interfacial Ekman layer. Observations revealed that at this stage the vortex had reached a state of equilibrium, with cyclonic velocities in both the upper layer and the lower layer. The flow is then in geostrophic (more precisely: cyclostrophic) balance, the velocities in each layer being depth-independent. A smooth transition between the upper-layer flow and the (weaker) lower-layer flow is accomplished by the thin Ekman layer at the interface.

As a result of frictional effects (confined to the Ekman layers) the baroclinic vortex thus produced appeared to decay gradually. This decay process proceeds at an extremely slow rate: in most experiments the vortex was still visible even after a few hundred rotation periods!

In experiments with $R < R_0$ the bottom front is unstable and wavelike distortions of the circular shape became visible soon after withdrawal of the inner cylinder. Although initially these distortions have the same appearance as the surface instabilities, after a short while they take a different appearance.

When the inner cylinder is lifted, one observes in the lower layer an *anticyclonic* motion which is confined to a narrow band, a few Rossby radii wide, at the front. At this stage waves may develop (see Fig. 12a), with a tendency to break backwards (thus producing cyclonic vortices, as at surface fronts). Within a few rotation periods, however, the anticyclonic motion in the bottom layer is reduced to zero (due to the spin-up mechanism caused by the Ekman layers mentioned before), and large radial excursions of lower-layer fluid are observed: the frontal distortions take on the appearance of 'bulges' (see Fig. 12b). As a result of the continuing spin-up action of the Ekman layers, these 'bulges' were after a few rotation periods seen to move in

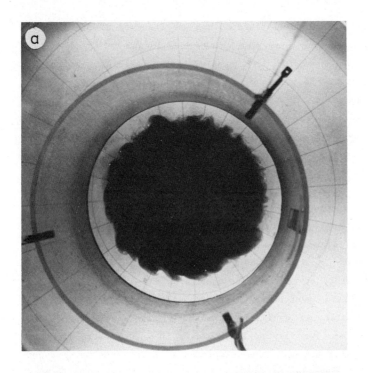

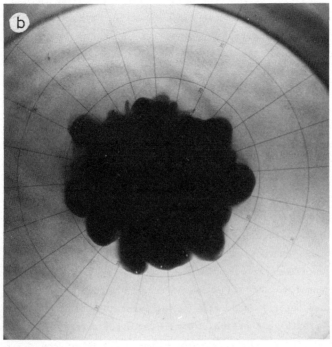

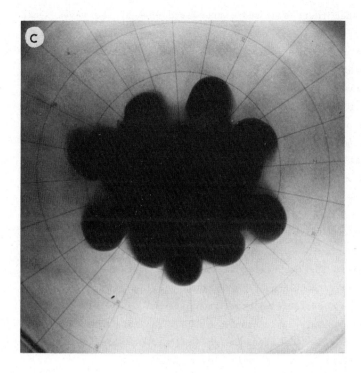

Fig. 12   A sequence of photographs showing the evolution of an unstable bottom front. The pictures are taken (a) 1.5, (b) 4 and (c) 6 rotation periods after the start of the experiment. Parameter values: $H_3 = H = 8.0$ cm, $f = 1.99$ s^{-1} and $g'_{23} = 1$ cm s^{-2}.

cyclonic direction. When their amplitudes are large enough, the waves again show a (weak) tendency to break backwards, thus producing (weak) anticyclonic vorticity in the tip of the protrusions. This can be seen in Fig. 12c. Note also the sharp leading edges, in contrast to the vagueness of the trailing edges – an effect caused by the bottom friction. Application of the earlier-mentioned visualization techniques revealed that the motion inside the 'bulges', as well as in the upper layer just above them, is cyclonic. As the amplitude growth continues, the waves can under certain conditions be pinched off ('vortex splitting', see Saunders, 1973). Each pinched off wave crest soon gets organized into a symmetric, cyclonic vortex and is accompanied by a 'Gaussian'-shaped dome of denser lower-layer fluid at the bottom. Fig. 13 nicely shows the evolution of an unstable bottom front into a symmetric pattern of four vortices. After they were pinched off, the vortices became perfectly circular.

In marked contrast to the observed behaviour of an unstable surface front (with its characteristic vortex-pair formation, leading to 'hammer-head'-shaped protrusions), an unstable bottom front seems to produce cyclonic *monopoles* rather than cyclonic-anticyclonic dipoles. The main reason for this different behaviour lies in the presence of the bottom Ekman layer, which tends to spin up the fluid in anticyclonic flow regions.

Before the frontal protrusions are pinched off, the instabilities usually have a very regular appearance, which allows accurate measurement of the 'final' wavelength of the fastest-growing instability mode. In order to present the observational results graphically, it is useful to define the geometric mean of the upper and lower-layer Rossby radii as

$$\hat{R} = \bar{R} \, [H/h - 1]^{\frac{1}{4}} \, ,$$

with $\bar{R}$ the Rossby radius of the upper layer based on its mean depth $\bar{h} = h_o R_o^2/\bar{R}^2$, $\bar{R}$ being the radius of the circular front before it becomes unstable. Following Griffiths & Linden (1982), the observed wavelengths λ, scaled by $\hat{R}$, have been plotted in Fig. 14 as a function of the Froude number F, defined as

$$F = (R_o/R)^2,$$

with R the Rossby radius of the upper layer, based on its initial depth. The broken lines indicate the band containing the data for unstable surface fronts obtained by Griffiths & Linden (1982, their figure 10). It appears that the observed wavelengths in the bottom front experiments lie well within this

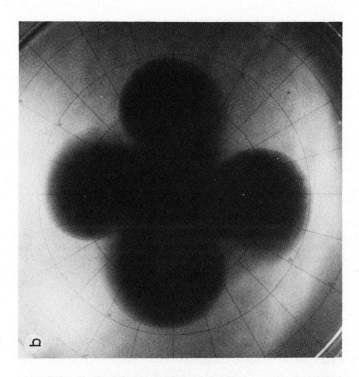

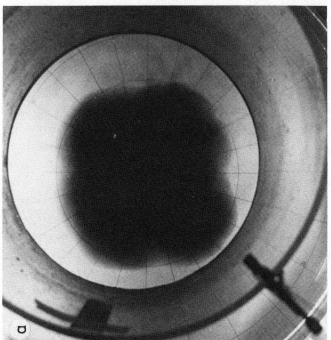

Fig. 13 Two photographs showing a central column of dense fluid breaking up into four cyclonic vortices. The pictures are taken approximately (a) 2 and (b) 4 rotation periods after withdrawal of the inner cylinder. Parameter values: $H_3 = H = 10.0$ cm, $f = 1.55$ s^{-1} and $g'_{23} = 5$ cm s^{-2}.

188

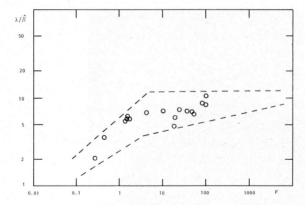

Fig. 14 The observed wavelength λ at unstable bottom fronts non-
dimensionalized by the mean Rossby radius $\hat{R}$ plotted against the
Froude number F. The broken lines indicate the band containing the
experimental data for surface fronts as obtained by Griffiths &
Linden (1982).

band. At larger Froude numbers (F > 1) the observed wavelengths take
approximately constant values. Averaging over 15 experiments with F > 1 one
finds $\lambda / 2\pi \hat{R} = 1.1 \pm 0.25$, which is exactly the value found by Griffiths &
Linden for unstable surface fronts. This suggests that bottom friction has
only a very limited effect (if any) on the wavelength of the most unstable
mode.

4.4 Tidal mixing fronts

By choosing the ambient fluid density ρ_2 in the range $\rho_1 < \rho_2 < \rho_3$ (see
Fig. 8c) one obtains a flow situation that can be regarded as an − admittedly
too simplifying − model of a tidal mixing front, provided that the appropriate
Rossby radii are sufficiently small that the upper and lower interfaces
intersect the free surface and the bottom, respectively. A number of such
experiments has been carried out, mainly in order to study the instability
behaviour of this type of fronts. In all cases the parameters were chosen such
that the upper and lower-layer Rossby radii had approximately equal values.
Figure 15 shows a sequence of photographs illustrating the evolution of the
unstable front. In this experiment the upper and lower-layer fluids in the
stratified central region were dyed with different colours which unfortunately
cannot be distinguished on the black-white photographs. Observation by eye and
photography from the side revealed the existence of an extremely strong
vertical coupling between the upper and lower layer, and also revealed that
the distortions of the surface front and the bottom front are identical.

Looking down into the tank, differences between the upper and lower frontal shapes are hardly visible, even when the flow eventually becomes irregular due to vortex 'splitting' and the subsequent interaction between the vortices. Careful observations made clear that in this particular type of experiment (with identical upper and lower-layer Rossby radii) the upper-layer flow is completely governed by the lower layer, as can be seen clearly in Fig. 15: the surface (and bottom) front visible on the photographs shows exactly the same instability behaviour as the bottom front shown in Fig. 12, which is significantly different from the surface fronts shown in Figs. 10 and 11.

The first photograph (Fig. 15a) shows the fluids immediately after withdrawal of the inner cylinder, the surface (and bottom) front still being circular then. When the next picture (Fig. 15b) was taken, approximately 2 rotation periods later, the upper and lower fluids in a narrow band near the fronts were observed to move in anticyclonic direction; the waves on the unstable fronts appear to break backwards, thus producing cyclonic vorticity. The third photograph (Fig. 15c) was taken 13 rotation periods after the start of the experiment, which roughly coincided with the stage at which the relative motion in the upper and lower-layer fluids was reduced to zero by the spin-up effect: the 'bulging' of the waves can be seen clearly (compare with Fig. 12b). The last picture (Fig. 15d) is taken some 7 rotation periods later. It clearly shows that the frontal waves move in cyclonic direction, as can be seen from the diffuse trailing edges of the bottom waves. Note the resemblance to the corresponding picture of the bottom front (Fig. 12c), on which similar features are visible.

An interesting aspect of the vertical coupling between the fluid layers is visible on photographs taken from the side (see e.g. Fig. 16), which all show 'threads' of coloured fluid between the 'blobs' in the upper and the lower layer. These 'blobs' correspond to cyclonic eddies in the frontal region; the vortices seem to extend over the full water column. The 'threads' are in fact cylindrical sheets of coloured fluid, and indicate vertical exchange between the upper and the lower layer. This vertical motion is likely to be driven by horizontal shear, similar to the vertical flow arising in the Stewartson shear layer encompassing a Taylor column in a rotating fluid. It might be expected that the instability behaviour is different in cases where the upper and lower-layer Rossby radii have dissimilar values. However, this problem has not been addressed here.

In shallow shelf seas the flow situation is much more complicated, as the stratification is continuous rather than discrete. Also, the occurrence of vigorous turbulent mixing in the individual layers is a complicating factor. These effects may be of significant importance to the vertical structure of

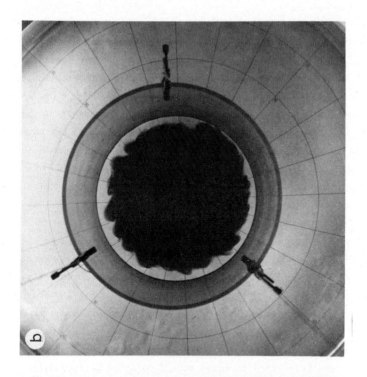

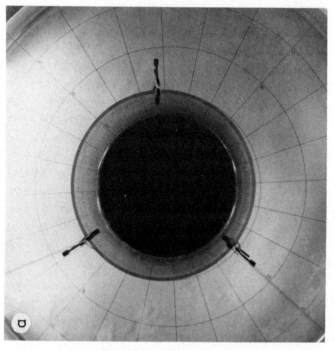

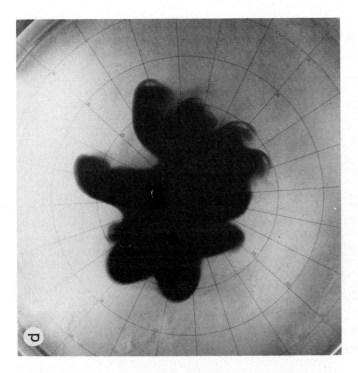

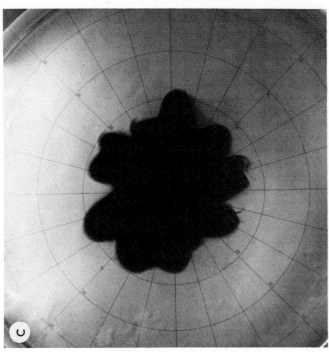

Fig. 15 A sequence of photographs showing the evolution of an unstable 'tidal mixing front' as produced by the collapse technique. The first picture (a) is taken immediately after withdrawal of the inner cylinder; the next photographs are taken (b) 2, (c) 13 and (d) 20 rotation periods after the start of the experiment. Parameter values: $H = 10.8$ cm, $H_3 = 6.2$ cm, $f = 2.16$ s^{-1}, $g'_{13} = 2$ cm s^{-2} and $g'_{23} = 1$ cm s^{-2}.

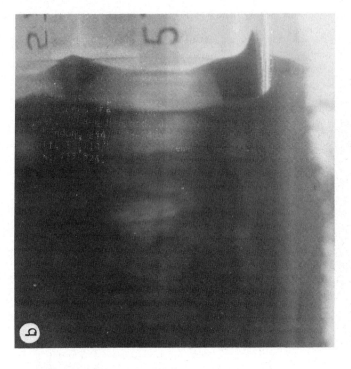

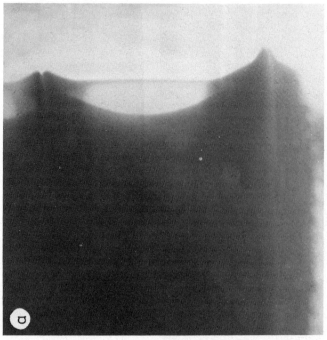

Fig. 16 Two photographs taken from the side, showing 'threads' of coloured fluid between the frontal eddies in the upper and the lower layer. The second picture (b) is taken during a later stage of the instability process, as can be seen from the abundancy of eddies. Note the reflections at the free surface and at the tank bottom.

tidal mixing fronts, and may change the picture obtained by laboratory experiments considerably. Nevertheless, it is thought that the simple laboratory model discussed here is helpful in gaining some insight into the intricate dynamics of unstable fronts.

As yet, too few laboratory experiments on tidal mixing fronts have been performed to draw reliable conclusions about the observed wavelengths. This work will be extended in the near future, however, and the results will be reported in due time.

5. CONCLUSION

Although the theoretical and laboratory models described in this paper are rather simplifying in that they are dealing with idealized flow situations, they give some useful insight into the frontal dynamics, particularly with respect to the along-frontal flow structure and the instability behaviour.

Unfortunately, detailed observations of the three-dimensional structure of tidal mixing fronts in continental shelf seas are scarce. An observational project with a number of thermistor chains or undulating bat fish to be towed simultaneously along different frontal cross-sections, in combination with a fine array of current meters moored in the frontal zone, backed up by air-borne and satellite observation of the sea surface is extremely costly and seems a utopian enterprise. Nevertheless, this paper concludes with a plea for such observations to be made, as these will provide valuable material for a better understanding of the various cross-frontal transport processes.

ACKNOWLEDGEMENTS

Some of the experiments described in section 4 were carried out in the Hydrodynamics Laboratories of the Twente University of Technology, The Netherlands, by kind permission of Professor van Wijngaarden. I wish to thank Lieuwe Seinstra for his skilful assistance with the laboratory experiments, and Leo Maas and Hendrik van Aken for providing the observational data plotted in Figs. 4 and 5. Financial support from the Working Group on Meteorology and Physical Oceanography (MFO) of the Netherlands Organization for the Advancement of Pure Research (Z.W.O.) is gratefully acknowledged.

REFERENCES

Van Aken, H.M., G.J.F. van Heijst and L.R.M. Maas, 1986. Observations of
    fronts in the North Sea. (In preparation).

Bowman, M.J. and W.E. Esaias (Editors), 1978. Oceanic Fronts in Coastal
    Processes. Springer, Berlin , 114 pp.

Garrett, C.J.R., and J.W. Loder, 1981. Dynamical aspects of shallow sea
    fronts. Phil. Trans. Roy. Soc. London, A302, 563-581.

Griffiths, R.W., and P.F. Linden, 1981. The stability of buoyancy-driven
    coastal currents. Dyn. Atmos. Oceans, 5, 281-306.

Griffiths, R.W., and P.F. Linden, 1982. Laboratory experiments on fronts.
    Part 1: Density-driven boundary currents. Geophys. Astrophys. Fluid Dyn.,
    19, 159-187.

Van Heijst, G.J.F., 1985. A geostrophic adjustment model of a tidal mixing
    front. J. Phys. Oceanogr. 15, 1182-1190.

James, I.D., 1978. A note on the circulation induced by a shallow-sea front.
    Estuarine Coastal Mar. Sci., 7, 197-202.

James, I.D., 1984. A three-dimensional numerical shelf-sea front model with
    variable eddy viscosity and diffusivity. Cont. Shelf Res., 3, 69-98.

Killworth, P.D., N. Paldor and M.E. Stern, 1984. Wave propagation and growth
    on a surface front in a two-layer geostrophic current. J. Mar. Res., 42,
    761-785.

Krause, G., G. Budeus, D. Gerdes, K. Schaumann and K. Hesse, 1986. Frontal
    systems in the German Bight and their physical and biological effects.
    (This issue).

Linden, P.F., and G.J.F. van Heijst, 1984. Two-layer spin-up and
    frontogenesis. J. Fluid Mech., 143, 69-94.

Pingree, R.D., 1978. Cyclonic eddies and cross-frontal mixing. J. Mar. Biol.
    Ass. U.K., 58, 955-963.

Pingree, R.D., 1979. Baroclinic eddies bordering the Celtic Sea in late
    summer. J. Mar. Biol. Ass. U.K., 59, 689-698.

Pingree, R.D, P.M. Holligan and G.T. Mardell, 1979.
    Phytoplankton growth and cyclonic eddies. Nature, 278, 245-247.

Saunders, P.M., 1973. The instability of a baroclinic vortex. J. Phys.
    Oceanogr., 3, 61-65.

Simpson, J.H., C.M. Allen and N.C.G. Morris, 1978. Fronts on the Continental
    Shelf. J. Geophys. Res., 83, 4607-4614.

Simpson, J.H., and J.R. Hunter, 1974. Fronts in the Irish Sea. Nature, 250,
    404-406.

Tomczak, G., and E. Goedecke, 1964. Die thermische Schichtung der Nordsee.
    Deutsche Hydrographische Zeitschrift, Ergänzungsheft Reihe B(4), Nr. 8,
    182 pp.

ZOOPLANKTON IN THE UPWELLING FRONTS OFF PT. CONCEPTION, CALIFORNIA

S.L. SMITH,[1] B.H. JONES,[2] L.P. ATKINSON,[3] and K.H. BRINK[4]

[1]Oceanographic Sciences Division, Department of Applied Science, Brookhaven National Laboratory, Upton, NY 11973

[2]Department of Biological Sciences, University of Southern California, Los Angeles, CA 90089

[3]Department of Oceanography, Old Dominion University, Norfolk, VA 23508

[4]Woods Hole Oceanographic Institution, Woods Hole, MA 02543

ABSTRACT

Surface maps of selected taxa of zooplankton were made off Pt. Conception, California, during three contrasting upwelling situations: moderate upwelling, strong upwelling, and downwelling. Number of taxa and number of individuals decreased with increasing upwelling intensity, indicating replacement of richer surface waters by relatively impoverished subsurface waters. The exception to this pattern was in copepodid stage V of Calanus pacificus which increased in numbers nearshore as upwelling strength increased. Since copepodid V is the deep-living, diapausing stage of this copepod, its increase in numbers is consistent with the upward movement of deep water and the life cycle of C. pacificus. Frontal zones always had more individuals than non-frontal zones, and these were primarily copepod nauplii. Estimated ingestion by copepod nauplii and Calanus pacificus in frontal zones during upwelling was twice the ingestion by these taxa outside the frontal zones, suggesting that the frontal zones associated with upwelling off Pt. Conception are sites of enhanced secondary production.

INTRODUCTION

The large and lengthy series of observations of plankton and hydrography in the California Current System and North Pacific Central Gyre has been analyzed primarily for large-scale correlations between physical forcing and planktonic response (Wickett, 1967; Bernal and McGowan, 1981; Bernal, 1981; Chelton, 1982). The generalities that have arisen are principally that maxima in zooplanktonic biomass in the California current region are associated with transport from the north (Wickett, 1967; Bernal and McGowan, 1981) and not with coastal upwelling (Bernal and McGowan, 1981; Chelton, 1982). Although Chelton (1982) proposes another offshore upwelling mechanism to explain the biomass peak observed approximately 100-200 km from the coast, he concedes the very important point that the CalCOFI surveys upon which these generalities are based did not include samples from the nearshore, coastal region. This narrow coastal zone,

where classical wind-driven coastal upwelling could be acting to increase zooplanktonic stocks, and located between Points Conception and Arguello (~34°-35°
N; 120°-121° W), is the area in which the present study was conducted.

The nearshore region off southern California is notorious for the spatial
variability in its plankton populations in spring and summer (Haury, 1976; Cox,
Haury and Simpson, 1982; Star and Mullin, 1981), seen clearly in satellite
coastal zone color scanner images such as those used to analyze the El Niño
phenomenon (Fiedler, 1984). Off San Diego, for example, zooplankton larger than
80 μm were most abundant and most patchy in a set of nearshore samples compared
with similar sample sets collected in the western part of the California Current
and the North Pacific Central Gyre (Star and Mullin, 1981). Patch size in the
California Current was likewise smaller than patch size in the central gyre
(Haury, 1976). Thus, there seems to be an onshore-offshore gradient in patch
size, with smallest patches nearshore. Surface waters in the Point Conception
area have been shown to contain small scale zooplankton patches, associated with
patches of chlorophyll a, and interpreted as topographically induced offshore
extensions of band-like or frontal coastal upwelling effects (Cox et al., 1982).

A prominent member of the nearshore and California Current plankton community in spring and summer is Calanus pacificus (Star and Mullin, 1981;
Fleminger, 1964) whose diel (Enright and Honegger, 1977) and ontogenetic
(Longhurst, 1967; Alldredge et al., 1984) vertical migrations may contribute
substantially to observed patchiness. Diel differences at 35 m during single,
short sampling intervals are not striking (Star and Mullin, 1981), but migration
by late copepodids (stage CV) and adults into the upper 20-25 m consistently
took place at night during spring and early summer, although the exact time of
ascent varied (Enright and Honegger, 1977). Thus during spring and summer,
migrating C. pacificus contribute to patchiness observed in the surface zooplankton community, while at other seasons when the resting stage (copepodid V)
is in large numbers at depth (Longhurst, 1967; Alldredge et al., 1984), it cannot contribute to observed shallow patchiness.

Two sets of observations have prompted the analysis of spatial patterns
herein. First, near Pt. Conception chlorophyll a at the surface is organized
into a frontal feature (Cox et al., 1982) which may also contain nauplii of
Calanus pacificus (Star and Mullin, 1981), assuming spatial correlations nearshore off San Diego are valid at Pt. Conception. Second, in a study of the
upwelling area off Peru, nauplii tended to be concentrated in the downstream
edges of plume-like features of chlorophyll a arising from upwelling processes
(Boyd and Smith, 1983). These "frontal" areas, which were potentially sites of

increased secondary production during upwelling, were much reduced during relaxation (Boyd and Smith, 1983). We therefore hypothesized that the upwelling area near Pt. Conception would have frontal zones where phytoplankton and copepod nauplii co-occurred. Furthermore, these spatial features arising from intermittent upwelling episodes might be critical to general production in the area because they were sites of increased herbivore activity (primarily due to Calanus pacificus) and secondary production. Spatial and temporal patterns in zooplankton at the surface in other upwelling areas have been shown to be dominated by copepods that have a strongly three-dimensional distribution during a life cycle containing a diapausing late juvenile stage (similar to C. pacificus; Binet and Suisse de St. Claire, 1975; Smith, 1982).

METHODS

Underway samples were collected from the R.V. New Horizon during rapid, uninterrupted steaming in the coastal area between Points Arguello and Conception (Fig. 1A). A PVC through-hull fitting (2 m depth) delivered seawater to the laboratory. No sea chest or other impediment obstructed the synoptic sampling of surface waters. Zooplankton were collected from the flow (average flow rate was 15 liters per minute) for 3 minutes each 15 minutes onto 70-µm mesh screen, washed into vials, and preserved in 5% buffered formalin. Each sample, therefore, represents an integration of approximately 0.7 km (0.4 nm) of horizontal distance covered during 3 minutes. Eleven such maps of zooplanktonic abundance at the surface were made during the OPUS study in April and May 1983. Three will be discussed in this paper. In the laboratory all zooplankton in each vial were counted, with total taxa equaling fifty. Calanus pacificus and Paracalanus parvus were staged for sizes collected quantitatively (to CI for C. pacificus; to CII/III for P. parvus), but most copepods were grouped into adults or copepodids of the particular species. No attempt was made to identify non-copepod taxa to species and thus groupings such as chaetognatha, ostracoda and so forth are common in the data. During sampling and data analysis, we noted a very sharp rise in the abundance of Calanus pacificus at sunset. Other diel problems were not evident, but most analyses were done on day and night portions of maps separately, and the "sunset spike" sample in each map was not used.

Temperature, chlorophyll a, and nutrient concentrations were measured from the surface stream by a thermistor in the stream, and by in-vivo fluorescence and continuous autoanalyzer sampling. Position was recorded directly from a

LORAN-C unit onto a computer every minute (see Jones et al., 1986, for details
of the maps of nonzooplanktonic variables). The thermistor, fluorometer, and
autoanalyzer were calibrated by XBT, bucket, and CTD casts, and data are
reported in Paluszkiewicz et al. (1984). Temperatures used in this analysis are
those recorded by XBT's launched during the 3-minute sampling for zooplankton.

RESULTS

   We have selected three surface maps for analysis because they represent three
distinct situations observed during the upwelling season off Pt. Conception and
Pt. Arguello. The first one was sampled on April 14-15, 1983, during moderate
upwelling when the winds were approximately 8 ms^{-1} from the north-northwest.
Upwelled water was from the Santa Barbara Channel and south of Pt. Conception
(Atkinson et al., 1986). The second example was from a period of downwelling on
April 19-20, 1983. Winds were also approximately 8 ms^{-1} but from the
south-southeast. The third map was made during strong upwelling on May 6-7,
1983, when upwelling extended north of Pt. Arguello also and winds were
sustained at 8 ms^{-1} from the north-northwest for a 3-day period before the
sampling exercise and at 13 ms^{-1} for four days subsequent to it (Atkinson et
al., 1986). The behavior of near surface (5 m and 10 m) currents nearshore
(at the 70 m isobath) between Pt. Conception and Pt. Arguello were consistent
with locally wind-driven upwelling dynamics (Brink and Muench, 1985); that
is, flow in the upper 15 m tended to go offshore (onshore) in response to
equatorward (poleward) winds. The apparent upwelling center was located
between the Points at about 34°30.5'N, 120°36.6'W (Atkinson et al., 1986).

   When maps of six zooplanktonic taxa were compared with maps of surface temperature and chlorophyll a, inspection suggested that there were important

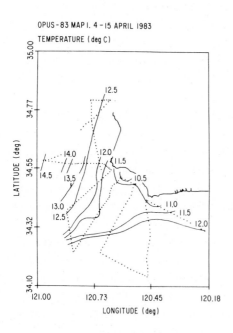

OPUS-83 MAP I. 4-15 APRIL 1983
TEMPERATURE (deg C)

Fig. 1A. Surface distribution of
temperature for map 4 (moderate)
upwelling) sampled on 14-15
April 1983. Maps of temperature
were constructed from 5-minute
averages of the data.

correlations between surface temperature and organism distributions. For example, during moderate upwelling (April 14-15) most of the copepod nauplii in the survey area were found on the southerly edge of what might be called a surface "streamer" of cool water extending offshore from between Points Arguello and Conception (Figs. 1A and 1B). Chlorophyll a was similarly highest along this southerly edge (Fig. 1C). During strong upwelling, abundances of copepod nauplii were greatly reduced, but highest concentrations were again in an area where isotherms were relatively closely spaced (Figs. 2A and 2B). Chlorophyll a was uniformly low (< 1 mg·m^{-3}) in concentration throughout the study area (Fig. 2C). In the downwelling period, surface temperature was uninformative (Fig. 3A), even though a distinct area of high abundance of copepod nauplii coincided with an area of elevated surface chlorophyll a concentration (Figs. 3B and 3C). Highest abundances of copepod nauplii were never adjacent to the coast, and were farthest removed from the coast during downwelling when they also achieved their maxima (Figs. 1B, 2B, and 3B). Maps of other taxa such as

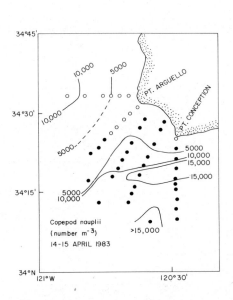

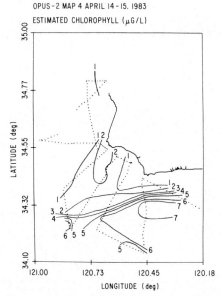

Fig. 1B. Surface distribution of copepod nauplii for map 4. Open circles indicate daytime samples; closed circles indicate night.

Fig. 1C. Surface distribution of chlorophyll a for map 4. Maps of chlorophyll a were constructed from 5-minute averages of the data.

euphausiid nauplii, appendicularians, total <u>Paracalanus</u> <u>parvus</u> and <u>Oithona</u> spp.
did not show such dramatic contrasts in abundance between upwelling and downwelling periods.

Because in both moderate and strong upwelling periods the copepod nauplii
seemed to be aggregated in areas that might be characterized as frontal zones
associated with the upwelling feature, we must define the "front" for each map
before analyzing its impact on the distributions of grazing organisms and the
food chain dynamics of the Pt. Conception area. To do this, we plotted surface
chlorophyll <u>a</u> and nitrate against temperature and chose the 1°C interval in
which chlorophyll <u>a</u> concentrations achieved their maximum value and nitrate
approached zero. We call this 1°C interval the frontal zone, and it corresponded to the area between the 12° and 13° isotherms in the moderate upwelling
and downwelling maps and the area between the 13° and 14° isotherms in the map

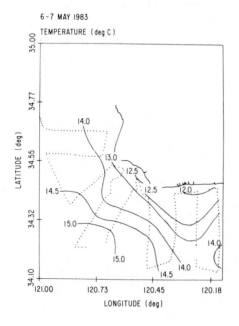

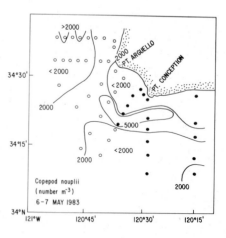

Fig. 2A. Surface distribution of
temperature for map 10 (strong
upwelling), sampled on 6-7 May 1983.

Fig. 2B. Surface distribution of
copepod nauplii for map 10.

made during strong upwelling (Figs. 4A-C). Such simplifications are necessary
in order to analyze events in this intensely heterogeneous area and attempt to
delineate effects associated with local upwelling from effects created by larger
scale advection (California Current) and oceanic influences.

The simplest question to ask is whether the total number of taxa and total
number of individuals varies in any identifiable way with increasing strength of
upwelling, and what the relationship is between frontal features and the surrounding waters. The fewest taxa and fewest individuals were found during
strongest upwelling (May 6-7) while the largest number of taxa and highest
animal abundances were observed during downwelling (April 19-20; Table 1). In
all three maps the number of taxa found in the frontal regions was similar to
the number found outside, but the number of individuals was always greatest
within the frontal zones (Table 1). The decreasing number of individuals with

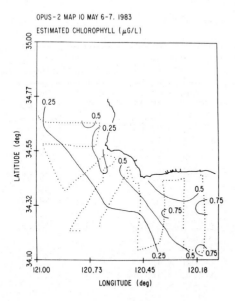

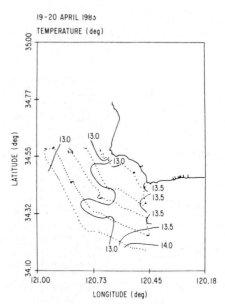

Fig. 2C. Surface distribution of
estimated chlorophyll a for map 10.

Fig. 3A. Surface distribution of temperature for map 5 (downwelling),
sampled on 19-20 April 1983.

TABLE 1

Mean characteristics of three maps made during varying upwelling conditions.

Characteristic	Downwelling		Moderate upwelling		Strong upwelling	
	In front	Out of front	In front	Out of front	In front	Out of front
Number of taxa	22(35)[+]	22(20)	21(21)	21(22)	13(14)	13(18)
Number of individuals·m^{-3}	20,034	17,514	12,135	9,654	5,421	3,474
Percentage copepod nauplii	87.0	81.4	77.6	63.2	70.4	45.7
Percentage C. pacificus	1.3	1.9	1.9	2.3	3.9	4.4
Percentage P. parvus	1.7	3.2	5.4	11.0	13.6	26.2
Percentage Oithona spp.	1.5	2.6	5.3	8.7	5.6	15.4
Ingestion by copepod nauplii*	6.2	5.3	3.8	2.1	1.5	0.5
Ingestion by C. pacificus*	1.9	0.7	3.8	1.6	0.9	0.7

[+](N)

*mg C·m^{-3}·day^{-1}

TABLE 2

Coefficients of variation of samples collected at night.

Taxon	Downwelling	Moderate upwelling	Strong upwelling
Copepod nauplii	In > Out	Out > In	In > Out
	17.63 > 15.85	47.26 > 40.42	97.41 > 51.45
Euphausiid nauplii	Out > In	Out > In	*
	162.16 > 125.66	151.52 > 97.03	*
Appendicularia	In > Out	Out > In	Out > In
	57.93 > 45.96	139.59 > 56.95	151.28 > 144.27
Oithona spp.	Out > In	Out > In	Out > In
	73.26 > 49.53	55.82 > 1.75	77.07 > 46.16
Paracalanus parvus	In > Out	Out > In	In > Out
	56.59 > 25.17	65.73 > 57.74	83.23 > 68.15
Calanus pacificus	Out > In	Out > In	In > Out
	228.85 > 81.21	172.78 > 146.08	170.74 > 134.52
Calanus pacificus C2	Out > In	Out > In	*
	170.78 > 159.85	374.17 > 16.66	*
C3	Out > In	Out > In	*
	235.74 > 97.24	176.03 > 164.95	*
C4	Out > In	Out > In	*
	264.58 > 169.39	343.00 > 147.37	*
C5	Out > In	Out > In	Out > In
	264.58 > 200.25	282.24 > 196.60	179.01 > 173.19

*None captured.

increasing strength of upwelling suggests that when surface waters are replaced
by deeper water during upwelling, the deeper water is relatively impoverished in
zooplankton. Surface fauna are presumably moved offshore and diluted by mixing
with the deep water rising during strong upwelling.

Further evidence of the surfacing of deep water nearshore, primarily during
strong upwelling events, can be seen in maps of copepodid stage V of Calanus
pacificus (Figs. 5A-C). It was found at the surface nearshore only during
moderate and strong upwelling. During downwelling it was found in greatest
abundance offshore in clusters suggesting origination north of Pt. Arguello and
in the Santa Barbara Channel (Figs. 5A-C). Copepodid stage V is the stage in
which diapause at depth is experienced as part of the life cycle of this copepod
(Longhurst, 1967; Alldredge et al., 1984). Presumably the copepodid stage V

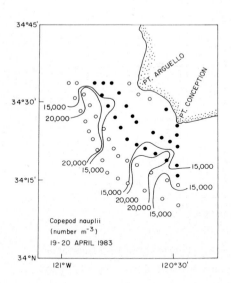

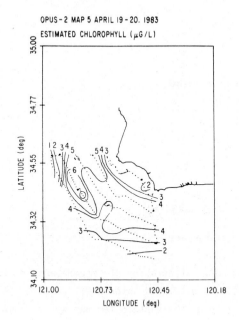

Fig. 3B. Surface distribution of
copepod nauplii for map 5.

Fig. 3C. Surface distribution of estimated chlorophyll a for map 5.

204

surfaces, molts, and reproduces during the upwelling season in this area since the CalCOFI atlas (Fleminger, 1964) shows no other taxon to be as abundant as C. pacificus in the Pt. Conception area, and the maxima in abundances of C. pacificus were noted in April and July. Strong upwelling moves surface plankton out of the Pt. Conception area and introduces subsurface organisms such as the stage CV of Calanus pacificus. The percentage of total numbers comprised by Calanus pacificus increases as strength of upwelling increases (Table 1), a further indication of its association with strong advection and surfacing of deep water.

During downwelling the importance of biological, rather than physical, effects becomes obvious. Numbers of individuals rise dramatically and the

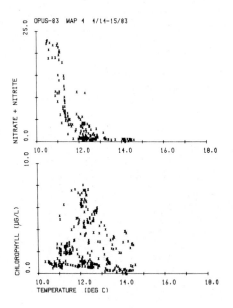 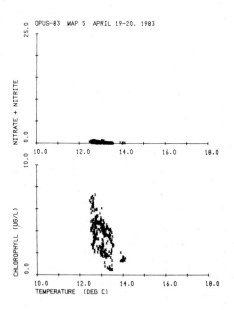

Fig. 4A. Correlation of temperature (°C) and estimated chlorophyll a and nitrate at the surface for map 4 (moderate upwelling) sampled on 14-15 April 1983.

Fig. 4B. Correlation of temperature (°C) and estimated chlorophyll a and nitrate at the surface for map 5 (downwelling) sampled on 19-20 April 1983. Note the very low concentrations of nitrate and relatively high concentrations of estimated chlorophyll a.

percentage of copepod nauplii reaches 87% of total numbers in the frontal zones (Table 1). In all three maps, copepod nauplii were more abundant in the frontal zones than in surrounding water (Table 1) indicating physical effects acting to aggregate these buoyant organisms. The aggregation of nauplii in the frontal zones suggests that these areas may be important in overall recruitment of copepods in this upwelling area. Total ingestion by copepod nauplii, calculated by assuming each nauplius ingested 0.4 µg C·day[-1] as has been shown for C. pacificus (Paffenhöfer, 1971; Mullin and Brooks, 1970), was always higher in the frontal zones than outside (recall the frontal zone is partially defined as the area where chlorophyll a concentrations achieve their maximum values) and was highest during downwelling and lowest during strong upwelling, following the abundance trends already noted (Table 1). Similarly, ingestion by Calanus pacificus, assuming ingestion by each copepodid stage was equivalent to that reported by Mullin and Brooks (1970), was always higher in the frontal zones than outside (Table 1).

The general characteristics show four interesting trends.

1. As strength of upwelling increased, number of taxa decreased, and there was no difference between frontal and nonfrontal zones.

2. Number of individuals decreased with increasing upwelling strength, and frontal zones always had higher abundances than nonfrontal zones.

3. Although total numbers decreased as upwelling strength increased, the percentage of total numbers comprised by copepodids and adults of three copepod taxa (Calanus pacificus, Paracalanus parvus, and Oithona spp.) increased as upwelling strength increased.

4. Abundance of copepod nauplii increased dramatically during downwelling, constituting one of the strongest contrasts observed.

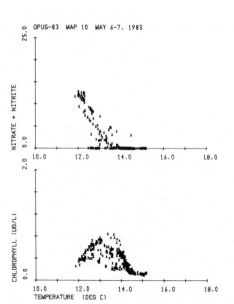

OPUS-83 MAP 10 MAY 6-7, 1983

Fig. 4C. Correlation of temperature (°C) and estimated chlorophyll a and nitrate at the surface for map 10 (strong upwelling) sampled on 6-7 May 1983.

The coefficient of variation in abundances (standard deviation as a percentage of the mean) is a simple way of assessing variability in a set of samples, and in the case of the maps is useful as a criterion for comparing the different upwelling situations and also in assessing diel variability within a given map. Of the samples collected at night, the trend was for variation to be less within frontal zones than outside (20 out of 26 cases; Table 2). Abundance of Calanus pacificus was less variable within frontal zones compared with outside in all cases (Table 2).

One would expect higher variability at the surface at night than during the day owing to the various patterns of diel vertical migration possible for different taxa. With the exception of Oithona spp., C. pacificus C3, and P. parvus C4, this was true in the frontal zones during the downwelling episode (Table 3). During moderate (and strong) upwelling, however, variation was always (generally) greater in the frontal zones in daytime, suggesting that advective effects predominated over biological effects (diel migration) in determining variability. For both a passive organism, such as a weakly swimming nauplius, and a more active organism, such as the ontogenetically migrating

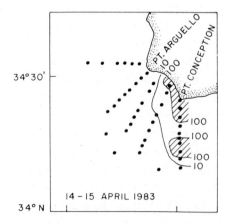

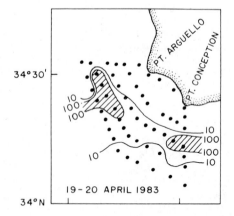

Fig. 5A. Abundance of copepodid stage V of Calanus pacificus at the surface during moderate upwelling.

Fig. 5B. Abundance of copepodid stage V of Calanus pacificus at the surface during downwelling. Note that this resting stage is found nearshore only during upwelling but is found offshore during downwelling.

TABLE 3

Coefficients of variation of samples collected in frontal areas.

Taxon		Downwelling[+]	Moderate upwelling	Strong upwelling
Copepod nauplii		N > D	D > N	D > N
		17.63 > 12.76	65.13 > 40.42	113.66 > 97.41
Euphausiid nauplii		N > D	D > N	*
		125.66 > 64.10	156.91 > 97.03	*
Appendicularia		N > D	D > N	N > D
		57.93 > 40.87	200.00 > 56.95	144.27 > 129.45
Oithona spp.		D > N	D > N	D > N
		91.30 > 49.53	91.21 > 1.75	114.38 > 46.16
Paracalanus parvus		N > D	D > N	N > D
		56.59 > 43.84	93.92 > 57.74	83.23 > 66.27
Calanus pacificus		D > N	D > N	N > D
		87.60 > 81.21	200.58 > 146.08	170.74 > 129.82
Calanus pacificus	C2	N > D	D > N	*
		159.85 > 109.03	200.01 > 16.66	*
	C3	D > N	*	*
		107.40 > 97.24	*	*
	C4	N > D	*	*
		169.39 > 109.84	*	*
	C5	N > D	*	D > N
		200.25 > 153.53	*	202.19 > 173.19
Paracalanus parvus		N > D	D > N	D > N
		131.49 > 105.16	160.93 > 92.76	115.38 > 16.29
	C4	D > N	D > N	D > N
		152.73 > 120.15	141.40 > 68.50	89.19 > 87.50
	C5	N > D	D > N	N > D
		75.51 > 65.02	71.13 > 52.91	120.04 > 100.31

[+]N denotes night, D denotes day.
*None captured.

TABLE 4

Coefficients of variation of samples collected at night.

Taxon	Upwelling condition	
	In Front	Out of Front
Copepod nauplii	SUP[+] > MUP > DWN	SUP > MUP > DWN
	97.41 > 40.42 > 17.63	51.15 > 47.26 > 15.85
Euphausiid nauplii	* DWN > MUP	SUP > DWN > MUP
	* 125.66 > 97.03	253.39 > 162.16 > 151.52
Appendicularia	SUP > DWN > MUP	SUP > MUP > DNW
	144.27 > 57.93 > 56.95	151.28 > 139.59 > 45.96
Oithona spp.	MUP > SUP > DWN	SUP > DWN > MUP
	49.53 > 46.16 > 1.75	77.07 > 73.26 > 55.82
Paracalanus parvus	SUP > MUP > DWN	SUP > MUP > DWN
	83.23 > 57.74 > 56.59	68.15 > 65.73 > 25.17
Calanus pacificus	SUP > MUP > DWN	DWN > MUP > SUP
	170.74 > 146.08 > 81.21	228.85 > 172.78 > 134.52

[+]SUP denotes strong upwelling, MUP denotes moderate upwelling, and DWN denotes
downwelling.
*None captured during strong upwelling.

Calanus pacificus, the least variability in frontal zones was observed during downwelling (Table 4). The most variability was often, but not always, observed in strong upwelling (9 out of 11 cases; Table 4). In general, increasing strength of upwelling was associated with increased variability in abundance of zooplankton captured at night, both in frontal zones and outside.

In daytime, variability in general was also least in the frontal zones during downwelling and greatest during moderate upwelling (Table 5). Areas outside the frontal zones showed greatest variation during strong upwelling (Tables 4 and 5), suggesting that overall in the Pt. Conception area, diel migration is not a very important contributor to variation in surface populations relative to the importance of the strength of advection associated with upwelling episodes.

Abundances within fronts and outside were compared statistically using log-transformed data. At night during moderate upwelling, copepod nauplii, euphausiid nauplii, and appendicularia were significantly more abundant in frontal zones than outside (t_{25} = 2.291; P $\leq$ 0.05, t_{25} = 4.115, P < 0.05, and t_{25} = 3.263; P $\leq$ 0.05). At night during strong upwelling, copepod nauplii were significantly more abundant in frontal zones than outside (t_{10} = 2.300; P $\leq$ 0.05). At night during downwelling, appendicularia and total Paracalanus parvus were significantly more abundant outside frontal zones than they were inside (t_{22} = 3.034; P $\leq$ 0.05; t_{21} = 4.171; P $\leq$ 0.05). Oithona spp. and total Calanus pacificus showed no statistically significant differences at night.

A final consideration is whether the rank ordering of taxa varies with respect to frontal and nonfrontal zones and according to strength of upwelling. Rank differences between frontal and nonfrontal zones showed greatest contrast in the downwelling situation and

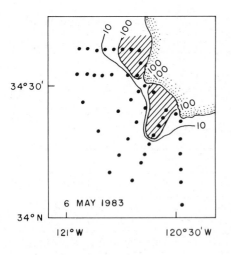

Fig. 5C. Abundance of copepodid Stage V of Calanus pacificus at the surface during strong upwelling.

TABLE 5

Coefficients of variation of samples collected in the day.

Taxon	Upwelling condition	
	In Front	Out Front
Copepod nauplii	SUP⁺ > MUP > DWN	SUP > MUP > DWN
	113.66 > 65.13 > 12.76	110.19 > 61.06 > 14.35
Euphausiid nauplii	* MUP > DWN	* MUP > DWN
	* 156.91 > 64.10	* 264.54 > 96.43
Appendicularia	MUP > SUP > DWN	MUP > SUP > DWN
	200.00 > 129.45 > 40.87	168.38 > 116.72 > 64.28
Oithona spp.	SUP > DWN > MUP	SUP > DWN > MUP
	114.38 > 91.30 > 91.21	54.29 > 47.22 > 39.51
Paracalanus parvus	MUP > SUP > DWN	SUP > DWN > MUP
	93.92 > 66.27 > 43.84	47.94 > 41.00 > 40.65
Calanus pacificus	MUP > SUP > DWN	DNW > MUP > SUP
	200.58 > 129.82 > 87.60	179.23 > 178.44 > 160.65

⁺SUP denotes strong upwelling, MUP denotes moderate upwelling, and DWN denotes downwelling.
*None captured during strong upwelling.

least contrast in the strong upwelling situation (Table 6). When ranks of taxa within the frontal zone are compared, greatest difference was between downwelling and moderate upwelling and the least difference was between moderate upwelling and strong upwelling (Table 6). In all situations copepod nauplii were ranked first, and within the frontal zones, copepod nauplii, Oithona spp. and appendicularia comprised the top three taxa regardless of upwelling strength (Table 6). During strong upwelling, no Calocalanus spp. were captured at all; euphausiid nauplii were not captured within the fronts; and no Mecynocera clausi were captured outside the fronts (Table 6). Except for copepodid stage I, no Calanus pacificus were found outside the frontal zones during strong upwelling (Table 6). The greatest differences in rank order in fronts between moderate upwelling and strong upwelling were in Oncaea spp. (higher in moderate upwelling) and Calanus pacificus stage CI (higher in strong upwelling); between downwelling and strong upwelling greatest differences were in Paracalanus parvus stages CII/III and CIV (higher in strong upwelling). The average rank of Calanus pacificus, when all copepodid stages were present, was highest in strong upwelling and lowest in moderate upwelling.

DISCUSSION

The nearshore zone influenced by coastal upwelling generally has been ignored in previous analyses of zooplankton in the California Current and coastal areas off southern California. In so doing, earlier studies identified offshore locations of increased biomass and found reasonable mechanisms to explain the

210

TABLE 6
Rank order of zooplankton taxa in frontal regions and outside during three different upwelling regimes.

Rank	DOWNWELLING In	DOWNWELLING Out	MODERATE UPWELLING In	MODERATE UPWELLING Out	STRONG UPWELLING In	STRONG UPWELLING Out
1	copepod nauplii	copepod nauplii	copepod nauplii	copepod nauplii	copepod nauplii	copepod nauplii
2	Appendicularia	Appendicularia	Oithona spp.	Oithona spp.	Oithona spp.	Oithona spp.
3	Oithona spp.	Oithona spp. ‡‡	Appendicularia*‡	Oncaea spp.	Appendicularia*	P. parvus C4
4	euphausiid nauplii	P. parvus ‡‡	Oncaea spp.	P. parvus C2+3	P. parvus C5	euphausiid nauplii
5	euphausiid calyptopes	P. parvus C5	P. parvus C2+3	P. parvus C5	P. parvus C4	P. parvus C5
6	P. parvus ‡‡	euphausiid nauplii	P. parvus C5	P. parvus ‡‡	P. parvus C2+3	P. parvus C2+3
7	P. parvus C5	Oncaea spp.	euphausiid nauplii*‡	euphausiid calyptopes	C. pacificus C1	Oncaea spp.
8	Oncaea spp.	C. pacificus C4	euphausiid calyptopes	P. parvus C4	P. parvus ‡‡	C. pacificus C1
9	Clausocalanus ‡‡*	euphausiid calyptopes	P. parvus C4	C. pacificus C5	C. pacificus ‡‡	Appendicularia
10	C. pacificus C1*	P. parvus C2+3	C. pacificus C5	Appendicularia	C. pacificus C2	Corycaeus spp.
11	Corycaeus	C. pacificus C5	P. parvus ‡‡	Mecynocera clausi	Corycaeus	P. parvus ‡‡
12	Mecynocera clausi	P. parvus C4	C. pacificus C5	Clausocalanus ‡‡	Mecynocera clausi	Acartia spp. ‡‡
13	C. pacificus C4	Acartia spp.	C. pacificus C4	Corycaeus	euphausiid calyptopes	Clausocalanus ‡‡
14	C. pacificus *+C2	C. pacificus C3	Calocalanus *	C. pacificus C4	C. pacificus	euphausiid calyptopes
15	C. pacificus C3	Mecynocera clausi	Corycaeus	euphausiid nauplii	C. pacificus	Chaetognatha
16	P. parvus *+C4	Corycaeus	Mecynocera clausi	C. pacificus C1	Oncaea spp.	
17	C. pacificus* C5	Calocalanus	Chaetognatha ‡‡	Chaetognatha	Clausocalanus ‡‡	
18	Acartia spp.	C. pacificus C1	C. pacificus ‡‡	Acartia spp.	C. pacificus C3	
19	P. parvus *+C4	Clausocalanus ‡‡	Acartia spp.	C. pacificus ‡‡	Chaetognatha	
20	Chaetognatha	C. pacificus C2	Clausocalanus ‡‡*‡	Calocalanus		
21	Calocalanus spp.	C. pacificus	C. pacificus C3	C. pacificus C3		
22	C. pacificus ‡‡	Chaetognatha	C. pacificus C1 / C. pacificus C2	C. pacificus C2		

*Largest differences in rank in frontal zone versus out of frontal zone (>6 units).
†Difference significant at P < 0.05.

TABLE 7

Occurrence of copepodid stage V (CV) of Calanus pacificus within ten kilometers
of the coast.

Map Number	Date	Upwelling State	Number of Samples	Percentage Containing CV's
1	April 6	Non-upwelling	9	0
2	April 9	Non-upwelling	15	20
3	April 11	Moderate upwelling	6	33
4	April 14-15	Moderate upwelling	12	42
5	April 19-20	Downwelling	8	0
7	April 24-25	Non-upwelling	20	55
10	May 6-7	Strong upwelling	12	92

observations (Bernal and McGowan, 1981; Chelton, 1982). We have identified a
nearshore source that may also contribute to the biomass peak observed 100-200
km from shore and that has considerable small-scale spatial variation. All of
our observations suggest a conceptual model in which frontal zones are sites of
highest secondary production and in which the upwelling process actively introduces a late subadult herbivore into the surface layer. Any ideas about how the
nearshore pelagic ecosystem functions are somewhat difficult to test because of
the three-dimensional distribution and seasonal appearance of Calanus pacificus
(and perhaps other herbivores as well) combined with a complicated near-surface
circulation (Davis and Regier, 1984; Brink, 1983).

The importance of the upwelling process combined with ontogenetic upward
migration of copepodid stage V of Calanus pacificus is obvious in the seven maps
that have been analyzed (Table 7). When the state of upwelling is assigned an
arbitrary number, one for downwelling through four for strong upwelling, the
correlation coefficient between upwelling strength and the percentage of nearshore samples containing copepodid stage V is $r^2 = 0.844$ (n=7). During strong
upwelling, 92% of samples within 10 km of shore contained stage CV, while during
downwelling none did (Table 7). Although a seasonal effect cannot be completely
ruled out as a possible contributor to this striking contrast (more animals were
migrating upward at the later date), the variation in upwelling strength and the
persistence of the trend over the five weeks of the study suggest that upwelling
does play a major role in bringing these animals to the surface nearshore.
Deeper water either in the Santa Barbara Channel or north of Pt. Conception can
be thought of as containing a seed population of stage CV Calanus pacificus, in
diapause perhaps from the previous upwelling season. A combination of upward
migration and upwelling processes introduce these CV's into the surface layer
where they molt and become reproductively active. Nauplii resulting from the
reproduction are aggregated in frontal zones where concentrations of food are

elevated also. The frontal zones associated with upwelling thereby become sites of high secondary production compared with surrounding waters.

Balances in population dynamics cannot be achieved locally in this upwelling area because of variability in sources of water for upwelling (Atkinson et al., 1986) and in the trajectories of water at the surface (Davis and Regier, 1984). During the strong upwelling event discussed here, surface current - following drifters covered the distance from Pt. Conception nearly to Santa Cruz Island (34°N, 120°W or approximately 106 km) in approximately 3 days. Thus, any copepodid stage V's surfacing during this event would have tended to be entrained in the direction of the Santa Barbara Channel.

During downwelling, on the other hand, net surface flow tended to be less (surface drifters tended to traverse small loops) in the Pt. Conception - Pt. Arguello area, variability was reduced and abundance of copepod nauplii increased dramatically. Frontal zones, however, were still the areas in which potential secondary production was highest. The life history of Calanus pacificus, its relationship to upwelling circulation and the potentially dominant role of frontal zones in focusing secondary (and primary) production, all require considerable additional exploration. It seems that Calanus pacificus, which dominates the crustacean plankton during the upwelling season, is "injected" into the surface layer during upwelling where its secondary production may be achieved primarily in frontal zones.

ACKNOWLEDGMENTS

We thank K. Devonald and the captain and crew of R.V. New Horizon for assistance at sea in collecting the zooplankton samples. Laboratory analyses were done by E.M. Schwarting, and her thorough and professional help is gratefully acknowledged. The field work was supported by N.S.F. grant OCE-82-15228 and the laboratory analyses by Department of Energy grant DE-AC02-76CH00016 and NSF grant OCE-85-07438. The manuscript was prepared with finanacial support from DOE grant DE-AC02-76CH00016 and N.S.F. grant OCE-8507438.

LITERATURE CITED

Alldredge, A.L., B.H. Robison, A. Fleminger, J.J. Torres, J.M. King, and W.M. Hammer, 1984. Direct sampling and in situ observation of a persistent copepod aggregation in the mesopelagic zone of the Santa Barbara basin. Mar. Biol., 80: 75-81.

Atkinson, L.P., K.H. Brink, R. Davis, B.H. Jones, T. Paluszkiewicz, and D. Stuart, 1986. Mesoscale variability in the vicinity of Points Conception and Arguello during April-May 1983: the OPUS 83 experiment. J. Geophys. Res., submitted.

Bernal, P.A., 1981. A review of the low-frequency response of the pelagic ecosystem in the California Current. Calif. Coop. Oceanic Fish. Invest. Rep., 22: 49-62.

Bernal, P.A. and J.A. McGowan, 1981. Advection and upwelling in the California Current. In: F.A. Richards (Editor), Coastal Upwelling. American Geophysical Union, Washington, D.C., pp. 381-399.

Binet, D. and E. Suisse de Sainte Claire, 1975. Le copepode planktonique
    Calanoides carinatus repartition et cycle biologique au large de la Cote
    d'Ivoire. Cahiers O.R.S.T.O.M., series Oceanographique, 13: 15-30.
Boyd, C.M. and S.L. Smith, 1983. Plankton, upwelling, and coastally trapped
    waves off Peru. Deep-Sea Res., 30: 723-742.
Brink, K.H., 1983. The near-surface dynamics of coastal upwelling. Prog.
    Oceanog., 12, 223-257.
Brink, K.H. and R.D. Muench, 1986. Circulation in the Point Conception-Santa
    Barbara channel region. J. Geophys. Res., in press.
Chelton, D.B., 1982. Large-scale response of the California Current to forcing
    by the wind stress curl. Calif. Coop. Oceanic Fish. Invest. Rep., 23:
    130-148.
Cox, J.L., L.R. Haury and J.J. Simpson, 1982. Spatial patterns of
    grazing-related parameters in California coastal surface waters, July 1979.
    J. Mar. Res., 40: 1127-1153.
Davis, R.E. and L. Regier, 1984. Current - following drifters in OPUS-83.
    Scripps Institution of Oceanography Ref. No. 84-12, 41 pp.
Enright, J.T. and H.-W. Honegger, 1977. Diurnal vertical migration: adaptive
    significance and timing. Part 2. Test of the model: details of timing.
    Limnol. Oceanogr., 22: 873-886.
Fiedler, P.C., 1984. Satellite observations of the 1982-1983 El Niño along the
    U.S. Pacific coast. Science, 224: 1251-1254.
Fleminger, A., 1964. Distributional atlas of calanoid copepods in the
    California Current region. Part I. Calif. Coop. Oceanic Fish. Invest.,
    Atlas 2, 313 pp.
Haury, L.R., 1976. A comparison of zooplankton patterns in the California
    Current and North Pacific Central Gyre. Mar. Biol., 37: 159-167.
Jones, B.H., K.H. Brink, D. Blasco and L.P. Atkinson, 1986. Asymmetric
    distributions of phytoplankton associated with a coastal upwelling center.
    Cont. Shelf Res., submitted.
Longhurst, A.R., 1967. Vertical distribution of zooplankton in relation to the
    eastern Pacific oxygen minimum. Deep-Sea Res., 14: 51-63.
Mullin, M.M. and E.R. Brooks, 1970. Production of the planktonic copepod
    Calanus helgolandicus. Bull. Scripps. Inst. Ocean., 17: 89-103.
Paffenhöfer, G.-A., 1971. Grazing and ingestion rates of nauplii, copepodids
    and adults of the marine planktonic copepod Calanus helgolandicus. Mar.
    Biol., 11: 286-298.
Paluszkiewicz, T., W. Chandler, A. Grindle, and L. Atkinson, 1984. OPUS:
    Preliminary Hydrographic Data Report. Skidaway Institute of Oceanography,
    Savannah. 218 pp.
Smith, S.L., 1982. The northwestern Indian Ocean during the monsoons of 1979:
    distribution, abundance and feeding of zooplankton. Deep-Sea Res., 29:
    1331-1353.
Star, J.L. and M.M. Mullin, 1981. Zooplanktonic assemblages in three areas of
    the North Pacific as revealed by continuous horizontal transects. Deep-Sea
    Res., 28A: 1303-1322.
Wickett, W.P., 1967. Ekman transport and zooplankton concentrations in the
    North Pacific Ocean. J. Fish. Res. Bd. Canada, 24: 581-594.

OBSERVATIONS OF FINESTRUCTURE FORMED IN A CONTINENTAL SHELF FRONT (SOUTHEASTERN
BERING SEA)

L.K. COACHMAN

ABSTRACT

Observations of finestructure formed in the mid-shelf front of a broad
continental shelf sea provide unique data for elucidating the formation and
subsequent behavior of distinct water parcels created by partial mixing of
shelf bottom water and basin water. The measurements include one week of
hourly vertical current profiles and grids of CTD stations. The partial
mixtures are identified by a cold temperature signal from the shelf bottom
water. They are extruded seaward from the mixing zone by pressure gradients
along isopycnals into the graded density structure of water columns to seaward.
If the mixtures have sufficient initial volume, they collapse into "blini"
(Russian pancakes) of order 10 to 20 m thickness and 3 to 5 km diameter, the
horizontal scale determined by the Rossby radius, and maintain a distinct
dynamical identity after leaving the frontal zone. Salt-finger activity is
associated with some of the initial mixtures but appears to be a transient
phenomenon. If the ocean environment to seaward has low energy (as on the
Bering shelf) the larger "blini" can have lifetimes of approximately one month.

INTRODUCTION

Continental shelf seas of mid- and higher latitudes contain semi-permanent
fronts (Fig. 1). The fronts are zones of enhanced horizontal property gradients
denoting the boundaries between water masses differing in T/S properties and/or
water column structure. Within these fronts, the water masses are interacting
"laterally", and a common product of their interaction is the formation of
finestructure features. Finestructure, correlated inversions in T/S on vertical
scales of 1 m to 100 m (Munk, 1981) due to an interleaving and layering of two
water masses with nearly the same densities, are common features of all oceanic
frontal zones (see, e.g., Roden, 1974; Gordon et al., 1977; Horne, 1978a;
Williams, 1981). The interleaving of two water masses juxtaposed laterally on
vertical scales significantly larger than microstructure scales (<1 m vertical
dimension) appears to be a predominant mode of their interaction (Joyce, 1977);
this preference for a dominant scale larger than microstructure has been
demonstrated in the elegant laboratory experiments of Turner (1978) and Ruddick
and Turner (1979).

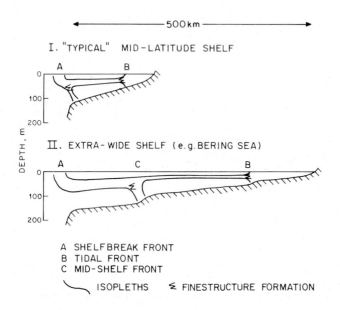

I. "TYPICAL" MID-LATITUDE SHELF

II. EXTRA-WIDE SHELF (e.g. BERING SEA)

A SHELFBREAK FRONT
B TIDAL FRONT
C MID-SHELF FRONT

ISOPLETHS   FINESTRUCTURE FORMATION

Fig. 1. Schematic cross-sections of a "typical" mid-latitude shelf (I) and one of the four World Ocean shelves 5 to 10 times broader (II), showing the semi-permanent fronts. The extra-wide shelves contain a third front in addition to the shelfbreak and inner tidal fronts at mid-shelf, which is an important site of finestructure formation.

The importance of the finestructure formed in the shelf fronts is that they modulate to an important extent the cross-shelf exchanges of properties (e.g., nutrients; Voorhis et al., 1976; Horne, 1978b; Houghton and Marra, 1983) and, where they are present, enhance the larger-scale vertical[1] fluxes of properties (Voorhis et al., 1976; Coachman and Walsh, 1981). The latter effect is due in part to a smaller-scale shearing of the property fields by the intrusive layering, and sometimes to the presence of "double-diffusive" processes (Turner, 1978) which may be associated with the upper and lower boundaries of the intrusive layers (Gregg, 1975; Gargett, 1976).

The finestructure are partial mixtures of the juxtaposed water masses which escape the formation area before complete mixing, thus retaining their identity as separate water parcels. These parcels are 3-dimensional, but the only dimension known with any certainty is the vertical; modern CTDs provide accurate resolution of T/S down to better than 1 m. What is their horizontal shape and extent? Are they ribbon or sheet-like? or rounder, more like

---

[1]No distinction is made in this paper between "vertical" and "cross-isopycnal" when considering quasi-vertical diffusion. Where this phenomenon is considered (e.g., section 5) isopycnals are quasi-horizontal and there is little distinction between them.

pancakes? Not only are the shapes of these features not known in detail, neither are important aspects of their behavior, such as relative motion, lifetime, and effect on fluxes. The problem is that the typical finestructure have small horizontal scales (order of a few kilometers) and are too variable spatially and temporally to be resolved by ordinary oceanographic sampling strategy.

The latest information providing details of these features for shelf sea fronts comes from the study of Houghton and Marra (1983). They occupied "microlines" of stations (1 km station spacing) across the shelfbreak front of the New York Bight (NYB), where the NYB "cold pool" water (Houghton, et al., 1982) extrudes as a nose-like feature ˜10 m to 40 m thick into continental slope water (cf. Voorhis, et al., 1976; Posmentier and Houghton, 1978, 1981; Welch, 1981). This is a zone of active finestructure formation (Fig. 1, I). Houghton and Marra reported 5 to 10 m scale finestructure to be prevalent in two regions, along the "nose" of the cold pool extrusion and on its underside. The features near the "nose" were not coherent over distances of 1 km, but this is a region of horizontal shear of the slope current (as suggested by the float measurements of Voorhis et al., 1976 and the current meter data referenced by Houghton and Marra (1983)). In contrast, the features under the cold extrusion had horizontal coherences of about 5 km. Likewise, similar finestructure associated with a slope water intrusion above the cold pool appeared to be coherent alongshore over 6-10 km and had cross-shelf dimensions of a few kilometers. These results suggest that not only is finestructure occurring on a hierarchy of vertical scales, the interleavings also exhibit a hierarchy of horizontal scales. This is implied in the models of intrusions of Toole and Georgi (1981), whose work identified a number of parameters important to setting vertical scales, such as effective vertical salt diffusivity and flux ratio of heat to salt, vertical density gradient and lateral salinity gradient, as well as the Coriolis acceleration for defining the along front scale.

Houghton and Marra (1983) also mapped the density ratio $R_\rho = \alpha \Delta T / \beta \Delta S$ [where $\alpha = -\rho^{-1}(\partial \rho / \partial T)$, $\beta = \rho^{-1}(\partial \rho / \partial S)$, and ΔT and ΔS are T and S differences over a vertical interval--they used 5-10 m], on their detailed "microline" to show that salt fingering, the areas where $1.0 < R_\rho < 1.5$, was active in the same areas where the 5-10 m finestructure was prevalent, viz. off the nose of the cold pool and on its underside. One finestructure intrusion under the cold pool had a vertical dimension of ˜4 m, on-off shelf dimension of ˜5 km, and a density ratio on the lower side of R_ρ ˜1.2-1.4, from which they estimated a lifetime of $T \lesssim 2$ days. This is the same lifetime as estimated for other similar features, e.g., 1-2 days for a ˜10 m thick layer with ΔT ˜ 2°C and ΔS ˜ 0.5 gm kg^{-1} by Posmentier and Houghton (1978), who noted further that without double-diffusive flux enhancement lifetimes would be considerably longer.

The extra-wide Bering Sea shelf contains a "cold pool" of water over the central shelf, for which the seaward boundary is the middle front (Fig. 1, II). This front is a prominent site of finestructure formation (Coachman and Charnell, 1979; Kinder and Schumacher, 1981); the finestructure are partial mixtures of cold pool water with warmer, more saline Bering Sea basin water moving landward as a bottom layer across the outer shelf. The only report providing any detail on this finestructure is Coachman and Charnell (1979). They analyzed a cross-shelf section of five stations at ~25 km intervals. Two adjacent stations (stas. 60, 61; Coachman and Charnell, 1979, Figs. 19-21) showed a complex hierarchy of finestructure very much like that of a station from the NYB (Posmentier and Houghton, 1978, Fig. 3) with nine distinctive layers of alternating warm, saline with cold, less saline; there was one large cold extrusion between 60 and 80 m depths with ΔT ~ 4°C and ΔS ~ 0.5 gm kg^{-1}, and the layers, 4 to 10 m thick with $0.2 < \Delta T < 0.6$°C and $0.2 < \Delta S < 0.5$ gm kg^{-1}, were distributed between 30 m and 90 m depths. Because the stations showed such precise similarity in layering, Coachman and Charnell concluded the features were sheet or ribbon-like with an on-offshore dimension >25 km. However, this may have been an unusual case or an incorrect interpretation; in many repeated sections of a standard cross-shelf section during the PROBES project with stations spaced at 25 km finestructure features could not be correlated between adjacent stations (Coachman, 1985); it thus seems likely that the usual cross-shelf scale for individual finestructure features is <25 km.

Two observations of Coachman and Charnell (1979) suggest that double-diffusive processes are associated with the finestructure. One was the downward progression of the T/S correlation at their stas. 60, 61 which showed a series of cyclonic whorls which, as analyzed by Posmentier and Houghton (1978), are characteristic of a more rapid vertical heat than salt flux. The other was that their layers definitely increased in density in a bulk sense, by about 0.2 σ_t, going off-shore. Where double-diffusion is active vertical fluxes are dominated by the salt-fingering phenomenon--thus the buoyancy flux is downward into cold, less saline intrusions which tend to gain density and sink as they move away from their formation area, and the opposite is true for warm, saline intrusions (Turner, 1978). In the Bering Sea case cold pool extrusions dominate the finestructure, and the observed increase in density is compatible with double-diffusion.

Thus there appears to be considerable similarity between the New York Bight and southeastern Bering Sea shelves in the nature of the lateral water mass interaction between cold pool shelf water and slope water. The gross interaction is an extrusion of cold pool water creating a large finestructure some tens of meters thick, colder and less saline than the slope water. Superimposed, and particularly along the upper and lower sides and at the nose of the larger finestructure, are many interleavings of 2 to 10 m vertical scale. With these intrusions there may be associated double-diffusive phenomena. The horizontal scales of these predominant finestructure are a few km, with life-times perhaps on the order of a few days.

There are differences between the regimes within which the finestructure disperses, as well as many unanswered questions about their dissipation and effect on horizontal and vertical fluxes. A major difference is the nature of the flow field. In the NYB the finestructure extrudes into a modest current; this undoubtedly is responsible for an along-shelf elongation of the features as suggested by Voorhis et al. (1976), but its role in their dissipation is unknown. In the Bering Sea the flow field is overwhelmingly tidal, with only slow (<5 cm/s) along- and across-shelf low-frequency currents (Schumacher and Kinder, 1983), which raises another question. A colder mid-depth layer (~50 to ~80 m) exists across the entire outer shelf >100 km from the middle front for which the source is the cold pool water; also, significant finestructure is frequently encountered in the outer part of the outer shelf (Coachman and Charnell, 1979; Kinder and Schumacher, 1981; Coachman, 1985), so the question is, if the lifetimes of the typical finestructure features is a few days, how do these features survive the transit of the outer shelf, a trip of >3 weeks at 5 cm/s?

In June-July 1982 a study was undertaken in the southeastern Bering Sea with the R/V Alpha Helix specifically to investigate the finestructure formed in the middle front. In addition to a Neil Brown CTD, the study employed a vertical-profiling current meter system (Cyclesonde: Van Leer et al., 1974). We encountered strong finestructure and watched its behavior for several days. This paper reports the results of the observations and analyses, which provide new data toward understanding these important features of oceanic mixing.

THE EXPERIMENT

The location of the experiment was chosen to coincide with the standard cross-shelf section established for the PROBES studies in 1978 (Goering and McRoy, 1981) and which has been occupied literally dozens of times since, providing considerable background on the general hydrography. Figure 2 shows the locations of stations 4-14 of the PROBES standard section. Also shown are the locations of two current meter moorings straddling the middle front (PR-1, PR-2) for which records are available from two different periods: one month in March-April 1980, analyzed by Coachman (1982), and $4\frac{1}{2}$ months Feb-June 1981, for which the records are as yet only partially analyzed (Coachman, 1985); thus there was also background on the flow field near the site of the finestructure formation. Moorings were in place for the experiment, but, unfortunately, the important mooring PR-1 was never recovered.

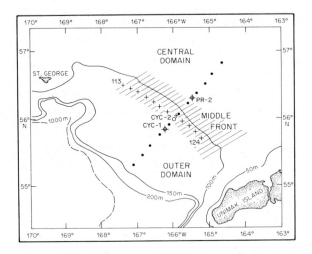

Fig. 2. Finestructure experiment over the outer part of the southeastern Bering Sea shelf. Stations 4-14 (solid circles) are the standard PROBES line; PR-1 and PR-2 current meter mooring locations previously occupied; CYC-1 and CYC-2 the Cyclesonde deployment locations; and stas. 113-124 the section along the 110 m isobath. Hachure indicates the middle front zone of finestructure formation.

The general scheme was to run the PROBES section to determine the location of the middle front and the T/S character of the basic interacting water masses in June, 1982, then deploy the Cyclesonde in a location with strong finestructure development. The first deployment, for $4\frac{1}{2}$ days, was close to PR-1 (CYC-1 in Fig. 2); it was then moved one station spacing onshore (CYC-2) where a full 7 days of 1 per hour vertical profiles of horizontal currents and temperature were obtained. For calibration purposes, CTDs were taken as close as we dared to the Cyclesonde timed to a vertical profile: 7 during CYC-1 and 19 during CYC-2. Also during the second deployment three grids of closely spaced (3-5 km) CTD stations surrounding CYC-2 were occupied to define the horizontal structure of the major finestructure. At the close of the experiment a section parallel to the face of the front along the 110 m isobath (stas. 113-124 in Fig. 2) was taken to provide data on the along-front nature of the finestructure intrusions.

The CTD was a Neil Brown, with accuracies better than 0.01 units in both T and S. Salinities and densities were calculated using the International Oceanographic Tables recently adopted by UNESCO. Because the focus of the work was on finestructure, the S and T data were averaged over 1-m intervals in the vertical.

Positions were determined using Loran-C, which provides excellent resolution in the southeastern Bering Sea.

The Cyclesonde was model PCM-1S from Marine Profiles, Inc., Miami, FL,
equipped with Savonius rotors (providing a redundancy in speed measurement),
a compass sensing orientation to within $\pm 3°$, a temperature probe (estimated
accuracy $\pm 0.01°C$), and a conductivity cell from which the data unfortunately
are useless. The instrument also records pressure and tilt from the horizontal.
We made a modification to the standard instrument: with the instrument running
continuously tape length limits the endurance to <2 days at the fastest sampling
rate, so a switch and timer were introduced which started tape transport at the
beginning of each profile, then stopped the tape at a selected time interval
later, e.g., 15 or 30 min. With hourly profiles of 15 min duration and the
fastest scan rate a full 7 days of measurements were achieved at CYC-2.

The Cyclesonde speed data were processed as follows. The mean vertical speed
of the instrument was about 10 cm s^{-1}, but varied a little between profiles
because it actually "flies" in the water and therefore its vertical speed
depends on horizontal current speed. The beginnings and ends of each profile
were estimated from the pressure and tilt data, and then the total number of
scans at 2-sec intervals allowed estimation of the mean vertical speed; this was
subtracted vectorically from the recorded speeds to give horizontal speeds.
The 2-sec sampling rate meant that only about 4 to 6 scans were obtained in
each vertical meter traversed, so the U and V values were smoothed with a 3-pt
running mean and averaged over 2-m intervals. Thus, the vertical shear data
are inadequate for examining microstructure scale phenomena (≤ 1 m) but suitable
for the finestructure which is the focus of the study (cf. Evans, 1982).

RESULTS

1. Large-scale interaction

Temperature and salinity sections for the initial occupation of the PROBES
line are shown in Fig. 3. Excellent conditions existed for the observation of
finestructure: the shelf water of T ~ 0°C and S ~ 31.7 gm kg^{-1} and basin water
of T ~ 3.8°C and S ~ 32.8 gm kg^{-1} provided temperature and salinity contrasts
of ΔT ~ 4°C and ΔS ~ 1 gm kg^{-1}.

The middle front lies between stas. 8 and 10 and finestructure formation is
clearly visible, especially in the T section. The layering of intrusions looks
very much like that in the NYB in summer: a warm, more saline layer aimed
onshore between ~35-55 m and a cold extrusion beneath, between ~55-80 m. There
is also a thinner cold layer ~5 m thick between the major warm layer and the
surface mixed layer. These features show nicely in the T/S correlations of the
stations (Fig. 4); from sta. 10 seaward to sta. 5 the curves each show two
temperature minima with the maximum from the warm, saline intrusion between.
These general features of the vertical water mass distributions are similar
across the outer shelf to beyond sta. 4, but in detail the adjacent stations do

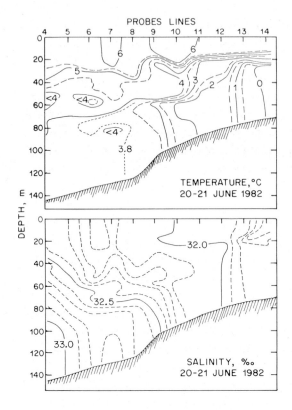

Fig. 3. Temperature (upper) and salinity (lower) sections along the PROBES
line at the beginning of the experiment. The middle front encompasses
stas. 9 and 10, within which can be seen two colder, less saline layers
extruding seaward with a warmer, more saline layer between. Note the
suggestion farther off-shore of isolated finestructure parcels.

not look very much alike. Looking also at the cross-sections (Fig. 3), the
finestructure does not appear to be sheets and layers of long lateral dimensions as was suggested from the section analyzed by Coachman and Charnell (1979),
but rather of numerous isolated features with horizontal dimensions less than
one station spacing (<25 km).

The T/S correlations (Fig. 4) show another result corresponding with that
reported by Coachman and Charnell (1979): the lower cold extrusion composed of
finestructure, which creates the temperature minimum in mid-water column across
the outer shelf to seaward of sta. 4, increases in density by >0.2 σ_t. Thus,
the finestructure features must on the average be losing buoyancy in their
>100 km transit of the outer shelf.

All the finestructure represents mixtures of three basic waters (cf. Fig. 3):
a surface water which covers the whole system down to 20-30 m depths, the cold,
deeper shelf water inshore, and the warm, saline basin water below ~80 m in the

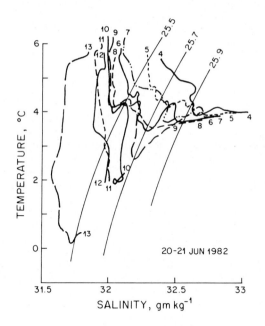

Fig. 4. T/S curves for stations 4-13 of the PROBES line, 20-21 June 1982.

outer part of the section. The three are arranged spatially such that only the
shelf bottom water mixes with both of the others--a quasi-lateral mixing with
basin water across the zone of the middle front, and vertical mixing with
surface water in the more landward part of the middle front. It is the very
cold (<0°C) signature of shelf bottom water that identifies the partial
mixtures of the finestructure. The shelf-surface water mixtures are a little
less dense and create the upper of the two cold finestructure extrusions, while
the lower, denser extrusions are products of the shelf-basin bottom waters
mixing "laterally" within the middle front. This hypothesis is depicted
schematically in the T/S plane in Fig. 5.

2. Cyclesonde observations

The Cyclesonde was deployed in two locations: CYC-1 was mid-way between
stas. 7 and 8 for $4\frac{1}{2}$ days, and CYC-2 mid-way between stas. 8 and 9 for 7 days
(Fig. 2). The instrument profiled every hour on the hour, at CYC-1 between
28 and 106 m and at CYC-2 between 28 and 85 m.

The vector-average current over the water column measured at CYC-1 is
shown in Fig. 6 in the form of a progressive vector diagram (PVD). We see the
dominant motion was a semi-diurnal tidal signal (with a large inequality)
superimposed on a quite steady set almost due north at 4.2 cm s^{-1}. As the
onshore direction is 043° (along the ordinate in Fig. 7), the flow was both

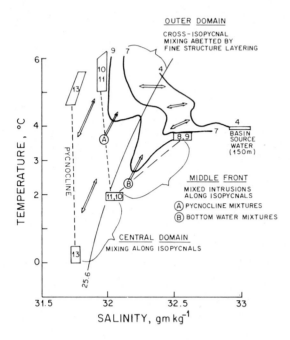

Fig. 5. Schematic diagram in the T/S plane of the mixing within the middle front which creates the two distinct cold, less saline finestructure: the shallower extrusions are shelf bottom water/surface layer mixtures and the deeper extrusions "lateral" mixings of shelf/basin water.

converging in the middle front and flowing northwest parallel with the bathymetry; this is a usual flow condition in the outer shelf domain (cf. Coachman, 1985).

The flow recorded at CYC-2 is shown in Fig. 7 in the same format. The same dominance of a tidal signal is evident, though it appears that the semi-diurnal component was somewhat stronger. Also, the low-frequency flow changed during the week: for the first three days it was directly onshore at ˜4 cm s^{-1} (the first day of record ˜5.8 cm s^{-1}), but then backed to the north, that is, turned more parallel to the bathymetry, and decelerated to <2 cm s^{-1}.

Temperature is the variable by which finestructure features can be positively identified, particularly the deeper cold extrusions that are mixtures of shelf bottom water (˜0°C) and basin water (˜3.8°C). Temperature profiles at CYC-1 showed some maxima and minima indicative of finestructure, particularly a tendency for a minimum between 50 and 65 m. But the greatest temperature contrast was <0.5°C, so the finestructure features were not obvious and amenable to more detailed analysis.

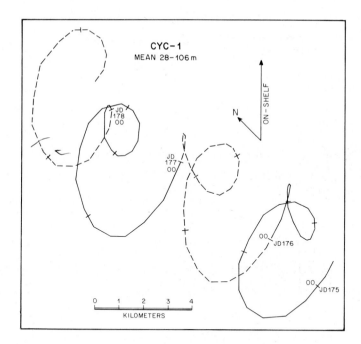

Fig. 6. The vector-mean flow over the measured water column (28-106 m) at CYC-1, for the period 23 June 2300 to 27 June 1600, presented as a PVD of the hourly values. Tic marks are at 6-hour intervals.

In contrast, the CYC-2 records contained observations of a very well developed finestructure, as can be seen in the time-history plot of the temperature field (Fig. 8). A cold core, 15 to 20 m thick centered at 50 to 55 m depth, with temperature colder by >1°C above or below appeared periodically at the station. It remained over the station typically 3 to 6 hours, then abruptly vanished only to reappear ~6 hours later, a timing strongly suggestive of tidal activity. The horizontal edge was observed to be quite sharp--it would on occasion completely appear or disappear between two profiles, which indicates a horizontal temperature gradient like that observed along the edge of the NYB cold extrusion, >1.5°C/km. For the first three days of record the core temperatures of the feature were cold, <2.5°C, but toward the end of the fourth day (JD 182) the cold signal became progressively less intense, and on the last two days the feature had nearly vanished from the Cyclesonde location, with only a weak indication of cold toward the ends of JD 184 and JD 185.

These observations can be explained in the following way. The strongly developed cold layer is a partial mixture of bottom shelf water and basin water formed in the middle front just landward of CYC-2, and which has extruded seaward in the form of a finestructure. The feature has a discrete lateral shape

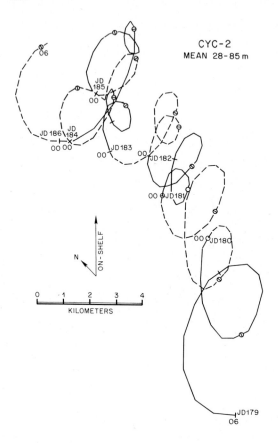

CYC-2
MEAN 28-85 m

Fig. 7. As in Fig. 6, but for CYC-2, 28 June 0600 to 5 July 0600.

with a fairly sharp boundary, at least along part of its perimeter, and is being advected about by the currents. As the flow field is predominantly oscillatory at tidal periods, the finestructure is being advected back and forth across the Cyclesonde location, reappearing at approximately 12-hour intervals. The finestructure has a very cold core, but is somewhat warmer towards its perimeter; sometimes only an edge of the feature reaches the CYC-2 location, particularly toward the end of the record period when the net advection toward the north (Fig. 7) has moved the core of the finestructure "out of reach" of CYC-2.

We now focus our attention on this distinctive feature of lateral water mass interaction, for which there are considerable water mass and horizontal flow field data.

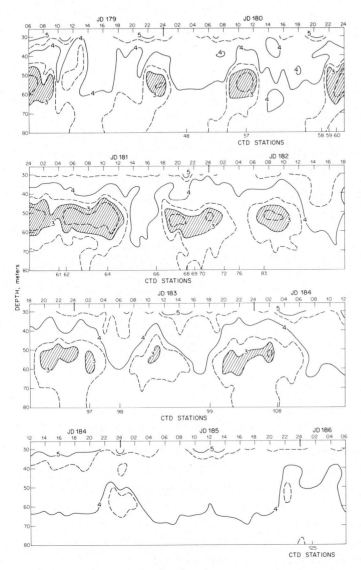

Fig. 8. Time-history of temperature (C) recorded by the Cyclesonde at CYC-2. CTD stations taken at CYC-2 are indicated beneath; T's <3°C are hachured.

3. The finestructure

3.1. Spatial reconstruction

The first objective is to map the size and shape of the finestructure. During the course of the Cyclesonde deployment 19 CTD stations were made very close to CYC-2, timed to a profile time, with the object of sampling the same water mass with the higher accuracy instrument: station times are indicated in Fig. 8. The T/S correlations for six stations which sampled five successive

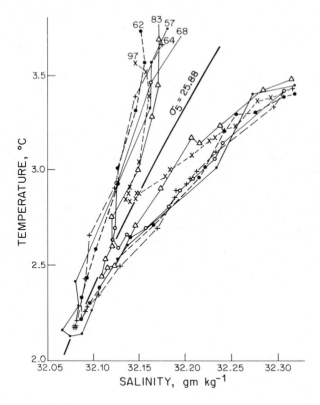

Fig. 9. T/S correlations for stations taken through five successive
re-appearances of the cold core at CYC-2 (cf. Fig. 8). Sigma-5 represents the
in situ density at p = 55 db, and the successive re-appearances are of the same
feature with the temperature minimum on the σ = 25.88 surface.

re-appearances of the cold core (cf. Fig. 8) are plotted in Fig. 9. We see

that the "core" lies precisely along the same density curve [in situ densities

are used in this analysis; thus, σ_5 represents $\rho_{S,T,P}$ at a pressure of 55 db,

the approximate mean depth of the core]. The conclusion is that the successive

re-appearances at CYC-2 are of the same finestructure feature--it is improbable

that separate, distinct partial mixings of bottom shelf and basin water would

produce features with precisely the same core densities.

The next step is to attempt to describe the spatial extent of the fine-

structure. At three times during the deployment week box-like grids of CTD

stations were occupied as rapidly as possible around CYC-2 (Table 1).

The temperature in the density band 25.87 < σ < 25.89 gives a distinctive

description of the feature. As CHASE provides the densest areal coverage,

these stations were dead-reckoned to JD 182, 0200, using the mean values of

current between 50 and 56 m depths observed by the Cyclesonde. The flow field

TABLE 1

Box experiments

Name	Date/Time	Stas	Nominal Spacing (n miles)
Box I	JD 180 0300-1100	48-57	4.5
CHASE	JD 182 0200-1230	72-95	3
Box II	JD 183-1800 JD 184-0630	99-111	5

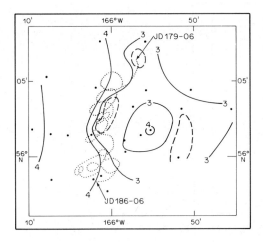

Fig. 10. Temperature on the σ = 25.88 surface, the density of the cold core of finestructure, constructed from the CTD stations of CHASE (dots) dead-reckoned to JD 182/0200. Hourly locations of water parcels determined from a PVD of the mean 50-56 m currents measured at CYC-2 were also dead-reckoned to JD 182/0200, creating the fine dotted line (cf. Fig. 11), and the 2.5, 3, and 4°C isotherms along this track were interpreted from Fig. 8. On this σ surface existed a cold feature (<3°C) ~20 x 20 km in extent which included 3 cold cores (<2.5°C) and one warm core (>4°C).

was not strong during CHASE (cf. Fig. 7) so maximum station displacement was <1 km. The dead-reckoned station locations are shown in Fig. 10.

Additional spatial definition of part of the T field was obtained as follows. The track of water motion indicated by the PVD diagram of mean 50-56 m currents was dead-reckoned to JD 182/0200--this reconstructed path is shown in Fig. 10 by the fine-dotted line. Then from the time-history of temperature at CYC-2 (Fig. 8) approximate times of appearance of specific temperature values (e.g.,

4°C, 3°C, 2.5°C, coldest T) were estimated and plotted along the reconstructed PVD track.

The reconstructed temperature field on σ = 25.88 is shown in Fig. 10. Unfortunately, the plan view of the finestructure features is not very precise; I was misled in the experimental design by my previous experience (viz. Coachman and Charnell, 1979) which had suggested horizontal coherence of the features of at least ~10 km, while the actual horizontal scale turns out to be ≦5 km. But one finestructure core is quite well defined when the extrapolated Cyclesonde data are included. A more detailed reconstruction is shown in Fig. 11. It was an ~3 km ovoid of T < 2.5°C centered, at JD 182/0200, just

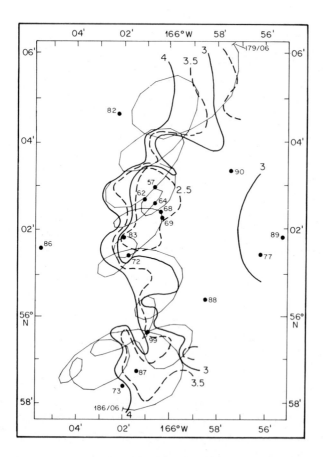

Fig. 11. Detailed reconstruction of the minimum temperature between 50 and 56 m along the Cyclesonde track. The straight segments creating the light line are hourly 50-56 m flow displacements (cf. Fig. 7) dead-reckoned to JD 182/0200 (sta. 72); numbered dots are dead-reckoned CTD stations. The isotherms near the Cyclesonde track were interpreted from the time-history of T at CYC-2 (Fig. 8). The cold finestructure seen repeatedly by the Cyclesonde was an ~3 km ovoid.

north of CYC-2. It had a very sharp horizontal temperature gradient along its western edge, >1.5°C km^{-1}, which accounts for the extremely abrupt appearances and disappearances of very cold temperature at CYC-2 (Fig. 8). Hereafter I refer to this well-defined cold core of the finestructure as CC.

We note that not all the appearances of <2.5°C temperature at the Cyclesonde were due to CC. The initial appearance, at JD 179/0600 (Fig. 8), was not connected with CC but a separate cold core, which by JD 182/0200 was located 10 km north of CYC-2. Likewise, the cold appearance at JD 184/0200 was a separate feature which, at JD 182/0200, lay 5 km SSE of CYC-2 (cf. Fig. 11). These two cold cores were never adequately sampled by either the Cyclesonde nor CTD stations, so only their presence within the larger cold finestructure field can be noted. Thus, the CTD stations which sampled CC were 57, 62, 64, 68, 83. [In Fig. 8 sta. 60 is shown as coinciding with the appearance of CC at CYC-2. However, its T profile did not match that of the Cyclesonde; it was apparently taken too far away from CYC-2 to be sampling the same water mass.]

To obtain a broader view of the cold finestructure field we can use all CTD stations taken within the general vicinity of CYC-2, including those of BOX I and BOX II, but to do so requires the assumption that time changes are very small compared to the spatial variability. This is the classical oceanographic sampling dilemma. However, the estimates obtained on vertical diffusion (see below), the most probable mechanism for "eroding" the temperature minimum, suggest only very slow warming and a possible long lifetime for these larger isolated water parcels. This is consistent with the contemporary discovery in the World Ocean of numerous isolated "lens" of water retaining their anomalous characteristics for months and located thousands of kilometers from their sources (see, e.g., Lindstrom and Taft, 1985; Riser et al., 1985).

With the above caveat in mind, Fig. 12 shows the temperature field on $\sigma = 25.88$ constructed from all stations and the Cyclesonde track (not shown) dead-reckoned to JD 182/0200. This is a less accurate presentation because of possible horizontal shear in the flow field, and maximum station displacement was 10 km; nevertheless, the result provides further insight into the spatial characteristics of the finestructure. The horizontal field appears to be composed as a hierarchy, with a large, cold feature, approximately 15 x 25 km, encompassing four much smaller and colder cores of order 3 to 5 km extent, and, in this case, also including a warm core of approximately the same dimension.

In Fig. 12, the small core including stas. 54, 94 is pictured as being distinct from the core measured at sta. 105. Fig. 13 shows the T/S diagram of these stations. Station 105 shows one very "active" cold layer, with a density inversion, centered on $\sigma = 25.88$, while the core ˜8 km farther north represented by stas. 54, 94 shows a two-layered cold structure, an indication

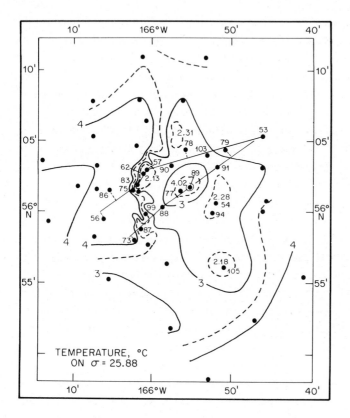

Fig. 12. Temperature on the σ = 25.88 surface constructed from all stations (BOX I,II, CHASE) and Cyclesonde data dead-reckoned to JD 182/0200. Some CTD stations are numbered, the sections of Figs. 14, 15 indicated, and T values given for four cold and one warm core. In horizontal space the finestructure shows a hierarchical structure, with a large (order 20 km) cold area encompassing four smaller (order 3-5 km) cold cores and one warm core.

of the cold on σ = 25.88 but a more strongly developed layer beneath on σ = 25.90. Data are inadequate for determining more exactly the relationship between these water masses, but they are definitely different cores.

In the vertical dimension the finestructure cores average between 10 and 20 m thickness, taking the 3°C isotherm as a boundary. This can be seen both from the measurements at CYC-2 (Fig. 8) and in the two sections constructed across the feature (Figs. 14, 15; see Fig. 12 for locations). In the coldest part of the cores it is a little thicker--maximum measured 3° isotherm separation was 19 m (JD 181/1000, Fig. 8). The edge of the feature tends to be blunt rather than tapered, and particularly so along its west and southwest side--when the cold core appeared at CYC-2 it was typically 8-10 m thick. When the orientation to the shelf topography is considered (see Fig. 12), there is the impression that the off-shore side of the finestructure tends to be blunter than the onshore side.

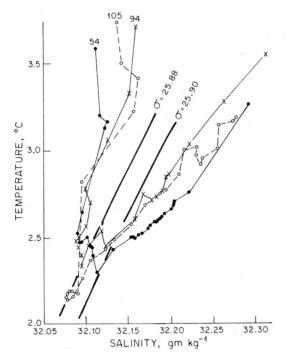

Fig. 13. T/S diagram of stations of two cold cores. The more southerly one at sta. 105 (cf. Fig. 12) had a single cold layer on σ = 25.88, while the core ~8 km farther north was double layered, on 25.88 and 25.90.

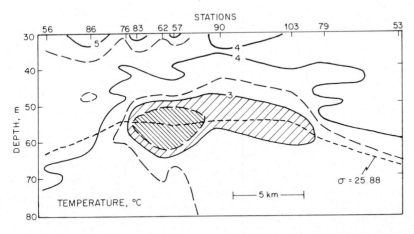

Fig. 14. Temperature section across the finestructure feature including CC (see Fig. 12). The density surfaces in the finestructure are elevated 10 m compared with those outside the feature.

234

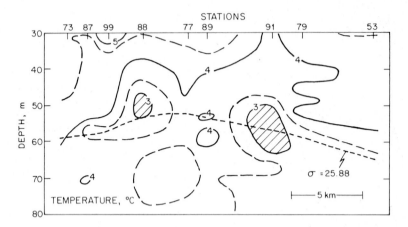

Fig. 15.   Temperature section across the finestructure (see Fig. 12), but including the warm core.

    The density surfaces in the general finestructure feature with its cores are elevated about 10 m above their depth in the water columns onshore and offshore of the feature.   This is shown in Figs. 14, 15 by the isopycnal $\sigma = 25.88$; in the plan view of the depth of the isopycnal (Fig. 16) we see the depth range

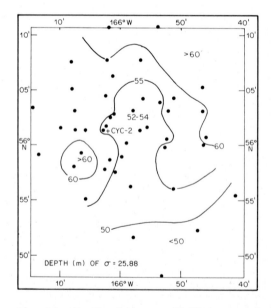

Fig. 16.   Depth of the $\sigma = 25.88$ isopycnal surface (station locations dead-reckoned as in Fig. 12).   Notice how the area of 52-55 m depths coincides with the general finestructure feature.   The even shallower band on the very south is a totally different water mass.

52-55 m essentially coincides with the feature, while on-shore and off-shore depths of this surface are ≥60 m. South and southeast of the feature at this time, i.e., in the along-shelf direction to the east, the σ = 25.88 surface lay even shallower, <50 m, with totally different water characteristics (T ˜ 4.5°C, S ˜ 32.4; cf. Figs. 9, 13). It is also notable that the warm core imbedded in the general finestructure feature also has elevated density surfaces (Fig. 15).

3.2 Field of motion

The finestructure, in addition to being swept around by the tidal currents, was being slowly advected northerly with the subtidal flow (Fig. 7). The dynamic topography of 55/110 db (Fig. 17) indicates a NNW set with a calculated

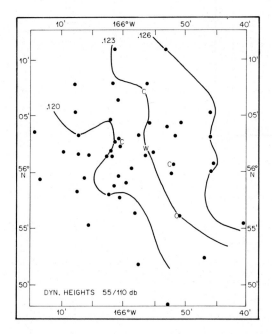

Fig. 17. Dynamic topography (dyn m) of 55/110 db, constructed from all stations dead-reckoned to JD 182/0200. C and W indicate locations of the cold and warm cores (cf. Fig. 12). The dynamic height difference across the general feature ΔD = 0.006 in 20 km is equivalent to 2.4 cm/s.

speed of 2.4 cm s^{-1}, in agreement with the subtidal set measured at CYC-2. Thus, the overall field seems to have been essentially in geostrophic balance, but the presence of the finestructure creates distortions in the baroclinic field which are not well resolved by the data.

We now examine more closely the motion associated with CC. Figure 18 shows vectors of the motion of the core layer of CC (50-56 m) relative to that in the

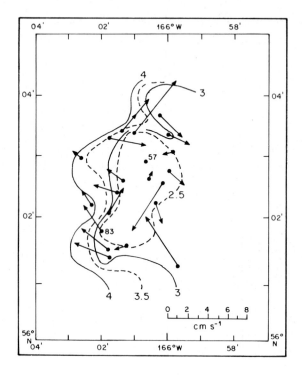

Fig. 18. Vectors of motion of the core layer (50-56 m) relative to that in the water column 10 m above for selected positions along the reconstructed Cyclesonde track, superimposed on the minimum temperature field. This depiction of relative motion suggests a clockwise rotation of CC at 2-3 cm/s with a 1-1$\frac{1}{2}$ km radius of curvature.

water column 10 m shallower for selected positions along the reconstructed Cyclesonde track. The vectors indicate a definite sense of clockwise rotation of the core relative to the water column, with a speed of ~2 cm s^{-1}. Geostrophic flow (over 100 db) was calculated for the section crossing the core (Fig. 19; cf. Figs. 12, 14). Geostrophy indicates a banded structure to the flow field, with a little stronger NW flow alternating with weak SE flow at the scale of the station spacing (~3 km). But right at the level of the cold core (50 to 60 m), indicated by the 2.5°C isotherm from the temperature section (Fig. 14), the flow between stas. 57 and 90 is ~2 to 3 cm s^{-1} less toward the NW than above or below, and between stas. 83 and 57 it is the same magnitude less to the SE. This is a clockwise motion of the core relative to the water column above and below of ~2 to 3 cm s^{-1}, in agreement with the direct measurements. The Rossby number (R_o = U/fr, where r = radius) for this relative motion is quite small, ~0.1. It thus appears that CC was rotating clockwise in quasi-geostrophic balance relative to the larger-scale flow field.

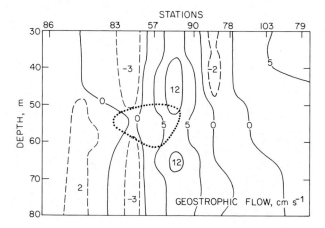

Fig. 19. Geostrophic flow (over 100 db) through the section crossing the finestructure including CC (see Figs. 12, 14). Speeds in cm/s, plus toward the
northwest. The dotted line is the 2.5°C isotherm from the temperature section
(Fig. 14).

Shear in the water column in relation to the cold core features was examined
in more detail. From the Cyclesonde data, temperature was used to define cores
as the layer encompassing water with T < 0.3°C above the minimum observed value.
The shear was determined as $S = (U^2 + V^2)^{1/2}$ where the U's and V's are the
differences in the U and V velocity components over a 6-m interval. For 54
profiles which registered a minimum T < 3°C, S was determined immediately above
and below the core and across the core itself. The results are plotted as
functions of core temperature in Fig. 20.

First, we note that in general the shears are not strong. The values across
the upper and lower interfaces of the cores (~ $1x10^{-2}$ s^{-1}) correspond to those
associated with thermohaline steps under the Mediterranean outflow (Simpson
et al., 1979), but are about one-half order smaller than reported from the base
of the mixed layer in the Atlantic (Evans, 1982).

The shear across the core for all cases was significantly less
(S = $0.28x10^{-2}$ s^{-1}) than the shear in the interfaces above (S = $0.50x10^{-2}$ s^{-1})
and below (S = $0.62x10^{-2}$ s^{-1}); the latter two values differ significantly at
the 98% level. When only the coldest cores are considered, the contrasts are
even greater: for all cores with T < 2.5°C (n = 27), core shear =
$0.23x10^{-2}$ s^{-1}, S(above) = $0.54x10^{-2}$ s^{-1} and S(below) = $0.69x10^{-2}$ s^{-1}.

In spite of the large scatter in the data of Fig. 20, there appear to be
trends in the distribution of water column shear with respect to the finestructure. The trends are suggested by least squares fits to the values from
above, below, and within cores. In very cold cores shear is less than above
and below (as noted above), but for warmer cores the values appear to converge.

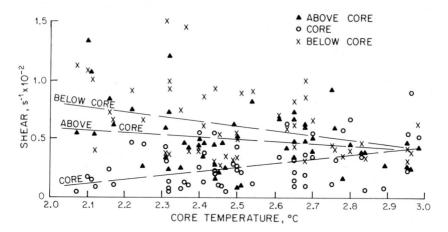

Fig. 20. Shears above, across, and below finestructure cold cores with T < 3°C, from 54 Cyclesonde profiles. Shear given as $S = [(\Delta U)^2 + (\Delta V)^2]^{1/2}$, calculated over 6-m intervals. The trend lines were fitted by least squares. When fine-structure cores are newly developed, shears across them are much less than above or below; when not well developed (or have been eroded) water column shears are more evenly distributed with respect to the finestructure feature.

The value of minimum temperature must to some degree reflect the age of the features--those with the coldest temperatures are much more likely to be more recently formed than those of higher minimum temperature. The only source of cold (T < 3°C) is the central shelf bottom water, and subsequently they can only become warmer. These "younger" features seem to be moving relative to the water column as a solid body, with little shear across (or within) them, while the relative motions are much greater and concentrated in the interfaces above and below the feature, particularly below (the double-diffusive interface). On "aging", the shears above and below definitely tend to decrease while that across the core increases (as shown by the trend lines in Fig. 20, fitted by least squares). Thus, as diffusion acts to reduce the temperature (and salt) contrast of the finestructure features, they also tend to lose their identity as distinct dynamical features in the water column.

4.   Formation of the finestructure

We are now in a position to discuss formation of the finestructure. Maxworthy (1984) recently presented a conceptual model for mixing at tidal fronts which considered the ambient stratification (Fig. 21, left). When the mixing of stratified fluid columns is localized in space, mixed products will escape from the mixing region into the stratified region as intrusions along the appropriate density surfaces. These may be as a single gravity current (Fig. 21b) or multiple fingers (Fig. 21c). When the stratification is weak and/or the mixing is strong, the zone will extend to the surface creating a

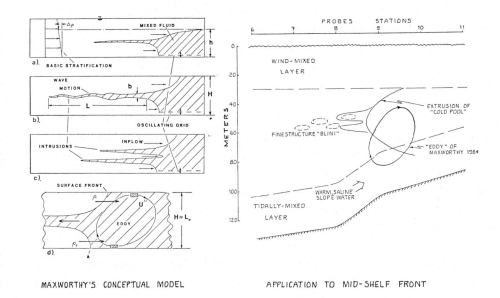

MAXWORTHY'S CONCEPTUAL MODEL   APPLICATION TO MID-SHELF FRONT

Fig. 21. Maxworthy's (1984) conceptual model of finestructure formation (left) and its application to finestructure formation in the middle front of the Bering Sea.

surface "front" (Fig. 21b); otherwise the mixing zone will remain subsurface, creating a "front" extending upward from the bottom over part of the water column (Fig. 21a). The limit of vertical extent will be the largest eddies produced by the tidal current kinetic energy that can overturn the stratification (Fig. 21d).

Maxworthy's conceptual model appears to fit precisely the observations of finestructure formation in the middle front of the Bering Sea shelf, and in Fig. 21, right, the model is scaled to the Bering Sea shelf situation. The cold bottom water of the central shelf domain inshore of PROBES station 11 is being extruded by the pressure field into the graded density structure of the outer shelf domain. Its density is a little greater than the waters of the surface wind-mixed layer but slightly less than those of the warmer, more saline slope waters working their way onshore as a bottom layer across the outer shelf domain (cf. Coachman and Charnell, 1979; Coachman and Walsh, 1981); therefore the extrusion is in the general depth range ~40 to 70 m.

The tidal currents encounter the somewhat more steeply sloping bottom near 100 m water depths; between PROBES stations 8 and 9 the bottom slope is 3 times greater than elsewhere in the central and outer domains, and the water columns shorten by ~20% going onshore. Turbulent kinetic energy from the tidal currents inshore from station 9 extends much farther up the water column creating partial mixtures of the "cold pool" water with the warmer, more saline slope water. The mixtures extrude seaward into the outer shelf ambient stratification as the finestructure features.

The reasonableness of this formation process for the Bering Sea case was investigated by comparing the increase of potential energy in water columns containing finestructure with the kinetic energy available from the tidal currents. The potential energy relative to the mixed state is (Simpson and Hunter, 1974):

$$V = \overline{Vh} = \int_{-h}^{o} (\rho - \overline{\rho})\, gz dz$$

with $\overline{\rho} = 1/h \int_{-h}^{o} \rho dz$, where ρ is density and h the column depth. For a vertically mixed system $V = 0$, and becomes increasingly negative for more stable stratification. As the finestructure was all observed deeper than 40 m, this was taken as the reference depth and a number of stations exhibiting well developed cold cores were integrated to bottom (~110 m); for these $\overline{V} = -15\, J\, m^{-3}$. Nearby stations with no evidence of finestructure were similarly averaged, and $\overline{V} = -30\, J\, m^{-3}$. It thus appears that formation of the finestructure calls for an increase in potential energy per unit volume of ~15 $J\, m^{-3}$.

The rate of decay of energy per unit mass from tidal currents is (Pingree and Griffiths, 1978):

$$\frac{C_D <|U^3|>}{h}$$

where C_D is the drag coefficient and U the tidal current speed; per unit volume this is:

$$\frac{\rho C_D <|U^3|>}{h}.$$

When multiplied by time, this expression gives an estimate of the amount of
kinetic energy per unit volume which has gone into mixing the water column of
length h. Taking a water column of h = 60 m (40 m to 100 m depths) and
C_D = 3.5 x 10^{-3} (Vincent and Harvey, 1976), Table 2 was prepared showing the
amount of KE expended on mixing for various time periods and speeds of tidal

TABLE 2
Energy per unit volume* from tidal currents for various speeds and durations

	Δt. hours		
U, cm s^{-1}	4	5	6
20	6.7	8.4	10.1
25	13.1	16.4	19.7
30	22	28	34

C_D = 3.5x10^{-3}; h = 60 m

$$*KE = \frac{\rho C_D <|U^3|>}{h} \Delta t, \ J/m^3$$

currents. The results suggest that tidal currents $\geqq$25 cm s^{-1} sustained for
periods >4 hours can provide an amount of energy for mixing which equals or
exceeds the potential energy involved in the observed finestructure formation.
The Cyclesonde current measurements and moored current meter records from the
same location (Coachman, 1982) show that incidences of U > 25 cm s^{-1} for
periods >4 hr are common, occurring on the average about every second tidal
cycle.

     Maxworthy (op. cit.) went on to point out that on initial formation
rotation is not important because stratification dominates, but following
collapse of the mixed fluid into a finestructure intrusion the internal
Rossby number

R_o = Nb/2ωL ,

where N = Väisälä frequency, b the thickness and L the length of the intrusion, will rapidly approach 1 and rotation will become equally important. Our
observations suggest that this is exactly what occurs; in fact, the horizontal
scaling of the "cold core" finestructure seems to be set by Coriolis acting
on the motion of the finestructure intrusions in the specific density gradients
of the water columns of the outer domain. For sta. 57 which defines the core
of the CC (Figs. 11, 12, 14), N = 1.12 x 10^{-2} s^{-1}, b = 19 m, and the internal
Rossby radius is ~1.8 km. This is not only the radius of this cold core, but
approximately that of the other three cold cores as well (cf. Fig. 12).

To summarize finestructure formation in the mid-shelf front of the Bering Sea: outer shelf water columns are stratified with density increasing downward to a warm, more saline slope water being "pumped" landward across the shelf, while central shelf columns contain a thick lower layer of colder, less saline water extruding seaward. These water masses are juxtaposed at about the 100 m isobath. At these depths the bottom slope is a little steeper than elsewhere, and on the shallower inshore side tidal current energy becomes sufficient much of the time to create mixtures of the two water masses to shallower depths (~40 m); thus ~100 m constitutes the boundary of the discrete mixing zone of Maxworthy's model. The partial water mixtures constituting finestructure collapse and extrude seaward initially as gravity currents, but quickly begin rotating clockwise under the influence of Coriolis' acceleration. The balance of Coriolis force with the pressure gradient force of the "foreign" water mass in the ambient stratification sets the scale size of these more or less circular "blini" (Russian pancakes with an aspect ratio of ~1:200) which continue to rotate in quasi-geostrophic balance slowly with respect to the water column for some time after their formation while being advected about by the ambient flow field.

5. Diffusion and lifetime

Following their formation and escape from the mixing zone into the stratified water columns of the outer shelf domain, decay of the features will be primarily through vertical diffusion filling in the temperature and salinity minima.

The finestructure are colder and less saline than the ambient water, so the interface above is the "salt finger" interface and that below the "double-diffusive" interface. Substantial evidence has now accumulated that double-diffusive convection (a more rapid flux of heat than salt) plays a significant role in ocean mixing; in the transition downward from warm, saline to colder, less saline water the differential diffusion can lead to the development of salt-finger activity. Schmitt (1979, 1981) showed that little salt-finger activity occurs unless $R_\rho < 2$, and that vigorous activity is present when R_ρ drops below 1.6 to 1.7.

The possibility of salt-finger activity associated with the finestructure was investigated by calculating R_ρ over 3 to 5-m intervals of steepest T and S gradients above the core for all CTD stations exhibiting a well-defined T minimum (n = 29). The values are plotted in Fig. 22 as functions of core temperature (A) and stability N^2 (B).

One-half of the stations had $R_\rho < 1.7$, with a minimum observed value of 0.93. In spite of the wide range of observed ratios, there is a definite trend

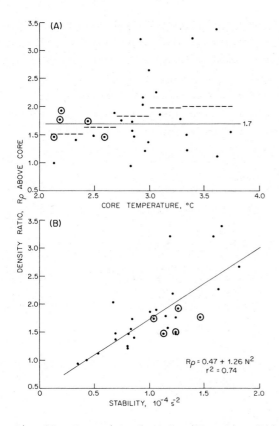

Fig. 22. Upper interface density ratios (R_ρ) for all CTD stations showing pronounced T and S minima indicative of finestructure, plotted as functions of core temperature (A) and stability N^2 (B). The broken lines in (A) are averages over increments of core temperature. Values in cold core CC (stas. 57, 62, 64, 68, 83) are circled.

of increasing R_ρ's with increasing core temperature (Fig. 22A); the colder ("younger") cores show a tendency for having values of $R_\rho < 2$ in their upper interface compared with warmer cores, where $\overline{R}_\rho \to 2$. But definitely not all cores exhibit evidence of salt fingering, only some of them. The density ratios are well correlated with stability of the upper interface (Fig. 22B), with the possibility of salt-fingering activity only when the stability is very low ($N < 1 \times 10^{-2} \, s^{-1}$). It thus appears that it is the degree of local in situ stratification into which a finestructure intrudes that determines whether or not there will be active "double diffusion" associated with it. If the newly-formed mixed parcel enters a strong density gradient, the possibility is less, and vice versa. Also, erosion of a core minimum over time tends to increase the stability of the upper (salt finger) interface, which adds to the good correlation of R_ρ with stability.

In Fig. 22, the stations of CC are identified separately, and their R_ρ's range between 1.5 and 1.9, values on the margin for salt-finger activity; their T/S curves (Fig. 9) also do not show any marked instabilities. It thus appears that this cold core finestructure was not the site of active salt-fingering. The only station of the ensemble that did show a marked density inversion was sta. 105 (Fig. 13). Vertical profiles of T, S, and in situ density through the minima are shown in Fig. 23. Over the depth interval 48 to 50 m R_ρ = 1.00, and

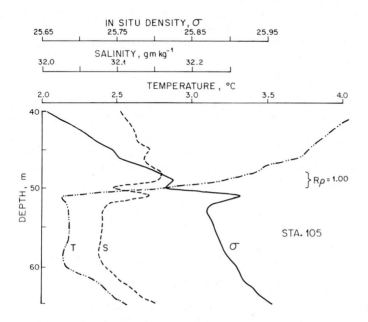

Fig. 23. Vertical profiles of T, S, and in situ density (σ) for sta. 105, the only station of the ensemble (n=29) showing marked density inversions. The R_0 = 1.00 and the "kinks" in the S (and σ) curves are suggestive of active salt-finger activity in the upper interface of the finestructure.

from the profiles one can visualize excess salt moving downward near the bottom of the thermocline, distorting (temporarily) the S and density curves. It appears that some of the finestructure formed in the middle front has double-diffusive activity (salt-fingering in the upper interface) but by no means all of the features do.

The question of possible Kelvin-Helmholtz instability of the interfaces, which could lead to a growth of internal waves and enhanced vertical mixing, is assessed through calculation of the gradient Richardson number $Ri = N^2/s^2$. When $Ri < 0.25$ the possibility of dynamic instability exists. The temperature profiles for all CTD stations taken at CYC-2 which sampled a cold finestructure between 50 m and 60 m were matched with their corresponding Cyclesonde T profiles; the six stations that showed close correspondence are plotted in Fig. 24 (the other stations do not correspond because of the large horizontal field gradients and our reluctance to take the CTD stations precisely at CYC-2). All the stations are within CC or on its edge (sta. 72). Also plotted are N^2, S^2, and Ri calculated over 2-m intervals.

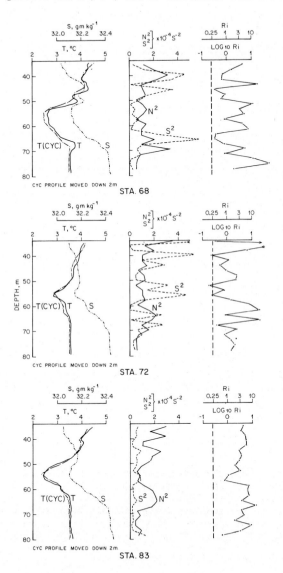

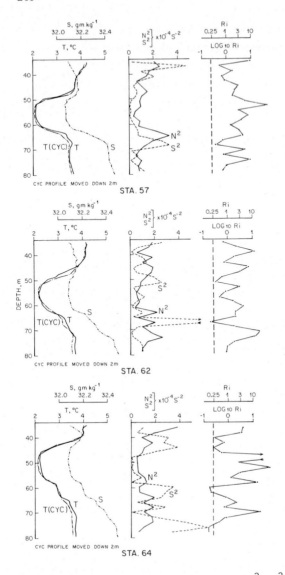

Fig. 24A,B. Vertical profiles of T, S, N^2, S^2, and Ri for the six stations with finestructure taken at CYC-2 for which the T profiles matched, indicating that the Cyclesonde and CTD were sampling the same water mass. Ri's <0.25 are rare, and do not seem to be associated with the main finestructure but with smaller-scale structure above and below.

Across the well-developed cold core the stabilities are not large, but the shears are even less, leading to large values of Ri. Only above and below the feature are there occasional large values of shear, and these don't seem to be associated with the upper and lower interfaces of the core but rather with step structure (stas. 62, 64) or much smaller scale secondary finestructure (sta. 68). Even so, values of Ri ≤0.25 are rare. When the cold core is less well developed, either at its edge (stas. 72, 83), or possibly "older" (sta. 83), shears are more uniformly dispersed through the water column (cf. Fig. 20); again Ri's ≤0.25 are seldom seen. We conclude that dynamic instabilities signified by small gradient Richardson numbers are not an important aspect of strongly developed cold core finestructure.

We can make an estimate of the vertical diffusion associated with CC, probably the major mechanism by which the T and S minima are reduced over time. With a constant eddy coefficient K^ϕ (ϕ = T,S), simple vertical diffusion is approximately

$$\frac{\Delta\phi}{\Delta t} = K^\phi \left\{ \left(\frac{\Delta^2\phi}{\Delta z^2} \right)_U + \left(\frac{\Delta^2\phi}{\Delta z^2} \right)_L \right\} ,$$

where subscripts U and L refer to the upper and lower interfaces of the finestructure minima. Five CTD sts. (57, 62, 64, 68, 83) over the course of two days registered strong minima in CC, and their minimum values of T and S on σ = 25.88 are plotted in Fig. 25. Station 64 differed from the others in that

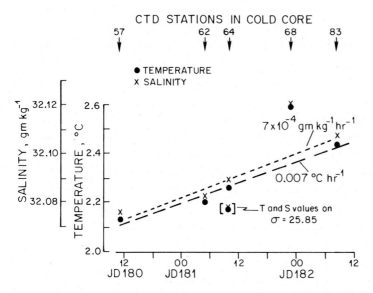

Fig. 25. Minimum values of T,S on σ = 25.88 at 5 stations in finestructure CC, as functions of time. Sta. 64 showed colder and less saline values on a slightly less dense surface. Sta. 68 (and probably also 83) were not in the "core" of the feature (cf. Fig. 11). The trend lines suggest maximum rates of heating and salting.

colder and less saline values were registered on a slightly less dense surface, a reflection of subtleties in the mixing and interleaving process giving rise to considerable variability in the small-scale features of finestructure (cf. also Fig. 13). Station 68 clearly was not in the "core" of CC. Likewise, it is also probable that sta. 83 was not actually in the "core", given its location at the extreme southern tip (Fig. 11). Therefore by including sta. 83 values we get an estimate of maximum rates of warming and salting, as shown in Fig. 25, of 1.9×10^{-6} °C s^{-1} and 0.19×10^{-6} gm kg^{-1} s^{-1}, respectively. Mean curvatures of T,S at the upper and lower boundaries of the minima were calculated over 4, 6, and 8 m intervals for the four stations. Then, using the diffusion equation, values of $K^T = K^S = 0.20$ m^2 s^{-1} were obtained. If the slightly elevated values of T,S at sta. 62 vis-a-vis sta. 57 represent more accurately the time change over 18 hours, then the rates are less. Using these rates and curvatures for the two stations gives $K^T = 0.12$ cm^2 s^{-1} and $K^S = 0.08$ cm^2 s^{-1}. The fact that the eddy coefficients for heat and salt flux are essentially the same reinforces the previous conclusion that CC was not the site of vigorous double-diffusive activity.

We can use the eddy coefficients to estimate the possible lifetime of CC. To blend in with the ambient T profile, the core temperature of CC must increase by >1.5°C. But as the core "fills", the curvatures in the upper and lower interfaces will decrease. Using a combined upper plus lower curvature value one-half that of the initial cold core (0.4×10^{-5} °C cm^{-2}) gives lifetimes between 22 and 43 days for K^T's of 0.2 and 0.1 cm^2 s^{-1}. Thus, it appears that the lifetimes of the coldest finestructure formed in the middle front that do not have vigorous double-diffusive activity associated with them is the order of a month.

It has been postulated (Coachman and Charnell, 1979) that the mid-depth layer (between the surface wind-mixed and bottom tidally-mixed layers) across the outer shelf contains very little mixing energy, which permits the persistence of the cold finestructure. In confirmation, comparison was made of the mean water column shear in the interval 40 m - 82 m between CYC-2 and CYC-1 located 25 km farther seaward (Fig. 2). Shears were calculated over 3-m intervals of the hourly profiles for a similar flow day at each mooring, with the following results:

CYC-1, JD 175 (Fig. 6): S = $0.57 \pm 0.37 \times 10^{-2}$ s^{-1}

CYC-2, JD 180 (Fig. 7): S = $0.73 \pm 0.46 \times 10^{-2}$ s^{-1}

The mid-layer shear is indeed less in the outer shelf away from the mid-shelf front. Thus the data support the hypothesis that the lifetimes of many of the cold finestructure features is sufficiently long, in a low-energy environment, for them to transit the outer shelf and cause the small temperature minima between 50 and 100 m frequently observed at stations taken near the shelfbreak.

CONCLUDING REMARKS

   This paper reports unique data from finestructure features formed in a shelf
sea front. Equipped with a good CTD and profiling current meter, and aided by
luck, we obtained new information for describing the vertical and horizontal
nature of these partial water mass mixtures and interpreting their formation
and subsequent behavior.

   The finestructure formed in mid-latitude shelf sea frontal zones where the
generally colder, less saline shelf water is mixing "laterally" with warmer,
more saline basin water appears to be generically similar, whether it is
formed in the shelfbreak front of "typical" shelves (New York Bight) or the
mid-shelf front of the extra-wide shelves (Bering Sea). Off-shore directed
pressure gradients push the colder, less saline shelf water seaward into the
graded density structure of the basin water, and the finestructure, partially
mixed products of the general "lateral" interaction process, are found on the
seaward side of the frontal mixing zone.

   The creation of the partially-mixed finestructure requires a spatial discontinuity in the mixing process. The major energy for mixing water masses on
these shelves comes from tidal currents, which stir the water columns upwards
from the bottom over some tens of meters. Partial mixtures of shelf and basin
water form where both water masses are within the tidal mixing zone, with
typically the colder, less saline shelf water over the warm, more saline basin
water; thus it is really a vertical mixing on a localized scale. The spatial
discontinuity is caused by changes in depth going across the shelf; on
"typical" shelves where tidal current action becomes strong enough and can
"reach" far enough up the water column to involve both water masses is close to
the shelfbreak. On the Bering Sea shelf, the outer shelf is too deep (~120 m)
for this to happen, and it occurs where the bottom shoals to <100 m. The
argument is reversible: the location of the frontal zone indicative of the
boundary between basic shelf water (filling the shelf sea beneath a surface
layer directly affected by mass and energy exchange) and basin water is
determined by this process.

   The finestructures are found only on the seaward side of the fronts, for
two reasons. (1) The pressure gradients are in general directed offshore, so
that the finestructures after creation (partial mixing) are on the average
pressed seaward, and (2) to survive for any length of time the features must
take up positions in water columns with much less mixing energy, and this is
only possible on the seaward (deeper) side of the fronts.

   Maxworthy's (1984) model appears to describe precisely the formation
process, and from his model and our results we can hypothesize further.
During some phases of some tidal cycles, shelf and basin waters are appro-
priately layered and stirring from the tidal current activity creates some

volume of partial mixture; from the T/S curves (Fig. 4) the larger features (e.g., CC) are perhaps 50 to 75% shelf water, with a total volume of order 10^8 m^3 (Figs. 11, 14). The larger volumes would be created when tidal excursions are fairly long, perhaps >3 km, and probably in the onshelf direction (flood) when water column motion is against the general offshore pressure gradient. This is a common occurrence (cf. Figs. 6, 7). Then during ebb the mixtures collapse and are pushed from the seaward side of the mixing zone, some actually escaping to produce the finestructures.

The finestructures are observed on a hierarchy of scales, both vertical and horizontal. In the vertical, Coachman and Charnell (1979) counted several hundred finestructures at 104 stations spread across the outer Bering Sea shelf during summer, 1976. The features ranged from 1 m up to 25 m vertical dimension, with a definite increase in numbers with decrease in thickness. In the horizontal, this study has documented a larger cold (the signature of the partially-mixed water) field of order 20 km across within which were four discrete colder cores of order 3 km. We can now hypothesize there is a cut-off scale size, a certain volume of partially-mixed water, such that when it collapses as a density current along its isopycnal into the graded density structure to seaward, it is sufficiently large to maintain itself as a separate entity against the subsequent ambient mixing. These become distinct dynamic features, more or less round because they rotate relative to the water column essentially as solid bodies and their scale size is set by the appropriate internal Rossby radius. These form the cold cores. Once formed as separate entities, they erode only very slowly because shears in the water column away from the front are small, and within the finestructure are even smaller. We note this will not be the case for finestructures formed in the more "typical" shelf sea fronts (Fig. 1, I) where the mid-layers of the offshore water columns are more energetic. Off the NYB, for example, the presence of horizontal shear distorts the features in the along-shelf direction and their lifetimes are much shorter than in the outer domain of the Bering Sea.

The partial mixtures creating finestructure smaller than the cut-off volume are transient and erode relatively rapidly. In areas and at times when lots of these mixtures are formed, their decay contributes to creation of a larger, generally colder mid-layer, as observed surrounding the cold cores during this study. The erosion of the cold cores themselves from their top and bottom surfaces and around their perimeters also contributes to the creation of the larger cold area.

The possible lifetimes of the larger finestructure, which take on a discrete dynamic identity, appears to be quite long (order of a month) because their decay is basically limited to simple diffusion in the vertical which is quite slow. There will also be some slow erosion of the perimeter by the formation of smaller finestructures through mixing of the core and ambient waters. Double diffusive phenomena, which enhances vertical diffusion and markedly reduces the lifetimes of the "bliny", does appear to be quite prevalent in the regions of finestructure. In this study about one-half of the calculated density ratios suggested salt-fingering, and substantial regions of low R_ρ's were found in the NYB finestructure formation region. But the occurrence of double-diffusive phenomena depends on the strength of the local density gradient; the lower the gradient, the more likely the occurrence, and vice versa. So as the finestructure erodes, the local density gradients increase and the double diffusive activity is damped out. Thus many of the smaller finestructures may be rapidly mixed away through this effect, but the larger parcels, bigger than the cut-off volume, are only reduced somewhat in thickness if double diffusion is present initially. The presence of instabilities, which could also cause more rapid erosion of the features is definitely not indicated. Even though the density gradients associated with the features are low, shears are even less, and Richardson numbers approaching 0.25 are rare.

The finestructure "blini" will also be eroded to some extent around their perimeters. The nature of this erosion is probably through the formation of much smaller finestructure (compare the hierarchical concept of Joyce, 1977), where the water masses involved are the core water and the ambient water of the adjacent stratification, and the energy for mixing is the local shear. This effect would continue throughout their lifetimes, i.e., while they transit the outer shelf domain. It seems probable that transient salt-finger activity would be associated with this process, which could be responsible for the observed bulk increase in density of the T and S minima across the outer shelf.

The finestructures formed in the mid-shelf front of the Bering Sea influence the water mass properties and structure across the whole outer shelf domain to the shelfbreak more than 100 km seaward (cf. Fig. 2). Water columns in this domain exhibit a temperature minimum at mid-depths, between ˜50 and 100 m (cf. Fig. 4). At the close of the study we occupied a 165 km CTD section along the 110 m isobath (see Fig. 2 for location), and the temperatures are plotted in Fig. 26. Westward from station 121 a number of larger finestructures were observed. The conclusion is that the minimum temperature layer results because of a sufficient production of larger finestructures whose lifetimes are long enough to survive transit of the outer domain before being completely dissipated. These features also transport seaward any other properties involved in the mixing, for example, higher levels of nutrients than are normal for the mid-depth layer. Fig. 27 shows the NO_3 distribution across CC. The source for high values of NO_3 is the bottom basin water, and clearly this nitrate has been mixed into the finestructure elevating its values vis-a-vis the ambient water, and will be transported by the finestructure. This

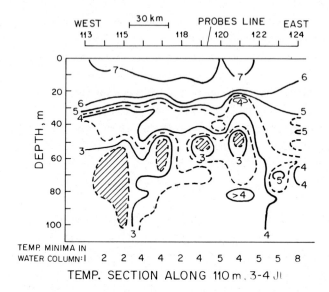

Fig. 26.   Temperature section along the 110 m isobath, the seaward side of the middle front (see Fig. 2 for location).   T's colder than 2.5°C are hachured. Numerous large finestructure were observed west of sta. 122.

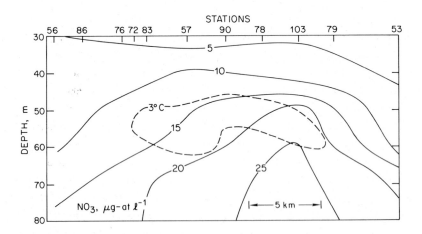

Fig. 27.   Nitrate distribution across finestructure CC (cf. Fig. 14).   Nitrate from the deeper basin water has been mixed up into the finestructure during its formation.

phenomenon accounts for the enhanced horizontal and vertical fluxes of nitrate in mid-depth layers observed in the outer domain of the Bering Sea (Coachman and Walsh, 1981).

ACKNOWLEDGEMENTS

Support for PROBES was provided by the National Science Foundation, Division of Polar Programs, which also supported the finestructure study, and the final stage of manuscript preparation was in part supported by the same organization under the ISHTAR program. I am indebted to my colleagues in these programs, afloat and ashore, for continuing congenial and fruitful scientific collaboration. I am also indebted to S. Riser for a critical evaluation of the manuscript.

REFERENCES

Beardsley, R.C. and C.N. Flagg 1976
    The water structure, mean currents, and shelf-water/slope-water front on the New England continental shelf. Mem. Soc. R. Sci. Liege 6 ser. X, pp. 209-225.

Coachman, L.K. 1982
    Flow convergence over a broad, flat continental shelf. Cont. Shelf. Res. 1: 1-14.

Coachman, L.K. 1985
    Circulation, water masses, and fluxes on the southeastern Bering Sea shelf. Cont. Shelf Res. (in press).

Coachman, L.K. and R.L. Charnell 1979
    On lateral water mass interaction - a case study, Bristol Bay, Alaska. J. Phys. Oceanogr. 9: 278-297.

Coachman, L.K. and J.J. Walsh 1981
    A diffusion model of cross-shelf exchange of nutrients in the southeastern Bering Sea. Deep-Sea Res. 28A: 819-846.

Cresswell, G.H. 1967
    Quasi-synoptic monthly hydrography of the transition region between coastal and slope water south of Cape Cod, Mass. WHOI Rep. 67-35.

Evans, D.L. 1981
    Velocity shear in a thermohaline staircase. Deep-Sea Res. 28: 1409-1415.

Evans, D.L. 1982
    Observations of small-scale shear and density structure in the ocean. Deep-Sea Res. 29: 581-595.

Gargett, A.E. 1976
    An investigation of the occurrence of oceanic turbulence with respect to finestructure. J. Phys. Oceanogr. 6: 139-156.

Goering, J.J. and C.P. McRoy 1981
    A synopsis of PROBES. EOS 62(44): 730-731.

Gordon, A.L. and F. Aikman III 1981
    Salinity maximum in the pycnocline of the Middle Atlantic Bight. Limnol. Oceanogr. 26: 123-130.

Gordon, A.L., D.T. Georgi and H.W. Taylor 1977
    Antarctic polar front zone in the Western Scotia Sea - summer 1975. J. Phys. Oceanogr. 7: 309-328.

Gregg, M.C.   1975
    Microstructure and intrusions in the California Current.   J. Phys. Oceanogr.
    5: 253-278.

Horne, E.P.W.   1978a
    Interleaving at the subsurface front in the slope water off Nova Scotia.
    J. Geophys. Res. 83: 3659-3671.

Horne, E.P.W.   1978b
    Physical aspects of the Nova Scotian shelfbreak fronts.   Chap. 7, pp. 59-68
    in Oceanic Fronts in Coastal Processes, M.J. Bowman and W.E. Esaias, eds.
    Berlin/Heidelberg/New York: Springer-Verlag.

Houghton, R.W. and J. Marra   1983
    Physical/biological structure and exchange across the thermohaline shelf/
    slope front in the New York Bight.   J. Geophys. Res. 88: 4467-4481.

Houghton, R.W., R. Schlitz, R.C. Beardsley, B. Butman and J.L. Chamberlain 1982
    The Middle Atlantic Bight cold pool: evolution of the temperature structure
    during summer 1979.   J. Phys. Oceanogr. 12: 1019-1029.

Joyce, T.M.   1977
    A note on the lateral mixing of water masses.   J. Phys. Oceanogr. 7: 626-629.

Kinder, T.H. and J.D. Schumacher   1981
    Hydrographic structure over the continental shelf of the southeastern Bering
    Sea.   In The Eastern Bering Sea: Oceanography and Resources, vol. 1,
    D.W. Hood and J.A. Calder, eds.   Seattle: Univ. Washington Press,
    pp. 31-52.

Lindstrom, E.J. and B.A. Taft   1985
    Small water transporting eddies: an analysis of statistical outliers in the
    POLYMODE Local Dynamics Experiment hydrographic data.   J. Phys. Oceanogr.
    (in press).

Maxworthy, T.   1984
    On the formation of tidal mixing fronts.   Ocean Modelling 55: 8-10
    (unpublished manuscript).

Mooers, C.N.K, C.N. Flagg and W.C. Boicourt   1978
    Prograde and retrograde fronts.   Chap. 6, pp. 43-58 in Oceanic Fronts in
    Coastal Processes, M.J. Bowman and W.E. Esaias, eds.   Berlin/Heidelberg/
    New York: Springer-Verlag.

Munk, W.   1981
    Internal waves and small-scale processes.   Chap. 9, pp. 264-291 in
    Evolution of Physical Oceanography, B.A. Warren and C. Wunsch, eds.
    Cambridge and London: The MIT Press.

Pingree, R.D. and D.K. Griffiths   1978
    Tidal fronts on the shelf seas around the British Isles.   J. Geophys. Res.
    83: 4615-4622.

Posmentier, E.S. and R.W. Houghton   1978
    Fine structure instabilities induced by double diffusion in the shelf/slope
    water front.   J. Geophys. Res. 83: 5135-5138.

Posmentier, E.S. and R.W. Houghton   1981
    Springtime evolution of the New England shelf break front.   J. Geophys. Res.
    86: 4253-4259.

Riser, S.C., W.B. Owens, H.T. Rossby and C.C. Ebbesmeyer 1985
The structure, dynamics, and origin of a small-scale lens of water in the
Western North Atlantic thermocline. J. Phys. Oceanogr. (in press).

Roden, G.I. 1974
Thermohaline structure, fronts, and air-sea energy exchange of the trade
wind region east of Hawaii. J. Phys. Oceanogr. 4: 168-182.

Ruddick, B.R. and J.S. Turner 1979
The vertical length scale of double-diffusive intrusions. Deep-Sea Res. 26:
903-913.

Schmitt, R.W. 1979
The growth rate of super-critical salt fingers. Deep-Sea Res. 26(1A): 23-40.

Schmitt, R.W. 1981
Form of the temperature-salinity relationship in Central Water: evidence for
double-diffusive mixing. J. Phys. Oceanogr. 11: 1015-1026.

Schumacher, J.D. and T.H. Kinder 1983
Low-frequency current regimes over the Bering Sea shelf. J. Phys. Oceanogr.
13: 607-623.

Simpson, J.H. and J.R. Hunter 1974
Fronts in the Irish Sea. Nature 250: 404-406.

Simpson, J.H., M.R. Howe, N.C.G. Norris and J. Stratford 1979
Velocity shear in the steps below the Mediterranean outflow. Deep-Sea Res.
26: 1381-1386.

Stern, M.E. 1967
Lateral mixing of water masses. Deep-Sea Res. 14: 747-753.

Toole, J.M. and D.T. Georgi 1981
On the dynamic- and effects of double-diffusively driven intrusions.
Prog. Oceanogr. 10: 123-145.

Turner, J.S. 1978
Double-diffusive intrusions into a density gradient. J. Geophys. Res. :
2887-2901.

Van Leer, J.C., W. Düing, R. Erath, E. Kennelly and A. Speidel 1974
The Cyclesonde: an unattended vertical profiler for scalar and vector
quantities in the upper ocean. Deep-Sea Res. 21: 385-400.

Vincent, C.E. and J.G. Harvey 1976
Roughness length in the turbulent Ekman layer above the sea bed. Mar. Geol.
22: M75-M81.

Voorhis, A.D., D.C. Webb and R.C. Millard 1976
Current structure and mixing in the shelf-slope water front south of New
England. J. Geophys. Res. 81: 3695-3708.

Welch, C.S. 1981
Mid-level intrusions at the continental shelf edge. J. Geophys. Res.
86: 11,013-11,019.

Williams, A.J. III 1981
The role of double diffusion in a Gulf Stream frontal intrusion.
J. Geophys. Res. 86: 1917-1928.

SOME ASPECTS OF THE LIGURO-PROVENCAL FRONTAL ECOHYDRODYNAMICS

J-H. HECQ[*], J-M. BOUQUEGNEAU[*], S. DJENIDI, M. FRANKIGNOULLE,
A. GOFFART[**] and M. LICOT

Department of Oceanography University of Liège
[*] Chercheur Qualifié FNRS
[**] Chercheur IRSIA

INTRODUCTION

In the oligotrophic sea water, the biological production is mainly influenced by zones of discontinuity such as fronts and thermoclines. In the Mediterranean Sea, the front at the edge of the continental shelf separates general and coastal circulations, limiting lateral diffusion. The vertical motions associated with this front bring nutrients to the surface, increasing the biological productivity (PRIEUR, 1981). The produced biological material is exported off-shore perpendicularly to the front.

Since a few years, interdisciplinary studies have been led in the Corsican area of the Liguro-Provençal front by the University of Liège at STARESO Station (Calvi - Corsica).

These studies are integrated in an international program with the collaboration of L. PRIEUR (Station zoologique of Villefranche-sur-Mer, France).

General circulation

The dynamics of the Mediterranean Sea is mainly dominated by the general and the local atmospheric forcings. Due to the climatology of the Mediterranean area, the water balance is negative, the evaporation being more important than the precipitations and the river supply. In this concentration basin, the deficit is compensated for by an important water flux through the Gibraltar Strait (LACOMBE, 1973; BETHOUX, 1980).

The Atlantic surface layer flows eastwards along the North African coast, forming the so called African current. In the Western part of the Mediterranean large cyclonic circulations are developed in the Thyrrhenian Sea and in the Liguro-Provençal basin (figure 1).

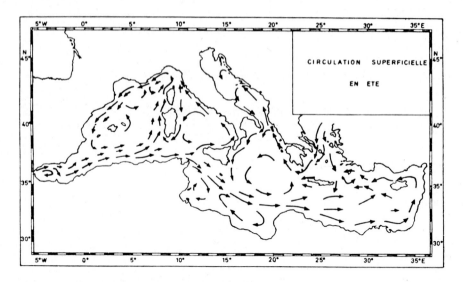

Fig. 1 : Summer superficial circulation (after SCHMIDT and NIELSEN)

The Summer heating of the surface layer leads to the formation of a seasonnal thermocline. During the winter, under the effect of dry and cold continental winds (Mistral, Tramontane, ...) and of the difference of tempe- rature between air and sea, the evaporation and heat transfer from the sea to the air become very important. Hence the density of the surface layer increases, inducing a baroclinic instability. The consequent vertical mixing and convection give rise to deep waters. This phenomenon of deep water for- mation is observed in the North-West Mediterranean (GASCARD, 1978), in the Levantin basin and in the Adriatic Sea.

The water formed during the winter in the Levantin basin crosses the Sicilian Strait and spreads in the Western basin, describing cyclonic circulations of a so-called intermediate water.

So, the Mediterranean Sea seems to be a three layer system : a surface layer of Atlantic water (~ 0 - 100 m) cold and little saline, an interme- diate levantine layer (~ 100 - 500 m) warmer and more salted and a typically mediterranean deep layer (under 600 m) colder and less salted. Above the summer thermocline, when it exists, the upper layer of the sea has higher temperature and salinity than the Atlantic water.

There is no doubt that this general scheme of circulation is highly perturbed and locally modified, at least in the surface layer, by the transient wind events.

From this general survey obtained by means of classical field measurements, one emphasizes the intense cyclonic gyre in the Liguro-Provençal basin carrying along surface and subsurface waters.

Since a few years, the new possibilities offered by the development of remote sensing techniques has allowed a better monitoring of oceanic phenomena like the frontal structures in the Western Mediterranean Sea.

The examination of a one year series of infrared thermographies provided by the Centre de Météorologie Spatiale (Lannion, France) shows for instance the seasonnal variability of the front separating the different water masses in the Liguro-Provençal basin (figure 2).

Spring images do not provide much information because of the weakness of the surface temperature gradients due to the strong mixing of the surface layer by the cold winds during the winter.

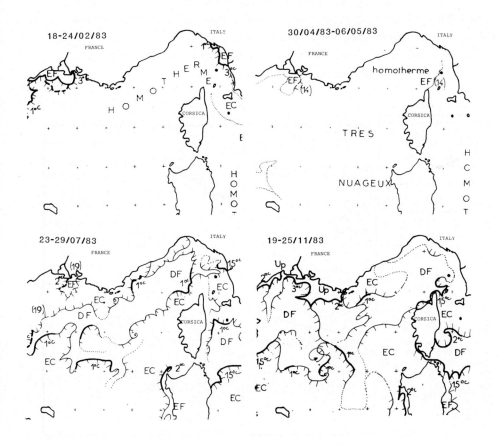

Fig. 2 : NOAA satellite infrared thermographies showing surface temperature gradient( )(from Centre de Météorologie spatiale, Lannion, France)

The heating of the surface water during the late spring progressively establishes a vertical stratification. The consequent seasonnal thermocline is affected by the water motions which disturb it, giving rise to horizontal temperature gradients visible form satellites in the form of more or less marked fronts. The most important front for us, in relation with the cyclonic circulation in the Liguro-Provençal basin, separates dense water of the cold core from the warm water running around it. The width of the warm water strip is variable, the front being sometimes firmly situated in the coastal zone, like it is the case off North Corsica from Calvi to Cape Corse. In spite of its variability in space and in time, the thermal front associated with this cyclonic loop is persistent throughout the summer and fall. However, during the autumn, it begins to weaken and to present instabilities.

The winter meteorological forcing removes the seasonnal thermocline and consequently the whole typical summer fronts. Then the winter and spring thermographies do not show the Liguro-Provençal gyre, although it is present in winter as shown in figure 3. During these seasons the density field appears as mainly controlled by the salinity.

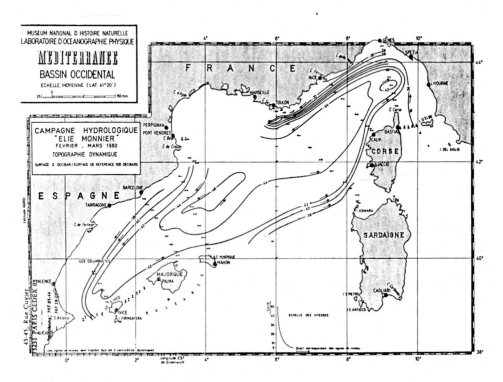

Fig. 3 : Dynamical topography in the North-West Mediterranean during the winter (from BOSCALS DE REAL).

Hydrological data

    The study of hydrological data (temperature, salinity) collected in March,
July and October 1984, during several oceanographic cruises across the
Ligure-Provençal front (Corsican area), provides a more detailed picture of
water masses distribution and seasonal fluctuations. These campaigns have
been carried out on board of the "Recteur Dubuisson", the oceanographic ship
of the University of Liège at Calvi (Corsica). Ten stations have been selected (figure 4) on the Calvi-Nice axis, from Calvi (station n°1) to 30 nautical miles offshore (station n°10).

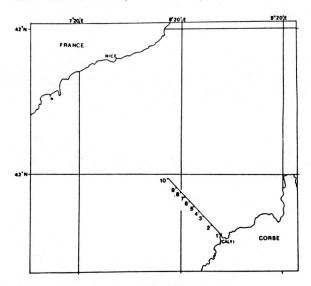

Fig. 4 : Position of sampling stations

    Temperature and salinity measurements have been carried out at every ten
stations, from the surface to 200 meters. The isotherms, isohalines and
isopycnals distributions are presented.
    In March, a strong gradient of salinity separates coastal waters
(S < 38.2 °/₀₀) from offshore waters (figure 5A) (S > 38.4 °/₀₀). The distribution of isotherms shows that the upper layers are not really homothermal.
Figure 5B shows that colder waters (T < 12.9°C) are merely situated beneath
the haline gradient while warmer waters (T > 12.9°C) are situated above it,
near the coast

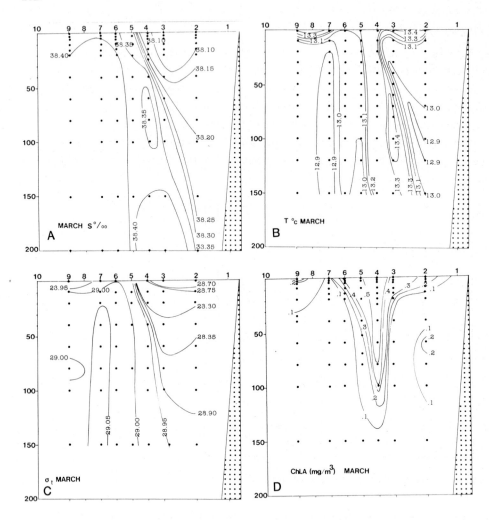

Fig. 5 : MARS 1984. Distribution of isohalines (A), isotherms (B),
isopycnals (C) and chlorophyll A concentration (D) across Corsican
front region from the coast (st 1) to 30 miles offshore (st 10)
and from surface to 200 meters deep (HECQ et al, 1985).

Density distribution (figure 5C) shows that the thermohaline front separated two areas :

- Offshore stations with waters of high density, characteristic of Levantine origin; this region is unstratified and seems to be a divergence area,
- Onshore stations with less dense waters, characteristic of Atlantic origin. This region is little stratified with a tendance of convergence close to the gradient. In that period, the front crosses the sea surface about 15 miles off the coast and the isopycnals slant with a 1.6 % slope from the front area to the coast (HECQ et al, 1985).

In June, a vertical stratification is initiated (figure 6B) with the heating of upper layers. This vertical gradient of temperature masks the horizontal gradient at least in the 75 upper meters. From the point of view of the salinity (figure 6A), the haline gradient separating offshore and coastal waters in more important below 50 m than above. The density diagram (figure 6C) summarizes the water masses distribution. In the upper layer (from the surface to the 50 m depth), the distribution of isopycnals is approximately horizontal. Below 75 m, both a coastal stratified area and an offshore unstratified zone with high density values are still observed. This suggests that the divergence does not reach the surface but only affects the waters below the thermocline.

In October (figure 7), the distribution of isohalines is quite similar to the situation in June and two regions are separated by a frontal discontinuity. The increase of isothermal and isopycnal slope suggests the onset of a destabilization.

The "in situ" measurements confirm fairly well the existence of different water masses as described above.

The water situated between the front and the Corsican coast has the characteristics of an Atlantic water. The water situated beyond the front has the characteristics of an intermediate Levantine water.

In winter and at the begining of spring, salinity influences the most the slope of isopycnals (from offshore areas to the coast). In that period, divergences reach surface layers.

In contrary, during summer, surface temperature influences horizontal stratification of the upper layers and frontal divergences seem to reach only water layers below 100 meters.

In addition to the measurements of salinity and temperature, total alkalinity analyses have been carried out on the same samples. Total alkalinity is independent of biological processes as far as reactions involving carbonates exchanges are negligible (FRANKIGNOULLE and BOUQUEGNEAU, 1985)

264

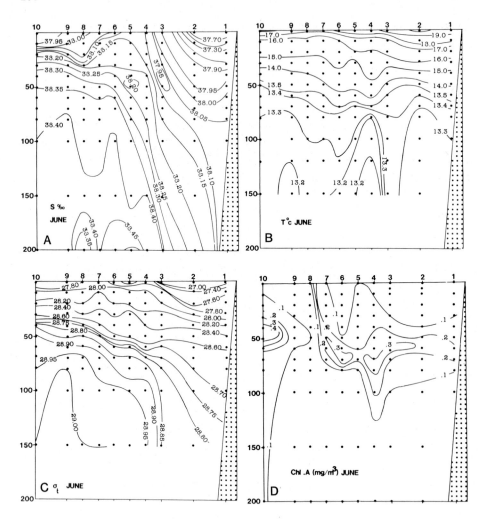

Fig. 6 : JUNE 1984. Distribution of isohalines (A), isotherms (B),
isopycnals (C) and chloropyll A concentrations (D) across Corsican
front region from the coast (st 1) to 30 miles offshore (st 10)
and from surface to 200 meters deep (HECQ et al, 1985).

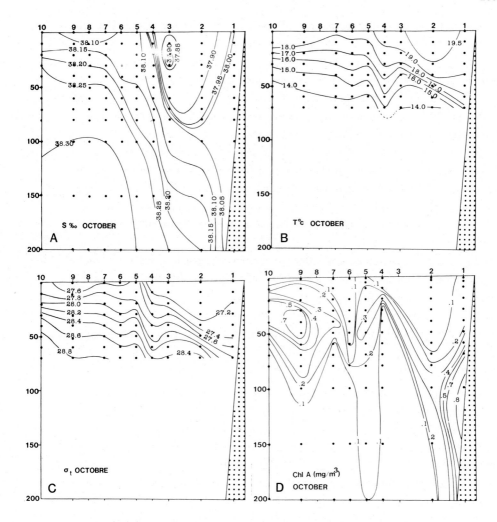

Fig. 7 : OCTOBER 1984. Distribution of isohalines (A), isotherms (B),
isopycnals (C) and chlorophyll A concentrations (D) across Corsican
front region from the coast (st 1) to 30 miles offshore (st 10)
and from surface to 200 meters deep (HECQ et al, 1985).

and we believe that this parameter can be used to discuss the origin and movements of water masses : deep waters are more alkaline than surface ones.

Figure 8 shows the result obtained in March, June and October 1984. In March, the thermohaline front separates less alkaline onshore waters ($\leqslant 2.60$ m Eq. L^{-1}) from alkaliner offshore ones ($\geqslant 2.65$ m Eq. L^{-1}). A divergence of offshore waters and a convergence of onshore ones are clearly suggested along the isopycnal discontinuity.

In June, once again, the haline front separates more and less alkaline waters. A divergence phenomenon appears in offshore waters but the maxima of alkalinity are found just beneath the thermocline : there is a divergence at station 7 up to 50 meters where the pycnocline has been detected; at station 10, at 40 meters depth, the highest observed alkalinity value is observed.

In October, the waters are relatively more homogeneous and no important divergence or convergence are suggested. The frontal discontinuity is till apparent and a gradient of alkalinity remains both from coastal to offshore waters and from the surface to the bottom.

These results fit well with the other hydrological data described above.

Phytoplankton data

Investigations performed in spring have shown a spatial heterogeneity in chlorophyll distribution in the Ligurian Sea (JACQUES et al, 1973).

Moreover, in the Bay of Calvi, vertical phytoplankton distribution exhibits a seasonal evolution related to the thermal structure of the water column (HECQ et al, 1981, 1985; LEGENDRE, 1981).

Oceanographic data collected in the area of the front during 1984 have lead to an accurate picture of the spatio-temporal phytoplankton distribution.

The area investigated and the techniques have been described earlier. Chlorophyll A has been analysed on water collected at twelve different depths (from surface to 200 m) (according to STRICKLAND and PARSONS, 1968).

In March, chlorophyll A concentration (figure 5D) is important in the frontal area : more than 0.3 mgr. chl. A/m^3 from station 2 to station 6, where the amount of nutrients is high.

The maximal concentrations are recorded at station 4 (> 0.5 mgr. chl. A/m^3). At this station, the living phytoplankton is found deeper than euphotic depth and the vertical distribution of chlorophyll A suggests a downwards transport of phytoplankton along isopycnals in the area of the front; e.g. at station 4, living chlorophyll content still reaches 0.4 mg/m^3 at 100 meters depth.

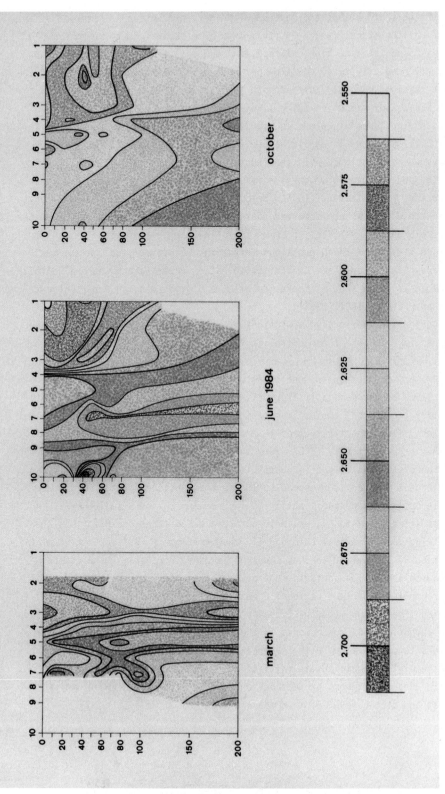

Fig. 8 : Distribution of alkalinity isolines across Corsican frontal region (number of stations.depth in meters) in March, June and October 1984. Alkalinity has been determined by electrochemical titration according to GRAN (1952)

During that period, the maximum of primary production is observed at the level of the front (figure 9) (from 0 to 25 m depth : 50 to 70 mg C m^{-3} D^{-1}) and is associated to high chlorophyll A concentrations. In the other hand primary productivity (production per unit biomass) reaches a maximum at the same stations but only just beneath the surface (200 mg C. mg chl A^{-1} B^{-1}). The biomass distribution being like a plume (figure 5D), we can conclude that phytoplankton is produced near the surface between stations 3 and 4 (LICOT, 1985), and is carried along the isopycnals, in relation with the convergence associated with the frontal system.

In June (figure 6D), when the value of the isopycnals slope is smaller than that found in March (0.5 % below the stratified layer) - maximal phytoplankton biomasses are observed just below the stratified layer (< 50 meters). The chlorophyll distribution seems to follow the general slope of isopycnals (σ_t 28.4 - 28.6). The highest concentrations (> 0.4 mgr. chl. A/m^3) are observed in the open sea at station 10, at 50 meter depth where alkalinity and density data show a divergence (figure 8) supplying an important nutrient concentration (LICOT, 1985).

During that period, accumulations of chlorophyll A are situated in the lowest level of the thermocline. At that level, the light intensity is ireduced but sufficiently high nutrients concentrations are present.

Primary production in June is also maximum at the coast and offshore at the level of the seasonal thermocline. At the level of the haline front, primary production remains important (> 25 mg C^{-3} D^{-1}) despite a poor algal biomass. The productivity profile are quite different : coastal stations (1 - 3) with high productivity (800 - 1000 mg. C mg chl A^{-1}, D^{-1}) are separated from offshore stations of poor productivity (< 200 mg. C mg chl A^{-1} D^{-1}) by an horizontal gradient corresponding to the thermohaline front.

In October, a destabilization begins between 10 and 15 miles away from the coast (stations 3 to 5), while in offshore areas (stations 8 to 10), the distribution of maximal chlorophyll concentrations (> 0.7 mgr. chl A/m^3 at station 9) is similar to that observed in June. However, in the coastal area, chlorophyll A contents show a completely different pattern which is characteristic of that period : a maximum of chlorophyll A (> 0.8 mg chl A/m^3) is found below 100 m depth. Probably this accumulation of chlorophyll below the euphotic layer corresponds to the general trend of the coastal upper waters to destabilize. With the breakdown of the stratification, a return to the winter conditions is initiated.

In October, highest productions and productivities are found near the coast and at the level of thermohaline front.

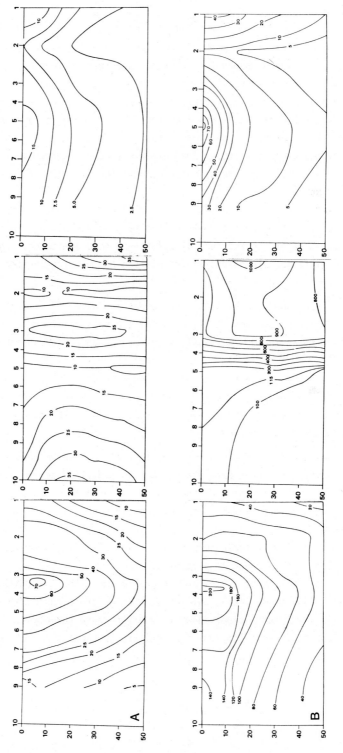

Fig. 9 : Distribution of primary production (mgC . m^{-3} . D^{-1}) (A) and primary productivity (mgC. mgChl A^{-1}.D^{-1}) (B) across Liguro-Provençal front in 1984 (LICOT, 1985).

CONCLUSIONS

Seasonal hydrological studies realized during 1984 show the presence of distinct water masses separated by fronts and thermoclines.

At these boundaries vertical movements occur as evidenced by temperature, salinity and alkalinity distributions. High alkalinity values characterize deep waters. This parameter has appeared to be an original and useful tool to study the water masses and movements.

The hydrological structures are relatively constant from year to year (LICOT, 1985) and their evolution throughout the year can be summarized as follows :

- In winter when the external forcing favours the mixing, an important haline front separates the coastal more stratified light pool from the waters of intermediate origin, nutrient-rich and undergoing high vertical mixing.

- In spring and summer, with surface heating a vertical stratification is induced, followed by the establishment of the seasonal thermocline, hence conducting to an increased stability. Hivernal water masses sink and the intermediate waters can not reach the surface. Destabilizing factors (such as winds, cold air masses) can locally generate divergences with rise of cold waters. Although vertical mixing is reduced by the presence of the thermocline, a shoaling of the isopycnals from the coast to the open sea is yet observed during that period.

- At the approach of the winter the thermal balance between the sea and the atmosphere reverses, leading to a destabilization of the water column and to the breakdown of the seasonal thermocline. Vertical mixing is enhanced and intermediate water masses can rise up to the sea-surface.

The description of the chlorophyllian pigments distribution in the upper layers along the transect is presented.

In spring, maxima of phytoplankton are found on the thermohaline front. In summer (June to October), chlorophyll maxima are situated below the stratified layer which slopes down from offshore to the coast.

The impact of frontal dynamics on primary production is well emphasized in our data.

ACKNOWLEDGEMENTS

The authors acknowledge the help of Captain X. BRUNEAU and the crewmembers of R.V. "Recteur Dubuisson" and D. BAY, co-director of the Station STARESO.

REFERENCES

Bethoux, J.P., 1980, Mean water fluxes accross sections in the Mediterranean sea evaluated on the basis of water and salt budgets and of observed salinities, Oceanologica Acta, **3**,1, 79-88.

Frankignoulle, M. and Bouquegneau, J.M., 1985, Ecohydrodynamic study of Liguro-Provençal front (Corsican Area), IV sea water CO_2 system data, Proc. 1st Cong. Oceanol. Belg. Acad. Sci., 4 - 6 March 1985 (in press).

Gascard, J.C., 1978, Mediterranean deep water formation. Baroclinic instability and oceanic eddies, Oceanologica Acta, **4**, 3, 315-330.

Gran, G., 1952, Determination of the equivalent point in potentiometric titrations - part II , Int. Congress Anal. Chem., 77 pp, 661-671.

Hecq, J.H., Gaspar, A. and Dauby, P., 1981, Caractéristiques écologiques et biochimiques de l'écosystème planctonique en baie de Calvi (Corse), Bull. Soc. Roy. Sci. Liège, **50**, 440-445.

Hecq, J.H., Licot, M., Goffart, A., Mouchet, A., Frankignoulle, M., Bouquegneau, J.M., Distèche, A., Godeaux, J. et Nihoul, J.C.J., 1985, Ecohydrodynamical study of the Liguro-Provençal front (Corsica), I. Hydrological Data, Proc. 1st Cong. Oceanol. Belg. Acad. Sci., 4-6 March 1985 (in press).

Jacques, G., Minas, H.J., Minas, M. et Nival, P., 1973, Influence des conditions hivernales sur les productions phyto- et zooplanctoniques en Méditerranée nord-occidentale, II, Biomasse et production phytoplanctoniques, Mar. Biol.,**23**, 251-265.

Lacombe, H., 1973, Aperçu sur l'apport à l'océanographie physique des recherches récentes en Méditerranée, Bull. Et. Commun. Médit., **7**, 5-25.

Legendre, L., 1981, Hydrodynamic control of marine phytoplankton production : the paradox of stability, Ecohydrodynamics : Proceedings of the 12th Int. Liège Colloquium on Ocean Hydrodynamics, Ed. J.C.J.Nihoul, Elsevier Oceanogr. **32**, 191-207.

Licot, M., 1985, Etude écohydrodynamique du front Liguro-Provençal au large de la Corse, Relation entre l'hydrodynamique, les paramètres physico-chimiques et la production primaire, P.h. Thesis University of Liege, 131 pp.

Prieur, L., 1981, Heterogénéité spatio-temporelle dans le bassin Liguro-Provençal, Rapp. Comm. Int. Mer Médit., **27**, 6, 177-179.

Strickland, J.D.H. et Parsons, T.R., 1968, A manual of sea water analysis, Bull. Fish. Res. Bd Can., **125**, 1-311.

PLANKTON DISTRIBUTIONS AND PROCESSES IN THE BALTIC BOUNDARY ZONES

M. Kahru[1], S. Nômmann[1], M. Simm[1], and K. Vilbaste[2]

[1]Institute of Thermophysics and Electrophysics, Paldiski St. 1, Tallinn 200031 (USSR)

[2]Institute of Zoology and Botany, Vanemuise St. 21, Tartu 202400 (USSR)

ABSTRACT

Due to their estuarine origin, the brackish water masses of the Baltic Sea result from complicated mixing processes. As a result, boundary zones span the whole spectrum of space scales in both the vertical and horizontal directions. The ecological effects of these boundary zones are exemplified by the results of recent surveys. Evidence of a striking increase in primary productivity in an offshore front, elevated levels of chlorophyll and phytoplankton biomass in the thermocline, offshore and in coastal boundary areas, at the periphery of synoptic eddies, and of a higher standing stock of zooplankton in or adjacent to frontal boundaries has been obtained. Flow-through particle counting has revealed abrupt changes in the size distribution of phytoplankton populations indicating different stages of the vernal bloom. It is stressed that the frequency of the boundary effects is poorly known in both space and time, thus the assessment of their overall significance is not possible at present.

INTRODUCTION

The Baltic Sea may be regarded as a large "overmixed" estuary (Shaffer, 1979). Its brackish waters result from mixing between different water masses. A variety of mixing processes are involved on the whole spectrum of scales. As the mixing is not a smooth, Fickian diffusion, a whole spectrum of boundary zones, both in the horizontal and vertical, is being generated. Most persistent are those in the vertical, i.e. the seasonal thermocline and the permanent halocline. As both are strong pycnoclines and restrict the extent of vertical mixing (Kullenberg, 1982), they exercise a profound influence on the ecology of the Baltic (e.g. Jansson et al., 1984). In the horizontal, the boundary zones are as ubiquitous though not so persistent. Sharp boundaries in larger scales are commonly known as fronts. These transition zones between different water masses are often characterized by the accumulation of biomass and the increased biological activity (e.g. Holligan, 1981), and are supposed to be important for the feeding and survival of larval and juvenile fish (Iles and Sinclair, 1982).

The ecological significance of boundary zones in the Baltic, apart from the restrictions to the vertical mixing by the pycnoclines, is virtually unknown. Compared to better studied oceanic shelf analogs (Holligan, 1981; Marra et al., 1983; Fasham et al., 1985), the fronts in the Baltic are expected to be smaller in scales and less persistent in time. Their shorter life-time and intermittent occurence makes their investigation extremely difficult. From almost any good-quality satellite image of the Baltic (see Horstmann, 1983; Gidhagen, 1984) it is readily evident that the whole Baltic is tightly packed with eddy-like features and has more or less distinct boundaries in between (Fig.1). Taken singularly, each boundary zone may have little ecological impact. However, to assess the impact of the whole pattern of locally increased gradients and boundary zones is a challenging and important task. Up to now only single ecological studies of fronts have been made in the Baltic (Kahru et al., 1984). It is our general feeling that due to the diversity of hydrological/hydrobiological situations and the complexity of interactions, a statistically large number of cases must be analysed before any reliable assessment of the overall ecological significance of fronts and other boundaries in the Baltic can be made. Here we report some results of our biologically oriented studies from both the offshore and coastal areas of the Baltic.

METHODS

Vertical profiling

Vertical distributions of temperature, salinity, density, and chlorophyll a fluorescence were obtained by a complex of a Neil Brown Mark III CTD and a submersible fluorometer. As fluorometer first a "Variosens" (Früngel and Koch, 1976) and later a "EOS" manufactured by Elektro Optik Juan F. Suarez (Henstedt-Ulzburg, FRG) and described by Astheimer and Haardt (1984) were used. The signals were interfaced to a computer and processed on-line to equispaced (Δz = 50 cm) vertical profiles.

"Discrete" measurements

Water samples were drawn from Niskin samplers and used for on-

Fig. 1. CZCS image (channel 3) of the western Baltic Sea. Aggregations of the blue-green algae serve as tracers of the near-surface current field. Well-defined eddies are visible south and west of the Bornholm island.

board analyses of the extracted chlorophyll a, nitrate, nitrite, phosphate, silicate, primary production and for later phytoplankton counts. Not all the parameters were measured on every survey. Carbon fixation tests were made in a controlled temperature and light bath using the ^{14}C teqhnique. The potential primary production was measured by incubating vials with 120 ml of unfiltered seawater in duplicate at a saturating light intensity. Zooplankton samples were taken by vertical net (mesh size = 0.09 mm) hauls through the whole water column or through the upper 10-m layer. Phytoplankton samples were processed with the Utermöhl method (Utermöhl, 1958). More details on the methods are found in Kahru et al. (1984).

Underway sampling

Since 1985 a new system has been developed, consisting of a submersible pump at 5 m depth, a bubble trap, an on-line particle size
analyzer Hiac/Royco PC-320 (Pugh, 1978), a Turner Designs 10-005R
fluorometer, and a computer/data logger with flexible disk storage.
Particles in 12 size classes with the equivalent diameter from 1 to
1000 micrometers were counted. The time interval between the full
cycles of measurements was 1 min. Depending on the ship speed the
corresponding space interval was between 180 and 400 m. Simultaneously the near-surface temperature and conductivity were recorded.

Near-shore surveys

Coastal areas were visited with smaller vessels and simpler
equipment (Nõmmann, 1985), leaving most of the samples to later laboratory analyses. In addition, algal assays were made with a culture of Scendesmus brasiliensis to assess the concentration of nutrients in the brackish water, directly utilizable by the phytoplankton (see Källqvist, 1973).

The principal component analysis was used to discover spatial
patterns in multivariate data of phytoplankton numbers, nutrient
concentrations and other environmental variables. The species-abundance curves were calculated for the empirical phytoplankton abundances to discover shifts in the dominance pattern within the phytoplankton assemblages (see Levich, 1980).

RESULTS

Vertical structure

The subsurface chlorophyll maximum is a conspicuous feature of
many oceanic and shelf areas (Cullen, 1982). In the Baltic, a close
association between the chlorophyll maximum layers and the depths of
the local maxima in the Brunt-Väisälä frequency has been ascertained (Kahru, 1981). In relation to the vertical distribution of density (or, temperature) four phenomenological types of the chlorophyll maximum layers may be distinguished (Fig.2). In case of a
well-mixed upper layer and bloom conditions the type H1 maximum may
extend to the surface. However, type H1 may also be a subsurface
maximum in an internal homogeneous layer. Type H2, i.e. a well-defined chlorophyll maximum without an adjacent vertical density gradient, is most probably caused by the active aggregation of vertically migrating phytoplankton. During the summer thermal stratification the chlorophyll maxima are most commonly associated with the

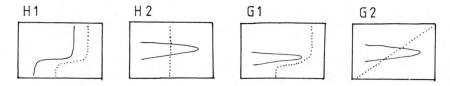

Fig. 2. Phenomenological typology of the chlorophyll maxima (continuous curves) in relation to the vertical density structure (dotted curve). The chlorophyll maxima reside either in a vertically homogeneous region (types H) or in a gradient region (types G). Both
types are further differentiated in relation to the association
with an extremum in the vertical density gradient. Types 1 are
bounded at least on one side by a density extremum. Type H1 may also be bounded from above. Types 2 are not bounded by extrema in the
density gradient.

vertical density gradient, belonging either to type G1 (with an extremum in the vertical density gradient) or type G2 (without an apparent extremum in the vertical density gradient). However, transitions between the types do occur. In the typical summer conditions in August, 1979 all the 62 vertical profiles observed had
chlorophyll maxima within the interval of 4 m from the nearest local maximum in the Brunt-Väisälä frequency with the standard deviation between the two depths of 1.3 m (Kahru, 1981). It should be
noted that the chlorophyll maxima considered here have the vertical
scale of a few meters and are reproduced on consequtive vertical
profiles. In this respect they differ from the "micropatches" discussed by Astheimer and Haardt (1984). As we have no rate measurements for the Baltic chlorophyll maxima, no firm connections between the phenomenological types and the functional types of Cullen
(1982) can be established.

In or near horizontal fronts the vertical structure is much more
complicated, showing various intrusions and interleavings (Fig. 3).
The conversion of fluorescence to chlorophyll a concentrations in
layers of different origin and with different phytoplankton populations may be problematical due to variations in the fluorescence
yield. However, in most cases with sufficient range of data points,
the calibration regressions (Fig. 4) are quite reliable (Kahru and
Aitsam, 1985).

Another problem, related to the potentials of remote sensing to
estimate the distributions of chlorophyll and primary production,
is the correlation strength between the near-surface chlorophyll
and the photic zone integral chlorophyll. On the basis of 570 vertical profiles from the Baltic (Kahru, 1985) it has been shown that

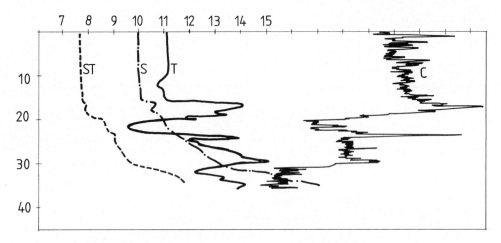

Fig. 3. Profiles of temperature (T, °C), salinity (S, °/oo), density (ST, sigma-t), and chlorophyll a fluorescence (C, rel. units) versus depth (m) near a front in the Arkona basin, 23 Sept 1979.

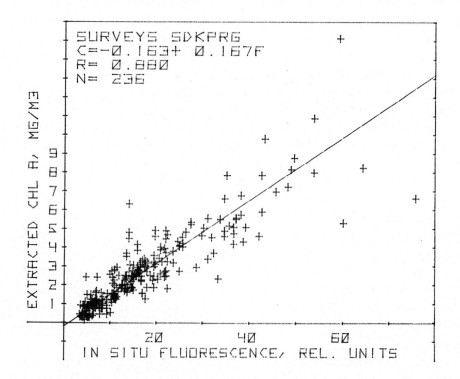

Fig. 4. Calibration regression of extracted chlorophyll a from discrete water samples versus in situ fluorescence for surveys in July-August, 1982.

only in case of a well-mixed upper layer serves the near-surface
concentration as a good estimate of the photic zone chlorophyll. On
the other hand, in case of a stratified photic layer the correlation between the near-surface and the photic zone chlorophyll is
low. Moreover, it has been documented that if there are stations
with a strong subsurface maximum and with a more or less uniform
vertical distribution (Fig. 5), or if the photic zone depth changes
primarily due to non-chlorophyllous material, the relation between
the near-surface and the photic zone chlorophyll is inverse. This
can be a source of error for the remote estimations of chlorophyll
and primary productivity.

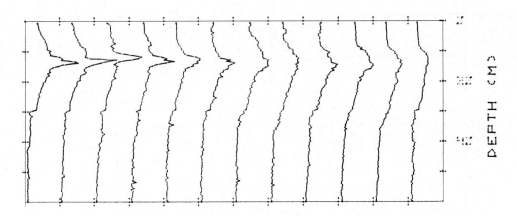

Fig. 5. Sequence of vertical chlorophyll a profiles along a transect line with a space step of 2.5 nautical miles (Gotland basin,
6 Aug 1982). Each ensuing profile is offset 3 mg m^{-3}. The shift in
the vertical distribution from a strong subsurface maximum to a
more or less uniform vertical distribution gave rise to an inverse
relationship between the surface and the photic zone chlorophyll
(only the central part of the transect is shown).

Offshore boundaries associated with fronts and eddies

Our first biological front study in the south-eastern Gotland
basin in 1982 yielded rewarding results (Kahru et al., 1984). A
30-m thick water mass with anomalously low salinity extended vertically across the horizontally uniform thermocline (Fig. 6). The
front was primarily a salinity front and not a density front (only
the isopycnals in the top 10-14 m were inclined). In the top layer
the higher salinity water had protruded onto the low-salinity water.
It was hypothesized that after the initial convergence of the two
water masses, a vertical stretching of the top layer was produced,
the top isotherms being lifted up, and a cyclonic circulation es-

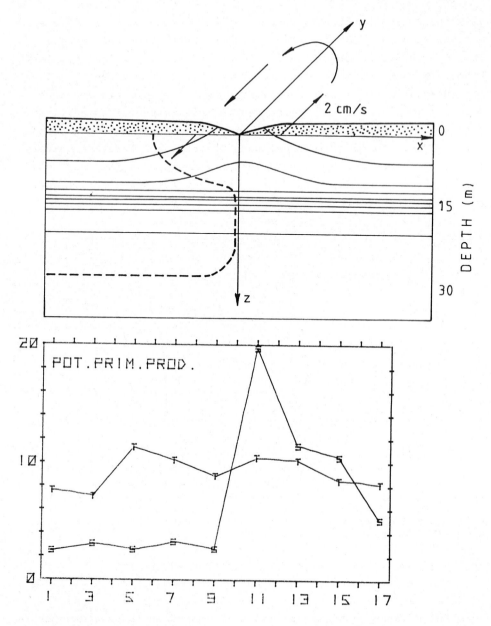

Fig. 6. (upper panel) Schematic description of the frontal structure and of the proposed mechanism for frontal upwelling (Gotland
basin, 6 Aug 1982). Left from the dashed curve: the low-salinity
anomaly; continuous curves: isopycnals (or isotherms); arrows: cyclonic circulation; stippled area: supposed deformation of the free
surface.
(lower panel) Horizontal distribution of the potential primary production (mgC $m^{-3}h^{-1}$) near the surface (S) and in the thermocline (T).

tablished. The limited upwelling of the upper thermocline water was
suggested to be responsible for the 7-fold increase in the potential primary productivity adjacent to the salinity interface. The
productivity values were higher in the higher-salinity side of the
front and levelled down further away from the front. The sharp thermocline where the upwelled waters could originate from obviously
accumulated sinking organic matter which released nutrients. The
changes in primary productivity were primarily caused by changes in
the assimilation numbers and not in the biomass as the chlorophyll
content had only a slight maximum adjacent to the frontal zone. However, the zooplankton biomass more than doubled in the higher-salinity side of the front.

Another type of fronts was under observation in the Bornholm basin in the summer of 1984. In a 4-day interval the transect was covered twice (2 and 6 August) and again after 4 days profilings were
made on the rectangular grid of stations (Fig. 7). While both of

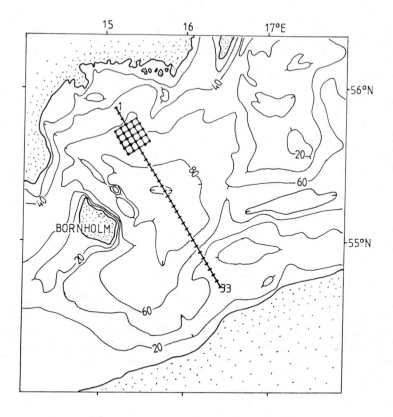

Fig. 7. Study area in the Bornholm basin in August, 1984. Station
spacing along the transect and on the grid is 2.5 nautical miles.

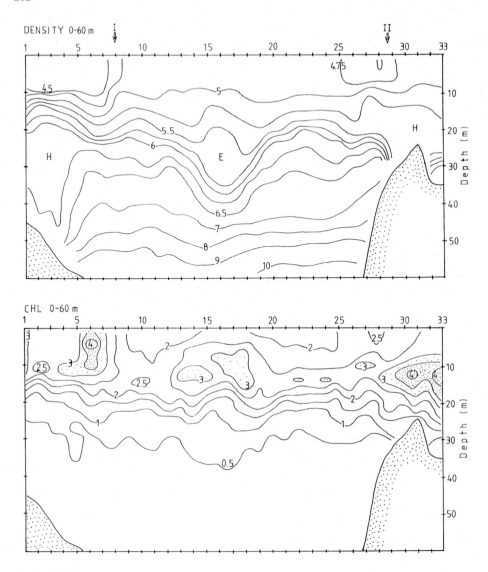

Fig. 8. Vertical distributions of density (sigma-t units) and chlorophyll a (mg m^{-3}) on the transect, 6 Aug 1984. Arrows point to the
frontal structures labeled I and II. An eddy-like deformation of the
thermocline (E) and two bottom-mixed homogeneous water masses (H)
near shallow banks are labeled. Note the chlorophyll maximum in the
low-salinity side of front I and another maximum above one of the
bottom-mixing areas. The contour interval is 0.5 for chlorophyll,
0.25 for the sigma-t range 4.5-6.5 and 1.0 for the sigma-t range
7-10.

the frontal structures found (Fig. 8) were primarily caused by
changes in salinity, they were also distinct density fronts. The
frontal structures were not stationary in space but were advected
(or meandered) in the NW direction (both fronts - 10 km in the 1st
4-day interval, front I - 5 km in the 2nd 4-day interval). The other
distributions revealed several features related to the frontal structures. The chlorophyll maximum was bound to the low-salinity edge of
front I on both of the transects. The two maxima in the distributions of the nonmigrating zooplankton most conspicuously responded
to the both frontal structures (Fig. 9A). The distribution pattern
of copepods could have been similar unless the upper layer samples
were overmasked by their diel vertical migration. The dominant species in the frontal maxima were <u>Bosmina coregoni maritima</u> for the
cladocerans and <u>Synchaeta monopus</u> for the rotifers. The nutrient
concentrations were quite variable; however, the nitrate maximum between the stas. 6-10 may be associated with the frontal region
(Fig. 9B). The phosphate maximum between the stas. 10-18 (not shown)
seems to be characteristic of the water mass on the SE side of
front I. Although the analytical error variance of the nutrient determinations might be considerable, the space resolution of 2.5 n.
miles is obviously too big to resolve the intense smaller scale
variability (see Hansen, 1985). The same was probably true of the
primary productivity distributions. The productivity values were
highly variable and did not show obvious connections with the other
variables. This may partly be attributed to the dominance of the
cyanobacteria (<u>Aphanizomenon, Microcystis, Nodularia</u>) forming large
aggregates and increasing therethrough the sample variance.

Unfortunately, no current measurements in the fronts were available but in general fronts are known as sites of active convergence/
divergence. The increased abundance of zooplankton in the frontal
regions might have resulted from the flow convergence in the fronts.
The organisms which can maintain their preferred depth can easily
accumulate in convergence zones (Olson and Backus, 1985). The same
mechanism, however, did not work for ichtioplankton which was more
abundant in the region between the two fronts. Although fronts have
increased gradients in the across-front direction, they are by no
means merely 2-dimensional (vertical and across-front) phenomena.
This was substantiated by mapping the temperature, salinity and
chlorophyll distributions on the 2-dimensional grid (Fig. 10).
Changes in the location of the frontal boundary, interpreted as the
meandering or advection of the front, could easily be followed in

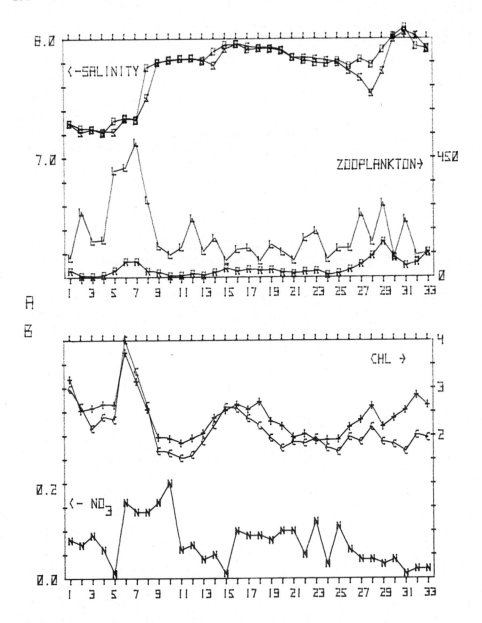

Fig. 9. Horizontal distributions along the transect through the Bornholm basin, 6 Aug 1984. (A) Salinity ($^o/_{oo}$) at the depths of 5 m (5) and 10 m (O), biomass (mg m^{-2}) of the cladocerans (L) and rotifers (R) in the upper 10-m layer.
(B) Surface (1m) concentrations of nitrate (N, μM), chlorophyll $\underline{a}$ as extracted from water samples (C), and the mean chlorophyll $\underline{a}$ concentration in the top 10-m layer as measured fluorometrically (+, mg m^{-3})

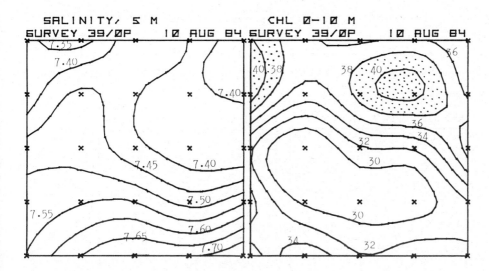

Fig. 10. Horizontal distributions on the 10 x 10 n. mile grid (for the position see Fig. 7): near-surface (5 m) salinity (‰) and the top 10-m chlorophyll concentration (mg m^{-2}).

the thermohaline structure. However, the biological and chemical distributions had undergone drastic changes between the surveys. As the dominant currents ought to be in the along-front directions, we suggest that the principal source of variations between the subsequent surveys was the advection of various scale patches across the station grid. This could account for the lack of resemblance between the consequent surveys for the chlorophyll (Figs. 8 and 10) and nutrient distributions.

We have presented evidence that the synoptic scale eddies contribute significantly to the distributions of biological and chemical variables in the Baltic (Kahru et al., 1981, 1982; Aitsam et al., 1984). Eddies may develop from frontal boundaries (Pingree et al., 1979) and can themselves produce smaller scale fronts by accentuating the existing gradients. The synoptic scale eddies contribute probably extensively to the vertical flux of nutrients in the interior of the Baltic (Kahru, 1982). However, the effect of an eddy field on the nutrient flux is highly dependent on the vertical thermohaline and nutrient stratification as well as on the energy available to overcome the buoyancy forces. Consequently, some eddies simply transport nutrient and chlorophyll anomalies over large distances (Aitsam et al., 1984). Apart from lifting or lowering the pycnoclines (and, possibly, nutricline) in eddy centers, the eddies

influence the mixing in their periphery, probably owing to the intensified shear zones at the eddy boundaries. Fig. 11 is a result of
a joint survey with the Institut für Meereskunde, Kiel (Dr. H.P.
Hansen) showing coherent patch-like anomalies in the hydrochemical

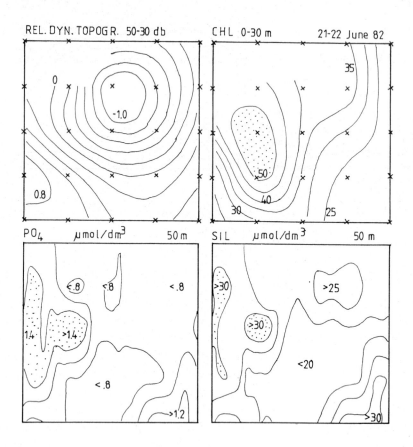

Fig. 11. Horizontal distributions on a 10 x 10 n. miles area in the
Bornholm basin, 21-22 June 1982. The phosphate and silicale data
were measured by Dr. H.P. Hansen (IfM, Kiel).

parameters and chlorophyll distribution. The anomaly with maxima in
the nutrient concentrations and minima in pH and oxygen was located
in the eddy periphery with maximal gradients in the geostrophic currents. The chlorophyll patch in the upper 30-m layer was located
exactly above the nutrient patch at 40-60 m depths. Some of the differences between the corresponding contour maps are not due to different distributions but to different sampling schemes. Whereas the

CTD and chlorophyll profiles were taken at the grid points with a
2.5 n. miles step, the hydrochemical data were obtained by horizontal scanning along the parallel grid lines.

Offshore fronts also delimit different plankton communities or
seasonal succession stages. This has been observed with the use of
the flow-through measurement system. As most of the data are still
under scrutiny, only one example will be presented. A sharp salinity
front was observed during the spring bloom near the opening to the
Gulf of Finland. Together with the drop in salinity an abrupt change
in the particle size structure was observed (Fig. 12). While the
channel 7 values remained on the same level, channels 8 and 9 as
well as the fluorescence showed a sharp increase. It was proved later that the changes were due to the increased mean size of the diatom Achnanthes taeniata "chains". The longer chains of A. taeniata
colonies were assigned by the counter into the higher size classes.
Together with the other changes (Table 1), e.g. the increased chlorophyll, pheopigment and primary production levels, phytoplankton
abundance and the increased numbers of the decaying cells of the
cold water species Melosira arctica and Nitzschia frigida this indicates that the front marked a boundary between different succes-
sional stages of the vernal bloom. Whether there were some specific
boundary effects, e.g. elevated values compared to the both sides,
is not clear but the waters of the Gulf of Finland were certainly
in a more advanced stage.

TABLE 1
Near-surface (5 m) hydrographic data and phytoplankton parameters
for stations representative of the either side of the front and
adjacent to the front (cf. Fig. 12).

	T5	F	T6
Salinity ($^o/_{oo}$)	7.33	6.77	6.85
PO_4-P (μM)	0.42	0.48	0.40
Chlorophyll a (mg m^{-3})	5.47	14.28	12.68
Phaeopigment (- „ -)	1.25	2.16	1.80
Light-saturated primary production (mg C m^{-3}h^{-1})	21.0	48.1	48.5
Total phytoplankton abundance (solitary cells and colonies/ml)	490	808	596
Achnanthes taeniata (colonies/ml)	225	388	339
Mean equivalent diameter of (A. taeniata colonies (μm)	41.9	45.7	48.3

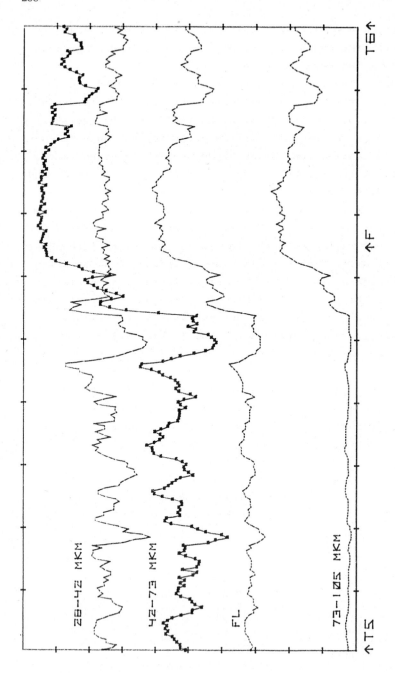

Fig. 12. Near-surface (5 m) horizontal distributions of chlorophyll fluorescence and particle concentrations in the equivalent diameter ranges 28-42, 42-73 and 73-105 μm. The interval between points T5 and T6 is 50 n. miles. Water samples were taken from points T5, F, T6. The full scale is equivalent to 12o particles/ml.

Boundary effects in the Moonsund area

Altogether 6 surveys were made in the study area (Fig. 13) in
May, July and October, 1982 and May, July and November, 1983 (Nõm-
man, 1985). In the space of the first two principal components a
boundary area between the waters of the Matsalu Bay and the Soela
Strait is distinguished (Figs. 13, 14). The boundary area is a site
of strong vertical stirring due to a strong, fluctuating current
over shallow areas, and of lateral exchange between different water
masses. The boundary discussed here is not a front in its usual
sense but rather an area of contact between different communities
and water masses, and has a scale of about 10 km. While some parameters (the water transparency) decreased gradually from the west
to the east (Fig. 15), several others showed pronounced maxima in
the boundary zone, e.g. the nutrients, algal assay values, phytoplankton biomass and species richness. The species-abundance curves
of the spring phytoplankton (Fig. 16) show that while the eutrophic
Matsalu Bay community is strongly dominated by a single species
(Diatoma elongatum v. tenuis), and the oligotrophic community has
very long abundances with a low number of species, the boundary
phytoplankton community has a rich variety of relatively uniformly
distributed species. The results of the 6 surveys varied to some
extent but at least some of the boundary effects mentioned were
present on all the surveys.

DISCUSSION

Using results from our recent studies we have presented examples
of various physical-biological couplings between the synoptic scale
and the smaller scale distributions of variables in the Baltic Sea.
We must admit that our abilities to interpret various details in
these distributions in terms of the dynamics (i.e., their generation, transport and decay) are limited and ambiguous. Although the
boundary zones described herein have often been characterised by
elevated values of plankton concentrations and/or rates of related
processes with respect to the surrounding waters, this is by no
means always the case. As rightly noted by Richardson et al. (1985),
there is a natural tendency to preferentially report those occasions with significant boundary effects, and to "forget" the cases
with no apparent differences. We have demonstrated that the ecological system in the Baltic can actively respond to the physical
forcings, such as the synoptic eddies, frontal dynamics, etc. We
believe that much of the variability in the response stems from

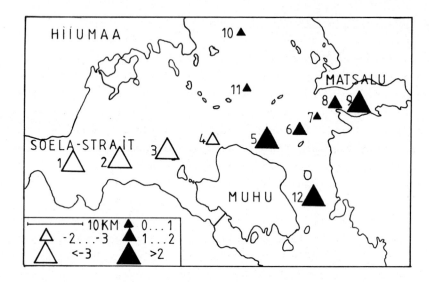

Fig. 13. Study area and horizontal pattern of scores of the first principal component in May, 1983.

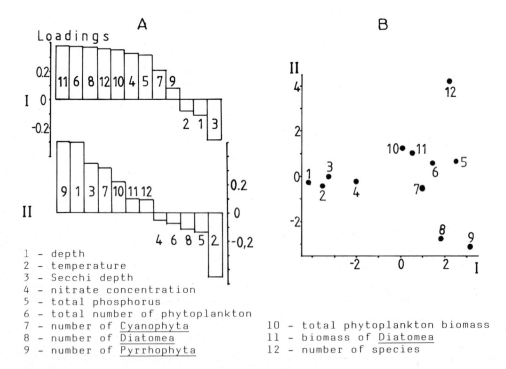

1 - depth
2 - temperature
3 - Secchi depth
4 - nitrate concentration
5 - total phosphorus
6 - total number of phytoplankton
7 - number of Cyanophyta
8 - number of Diatomea
9 - number of Pyrrhophyta

10 - total phytoplankton biomass
11 - biomass of Diatomea
12 - number of species

Fig. 14. (A) Structure of the first two principal components with the ingredient variables listed below;
(B) Ordination of stations in the space of the first two components.

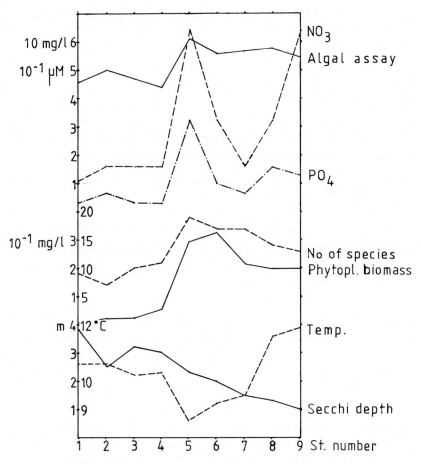

Fig. 15. Horizontal profiles of some biological, chemical and physical variables across the Moonsund boundary area in May, 1983.

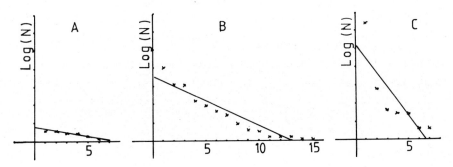

N - number of individuals per species

Fig. 16. Species-abundance curves of three phytoplankton communities:
A - Soela Strait, B - boundary area, and C - Matsalu Bay.

the intricate hydrodynamics of the forcings themselves. It seems that in the face of the ubiquitous occurrence of fronts and other boundary phenomena in the Baltic and in the face of the results from this and other areas, it is not the question any more whether the boundaries have biological effects but what is the density of occurrence in space and time of the boundaries with certain characteristics. The next task then should be the determination of the overall importance of the Baltic boundary zones to its ecology, to the local fisheries, etc. A major task for the coming years would be to accumulate representative statistics on biological effects of fronts and other boundary phenomena, and to reveal the mechanisms which govern the biological response to the physical forcing. To accomplish this task, a major improvement in the sampling strategy and techniques should take place. First, more extensive use of the remote sensing (see Campell and Esaias, 1985) and high resolution in situ methods should be made. The temporal dynamics of the phenomena should be studied in their full dynamics, i.e. from their generation till their decay. In a system which is not at equilibrium, the temporal sequence of events is crucial (Harris, 1983). Therefore the correlations between planktonic and environmental variables as measured at a single instant need not yield much insight. It is also evident that the variability at the level of integral parameters such as biomass or in vivo fluorescence obscures the inherent variability of a plankton community at the level of populations. It is at the level of populations or even their developmental stages that the significant interactions take place. We hope that the use of the high-resolution, flow-through particle counting system with the accompanying measurements will provide useful insights in the near future.

Acknowledgements. We are greatly indebted to a number of persons among whom J. Elken, I. Kotta, L. Käärman, J. Lilles, J. Lokk, M. Püvi, T. Raid, A. Randveer, H. Salandi and A. Siht have been most helpful. The CZCS image on Fig. 1 was kindly supplied by Dr. U. Horstmann (Institut für Meereskunde, Kiel).

REFERENCES

Aitsam, A., Hansen, H.P., Elken, J., Kahru, M., Laanemets, J., Pajuste, M., Pavelson, J. and Talpsepp, L., 1984. Physical and chemical variability of the Baltic Sea: a joint experiment in the Gotland basin. Cont. Shelf Res., 3: 291-310.

Astheimer, H. and Haardt, H., 1984. Small-scale patchiness of the chlorophyll-fluorescence in the sea: aspects of instrumentation, data processing, and interpretation. Mar. Ecol. Prog. Ser., 15: 233-245.

Campbell, J.W. and Esaias, W.E., 1985. Spatial patterns in temperature and chlorophyll on Nantucket Shoals from airborne remote sensing data, May 7-9, 1981. J. Mar. Res., 43: 139-161.

Cullen, J.J., 1982. The deep chlorophyll maximum: comparing vertical profiles of chlorophyll a. Can. J. Fish Aquat. Sci., 39: 791-803.

Fasham, M.J.R., Platt, T., Irwin, B. and Jones, K., 1985. Factors affecting the spatial pattern of the deep chlorophyll maximum in the region of the Azores front. Prog. Oceanog., 14: 129-165.

Früngel, F. and Koch, C., 1976. Practical experience with the Variosens equipment in measuring chlorophyll concentrations and fluorescent tracer substances, like Rhodamine, Fluorescein, and some new substances. IEEE J. Ocean. Engn., 1: 21-32.

Gidhagen, L., 1984. Coastal upwelling in the Baltic - a presentation of satellite and in situ measurements of sea surface temperatures indicating coastal upwelling. Parts I, II. SMHI-reports RHO-37, Norrköping, 40+60 pp.

Hansen, H.P., 1984. The significance of discrete Baltic nutrient samples discussed on the basis of continuous measurements. D. hydrogr. Z., 37: 245-258.

Harris, G.P., 1983. Mixed layer physics and phytoplankton populations: studies in equilibrium and non-equilibrium ecology. Prog. Phycol. Res., 2: 1-52.

Holligan, P.M., 1981. Biological implications of fronts on the northwest European continental shelf. Phil. Trans. R. Soc. Lond., A302: 547-562.

Horstmann, U., 1983. Distribution patterns of temperature and water colour in the Baltic Sea as recorded in satellite images: indicators for phytoplankton growth. Ber. Inst. Meeresk. Univ. Kiel, 106 (1), 147 pp.

Iles, J.D. and Sinclair, M., 1982. Atlantic herring: stock discreteness and abundance. Science, 215: 627-633.

Jansson, B.-O., Wilmot, W. and Wulff, F., 1984. Coupling the subsystems - the Baltic Sea as a case study. In: M.J.R. Fasham (Editor), Flows of energy and materials in marine ecosystems. Plenum Press, New York, pp. 549-595.

Kahru, M., 1981. Relations between the depths of chlorophyll maxima and the vertical structure of density field in the Baltic Sea. Oceanology, 21: 76-79.

Kahru, M., 1982. The influence of hydrodynamics on the chlorophyll field in the open Baltic. In: J.C.J. Nihoul (Editor), Hydrodynamics of semi-enclosed seas. Elsvier, Amsterdam, pp. 531-542.

Kahru, M., 1985. Remote sensing and the vertical distribution of the Baltic chlorophyll. Proc. 14th Conf. Baltic Oceanographers, Gdynia, in press.

Kahru, M. and Aitsam, A., 1985. Chlorophyll variability in the Baltic Sea: a pitfall for monitoring. J. Cons. int. Explor. Mer., in press.

Kahru, M., Aitsam, A. and Elken, J., 1981. Coarse-scale spatial structure of phytoplankton standing crop in relation to hydrography in the open Baltic Sea. Mar. Ecol. Prog. Ser., 5: 311-318.

Kahru, M., Aitsam, A. and Elken, J., 1982. Spatio-temporal dynamics of chlorophyll in the open Baltic Sea. J. Plankton Res., 4: 779-790.

Kahru, M., Elken, J., Kotta, I., Simm, M. and Vilbaste, K., 1984. Plankton distributions and processes across a front in the open

Baltic Sea. Mar. Ecol. Prog. Ser., 20: 101-111.

Kullenberg, G., 1982. Mixing in the Baltic Sea and implications for the environmental conditions. In: J.C.J. Nihoul (Editor), Hydrodynamics of semi-enclosed seas. Elsvier, Amsterdam, pp. 399-418.

Källqvist, T., 1973. Use of algal assay for investigating a brackish water area. In: S.G. Lönnberg (Editor), Algal assay of water pollution research. Oy Kirjapaino, Helsinki, pp. 111-124.

Levich, A.P., 1980. The structure of ecological communities (in Russian). Moscow State University, Moscow, 181 pp.

Marra, J., Houghton, R.W., Boardman, D.C. and Neale, P.J., 1982. Variability in surface chlorophyll a at a shelf-break front. J. Mar. Res., 40: 575-591.

Nômmann, S., 1985. Mesoscale spatial variability of the phytoplankton distribution in the Moonsund area of the Baltic Sea. Proc. 14th Conf. Baltic Oceanographers, Gdynia, in press.

Olson, D.B. and Backus, R.H., 1985. The concentrating of organisms at fronts: a cold-water fish and warm-core Gulf Stream ring. J. Mar. Res., 43: 113-137.

Pingree, R.D., Holligan, P.M. and Mardell, G.T., 1979. Phytoplankton growth and cyclonic eddies. Nature, Lond., 278: 245-247.

Pugh, P.R., 1978. The application of particle counting to an understanding of the small-scale distribution of plankton. In: J.H. Steele (Editor), Spatial pattern in plankton communities. Plenum Press, New York, pp. 111-129.

Richardson, K., Lavin-Peregrina, M.F., Michelson, E.G. and Simpson, J.H., 1985. Seasonal distribution of chlorophyll a in relation to physical structure in the western Irish Sea. Oceanol. Acta, 8: 77-86.

Shaffer, G., 1979. On the phosphorus and oxygen dynamics of the Baltic Sea. Contr. Askö Lab. Univ. Stockholm, No. 26, 90 pp.

Utermöhl, H., 1958. Zur Vervollkommnung der quantitativen Phytoplankton-Methodik. Mitt. Int. Ver. Theor. Angew. Limnol., 1-38.

THE ROLE OF THE LOOP CURRENT IN THE GULF OF MEXICO FRONTS

D.A. SALAS de LEON and M.A. MONREAL-GOMEZ

CONACYT, Consejo Nacional de Ciencia y Tecnologia, México

GHER, Université de Liège, B-5 Sart Tilman, B 4000 Liège, Belgium

ABSTRACT

The role of the Loop Current on the fronts of the Gulf of Mexico is studied using the results of the Monreal-Gómez and Salas de León (1985) Gulf of Mexico numerical model. The horizontal thermal structure can be depicted with the pycnocline anomaly. Assuming that biological processes take place in the frontal areas, the thermal boundaries at the cyclonic and anticyclonic eddies, as well as the Loop Current have an effect on the chlorophyll-a concentration in the Gulf of Mexico.

INTRODUCTION

Differences between the waters of the Gulf of Mexico and waters of the Loop Current are rather small, showing a slightly horizontal thermal gradient (Ichiye, 1962). Nevertheless, the anticyclonic and cyclonic rings separated from the Loop Current appear as warm salty bodies which advect into the Mexican Current, balancing the salt and heat budget of the western Gulf (Elliot, 1982). Fronts associated to surfacing of cooler waters with high levels of phytoplankton pigment and the path of the warm Loop Current, with low concentration of phytoplankton pigment were observed by Yentsch (1984) from Coastal Zone Color Scanner (CZCS) images of the west Florida coast. In the Bay of Campeche, a high biological productivity has been observed (Licea-Duran, 1977) as well as on the west Florida continental shelf. Using Earth Resources Technology Satellite (ERTS) and shipboard measurments, Maul (1975) depicted the Loop Current and demonstrated that, the cyclonic boundaries are associated with fronts and changes in chlorophyll-a concentration and temperature.

Using the above consideration, the pycnocline anomaly, and the cyclonic and anticyclonic eddy boundaries, will be related to the behavior of the major frontal zones in the Gulf of Mexico.

DISCUSSION

The role of the Loop Current and the detached eddies in the Gulf of Mexico frontogenesis can be inferred from the numerical model developed by Monreal-Gómez and Salas de León (1985). In this model, the driving mechanism is a jet through the Yucatan Channel and the Florida Strait. The Loop Current generated only with the Yucatan and Florida jet show lenses coinciding with the observed

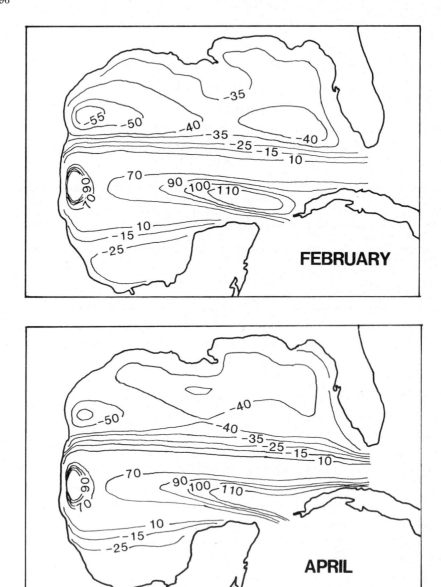

Fig. 1a. : Results of the linear baroclinic model. Pycnocline height anomaly in m.

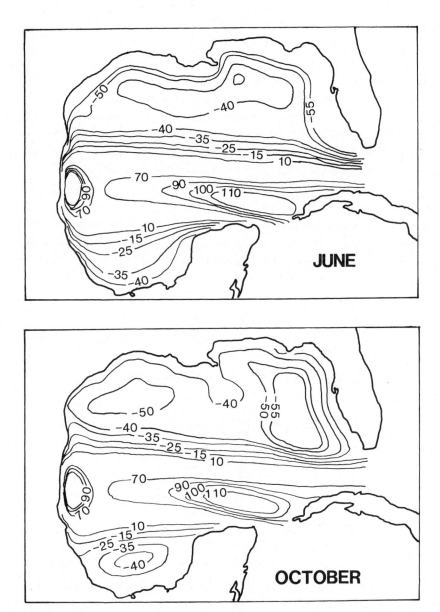

Fig. 1b. : Results of the linear baroclinic model. Pycnocline height anomaly in m.

thermal structures (Fig. 1 a, b).

Assuming that phytoplankton patches reflect the density or thermal field (Yentsch, 1984), the influence of eddies in the spatial patterns of growth can be explained in terms of what can be observed in Fig. 2. The anticyclonic eddies with a diving pycnocline are associated to warm rings in which lighter waters are located at the center and heavier waters at the external border of the eddies. The Loop Current, the anticyclonic and the cyclonic eddies represent different water masses, the last one colder than the others. The boundary between the Loop Current and the anticyclonic and the cyclonic eddies will be an active frontal zone. The anticyclonic eddies and the Loop Current located in the middle of the Gulf of Mexico will work as a natural boundary for the phytoplanktonic growth. On the other hand, with a cyclonic circulation in the west of Florida, the Texas-Louissiana shelf, and the Bay of Campeche, one will expect high concentration of phytoplankton.

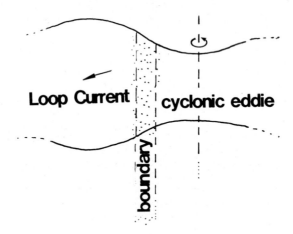

Fig. 2. : Geostrophic relationships in the Loop Current and cyclonic eddies. The boundary of the cyclonic eddies with the Loop Current is a zone of frontogenesis.

One can imagine three different mechanisms interacting between the Loop Current and the cyclonic eddies. The first one corresponds to a divergence zone in which the lower layer is surfacing. The second one represents a convergence zone produced by the interaction of the cyclonic transport into the Loop Current when the densities are almost similar. The last mechanism is similar to the second one, but in this case the density of the Loop Current waters are markedly lighter than eddy waters. Here the Loop Current waters will be pushed up, and frontal conditions will result (Fig. 3).

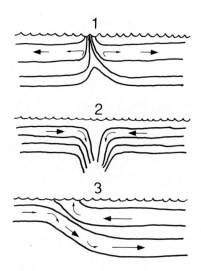

Fig. 3. : Interacting mechanisms between the Loop Current and the cyclonic eddies. The first one correspond to a divergence zone and the second and third ones to a convergence zone.

By comparing transport predicted by the model and distribution of mean concentration of organisms observed in the Bay of Campeche by Licea-Duran (1977), one can see high concentrations in the northeast Bay of Campeche for the month of September coinciding with a well defined cyclonic eddy and a high transport in the Yucatan Channel. This effect can be explained in terms of mechanism number 3 showed in Fig. 3.

Finally associating Loop Current and cyclonic eddies, one can identify the semi-permanent frontal zones in the area. Fig. 4 gives an example for the month of October.

CONCLUSION

Despite the simplifying hypothesis of the Monreal-Gomez and Salas de Leon (1985) model, many features of the Gulf of Mexico fronts can be inferred from the Loop Current dynamics. In particular the strong anticyclonic eddy in the middle of the Gulf appears as a natural boundary for the phytoplanktonic growth. Areas of possible frontal activity can be deduced in zones of convergence and divergence between the Loop Current and the cyclonic eddies.

300

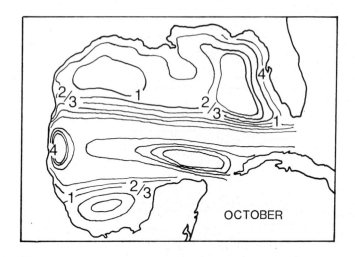

Fig. 4. : Semi-permanent front induced by the Loop Current. Numbers 1, 2 and 3 represent the mechanisms showed in Fig. 3. Numbers 4 represent a sensible area to frontal activity induced by a cyclonic or anticyclonic eddy and a closed boundary.

REFERENCES

Elliot, B.A., 1982. Anticyclonic rings in the Gulf of Mexico. J. Phys. Oceanogr. 12 : 1292-1309.
Ichiye, T., 1962. Circulation and water mass distribution in the Gulf of Mexico. Geofis. Int. Mex., 2 : 47-76.
Licea-Duran, S., 1977. Variacion estacional del fitoplancton de la Bahia de Campeche, Mexico (1971-1972). FAO Fish. Report, 200 : 253-273.
Maul, G.A., 1975. An evaluation of the use of the earth resources thecnology satellite for observing ocean current boundaries in the Gulf Stream. NOAA Technical Report ERL. 335-AOML 18, 125 pp.
Monreal-Gomez, M.A. and D.A. Salas de Leon, 1985. Barotropic and baroclinic modes in the Gulf of Mexico. (To be published) On Proc. Symp. on Oceanology, Bruxelles, 10 pp.
Yentsch, C.S., 1984. Satellite representation of features of ocean circulation indicated by CZCS colorimetry. In J.C.J. Nihoul (Editor), Remote Sensing of Shelf Sea Hydrodynamics. Elsevier Oceanography Series, 38. Amsterdam, pp. 337-354.

PRELIMINARY STUDY OF A FRONT IN THE BAY OF CAMPECHE, MEXICO

S.P.R. Czitrom, F. Ruiz, M.A. Alatorre and A.R. Padilla
Instituto de Ciencias del Mar y Limnología, Apartado Postal 70-305,
Ciudad Universitaria, U.N.A.M., 04510 México D.F., MEXICO.

ABSTRACT

Data collected during three surveys of the Bay of Campeche are used to describe the thermohaline field at different times of the year. During one of these surveys, a river discharge induced front and associated stratification were observed. A one dimensional model of vertical mixing is used to analyse the effect of wind on the frontal stratification. It is likely that layering will survive a typical storm for the area while the front appears to have greater persistence during windy weather than the associated stratification. Results from a computer model of the residual circulation suggest that attention must be payed to the general circulation of the Gulf of Mexico in the design of future data collection and in its interpretation.

INTRODUCTION

The Bay of Campeche in the Gulf of Mexico is an area where much human economic activity is carried out. Fishing grounds are highly exploited, substantial quantities of industrial wastes are dumped into the rivers that discharge in the area and, more recently, offshore oil exploration and exploitation on a grand scale has played an increasing part in financing the Mexican economy. The physical processes occurring in these waters must therefore be studied to provide the framework within which a rational exploitation of the area can be carried out.

In this context, the study of fronts - which may be defined as the boundaries between water masses of different physical, chemical or biological characteristics - acquires a special relevance. It has been known for many years that fish tend to congregate near fronts in the seas of Japan (Uda, 1938) while other authors (see Murray, 1975; Klemas, 1980) have observed that the size and distribution of oil slicks and other contaminants can be strongly affected by the presence of fronts.

Sightings of a sharp boundary between river discharge and oceanic waters along which bands of floating debris accumulate

302

have been made in the Bay of Campeche (I. Emilsson, L. Lizarraga, personal communication). This front may have helped to protect the Mexican shoreline from greater damage during the Ixtoc-1 oilwell blowout in 1979. The results presented in this paper are a preliminary study of fronts in the Bay of Campeche as part of a longer term effort to understand their dynamics.

THE DATA

Three surveys of the area shown in fig. 1 were carried out in september 1977, June 1978 and April 1984. In the first two, Niskin bottles fitted with reversing thermometers were used to take samples for the determination of salinity and density. In the third survey, a Neil Brown CTD system gave readings of conductivity and temperature in the water column at the stations marked with dots in fig. 1. Also, continuous measurements of surface salinity and temperature were made while the ship steamed between stations. A satellite positioning system was used for navigation.

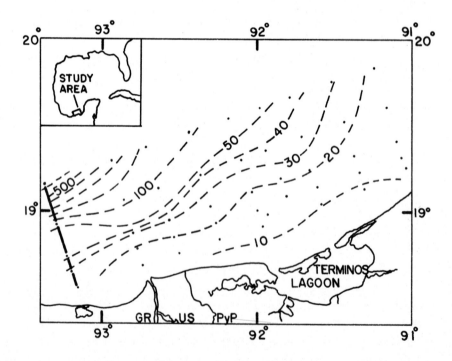

Figure 1. Bathymetry of the area of study based on soundings taken during the April 1984 survey. The transect referred to in the text is marked with a heavy line and the station positions with dots. The rivers Grijalva, Usumacinta and San Pedro y San Pablo are marked GR, US and PyP respectively.

Figure 2 shows contour maps of surface salinity, temperature and sigma-t during the CTD survey of April 1984. A marked front characterized by intensified gradients can be seen in the surface density field which is clearly the result of a frontal structure in the distribution of salinity. No frontal structure was apparent in the temperature field. The front was observed 25 nautical miles from the coast and ran approximately parallel to it for at least 70 nm, turning landwards off the western inlet to the Terminos Lagoon. The band of strongest surface gradients had a width of about 5 nm within which a change of 1 salinity unit and 1 sigma-t unit occurred. The location of this band between stations was pinpointed using the continuous surface measurements of salinity and temperature.

Contours of salinity, temperature and sigma-t on a section perpendicular to the coast which is typical of the western half of the grid can be seen in fig. 3. The section is marked with a heavy line in fig. 1. Two areas of vertical stratification can be distinguished. One is associated to the front and lies in shallow water to the south of it. The other one, apparently not related to the front, lies below 30 m and to the north. In this study we have focused our interest in the upper 30 m of water where the front and associated stratification are confined. In the eastern half of the grid, the water column exhibited little or no layering.

The strength of stratification for a surface layer 30 m thick was estimated by computing

$$\phi = 1/h \int_{-h}^{0} (\rho - \overline{\rho})gz \, dz \qquad \text{where} \qquad \overline{\rho} = 1/h \int_{-h}^{0} \rho \, dz$$

(see Simpson et al., 1977) for all CTD casts. Here ρ is density, h is depth and z is the vertical coordinate positive upwards. ϕ can be interpreted as the amount of work per m^3 needed to vertically mix a stratified column of water. The parameter Rs = $\phi s/\phi$ (Czitrom, 1982) was also computed for all stations. Here,

$$\phi s = 1/h \int_{-h}^{0} (\rho(\overline{T},S) - \overline{\rho}(T,S))gz \, dz \qquad \text{where} \qquad \overline{T} = 1/h \int_{-h}^{0} T \, dz$$

and T and S are temperature and salinity respectively. ϕs can be interpreted as the amount of work per m^3 that would be required to vertically mix only the observed salinity profile. Thus Rs is the proportion of density layering that is due to haline stratification.

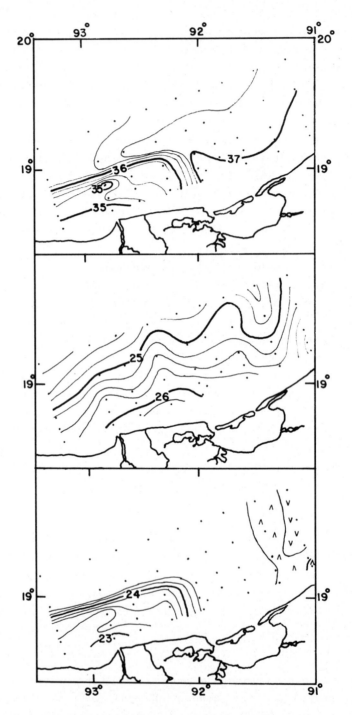

Figure 2. Isolines of a) Salinity, b) Temperature and c) Sigma-t at the surface in the area during the April 1984 survey. A density front is clearly the result of the salinity distribution.

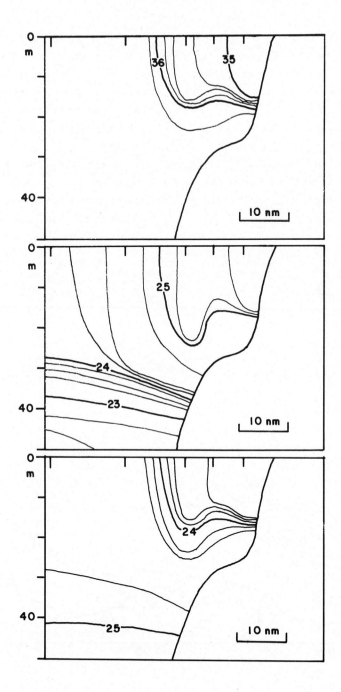

Figure 3. Isolines of a) Salinity, b) Temperature and c) Sigma-t on the transect marked on fig. 1.

306

Figure 4 shows the distribution of ϕ for a surface layer 30 m thick during the survey of April 1984. Most of the bay exhibited little or no layering except for an area immediately landward of the front (compare with fig. 2). At these stations ($\phi > 5$ Joules/m^3), nearly 80% of the stratification was due to the salinity profile (Rs = 0.79 ± 0.09) which points to the river discharge as the main stratifying agent.

At stations marked "*" in fig. 4, stratification below 30 m was nearly all thermal (Rs = 0.00 ± 0.06 for a layer 100 m deep). At those marked "o", significant salinity layering worked against vertical stability (Rs = -0.39 ± 0.12 for a layer 100 m deep). Overturning of the water column at these stations was prevented by the temperature profile.

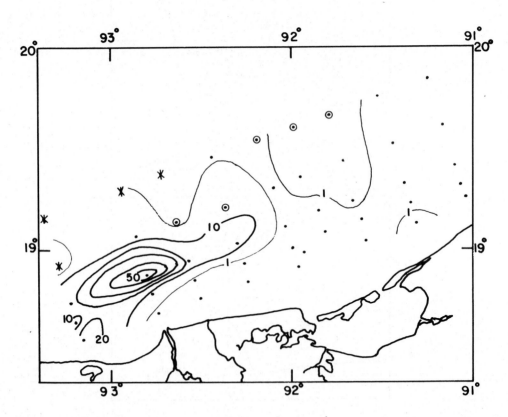

Figure 4. Distribution of ϕ computed for a layer 30 m deep. ϕ is the amount of work per m^3 needed to vertically mix the water column. At stations where $\phi > 5$ Joules/m^3, Rs = 0.79 ± 0.09. Rs is the proportion of density stratification that is due to salinity layering. Stations marked "*" and "o" had values of Rs of 0.00 ± 0.06 and -0.39 ± 0.12 respectively for a layer 100 m deep.

During the September 1977 and June 1978 surveys of the area, the thermohaline structure was entirely different to that described above. In 1977, no haline front and associated stratification was present. Further offshore (40 nm), the thermocline struck the surface forming a strong thermal front with associated stratification seaward of it. In June 1978 there was some indication of a haline front displaced westwards about 60 nm of where it was observed in April 1984.

DISCUSSION

The above observations show that the haline front and associated stratification are not permanent features in the shallow waters of the Bay of Campeche.

Rough estimates of the persistence of stratification during storms of various intensities can be made using a model given by Simpson et al. (1978) which describes the time variation of $\emptyset$ for the Irish Sea. This model considers the changes in stratification induced at a column of water by surface heat input, wind stirring and tidal mixing. If we assume that the wind mixing constants given by these authors apply to the Bay of Campeche, we can estimate the rate of change of $\emptyset$ due to this agent.

$$(\partial \emptyset / \partial t)_{WIND} = - \delta \ Ks \ \rho_A \ \overline{|W|}^3 / h \qquad (1)$$

where δ is the wind mixing efficiency coefficient (0.023), ρ_A is the density of air (1.216 kg/m^3), Ks is the surface drag coefficient (0.00215) x γ and γ is the ratio of the wind induced surface current to wind speed (0.03). $\overline{|W|}^3$ is the time average of the wind speed cubed and h is depth. During a typical storm in the area, $\overline{|W|}^3$ is likely to be of the order of 600 m^3/sec^3 as computed using wind speed and direction data taken during a storm in December 1982 (unpublished data). Assuming h is 30 m, eq. 1 yields a value of 3 Joules m^{-3} day^{-1}. Since the observed values of $\emptyset$ associated to the front were of about 30 Joules m^{-3}, it would take such a storm a minimum of 10 days to destroy layering. Typical storms in this area last between 3 and 5 days so that stratification associated to a front may be expected to survive such a storm. During a hurricane, however, $\overline{|W|}^3$ may be expected to reach 7000 m^3/sec^3 as computed using data published by Martin (1982) so that eq. 1 yields 36 Joules m^{-3} day^{-1}. Such a hurricane could destroy the stratification associated to a front in less than one day.

308

The destruction of stratification however does not necessarily imply the dissappearance of the front. This may be inferred from the presence of a frontal structure in fig. 5, which shows the distribution of depth averaged density during the survey of April 1984. Further work would have to be carried out to obtain more information on the time scales of evolution of the front and associated phenomenae.

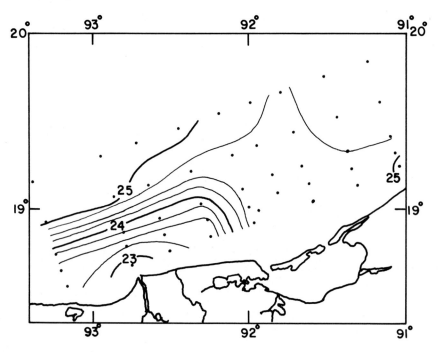

Figure 5. Distribution of depth averaged density in the area during the April 1984 survey. The presence of strong gradients in this figure indicates that the front may still be present even if the water column were vertically mixed by a storm (compare with fig. 2.c).

It would appear that the haline front is a consequence of discharge into the Bay of Campeche by the various rivers in the area. Fig. 6 shows averages over more than 10 years of mean monthly discharges into the bay for some of these rivers (S.R.H., 1969; S.R.H., 1976). It is interesting to note that while the haline front was clearly present in April of 1984 at the time of lowest discharge, it was not apparent in the area in September 1977 when rivers were probably at their highest level. One of the aims of future work in the area would be to study the mechanisms of frontogenesis and the relationship, if any, between the amount of river discharge and its evolution.

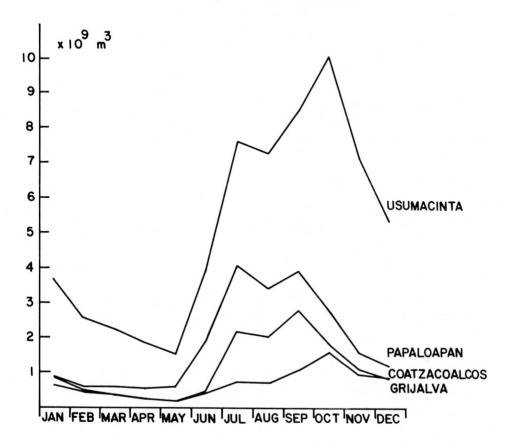

Figure 6. Mean monthly discharges averaged over more than 10 years for rivers in the area of study.

The rivers Papaloapan and Coatzacoalcos, some 150 nm to the west of the area of study, discharge comparable volumes of water into the Gulf of Mexico (see fig. 6). It remains to be seen whether these rivers also generate fronts and, if so, whether they form part of a continuous feature extending from the Terminos Lagoon or if they form separate frontal systems.

At other fronts of similar spatial scales, the density struc-ture induces a circulation at the surface which converges on the front line where foam and other floating objects are trapped (see Garvine, 1974). Whether such a circulation is present at the front in the Bay of Campeche has more than a academic interest only. This information would be valuable to decision makers during an emergency if indeed the floating hydrocarbons spilled during the Ixtoc-1 oilwell blowout were inhibited from reaching the coast by the circulation associated to a front. Computer modellers would have to make use of this information in the prediction of the behaviour of future oil spills.

Being such a shallow area which is not sheltered from the deep ocean in the Gulf of Mexico, it is likely that the evolution of the front, and in general the thermohaline structure in the Bay of Campeche, is highly vulnerable to physical processes occurring at the surface and at the landward and seaward boundaries. Evidence that the changing residual circulation in the Gulf of Mexico throughout the year may influence the evolution of the front is provided by a computer model of the residual circulation driven by wind and by the Yucatan current (Salas & Monreal, this volume). The circulation patterns predicted by this model show strong currents along the coast of the Bay of Campeche in September and weak ones in April. These results would be consistent with the observations presented in this study if the river fresh water were swept away by the strong residual currents in September despite the high level of discharge while being allowed to induce a well developed frontal structure in April when the currents are weak and river discharge is low. This example shows that careful consideration must be given to this type of process in the design of future data collection for the study of fronts in the Bay of Campeche while highlighting the potential usefulness of computer models for the interpretation of these data. This data acquisition could also assist computer simulators in the calibration of their models.

ACKNOWLEDGEMENTS

We wish to thank the members of CINVESTAV, Mérida, for allowing us to share their ship time. This research was sponsored by the Consejo Nacional de Ciencia y Tecnología (CONACYT), México with grant No. PCMABNA-005112.

Contribution No.408 of the Instituto de Ciencias del Mar y Limnología, U.N.A.M., México.

REFERENCES

Czitrom, S.P.R., 1982. Density Stratification and an Associated Front in Liverpool Bay. Ph.D. Thesis, University College of North Wales, Bangor, U.K.

Garvine, R.W., 1974. Dynamics of Small-Scale Oceanic Fronts. Journal of Physical Oceanography, Vol. 4, No. 4, pp. 557-569.

Klemas, V., 1980. Remote sensing of coastal fronts and their effects on oil dispersion. International Journal of Remote Sensing, Vol. 1, No. 1, pp. 11-28.

Martin, J., 1982. Mixed-layer simulation of buoy observations taken during hurricane "Eloise". Journal of Geophysical Research, 87C, pp. 409-427.

Murray, S.P., 1975. Wind and Current Effects on Large-Scale Oil slicks. Proceedings of the Seventh Annual Offshore Technology Conference, Houston, Texas, May 5-8, 1975. pp. 523-533.

S.R.H., 1969. Boletín Hidrológico No. 38 de la Secretaría de Recursos Hidráulicos, México.

S.R.H., 1976. Atlas del Agua de la República Mexicana, Secretaría de Recursos Hidráulicos, México.

Simpson, J.H., D.H. Hughes and N.G.C. Morris, 1977. The Relation of seasonal Stratification to Tidal Mixing on the Continental Shelf. In: M. Angel (Ed.), A Voyage of Discovery. Pergamon Press, Oxford, pp. 327-340.

Simpson, J.H., C.M. Allen and N.G.C. Morris, 1978. Fronts on the Continental Shelf. Journal of Geophysical Research, Vol. 83, No. C9, pp. 4607-4614.

Uda, M., 1938. Researches on "siome" or current rip in the seas and oceans. Geophysical Magazine, Vol. 11, No. 4, pp. 307-372.

THE INTERACTION OF PHYSICAL AND BIOLOGICAL PROCESSES IN A MODEL OF THE VERTICAL
DISTRIBUTION OF PHYTOPLANKTON UNDER STRATIFICATION

A.H. TAYLOR, J.R.W. HARRIS and J. AIKEN

Institute for Marine Environmental Research

Prospect Place, The Hoe Plymouth, PL1 3DH, UK.

ABSTRACT

A time-dependent advection-diffusion model of the vertical distribution of
phytoplankton and a single nutrient in a stratified hydrodynamic regime is
described. Its properties under summer conditions typical of a temperate shelf
sea or a deep ocean situation are analysed.

In the shelf case the system tends rapidly to a steady state with a
thermocline peak of phytoplankton. The model allows for partial nutrient
recycling; the effect of this is found to have a generally antisymmetric
relation to that of nutrient uptake by phytoplankton. Increasing recycling
increases steady state phytoplankton concentrations but has negligible effect on
nutrient, increasing uptake reduces steady-state nutrient but leaves
phytoplankton unaltered. As either nutrient uptake or recycling is reduced a
tendency to oscillate appears. Explicit relations for steady state
concentrations and conditions for oscillations are derived for a simplified
model of phytoplankton over an infinite nutrient pool.

Increased incident light is found to reduce surface phytoplankton, whose
nutrient supply is pre-empted by increased growth in the thermocline. This
effect is demonstrated in a simple two-layer model, and the determinants of
vertical distribution discussed. The diurnal light cycle is shown to be equivalent to a reduction in mean light intensity. Day-night light variation is only
weakly reflected in the phytoplankton but produces around 25% nutrient variation
in the mixed layer. Diurnal variation in turbulence due to reduced day time
convection has negligible effect on steady state mean levels, but somewhat
enhances this day-night nutrient variation.

INTRODUCTION

The abundance of phytoplankton at any depth changes in response to growth,
sinking, diffusion, nutrient limitation and grazing, as well as other processes
such as lateral advection. The earliest attempt to model vertical profiles of
phytoplankton was that of Riley et al. (1949) in the Western North Atlantic,
since which, numerical models with increasing degrees of complexity have been
constructed, some of these being discussed and compared by Patten (1968), Steele
and Henderson (1976), Platt, Denman and Jassby (1977), Steele and Mullin (1977)
and Fasham et al. (1983). Current limits to this development are perhaps represented by the work of Woods and Onken (1982) who have developed a Lagrangian
model in which a large number of individual phytoplankters are tracked in a
diurnally varying, vertically stratified sea; and Pace et al. (1984) who have

produced a continental shelf food web without stratification that is composed of 17 state variables.

Such models may describe or predict phytoplankton changes, but their very complexity may impede understanding of the way biological and physical processes interact in stratified systems, a perception which motivates the continuing theoretical analyses of simple systems. Criminale and Winter (1974) examined the dynamical behaviour of small perturbations from equilibrium in the model of Riley et al. (1949), and Criminale (1980) has recently extended this analysis to include a current and a current-shear. These studies investigated the vertical distribution of a single phytoplankton species without considering the dynamics of nutrients, grazers or other phytoplankton species. Following earlier work by Peterson (1975), Roughgarden (1978) and Tilman et al (1978), Powell and Richerson (1985) have examined the competition between two species of phytoplankton for two nutrients using simple models based on Michaelis-Menten expressions. Their models included nutrient recycling (without any nutrient loss) but did not consider vertical variation. The present paper is situated between these two extremes in that it treats a one dimensional stratified system with a single phytoplankton and a single nutrient. The model used is similar to those of Jamart et al. (1979) and Fasham et al. (1983). However, in the present case, the model has an open bottom so that cells can sink and be returned as nutrient, nutrient recycling may be imperfect, and daily variation of convectional mixing (Woods, 1980; Woods and Onken, 1982) is included. An understanding of the roles played by recycling and diurnal changes are objectives of this investigation. In particular, the daily variation in the turbulence of the mixed layer is a factor that has not been studied in simple models.

We concentrate on the mid-summer situation for a stably stratified shelf sea, typified by a shallow (25m) mixed surface layer, and a strong, narrow thermocline (temperature gradient up to $1^\circ C.m^{-1}$; 5m thick), above a tidally mixed bottom layer (typical depth 100m). These conditions are typical of the central Celtic Sea in mid-summer, where the upper mixed layer characteristically exhibits nutrient depletion and a peak concentration of phytoplankton (ca. 3 mg m^{-3}) occurs in the thermocline (Joint and Pomroy, 1984, Aiken 1985). Such areas have laterally homogenous vertical structures of temperature and chlorophyll concentration over distances of 50km (Aiken, 1985), it might be inferred that a quasi-steady state balance exists between the biological and physical processes. In the model there is no attempt to simulate the development of the seasonal cycle but attention is focused on the dynamics of the coupled phytoplankton and nutrient system. With this restriction it is unnecessary to treat the grazers dynamically. By way of a contrast to the shelf case, we consider a situation, typical of the summer conditions in a temperate ocean, in which the surface mixed layer overlies a broad deep thermocline (eg Williams and Hopkins, 1975,

1976). Our model contains several of the processes considered to be important in the more realistic simulation of the system, but is sufficiently simple that analysis allows an assessment of their relevance to its overall dynamics, and how this depends on the particular parameter values used. After outlining the structure and general behaviour of the model, the sensitivity to its parameters is explored by both numerical experiment and theoretical analysis.

MODEL

The model describes the nutrient-dependent growth of a single phytoplankton species, allowing for sinking and variable vertical mixing. The specific growth rate of the phytoplankton is taken to be the product of two components, a dependence (α) on incident light (I) and a function (ϕ) of the concentration (N) of a single nutrient. The form used for ϕ is the Michaelis-Menten relation. The phytoplankton are subject to a specific loss rate (m), due to death or direct respiration, which regenerates the nutrient with an efficiency, ε. Vertical profiles of phytoplankton and nutrients are calculated as functions of depth z and time t. Phytoplankton cells sink at a rate (v) within a hydrodynamic regime defined by a spatially (and on occasion temporally) variable turbulent diffusion coefficient, K(z,t). The changes in phytoplankton concentration (P) and nutrient concentration are then:

$$\partial P/\partial t = [\alpha(I)\phi(N)-m]P - v\partial P/\partial z + \partial[K(z,t)\partial P/\partial z]/\partial z \qquad (2.1)$$

$$\partial N/\partial t = -\gamma[\alpha(I)\phi(N)-\varepsilon m]P + \partial[K(z,t)\partial N/\partial z]\partial z \qquad (2.2)$$

Boundary conditions at the surface and bottom are also needed to constrain the model. As there can be no flux across the sea-surface:

$$K(z,t)\partial P/\partial z = vP \quad \text{and} \quad K(z,t)\partial N/\partial z = 0 \qquad (2.3)$$

must hold at the surface. At the bottom P(z,t) and N(z,t) are assigned constant values, since this allows phytoplankton to sediment to the sea-bed and nutrients to return to the water column from below. However, if at any time a flow of phytoplankton up from the bottom should occur the flux of phytoplankton at the sea-bed is set to zero.

Specific Growth Rate

The specific growth rate for excess nutrient, $\alpha(I)$, was obtained from the light intensity I at any depth by the 'tanh' formula of Jassby and Platt (1976) and Platt and Jassby (1976) which takes account of photosaturation at high light intensities:

$$\alpha(I) = \alpha_0 P_m \tanh (\alpha'I/P_m) \tag{2.4}$$

The light intensity, I, is the irradiance (Wm^{-2}) in the photosynthetically active region (400-700 nm), α' is the initial slope of the light saturation curve (0.12 mg C (mg Chl)$^{-1}$ h^{-1} W^{-1}m^2) and P_m is the assimilation number (2.5 mg C (mg Chl)$^{-1}$ h^{-1}). The constant α_0 was chosen to give a doubling time of 0.8 days ($\alpha(I) = 0.036$ h^{-1}) when I is 170 Wm^{-2}.

The surface light intensity was either kept constant at 170 Wm^{-2} or else approximated on the positive half of a diurnal sine wave of amplitude 534 Wm^{-2} which has the same daily average. The light irradiance profile was determined from

$$\partial I/\partial z = \kappa I \tag{2.5}$$

with the attenuation coefficient, κ, being dependent on phytoplankton concentration (P in mg (Chl m^{-3}) and given by:

$$\kappa = 0.10 + 0.018 P(z,t) \tag{2.6}$$

A carbon to chlorophyll ratio of 40 was used so that, assuming the nutrient to be nitrogen, the conversion factor Y is 0.503 mM N (mg Chl)$^{-1}$.

The extent of reduction in growth rate by nutrient limitation is determined by a Michaelis-Menten relation:

$$\phi(N) = N/(N+\nu) \tag{2.7}$$

Values of 0.05, 0.2 and 0.8 m M N m^{-3} were used for the half-saturation constant, ν, independent of depth (MacIsaac and Dugdale, 1969).

Loss, Recycling and Sinking

The specific loss rate, m, includes the mortality of cells due to death and grazing and the loss due to respiration. Two constant values are used for the specific mortality 0.05 or 0.2 x 10^{-5} s^{-1} and the respiration rate was taken to be 0.04 x 10^{-5} s^{-1} (ie 0.043, 0.173 and 0.035 day^{-1}, respectively). The recycling efficiency, ε, allows for the possibility that not all of the nutrient from this loss of cells is returned to solution. Three constant values have been used for the fraction returned: $\varepsilon = 0$ (no recycling), $\varepsilon = 0.2$ or $\varepsilon = 1$ (perfect recycling).

Although the sinking speed of phytoplankton cells depends on species, time of year, ambient nutrient concentration etc., such variations in sinking speed are probably not large in their effect in most cases and a constant value of 10^{-3} cm s^{-1} (0.86 m day^{-1}) was used throughout.

Turbulent Diffusion

The surface mixed layer was taken to be 25m deep and in the absence of diurnal changes the turbulent diffusion coefficient, K, was a constant 80 cm^2 s^{-1} within it. This is higher than the values computed by Fasham et al. (1983) for an area of the Celtic Sea when the thermocline was becoming established but is in agreement with values reported elsewhere (Kullenberg, 1971; James, 1977; Sundby, 1983). However, turbulent mixing in the surface layer follows a daily variation (Woods, 1980, 1985) while mixing in the immediate vicinity of the surface is by wind and wave action and is continuous, below this the convection which occurs at night is the main source of mixing. To explore the effect of this, in some instances K was dropped to 1 $cm^2 s^{-1}$ in the lower half of the mixed layer during the day. Below the mixed layer no diurnal variation was included in the turbulent diffusion coefficient. In this region two profiles of K were used: a 'shelf' case when a value of 0.1 cm^2 s^{-1} was used between 25 and 30 m and a value of 100 cm^2 s^{-1} was used in a tidally mixed layer between 30 m and the bottom, and an 'oceanic' case where K was set at 1 cm^2 s^{-1} at all depths below the mixed layer. These cases are in general agreement with estimates of 0.05 to 1 cm^2 s^{-1} for the thermocline (James, 1977; King and Devol, 1979; Jassby and Powell, 1975) and of 10^2 to 10^3 cm^2 s^{-1} from bottom tidal mixing (James, 1977 or using equation 4 of Pingree and Griffiths, 1977).

Numerical Solution

If the values of P(z,t) and N(z,t) are defined on a vertical grid of J points from j = 1,2,.....,J with spacing Δz, equations (2.1) and (2.2) can be approximated by the finite difference equations:

$$(N_j^{n+1} - N_j^n)/\Delta t = -\gamma \, \alpha_j^n \, N_j^{n+1} \, P_j^n/(N_j^n + \nu) + \varepsilon \gamma m P_j^n$$

$$+ [K_{j+\frac{1}{2}}^n(N_{j+1}^{n+1} - N_j^{n+1}) - K_{j-\frac{1}{2}}^n(N_j^{n+1} - N_{j-1}^{n+1})]/(2\Delta z^2)$$

$$+ [K_{j+\frac{1}{2}}^n(N_{j+1}^n - N_j^n) - K_{j-\frac{1}{2}}^n (N_j^n - N_{j-1}^n)]/(2\Delta z^2)$$

$$(P_j^{n+1} - P_j^n)/\Delta t = \alpha_j^n \, N_j^{n+1} \, P_j^n \, / \, (N_j^n + \nu) -m \, P_j^n - v \, (P_j^n - P_{j-1}^n)/\Delta z$$

$$+ [K_{j+\frac{1}{2}}^n (P_{j+1}^{n+1} - P_j^{n+1}) - K_{j-\frac{1}{2}}^n (P_j^{n+1} - P_{j-1}^{n+1})]/(2\Delta z^2)$$

$$+ [K_{j+\frac{1}{2}}^n (P_{j+1}^n - P_j^n) - K_{j-\frac{1}{2}}^n (P_j^n - P_{j-1}^n)]/(2\Delta z^2)$$

$$(2.8)$$

P_j^n, N_j^n are the values at the jth grid-point after the nth time-step of length Δt. The growth term has been treated implicitly and the Crank – Nicholson scheme is used for the diffusion terms. The growth rates, α_j^n, are calculated at the same grid-points as P_j^n and N_j^n while the diffusion coefficients $K_{j+1/2}^n$ are evaluated at points mid-way between the grid points. These equations were solved, incorporating the boundary conditions, by means of the tri-diagonal matrix algorithm (eg Richtmyer and Morton, 1957, p 189), using depth increment of 1 metre and a 5 minute time-step.

RESULTS WITHOUT DIURNAL CHANGES

Each run of the model was for a period with a shallow (25 m) mixed layer in which the nutrient was initially already considerably depleted ($N(z,0) = 0.1$mM N m^{-3}). Beneath this, initial concentrations were higher; 3 mM N m^{-3} in the thermocline and 6 mM N m^{-3} below that. The initial concentration of phytoplankton, $P(z,0)$ was 0.1 mg Chl m^{-3} at all depths. From these initial conditions both phytoplankton and nutrient concentrations generally tend towards a steady state exhibiting further nutrient depletion in the surface mixed layer and a peak phytoplankton concentration in the thermocline (Fig. 1). The mixed layer phytoplankton concentrations achieved in the shelf case are higher than in the ocean, and the peak in the thermocline larger but narrower, a difference in structure which is frequently observed. The shelf turbulence profile produces a more rapid approach to the steady state (within about 50 days) than is found in the ocean case (Fig. 2), which may be attributable to the greater disparity between the initial and final nutrient profile in the latter. In most cases development towards the steady state in the mixed layer is characterised by an initial overshoot of phytoplankton concentration in response to the somewhat high initial nutrient concentrations, followed by its more or less damped return towards the steady state. The rise towards the overshoot is largely exponential because the nutrient level is almost unchanged. Since the rate of increase

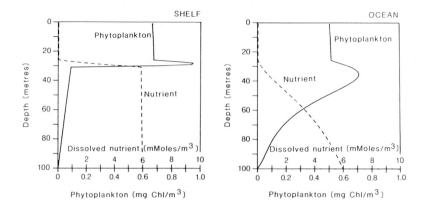

Fig. 1 Vertical profiles of phytoplankton and nutrient after 100 days under shelf or oceanic turbulence conditions (see text), with $\nu = 0.2$ mM per cubic m, $m = 0.078$ per day and $\varepsilon = 0.2$.

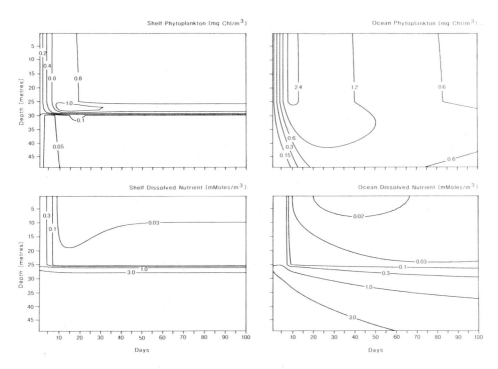

Fig.2 The temporal development towards the profiles of Fig. 1 from the initial conditions described in the text (ν = 0.2 mM per cubic m, m = 0.078 per day, ε = 0.2)

depends on the difference between growth and losses (including diffusion and sinking), a small change either in growth or mortality is amplified in the rate.

After an initial slight increase, reflecting upward diffusion unbalanced by phytoplankton uptake, changes in the dissolved nutrient concentration of the mixed layer mirror those of the phytoplankton; the total nutrient concentration temporarily increases while upward diffusion of nutrients is greater than downwards loss of phytoplankton.

Clearly the rate of the initial increase will be determined largely by the net growth rate of the phytoplankton and so can be markedly raised by reduction in loss (m) or the Michaelis constant (ν); (Figs. 3 and 5). Although the steady state nutrient concentration also shows a marked positive relation to ν (Figs 3 and 5), this is not mirrored by a corresponding decline in steady state phytoplankton levels, which are relatively insensitive to the Michaelis constant. Conversely, and equally contrary to naive expectations, recycling of nutrient seems largely irrelevant to its final concentration (Fig. 3) but when highly efficient it may considerably increase the concentration of phytoplankton which this supports (Fig. 3). Thus, for different reasons, increasing the recycling efficiency (ε) and reducing the Michaelis constant (ν) both increase the propor-

320

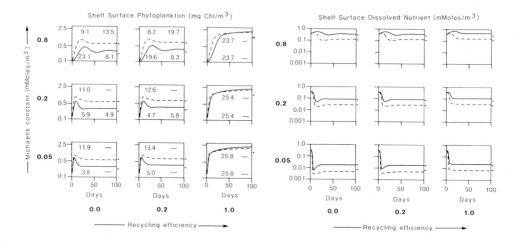

Fig. 3   Effects of the Michaelis constant,   recycling efficiency and loss rate
(dashed lines,   m = 0.078 per day;   continuous lines,   m = 0.208 per day) on the
calculated   temporal   development of surface phytoplankton   and   nutrient   under
shelf conditions.   Numerical values within each sub-figure indicate decay times
and   oscillation periods (days) calculated using the single-layer   approximation
(see text).   Arrows   indicate   steady-state   values expected   under   the   same
approximation (parameter values as Fig. 6)

tion of the nutrient in the mixed layer which is locked in the phytoplankton, an

effect   which,   although present,   is less marked in the thermocline   (Fig.   4).

Another   feature which may be seen in Fig.   3 is that,   under conditions of   low

recycling   efficiency,   as   the   net growth rate $(\alpha\phi - m)$ declines   and   hence   the

biological response time of the system increases relative to its fixed   physical

characteristics, a tendency to oscillate appears.

The source of much of this behaviour,   difficult to discern in the full model

(simple   though   it is) is elucidated by consideration of a yet   more   schematic

view of   reality.   As the thermocline and mixed layer show similar   responses,

consider, as a model, a layer of phytoplankton growing over a pool of nutrients.

Suppose   the   layer   is of thickness h and the concentrations   of   nutrient   and

phytoplankton are $\bar{N}$ and $\bar{P}$,   respectively,   with $\bar{N}_0$, $\bar{P}_0$ being the   corresponding

values below the layer.   Equations (2.1) and (2.2) may be approximated by:-

$$d\bar{P}/dt = (\beta\bar{N} + \theta - m - v/h)\bar{P} + k\,(\bar{P}_0 - \bar{P})/h \qquad\qquad (3.1)$$

$$d\bar{N}/dt = -\gamma(\beta\bar{N} + \theta - \epsilon m)\bar{P} + k\,(\bar{N}_0 - \bar{N})/h \qquad\qquad (3.2)$$

$\alpha N/(N+v)$ has been approximated in these equations by $\beta N + \theta$ and,   since on   most

occasions $\bar{N} \ll v$, $\beta \approx \alpha/v$ and $\theta \approx 0$.   The last term in each equation represents a

simple   approximation   to   vertical diffusion.   The coefficient k will   be   the

turbulent   diffusion coefficient in the thermocline divided by the square of   an

appropriate vertical distance (of the order of h).   If $\bar{P}_0$ is negligible compared

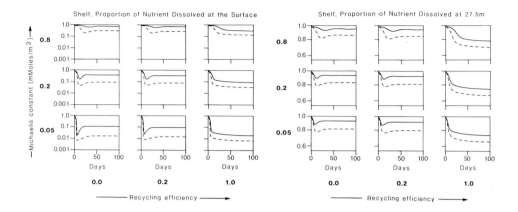

Fig. 4   The proportions of the total nutrient dissolved at the surface and in
the thermocline in the situations depicted in Fig. 3.

to $\bar{P}$, the steady-state solutions to (3.1) and (3.2) are:-

$$\hat{N} = (m + v/h + k/h - \theta)/\beta \qquad (3.3)$$

$$\hat{P} = \frac{k(\bar{N}_0 - \hat{N})}{hY(\beta\hat{N} + \theta - \varepsilon m)}$$

$$\text{or} \quad \hat{P} = \frac{k(\beta\bar{N}o - m - v/h - k/h + \theta)}{h\beta Y (m + v/h + k/h - \varepsilon m)} \qquad (3.4)$$

In these equations $\hat{N}$ does not depend on the recycling efficiency, even
though ε only appears in the equation for $d\bar{N}/dt$. The phytoplankton abundance $\bar{P}$
increases with ε, so that the proportion of the total nutrients that is dissolved decreases with increasing recycling efficiency as in Fig. 4. The ratio of
dissolved nutrients to the nutrients in the phytoplankton is:

$$\frac{\hat{N}}{Y\hat{P}} = \frac{h(m + v/h + k/h - \varepsilon m)(m + v/h + k/h - \theta)}{k(\beta\bar{N}_0 - m - v/h - k/h + \theta)} \qquad (3.5)$$

an expression which shows that the relative decline of dissolved nutrients with
increased recycling efficiency is due to a balance of the three processes: cell
loss, recycling and upward diffusion of nutrients. Further, since a reduction
in the Michaelis constant is equivalent to an increase in both β and θ this will
also reduce the ratio.

The predictions of equations (3.3) and (3.4) are shown on Fig. 3. The
agreement is generally close, in particular, the insensitivity of the phytoplankton abundance to the Michaelis-Menten parameter is reproduced.

The temporal behaviour of this single-layer model in the vicinity of its
steady state also reproduces the changes in Fig. 3. If $\hat{P}$ and $\hat{N}$ are

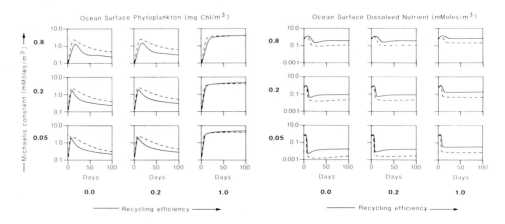

Fig. 5   The dependence of the temporal development of surface phytoplankton and nutrient on nutrient uptake, recycling and phytoplankton loss in an oceanic turbulence regime (dashed lines, m = 0.078 per day; solid lines, m = 0.208 per day).

equilibrium solutions of (3.1) and (3.2), $\overline{P} = \hat{P} + p$ and $\overline{N} = \hat{N} + n$ then in the vicinity of the steady-state:

$$dp/dt = (\beta\hat{N} + \theta - m - v/h - k/h)p + \beta\hat{P}n \qquad (3.6)$$

$$dn/dt = -\gamma (\beta\hat{N} + \theta - \epsilon m)p - (\gamma\beta\hat{P} + k/h)n \qquad (3.7)$$

When solutions $p = Ae^{\lambda t}$, $n = Be^{\lambda t}$ are sought, λ is given by a quadratic equation and λ may be either <u>real</u>, representing growth or decay, or <u>complex</u>, representing growing or decaying waves.   These solutions are expressed on Fig. 3 as a decay-time defined as $2\pi/(\text{real part of } \lambda)$ and an oscillation period which is $2\pi/(\text{imaginary part of } \lambda)$.   Oscillating solutions are predicted in the region of parameter values where these occurred in the full numerical model, although the predicted periods tend to be too short.   There is general agreement between the decay-time and the time to reach equilibrium after the initial increase (at the start conditions are too far from equilibrium for (3.6) and (3.7) to be applicable); in particular, very slow decay is predicted at high recycling efficiency and this is where the steady-state is barely reached.   The simple model (3.6) and (3.7) therefore allows the range of parameters giving rise to oscillation to be determined (Fig. 6).

   If P_o is negligible, the coefficient of p disappears from (3.6) and λ is a solution of:

$$\lambda^2 + \lambda(k/h + \beta\gamma\hat{P}) + \beta k(\overline{N}_o - \hat{N})/h = 0 \qquad (3.8)$$

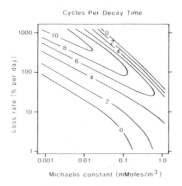

Cycles Per Decay Time

Michaelis constant (mMoles/m³)

Fig. 6   The effects of the Michaelis constant (using $\beta = \alpha/\nu$, see text) and phytoplankton loss rate on the dynamics of the single-layer model of nutrient-dependent growth (Eq. 3.1 and 3.2) $\alpha = 0.583$ per day, $\theta = 0$, $\varepsilon = 0.2$, $\hat{P}_0 = 0.1$ mg Chl per cubic m, $\bar{N}_0 = 6$ mM per cubic m, $k = 0.102$m per day, $h = 25$m.

Oscillating solutions to (3.8) occur if and only if

$$(k/h + \beta\hat{\gamma P})^2/4\beta < k(\bar{N}_0 - \hat{N})/h \qquad (3.9)$$

The right hand side represents the vertical diffusion of nutrients, and is independent of recycling efficiency. However, increasing ε raises $\hat{P}$ (Eq. 3.4) and so the propensity to oscillate is greater. On the other hand, low values of $\bar{N}_0$, as occur in the oceanic case, tend to eliminate oscillations (Fig. 5).

RESPONSE TO LIGHT INTENSITY

As expected, a reduction in light intensity reduces the initial rate of increase of the phytoplankton. However, unlike the parameter variations described in the previous section, which all produced similar changes in the steady-state phytoplankton in the thermocline to those in the mixed layer, a change in light intensity causes opposite changes in the two regions. Thus, while the equilibrium concentration of phytoplankton in the thermocline increases as the light intensity is raised (Fig. 7) the concentration in the mixed layer declines. The concentration of dissolved nutrient in the mixed layer also declines with increasing light and so it appears that, when more light penetrates below the mixed layer, there is more growth at these depths reducing the supply of nutrients to the surface layer. This process can be thought of, by analogy with internal waves, as an internal mode of the system. The process can be described if our earlier analysis is extended to consider two layers: a mixed layer with nutrient and phytoplankton concentrations $\bar{N}_M$ and $\bar{P}_M$ and a thermocline in which these values are $\bar{N}_T$ and $\bar{P}_T$, respectively. The corresponding values in the bottom layer are $\bar{N}_0$ and $\bar{P}_0$ (=0). The equations relating these variables are then:-

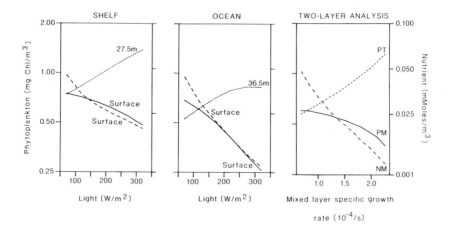

Fig. 7 The dependence of phytoplankton (dashed lines) and nutrient (solid lines) concentrations after 100 days on light incident at the surface (ν = 0.2 mM per cubic m, m = 0.078 per day, ε = 0.2). Values at 27.5m and 36.5m depth are representative of the thermocline in shelf and ocean respectively. The right-hand plot was calculated using the two-layer model (Eq. 4.1 to 4.4), with parameters adjusted to provide general agreement with the shelf case (α_M = 0.618 per day, N_{TO} = 3 mM per cubic m, k_1 = 0.216m per day, k_2 = 0.173m per day, h_M = 25m, h_T = 5m).

$$d\bar{P}_M/dt = (\beta_M \bar{N}_M + \theta_M - m - \nu/h_M)\,\bar{P}_M + k_2\,(\bar{P}_T - \bar{P}_M)/h_M \qquad (4.1)$$

$$d\bar{N}_M/dt = -\gamma(\beta_M \bar{N}_M + \theta_M - \varepsilon m)\bar{P}_M + k_2\,(\bar{N}_T - \bar{N}_M)/h_M \qquad (4.2)$$

$$d\bar{P}_T/dt = (\beta_T \bar{N}_T + \theta_T - m)\bar{P}_T - \nu(\bar{P}_T - \bar{P}_M)/h_T - k_1\,\bar{P}_T/h_T - k_2\,(\bar{P}_T - \bar{P}_M)/h_T \qquad (4.3)$$

$$d\bar{N}_T/dt = -\gamma[\beta_T \bar{N}_T + \theta_T - \varepsilon m]\bar{P}_T + k_1\,(\bar{N}_0 - \bar{N}_T)/h_T - k_2\,(\bar{N}_T - \bar{N}_M)/h_T \qquad (4.4)$$

The coefficient k_2 represents diffusion between the mixed layer (depth h_M) and the thermocline, k_1 between the thermocline (thickness h_T) and the bottom layer. In each layer the Michaelis-Menten relation has been linearised about a point appropriate to the layer. Although the mixed layer the values $\beta_M = \alpha_M/\nu$ and θ_M = 0 can be used as before, in the thermocline:

$$\beta_T = \alpha_T[1 - N_{TO}/(N_{TO} + \nu)]/(N_{TO} + \nu) \qquad (4.5)$$

$$\theta_T = \alpha_T[N_{TO}/(N_{TO} + \nu)]^2 \qquad (4.6)$$

where N_{TO} is a typical nutrient concentration in the thermocline. Therefore, if $\alpha_T = \alpha_M$, β_T is smaller than β_M and θ_T is larger than θ_M reflecting the non-linearity of the Michealis-Menten relation.

Generally, N_{TO} will be sufficiently large compared to ν that β_T will be effectively zero. Under these circumstances a solution is readily available; its properties are encapsulated by the following relations:

$$\hat{N}_M = [m - \theta_M + q_1 + s_3 q_2/(\theta_T - m - q_3)]/\beta_M \tag{4.7}$$

$$\hat{P}_M = \frac{1}{\gamma} s_1 s_3 (N_0 - N_M)/[(s_1 + s_2)(\beta_M N_M + \theta_M - \epsilon m) + q_2 s_3 (\theta_T - \epsilon m)/(m - \theta_T + q_3)] \tag{4.8}$$

$$\hat{N}_T/\hat{P}_T = (m - \theta_T + q_3)[N_M/P_M + \gamma(\beta_M N_M + \theta_M - \epsilon m)/(s_2 q_1)]/q_2 \tag{4.9}$$

$$\hat{P}_M/\hat{P}_T = (m - \theta_T + q_3)/q_2 \tag{4.10}$$

In these, specific transfer terms have been defined by $s_1 = k_1/h_T$, $s_2 = k_2/h_T$, $s_3 = k_2/h_M$, $q_1 = (\nu + k_2)/h_M$, $q_2 = (\nu + k_2)/h_T$ and $q_3 = q_2 + s_1$. Equations 4.8 to 4.10 have been simplified by the recognition that N_M may be neglected in comparison with N_T. As with the single layer analysis, in the mixed layer the nutrient concentration depends on the recycling efficiency and the phytoplankton concentration increases with ϵ so that the proportion of the total nutrients that is dissolved decreases with ϵ. The proportion of nutrient dissolved behaves similarly in the thermocline (Eq. 4.9) and this is reflected in the numerical results (Fig. 4).

Increasing the surface light intensity will make α and hence β_M, θ_M and θ_T larger in these equations. When β_M, θ_M and θ_T are all increased in the same proportion both $\hat{N}_M$ and $\hat{N}_T$ are decreased and the ratio of $\hat{P}_M$ to $\hat{P}_T$ declines, as was found in the numerical model. Note that the ratio of $\hat{P}_M$ to $\hat{P}_T$ does not depend on ϵ or N_0. In equations (4.6) to (4.9) only the non-linearity of the Michaelis-Menten relationship has been included, there is no need of self-shading or light saturation although these will cause some modification of the process. This response to variations in α is not the result of our simplifying assumptions, for it applies to the steady-state solutions of (4.1) to (4.4) which can be determined numerically (Fig 7). In these calculations β_T and θ_T have been evaluated using equation (4.5) and $\alpha_T = \alpha_M/4$ to allow for the reduced growth due to lower light levels in the thermocline.

The total phytoplankton abundance, $h_M \hat{P}_M + h_T \hat{P}_T$, declines slightly with decreasing light. However in the full numerical model the total phytoplankton in the mixed layer and thermocline increases as the light is reduced and only if the vertical integration includes the population below the thermocline does a decrease occur.

EFFECT OF DIURNAL VARIATION

There are two major physical sources of daily variation: the day-night cycle of irradiance and the cycle of convective mixing.   These processes are most easily examined using the shelf case where a steady state is more rapidly acheived.

Diurnally varying light

The daily cycle of light concentrates the irradiance into intervals of higher intensity, and some of this radiation is then not utilised because phytoplankton growth is related to light intensity by a law of diminishing returns.   For this reason a diurnally varying incident light has an effect similar to a reduction in mean light level.   That is, there is an increase in both steady state phyto-plankton and nutrient in the mixed layer and a decline in the thermocline peaks of phytoplankton (Table 1, column 2; cf. Fig. 7). The effect is most pronounced for the surface nutrient concentration which is almost trebled whereas the change in phytoplankton is only slight (equivalent to a change in mean light of 10 - 20%).   However, a marked slowing of the initial increase is produced by amplification of the small reduction in growth rate in the manner noted in section 2.

Table 1 Effect of the diurnal cycle.   Phytoplankton is in mg Chl per cubic m and nutrient is in mM per cubic m.   Values given are for the shelf case at the end of 100 days. Effects on the daily averages are expressed as deviations from the averages without a daily cycle.

	Mean with no cycle	DAILY AVERAGES		
			Additive effects	
		Light cycle	Turbulence cycle	Light and turbulence
Surface phytoplankton	0.67	0.02	-0.02	0.00
Surface nutrient	0.028	0.043	-0.004	0.040
Thermocline phytoplankton	0.95	-0.08	0.01	0.08
Thermocline nutrient	1.960	0.145	0.030	0.175

	PERCENTAGE INCREASE FROM NIGHT TO DAY		
	Light cycle	Turbulence cycle	Light and turbulence
Surface phytoplankton (day)	6	-2	3
Surface nutrient (day)	-23	-37	-38
Thermocline phytoplankton (day)	14	0	12
Thermocline nutrient (day)	-2	0	-2

	TIME TO FIRST MAXIMUM OR MINIMUM (DAYS)			
	No daily cycle	Light cycle	Turbulence cycle	Light and turbulence
Surface phytoplankton max	8.75	19.0	8.75	18.5
Surface nutrient min	13.5	24.0	13.5	24.5

In addition to such effects averaged over the daily cycle, phytoplankton concentrations vary only slightly diurnally but nutrient concentrations are noticably higher at night, around 30% in the example shown in Table 1. Because the time-scale for vertical mixing between the surface and the thermocline (~(mixed layer depth)^2/K) is at least one day, the day-night cycle does not show the coupling between the thermocline and mixed layer which is the equilibrium response to changing light.

Diurnally varying turbulence

During the daytime, mixing in the lower part of the mixed layer is greatly reduced because of the absence of convection; this reduces the flow of nutrients up through the thermocline and may affect the downward flux of phytoplankton cells. As a result, the effect of the daily cycle of turbulence is to reduce the daily average concentration of both phytoplankton and nutrient at the surface (Table 1, column 2). About half of this change appears attributable to a reduction in the mean turbulence of the mixed layer. Surface phytoplankton abundance is slightly higher at night than in the day (Table 1) which is the result of diffusion from the lower part of the mixed layer where there is an enhancement by day. There is nearly twice as much surface nutrient at night as in the day. Adding a diurnal variation in turbulence did not affect the response time of the model, the times of the first phytoplankton maximum and nutrient minimum remaining unchanged (Table 1).

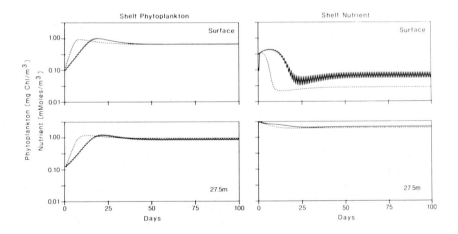

Fig. 8   The effects of diurnally varying turbulence and light on the temporal evolution of phytoplankton and nutrient concentrations at the surface and in the thermocline, subject to a shelf turbulence profile.   The dashed line is under constant conditions. For the solid line the system was subject to the same mean surface light, but concentrated in the positive part of a diurnal sine wave; turbulence was simultaneously reduced in the lower half of the mixed layer during the day. In each case the parameter values of Figs. 1 and 2 were used.

Diurnally varying light and turbulence

When combined, the effects of the cycles of light and turbulence appear additive and their effects on the phytoplankton cancel at the surface leaving only the increase in nutrients due to light variation. Particularly noticable is the resulting increase in the amplitude of the day-night cycle in the surface nutrient. The overall effect of the cycle of day and night is to decrease the average proportion of nutrient incorporated in the phytoplankton (Fig. 8).

DISCUSSION

The phytoplankton and nutrient in the shelf case reach a dynamical equilibrium within 10-20 days, even in the presence of diurnal forcing. The ocean case also seems to be moving towards an equilibrium. While fluctuating weather patterns will generally prevent conditions in the sea being stable for this length of time, the system will probably be displaced from equilibrium less than the initial conditions used here, so that many properties of the steady state may be relevant to conditons in the sea.

The approach to the steady state sometimes involves oscillations; these being induced by an imbalance between the processes of growth and decay, and the nutrient supply by diffusion (c.f eq. 3.9). Such oscillations represent the near equilibrium response of Lotka-Volterra systems (eg May, 1974), and were present in the two-species, two-nutrient competition model of Powell and Richerson (1985), although with much longer periods than given here (100 days or longer).

When the system has reached a steady state the total nutrient content of the mixed layer is maintained by a balance of three flows: the net downward transport of phytoplankton cells, the loss during their incomplete recyling and the upward diffusion of dissolved nutrient through the thermocline. We have shown that varying the efficiency of recycling has negligible effect on the upward nutrient flux and merely redistributes the two losses.

A limitation of the present model may be that the recycling occurs instantaneously, although the model does allow phytoplankton to leave at the bottom and return as nutrient. Any time delay is likely to be most important when it leads to a transport of the nutrients. In particular, zooplankton grazing and excretion can result in the vertical transfer of nutrients. If the zooplankton abundance is only weakly affected by changes in the phytoplankton concentration the model results will still apply, for the effect can be treated by making adjustments to the diffusion rate or the recycling efficiency.

Because the irradiance is depth dependent, changing the light intensity causes a higher order response of the system in which the mixed layer and thermocline each behave differently. The two-layer analysis shows that the influence of growth in the thermocline on the flux of nutrients across it is sufficient to explain the model results, without considering such processes as self-shading. However, the addition of self-shading introduces the possibility of a positive feedback which may be important under certain conditions, producing surface blooms. The day - night cycle of light is equivalent to a reduction in light intensity and so the response of the model to changing light is important in considering the effects of a diurnal cycle.

Diurnal light variations do not qualitatively change the results from the model and the quantitative effects are slight. Including the daily cycle of turbulence tends to cancel even these so that it may not always be necessary to consider diurnal variations in models of this type. Only in the nutrient or in the transient response is the absence of a diurnal variation likely to be at all noticeable. Day-night differences in phytoplankton abundance were negligible in agreement with Woods and Onken (1982) even if the doubling time was reduced by an order of magnitude. The effects due to the light and turbulence cycles again tended to cancel. The daily signal was stronger in the surface nutrient where it was almost a factor of two, the processes reinforcing each other in this case. Of course, some of the biological processes not treated might make the daily cycle more important (eg diel migration of grazers), but then the general buffering systems of phytoplankton physiology (eg photoadaptation and nutrient reservoirs) could lessen its importance. Inclusion of additional trophic levels may change the properties of the system we have described. Further analysis of simplified models can be used to explore the extent to which such additions define the limits of applicability of our system.

REFERENCES

Aiken, J., 1985. The Undulating Oceanographic Recorder Mark 2, a multiple oceanographic sampler for mapping and modelling the biophysical marine environment. In: Zirino, A. (ed), Mapping Strategies in Chemical Oceanography. American Chemical Society, 209, pp 315-332.
Criminale, W.O., 1980. Effects of mean current and stability on depth distribution of marine phytoplankton. J. Math. Biol., 10: 33-51.
Criminale, W.O. and Winter, D.F., 1974. The stability of steady-state depth distribution of marine phytoplankton. Am. Nat., 108: 679-687.
Fasham, M.J.R., Holligan, P.M. and Pugh, P.R., 1983. The spatial and temporal development of the spring phytoplankton bloom in the Celtic Sea, April 1979. Prog. Oceanog., 12: 87-145.
Jamart, B.M., Winter, D.F. and Banse, K., 1979. Sensitivity analysis of a mathematical model of phytoplankton growth and nutrient distribution in the Pacific Ocean of the northwestern U.S. coast. J. Plankt. Res., 1: 267-290
James, I.D., 1977. A model of the annual cycle of temperature in a frontal region of the Celtic Sea. Est. Coastal Mar. Sci., 5: 339-353.
Jassby, A.D. and Platt, T., 1976. Mathematical formulation of the relationship between photosynthesis and light for phytoplankton. Limnol. Oceanogr., 21: 540-547.
Jassby, A.D. and Powell, T., 1975. Vertical patterns of eddy diffusion during stratification in Castle Lake, California. Limnol. Oceanogr., 20: 530-542.
Joint, I.R. and Pomroy, A.J., 1983. Production of picoplankton and small nanoplankton in the Celtic Sea. Mar. Biol., 77: 19-27.

King, F.D. and Devol, A.H., 1979. Estimates of vertical eddy diffusion through the thermocline from phytoplankton nitrate uptake rates in the mixed layer of the eastern tropical Pacific. Limnol. Oceanogr., 24: 645-651.

Kullenberg, G., 1971. Vertical diffusion in shallow waters. Tellus, 23: 129-175.

MacIsaac, J.J. and Dugdale, R.C., 1969. The kinetics of nitrate and ammonium uptake by natural populations of marine phytoplankton. Deep-Sea Res., 16: 45-57.

May, R.M., 1973. Stability and complexity in model ecosystems. Princeton, Princeton University Press, 265pp.

Pace, M.L., Glasser, J.E. and Pomeroy, L.R., 1984. A simulation analysis of continental shelf food webs. Mar. Biol., 82: 47-63.

Patten, B.C., 1968. Mathematical models of plankton production. Int. Revue. ges. Hydrobiol., 53:357-408.

Peterson, R., 1975. The paradox of the plankton: an equilibrium hypothesis. Am. Nat., 109: 35-49.

Pingree, R.D. and Griffiths, D.K., 1977. The bottom mixed layer on the continental shelf. Est. Coast. Mar. Sci., 5: 399-413.

Platt, T., Denman, K.L. and Jassby, A.D., 1977. Modelling the productivity of phytoplankton. In: Goldberg, E.D., McCase, I.N., O'Brien, J.J. and Steele, J.H. (Editors), The Sea Vol. 6: Marine Modelling, John Wiley and Sons, New York, London, Sydney, Toronto, pp.857-890.

Platt, T. and Jassby, A.D., 1976. The relationship between photosynthesis and light for natural assemblages of coastal marine phytoplankton. J. Phycol., 12: 421-430.

Powell, T and Richerson, P.J., 1985. Temporal variation, spatial heterogeneity, and competition for resources in plankton systems: a theoretical model. Am. Nat., 125: 431-464.

Richtmyer, R.D. and Morton, K.W., 1957. Difference methods for initial-value problems. Interscience publishers, New York, London Sydney., 405 pp.

Riley, G.A., Stommel, H. and Bumpus, D.F., 1949. Quantitative ecology of the plankton of the western North Atlantic. Bull. Bingham Oceanogr. Coll., 12: 1-169.

Roughgarden, J., 1978. Influence of competition on patchiness in a random environment. Theor. Popul. Biol., 14: 185-203.

Steele, J.H. and Henderson, E.W., 1976. Simulation of vertical structure in a plankton ecosystem. Scot. Fish. Res. Rep., No. 5.

Steele, J.H. and Mullin, M.M., 1977. Zooplankton dynamics. In: Goldberg, E.D., McCave, I.N., O'Brien, J.J. and Steele, J.H. (Editors), The Sea Vol. 6: Marine Modelling, John Wiley and Sons, New York, London, Sydney, Toronto, pp. 857-890.

Sundby, S., 1983.. A one-dimensional model for the vertical distribution of pelagic fish eggs in the mixed layer. Deep-Sea Res., 30: 645-661.

Tilman, D., Kilham, S.S. and Kilham, P., 1982. Phytoplankton community ecology: the role of limiting nutrients. Ann. Rev., Ecol., Syst., 13: 349-373.

Williams, R. and Hopkins, C.C., 1975. Sampling at ocean weather station INDIA in 1973. Annales Biologiques, 30: 60-62.

Williams, R. and Hopkins, C.C.,1976. Sampling at ocean weather station INDIA in 1974. Annales Biologiques, 31: 56-60.

Woods, J.D., 1980. Diurnal and seasonal variation of convection in the wind-mixed layer of the ocean. Quart. J., R. Met. Soc., 106:379-394.

Woods, J.D., 1985. The physics of thermocline ventilation. In: Nihoul, J.C.J. (Ed.), Coupled ocean-atmosphere models. Elsevier Oceanography Series, 40, Amsterdam, Oxford, New York, Tokyo, pp 543-590.

Woods, J.D. and Onken, R., 1982. Diurnal variation and primary production in the ocean - preliminary results of a Lagrangian ensemble model. J. Plankt. Res., 4: 735-756.

ESTIMATES OF THE NITROGEN FLUX REQUIRED FOR THE MAINTENANCE OF
SUBSURFACE CHLOROPHYLL MAXIMA ON THE AGULHAS BANK.

R.A. CARTER, P.D. BARTLETT and V.P. SWART

National Research Institute for Oceanology, Box 320,

Stellenbosch, 7600 RSA

ABSTRACT

The waters overlying the Agulhas Bank are dominated by strong
thermal stratification in the austral summer. Chlorophyll distributions associated with the stratification show well developed maxima at the base of the thermocline. Nitrate-nitrogen
profiles generally show that the chlorophyll maxima are situated in the region of the water column where nitrate-nitrogen
concentrations are limiting for phytoplankton growth. Estimates
of mesozooplankton grazing and the partition of phytoplankton
production to grazing, decomposer pathways, etc., indicate a
phytoplankton production rate of c. 630 mgC/metre squared/day as
being necessary to maintain the observed chlorophyll maxima. This
estimate converts to a nitrogen requirement of 8.2 mM N/metre
squared/day. 24% of this requirement can be supplied by zooplankton excretion. The remainder (6.2 mM N/metre squared/ day)
can be supplied by vertical flux from the bottom waters, if the
conditions pertaining in the stratified waters of the English
Channel apply and the differences in nitrate-nitrogen concentration gradients in the nitraclines in English Channel and Agulhas
Bank waters are taken into account. Additional phytoplankton
grazing by other size fractions in the zooplankton and micronekton will require either increased vertical diffusion rates or
turbulent breakdown of the thermocline to meet the increased
nitrogen demand.

INTRODUCTION

The Agulhas Bank is an extensive, irregularly shaped continental shelf area situated south of the south African sub-
continent (Fig. 1). Surveys (eg. Shannon, 1966) have shown that
the waters overlying the Agulhas Bank are strongly thermally
stratified in summer and the sparse plankton literature from the
area (eg. De Decker, 1973; Brown, 1981) indicates that subsurface chlorophyll maxima develop in association with the thermoclines. Shannon et al. (1984) have discussed the likely connection between the subsurface chlorophyll maxima, their associated zooplankton populations, and the intense spawning by the

anchovy, _Engraulis capensis_, that occurs on the Agulhas Bank (cf. _Engraulis mordax_ off southern California (Lasker, 1975)).

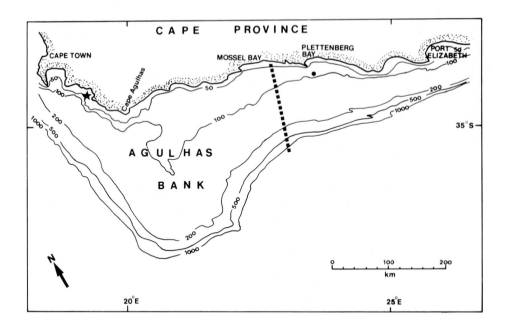

Fig. 1. Map showing the location and dimensions of the Agulhas Bank, south of South Africa. Depths are in metres. The dashed line indicates the position of the transect illustrated in Fig. 2; the closed circle shows the station location for which data are presented in Fig. 4 and the star indicates the location of the measurements made by Tromp and Horstman (1979).

Recent surveys by the R.V. Meiring Naudé on the Agulhas Bank have confirmed the existence of thermal stratification and associated subsurface chlorophyll maxima. Fig. 2 illustrates representative distributions in the form of vertical sections taken from a cross-shelf transect (for location see Fig. 1) made in February, 1984. Features of note are the strengths of the structures as well as their large spatial extent. This latter point has a parallel in the temporal persistence of these structures, research carried out west of Cape Agulhas (Fig. 1) indicating life spans of three to five months in a geographically restricted area (Tromp and Horstman, 1979).

Fig. 3 depicts typical summer vertical profiles of temperature, chlorophyll _a_ and nitrate-nitrogen from the Agulhas Bank. Similar to northern hemisphere temperate continental shelf waters

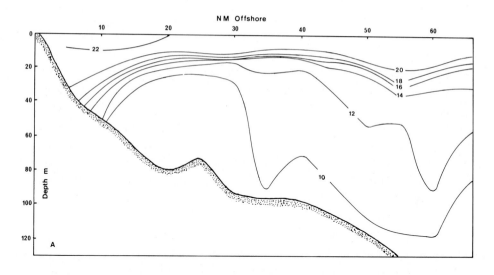

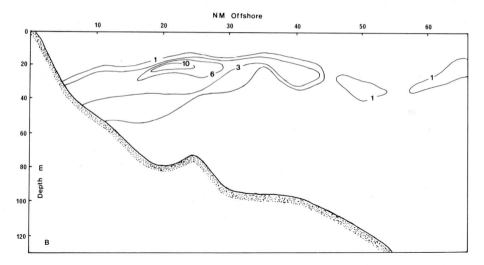

Fig. 2. Vertical sections from a cross-shelf transect on the Agulhas Bank, February, 1984. A: temperature; B: chlorophyll a (units are μg/ℓ).

(e.g. Cullen, 1982: Holligan et al.,1984a, b), the chlorophyll maxima are situated in the thermocline but generally above (e.g. Fig. 3b, c) the maximum gradient in nitrate-nitrogen concentrations. These features suggest that phytoplankton production in and above the chlorophyll maxima may be nitrogen limited and consequently nitrogen supply to the phytoplankton in these circumstances has received some attention. Cullen and Eppley (1981) showed that southern Californian subsurface chlorophyll maxima were dominated by dinoflagellates (as are Gulf of Maine maxima (Holligan et al. 1984a)) and that these organisms migrated vertically to replenish cell nutrient levels. Holligan et al. (1984c) demonstrated that Celtic Sea chlorophyll maxima were dominated by naked flagellates and that vertical diffusion supplied their nutrient requirements. The Agulhas Bank chlorophyll maxima differ from the above examples in that they are dominated by diatoms (De Decker, 1973; Carter, unpublished data), and also generally attain higher chlorophyll a concentrations. Nitrogen supply in this case must therefore either be through vertical diffusion or by turbulent disruption of the thermocline.

As a preliminary to further research on the dynamics of Agulhas Bank chlorophyll maxima here we calculate the nitrogen supply rates required to maintain observed phytoplankton (chlorophyll) distributions in the face of estimated grazing pressure by the associated mesozooplankton. This method is used because of the paucity of reliable primary production estimates in the area and the longer time base the approach provides. Our major purpose is to determine whether thermocline disruption processes have to be invoked to provide nitrogen at the required rate to the euphotic zone.

DATA

The calculation described below is based on the vertical distributions of phytoplankton (as chlorophyll a) and zooplankton depicted in Fig. 4. The data on which this figure is based were collected in the Plettenberg Bay region (Fig. 1) during April, 1984. Chlorophyll a and phaeopigments were measured on 90% acetone extracts of filtered 1ℓ samples on a Turner fluorometer calibrated against extracts of pure chlorophyll (Sigma Corporation) as well as parallel spectrophotometric determinations (Strickland and Parsons, 1972). Mesozooplankton samples were obtained by means of a high yield (2000 ℓ/min) submersible pump system operated at the same depths as those used for the phytoplankton pigment samples. Only zooplankton larger than 105 μm were collected with subsamples being taken for biomass estimates (dry weight; Lovegrove, 1966), and species identification. Zooplankton dry weights were converted to carbon equivalents by

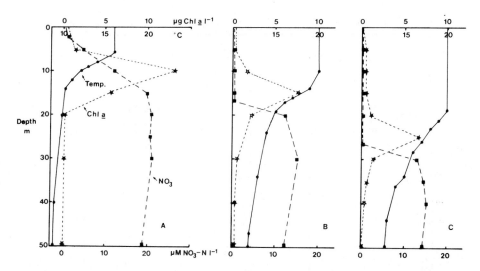

Fig. 3. Vertical profiles measured on the Agulhas Bank during February, 1984, showing typical distributions in shallow (A), moderately deep (B) and deep mixed layer (C) conditions.

multiplying by a factor of 0,45 (Hutchings, 1979). Both sets of subsamples were fractionated into 105-297 μm and >297 μm size ranges by filtration through appropriate meshes. The nitrate- and ammonia-nitrogen concentrations shown in Fig. 4 were measured according to Mostert (1983) on filtered samples on an onboard Technicon autoanalyzer system.

CALCULATION

The calculation is based on the following two assumptions:

1) Phytoplankton production below the chlorophyll maximum is negligible relative to rates at and above the chlorophyll maximum.

2) Steady-state conditions apply, i.e. there is no isopycnal transport of either phytoplankton or nitrogen into the area.

The consequences of these assumptions are that all the herbivorous zooplankton in the water column are dependent on phytoplankton production at and above the chlorophyll maximum and that the only source of 'new' nitrogen to the euphotic zone is diffus-

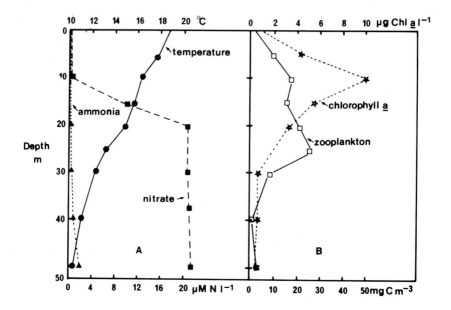

Fig. 4. Vertical profiles measured on the Agulhas Bank in April, 1984. A: Temperature, nitrate- and ammonia-nitrogen; B: Plankton.

ion or advection from below.

The integrated combined zooplankton biomass estimate for the profile presented in Fig. 4 is 494 mgC/metre squared. The zooplankton was dominated by the Calanoid copepods Calanoides carinatus, Paracalanus parvus, Paracalanus sp. and Centropages brachiatus. All of these species were reported from the area by De Decker (1973). Carter (1983) measured respiration rates (R) versus temperature (T) for similar sized species from southern Benguela kelp beds and obtained regressions of

$$R \ (\ell \ O2/mgDW/hr) = 3,45 + 0,17T$$

for copepods in the size range 105-297 μm, and

$$R(\ell \ O2/mgDW/hr) = 3,11 + 0,16T$$

for copepods larger than 297 μm. Carter (1983) used these regressions and those for production/respiration given by Humphreys (1979) to calculate consumption rates for the two size ranges of copepods. Applying these rates and the appropriate temperature corrections to the two size range biomass estimates used to construct Fig. 4 yields a zooplankton community consumption estimate of 126 mgC/metre squared/day.

Recent analyses have indicated that only some 20-30% of phytoplankton production is channelled through zooplankton; the rest going through decomposer processes (e.g. Holligan et al., 1984c; Peterson, 1983). If a figure of 20% is assumed for the purposes of this calculation the net photosynthesis required to satisfy the zooplankton ration is 630 mgC/metre squared/day. According to the integrated chlorophyll a concentration (207 mg/metre squared) this implies a phytoplankton assimilation number of 3.0 mgC/mgChla/metre squared/day which, using a measured carbon/chlorophyll a ratio of 55,85, is equivalent to an overall phytoplankton community doubling rate of 0,05/day. But, as stated above, it is assumed that most of the phytoplankton production occurs at and above the chlorophyll a maximum; the integrated chlorophyll concentration for this section of the water column is 52 mg/metre squared and consequently the required doubling rate becomes 0,22/day and the assimilation number 12,1 mgC/mgChla/metre squared/day. These values are not extraordinarily high, Carter (1983) reporting phytoplankton doubling rates between 0,23 and 1,48/day in southern Benguela kelp beds, whilst the assimilation number is well within the theoretical maximum (eg. Williams et al., 1983).

The mean carbon/nitrogen ratio for the phytoplankton sampled on the Agulhas Bank in April, 1984 was 6,07 (sd=0,15, n=12). The nitrogen requirement for the above estimated production rate is therefore 8,2 mM N/metre squared/day.

Holligan et al. (1984c) measured excretion rates for a range of different sized copepods in the English Channel. Paracalanus parvus, a species important on the Agulhas Bank, was included in their suite of measurements; an excretion rate of 19,1% of body nitrogen/day being obtained for this species. Generalizing this to the zooplankton biomass estimate derived from Fig. 4 yields a population excretion rate of c. 2 mM ammonia-N/metre squared/day. This is equivalent to 24% of the estimated nitrogen requirement of the phytoplankton. Thus 76% or 6,2 mM N/metre squared/day must be supplied by vertical flux of nitrogen from below the thermocline if the phytoplankton is to maintain itself without episodic or periodic disruption of the thermocline.

Holligan et al. (1984c), in thermally stratified water columns in the Celtic Sea, calculated a vertical diffusion rate of 2,3 mM N/metre squared/day on the basis of the heat flux into the bottom waters. This is far less than the calculated required flux for the Agulhas Bank. However, nitrate-nitrogen concentration gradients in the nitracline in English Channel waters, displayed in Holligan et al. (1984c), attain c. 0,75 μM N/metre whereas the gradient apparent in Fig. 4 is c. 2 μM N/metre. As diffusion rate is proportional to the concentration gradient (Fick's Law), a proportionally higher rate is thus expected for Agulhas Bank waters compared to that calculated for the English Channel. On this basis a vertical diffusion rate of c. 6 mM N/metre squared/day can be calculated for stratified Agulhas Bank waters which is close to the rate required by the phytoplankton.

Therefore, when viewed in the light of the above calculations and the analyses of Holligan et al. (1984a, b and c), nitrogen recycling by the zooplankton and vertical nitrogen diffusion appears to be just sufficient to maintain the observed chlorophyll maximum (Fig. 4). However, this conclusion is based on the fact that only mesozooplankton biomass was used in the estimation of the zooplankton ration. Larger zooplankton (500 μm) and also micronekton, eg. fish larvae, will also graze on the phytoplankton thus increasing the required phytoplankton production rate. Consequently higher vertical diffusion rates or turbulent disruption of the thermocline and associated enrichment of the waters above the chlorophyll maximum must thus be invoked to meet the higher nitrogen demand. Processes that can operate here are double diffusion, upwelling (e.g. Swart, 1983; Schumann et al., 1982), storms or breaking internal waves. These phenonema are currently being investigated.

REFERENCES

Brown, P.C., 1980. Phytoplankton production studies in the coastal waters off the Cape Peninsula, South Africa. M.Sc thesis, University of Cape Town, 98pp.

Brown, P.C., 1981. Pelagic phytoplankton, primary production and nutrient supply in the southern Benguela region. Trans. roy. Soc. S. Afri., 44: 347-356.

Carter, R.A., 1983. The role of plankton and micronekton in carbon flow through a southern Benguela kelp bed. Ph.D thesis, University of Cape Town, 174pp.

Cullen, J.J., 1982. The deep chlorophyll maximum: comparing vertical profiles of chlorophyll a. Can. J. Fish. Aquat. Sci., 39: 791-803.

Cullen, J.J. and Eppley, R.W., 1981. Chlorophyll maximum layers of the Southern California Bight and possible mechanisms of their formation and maintenance. Oceanol. Acta, 4: 23-32.

De Decker, A.H., 1973. Agulhas Bank Plankton. In: B. Zeitschel (Editor), Biology of the Indian Ocean. Ecological studies. Analysis and synthesis, vol. 3, Springer Verlag, Berlin, pp. 189-219.

Holligan, P.M., Balch, W.M. and Yentsch, C.M., 1984a. The significance of subsurface chlorophyll, nitrite and ammonia maxima in relation to nitrogen for phytoplankton growth in stratified waters of the Gulf of Maine. J. Mar. Res., 42: 1052-1073.

Holligan, P.M., Harris, R.P., Newell, R.C., Harbour, D.S., Head, R.N., Linley, E.A.S., Lucas, M.I., Tranter, P.R.G. and Weekley, C.M., 1984b. Vertical distribution and partitioning of organic carbon in mixed, frontal and stratified waters of the English Channel. Mar. Ecol. Prog. Ser., 14: 111-127.

Holligan, P.M., Williams, P.J.LeB., Purdie, D. and Harris, R.P., 1984c. Photosynthesis, respiration and nitrogen supply of plankton populations in stratified, frontal and tidally mixed shelf waters. Mar. Ecol. Prog. Ser. 17: 201-213.

Humphreys, W.E., 1979. Production and respiration in animal populations. J. Animal Ecol., 48: 427-453.

Hutchings, L., 1979. Zooplankton of the Cape Peninsula upwelling region. Ph.D thesis, University of Cape Town, 240pp.

Lasker, R., 1975. Field criteria for survival of anchovy larvae: the relation between inshore chlorophyll maximum layers and successful first feeding. Fish Bull. (U.S.), 73: 453-462.

Lovegrove, T., 1966. The determination of the dry weight of plankton and the effects of various factors on the values obtained. In: H. Barnes (Editor), Contemporary Studies in Marine Science. George Allen and Unwin Ltd., pp 429-467.

Mostert, S.A., 1983. Procedures used in South Africa for the automatic photometric determination of micronutrients in seawater. S. Afr. J. mar. Sci., 1: 189-198.

Peterson, B.J., 1983. Synthesis of carbon stocks and flows in the open ocean mixed layer. NATO Doc. No. 1873A-03/28/83, 10pp.

Schumann, E.H., Perrins, L.A. and Hunter, I.T., 1982. Upwelling along the south coast of the Cape Province, South Africa. S. Afr. J. Sci., 78: 238-242.

Shannon, L.V., 1966. Hydrology of the south and west coasts of South Africa. Invest. Rep. Div. Seafish. S. Afr., 58: 52pp.

Shannon, L.V., Hutchings, L., Bailey, G.W. and Shelton, P.A., 1984. Spatial and temporal distribution of chlorophyll in Southern African waters as deduced from ship and satellite measurements and their implications for pelagic fisheries. S. Afr. J. mar. Sci., 2: 109-130.

Strickland, J.D.H. and Parsons, T.R., 1972. A practicl handbook of seawater analysis. 2nd Edition, Bull. Fish. Res. Bd. Canada, 167: 310pp.

Swart, V.P., 1983. Influence of the Agulhas Current on the Agulhas Bank. S. Afr. J. Sci., 79: 160-161.

Tromp, B. and Horstman, D., 1979. Cape south coast upwelling measurements. Unpublished MS., Sea Fisheries Research Institute, 44pp.

Williams, P.LeB., Heinemann, K.R., Marra, J. and Purdie, D.A., 1983. Comparisons of 14C and O2 measurements of phytoplankton production in oligotrophic waters. Nature, 305: 49-50.

MODELLING THE TIME DEPENDENT PHOTOADAPTATION OF PHYTOPLANKTON TO FLUCTUATING LIGHT*

KENNETH L. DENMAN[1] and JOHN MARRA[2]

[1]Institute of Ocean Sciences, P.O.Box 6000, Sidney B.C. Canada, V8L 4B2

[2]Lamont-Doherty Geological Observatory of Columbia University, Palisades, N.Y., U.S.A. 10964

ABSTRACT

   Phytoplankton in the pycnocline and in the upper mixing layer of the ocean are subjected to vertical displacements over a range of time scales characteristic of fluid motions resulting from internal waves and turbulent processes. Because underwater light decreases approximately exponentially with depth, the phytoplankton thus experience variations in irradiance over the same range of time scales. In general, because of photoadaptation, the instantaneous rate of photosynthesis does not depend only on the instantaneous irradiance but rather on some function of the cells' light history. We present a linear response model for photoadaptation of the rate of photosynthesis to varying irradiance. Results of laboratory experiments under controlled lighting are used to determine time scales for photoadaptation and to evaluate functional relationships and coefficients for the model. The model is used to simulate the generation of photosynthesis- irradiance curves from incubation experiments, and it is tested with data from a culture grown in a greenhouse under natural sunlight. For an idealized case, the frequency response function for adaptation to internal wave displacements is calculated. Waves with periods equal to or shorter than the adaptive response time allow little adaptation.

INTRODUCTION

   Successful survival and photosynthetic production of phytoplankton require that they find and stay in regions of the ocean with both sufficient light and sufficient nutrients. This admittedly simplistic view of phytoplankton dynamics emphasizes the importance to primary production of light availability and utilization by the cells. Significant research effort has gone into understanding and quantifying the functional dependence of the rate of photosynthetic production P on available light I (e.g. Platt, Gallegos and Harrison, 1980). The so-called 'P-I curves' that result are usually obtained from ^{14}C incubation experiments where samples of living phytoplankton are inoculated with ^{14}C, and split into subsamples which are incubated simultaneously at different constant irradiances for a set time period (1 h to 1 day, usually several h). The net carbon uptake by the cells is determined and divided by the incubation time to give an average rate of photosynthetic carbon production at each light intensity. The individual data values are then plotted, usually giving a relatively

* L-DGO contribution no. 3891.

smooth P-I curve. In general, P-I curves are not repeatable between experiments, and determining the sources of this variability is a research problem currently of great interest.

That one source of this variability may be the adaptation of the cells' physiology to the varying irradiance itself was clearly demonstrated by Harris and Lott (1973) and Harris (1973). They showed from short term photosynthetic oxygen production rate measurements with laboratory cultures and field samples that the 'instantaneous' rate values presented on a P-I phase plot for a full day did not trace out a single P-I curve. There was hysteresis: a closed curve was drawn out with high rates in the morning, a midday depression, and values in the afternoon recovering but lower than at the same irradiances observed in the morning. The extent of the hysteresis varied with season and with the degree of cloudiness over the previous few days. These findings were repeated and extended by Marra (1978a, b).

More recent studies by Marra (1980b), Falkowski (1980, 1983), Prézelin and Matlick (1980), Rivkin et al. (1982), Slagstad (1982), Lewis and Smith (1983), Lewis, Cullen and Platt (1984), Post et al. (1984), Geider and Platt (submitted), and others have established that this photoadaptation of the phytoplankton to different preconditioning or light histories occurs in a variety of different forms and a variety of different time scales. In addition, several of these studies have been successful in modelling the adaptive responses by first order reaction kinetics, suggesting a possible form for a more detailed model.

MODEL FORMULATION

Linear response model

Previous work shows that the functional relationship between the rate of primary production and the irradiance (the P-I curve) is clearly nonlinear and not unique. However, the coefficients that describe the P-I curve may themselves adapt to irradiance conditions in a linear fashion. Hence, we will start with the assumption that each coefficient obeys the principles of linear systems theory and can be described by a linear response model as outlined in, for example, Jenkins and Watts (1968). Apart from the solutions being known and straight forward, other advantages are that solutions with different response times can be superposed and that each coefficient can have independent response times.

We assume that the phytoplankton have two limiting P-I curves, a fully dark adapted curve (subscript D) and a fully high light adapted curve (subscript L), such that the instantaneous rate of production would always fall between these two curves. Furthermore, the degree of adaptation of the cells at any time from the dark adapted curve to the light adapted curve is a linear function of recent 'inhibiting' irradiances. If $A(t)$ is a time dependent coefficient of the curve describing the functional dependence on irradiance $I(t)$ of the instantaneous photosynthesis rate $P(t)$, then, after Jenkins and Watts (1968),

$$A(t) = Y(t) A_L + (1 - Y(t)) A_D = Y(t) (A_L - A_D) + A_D. \qquad (1)$$

The degree of adaptation $Y(t)$ of $A(t)$ from curve D to curve L is given by the convolution integral

$$Y(t) = \int_0^\infty h(t_s) \, X(t - t_s) \, dt_s. \qquad (2)$$

Formally, $h(t)$ is the response of the linear system to an impulse function input, and $X(t)$, the input function, is the inhibiting or adaptive capability of a given irradiance $I(t)$. Marra (1980b), Falkowski (1980, 1983) and others have used with some success first order kinetics to fit data from photoadaptation experiments, suggesting a simple exponential response function form for $h(t)$:

$$h(t) = \frac{1}{T} e^{-t/T}. \qquad (3)$$

In particular, for a step function in $X(t)$, we get $Y(t) = 1 - e^{-t/T}$. The functions Y, h, X and T are specific to the coefficient A and would be so subscripted when more than one coefficient was being modelled. In addition, there can be multiple response times T for each coefficient, for example, one with a value of 1 or 2 h and one with a value of several days.

Before the adaptation model described by equations (1)-(3) can be of use, we must choose explicit functional forms for the instantaneous rate of photo-synthetic production $P(t)$ and the inhibiting or adaptive capability $X(I(t))$ of a given irradiance, and a value for the adaptation response time T that scales the exponential response. Since the experiments to be used measured $P(t)$ rather than $A(t)$ (e.g. Marra, 1978a, b), simpler functional forms that can be inverted to solve for the coefficients $A(t)$ are desired.

Choice of functional forms

Substantial effort has been expended on fitting various functional forms of the P-I curve to experimental data, e.g. Platt et al., 1980; Lederman and Tett, 1981; and Gallegos and Platt, 1981. These curves do not represent instantaneous relationships since the productivity rates obtained are time averages from ^{14}C incubation experiments lasting from about 1/2 to several h. Nevertheless, we shall assume similar forms for the instantaneous P-I curves. We shall use two expressions that fit the incubation data well where photoinhibition does not occur:

$$P^a(t) = P_s(t) \tanh \frac{a \; I(t)}{P_i(t)} \tag{4}$$

$$P^b(t) = P_s(t) \; [1 - e^{-a \; I(t)/P_i(t)}] \tag{5}$$

where $P^a(t) = P_s(t)$ for constant slope at $I = 0$ and $P^b(t) = P_D$ for constant $I_k = P_D/a$. The two cases of constant initial slope and constant I_k usually will only yield small differences in the results. The latter case represents the situation often found with fitted coefficients from a large number of incubations with natural assemblages of phytoplankton: that the coefficients a and P_s are positively correlated implying that their ratio I_k is roughly constant. Expressions (4) and (5) are plotted in Fig. 1.

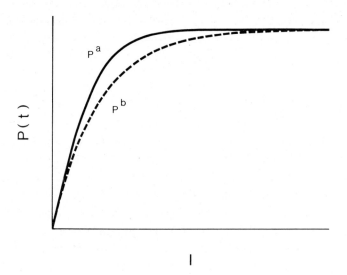

Fig. 1. Expressions for the instantaneous rate of production P as a function of irradiance I.

We have made two implicit assumptions: (i) the coefficient A(t) in (1) is now in fact the asymptotic rate of production $P_s(t)$, and (ii) there is no explicit photoinhibition factor b. Rather, adaptation occurs only by varying $P_s(t)$. In principle of course, the other coefficients a and b could also vary.

There is little information on the inhibiting or adaptive capability of a given irradiance I, other than the assumptions that high light somehow damages the cellular photosynthetic apparatus at a rate faster than recovery processes occur but that at lower light recovery processes may keep pace with photo-destruction. For mathematical expediency, we try two expressions for X(I), one

with adaptive inhibition increasing from I = 0 and one with a threshold for
inhibition at I_b (see Platt and Gallegos, 1980):

$$X^a(I) = 1 - e^{-(I/I_b)^2} \tag{6}$$

$$X^b(I) = 0 \qquad\qquad (I < I_b)$$

$$= 1 - e^{-[(I - I_b)/I_b]^2} \qquad (I \geq I_b). \tag{7}$$

The expressions are plotted in Fig. 2; the sharpness of the onset of inhibition
can be increased by increasing the exponent in (6) and (7). We shall choose I_b
(which also scales the rate of increase of X) from experimental data.

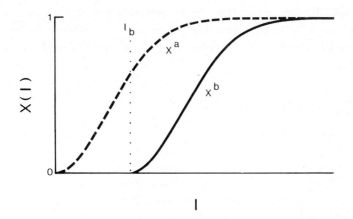

Fig. 2. Expressions for the inhibiting or adapting capability X(I) as a function
of irradiance.

Once all the required functions and coefficients are estimated, the model is
solved in the following manner. At each time t, the degree of adaptation Y(t)
is calculated from (2) for each varying coefficient based on the history of
inhibiting irradiance received by the cells X(I) (calculated from (6) or (7)).
The instantaneous value of each adapting coefficient is calculated from (1) and
the instantaneous rate of photosynthetic production is calculated from (4) or
(5). Given an appropriate initial condition, the convolution integral (2) need
only be calculated back one time step and the previous value adjusted by a
constant factor. For special simple cases (e.g. constant irradiance), the model
can be solved exactly as a function of time.

Estimation of coefficients from experiment

Marra (1978b, 1980a) carried out laboratory experiments with phytoplankton cultures grown under controlled light conditions that provide data suitable for testing the adaptation model. Pure cultures were subjected to a 12-h dark: 12-h light cycle (12 D: 12 L), where the light was constant during each day but the intensity was set randomly at one of several levels between 60 and 1500 μE m^{-2} s^{-1} of photosynthetically active radiation (PAR). The rate of photosynthetic oxygen production was measured at 0.5 h intervals with an oxygen electrode. The results shown in Fig. 3 for the diatom Lauderia borealis were similar for several species: initial high oxygen production at high light intensities falling off rapidly to roughly constant low values after several hours, and small changes in production at low light intensities. The effect was not strictly a diurnal feature as similar curves were obtained by varying the start time by several hours. The suppression at high light intensities may be the result of increased respiration with exposure to high light (Falkowski, Dubinsky and Wyman, 1985).

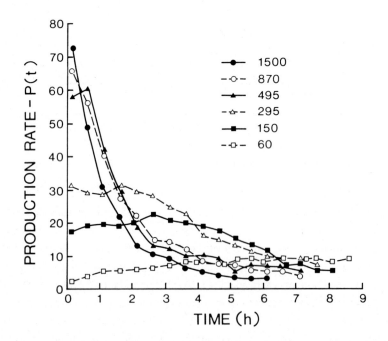

Fig. 3. Measurements of oxygen production rate P (in pg-at. O$_2$ h^{-1} cell^{-1}) against time at various constant irradiances (in μE m^{-2} s^{-1} PAR), after 12 h of dark conditioning. Redrawn from Marra, 1978b (we have corrected units for P).

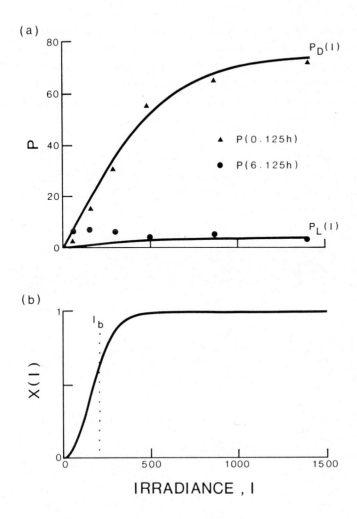

Fig. 4. (a) Data from Fig. 3 replotted against irradiance for two times. The solid lines are estimated P-I curves for equilibrium dark adapted $P_D(I)$, and high light adapted $P_L(I)$, cells. (b) Estimate of $X(I)$, the adapting capability of a given irradiance, from the fractional distance between $P_D(I)$ and $P_L(I)$ of the solid circles in panel (a).

The curve in Fig. 3 for the highest irradiance approximates the adaptation from fully dark adapted cells to fully light adapted cells. The other curves represent the adaptation from dark adapted cells to cells adapted to some intermediate light intensities. In Fig. 4 we have replotted the oxygen production rates against irradiance, for the initial sample time (t = 0.125 h) and the final common time (t = 6.125 h). The two curves are estimated P-I curves (using (4)) for dark adapted cells, $P_D(I)$, and for high light adapted cells, $P_L(I)$. Curve $P_L(I)$ would occur for cells that were subjected to high irradiances such

348

that $X(I) \approx 1$ for sufficient time that $Y(X(I)) \approx 1$. The production rates at t = 6.125 h in Fig. 4 are between the two curves because the cells had adapted to lower irradiances such that $X(I) < 1$. The fraction of the production rates at t = 6.125 h between $P_D(I)$ and $P_L(I)$ is an approximation to $X(I)$, the inhibiting strength of an irradiance I. In the bottom panel we have plotted an estimate of $X(I)$ using (6) from the data in the top panel. This curve will be used in the model.

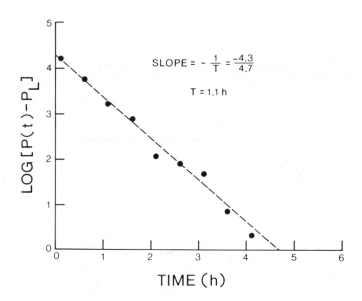

Fig. 5. Log-linear plot of data from Fig. 3 (for I = 1500). The asymptotic production rate at large time was taken as $P_s = P_L = 4$ according to Fig. 2a. Units are as before.

The adaptation time T for the data of Fig. 3 can now be estimated. The response for a step function increase in I (and X) from 0 is $Y(t) = x_2(1 - e^{-t/T})$ where x_2 is the value of X after the step increase in I. For large I, $P(t) \rightarrow P_s(t)$ in either (4) or (5), and substituting in (1) where $A(t) = P_s(t) = P(t)$

$$P(t) = x_2 (P_L - P_D) (1 - e^{-t/T}) + P_D \qquad (8)$$

Rearranging, taking logarithms, and setting $x_2 = 1$ for large I, we get the expression

$$\log (P(t) - P_L) = -t/T + \log (P_D - P_L). \qquad (9)$$

Thus a plot of log $(P(t) - P_L)$ as a function of time t should yield a straight line with slope $-1/T$ if this model is a reasonable representation of the adaptation. We have plotted these data for $I = 1500$ μE m^{-2} s^{-1} in Fig. 5. The curve is roughly linear with a slope that gives a response time estimate of $T = 1.1$ h. Beyond a time $t = 4$ h, the production rates were so near P_1 that there was a large scatter on the logarithmic plot.

From the data of Marra's (1978b, 1980a) 12 D: 12 L experiments, we have been able to estimate all the coefficients required to complete the linear adaptation model for those data. The model should now be able to simulate those data.

SOLUTIONS FOR BATCH CULTURE DATA
Light-Dark Step Functions

The analytic solution (8) to the model for a step function in irradiance can be extended to the general case of a step at time $t = 0$ from any constant preconditioning irradiance I_1 to another constant irradiance irradiance I_2. The solution for the adaptation of $P_s(t)$ is

$$P_s(t) = P_D + (P_L - P_D) [x_2 - (x_2 - x_1) e^{-t/T}] \tag{10}$$

where $x_1 = X(I_1)$ and $x_2 = X(I_2)$. Equation (10) can then be substituted into (4) or (5).

The results from (10) for Marra's experiments (Fig.3) are shown in Fig. 6. We used (4) with constant I_K and (6) with $I_b = 200$ μE m^{-2} s^{-1}. Qualitatively, the model has captured the main features of the data. It does not duplicate the slow increase (or initial plateau) of the rate of production at the lower light intensities. That would probably require separate adaptation of the slope a with a different response time and a different functional form for $X(I)$. Clearly from Fig. 3, in the usual experiments to obtain P-I relations, adaptation can occur during the incubation. Each estimate of production is then only a time average for the duration of the experiment and is in general a poor estimate of the instantaneous rate at any time during the incubation. By integrating along the curves in Fig. 3, Marra (1978b) simulated P-I curves from incubations by calculating the average production rate at each irradiance for several different incubation times T_i:

$$P_{av}(I) = \frac{1}{T_i} \int_0^{T_i} P(I(t)) \, dt \tag{11}$$

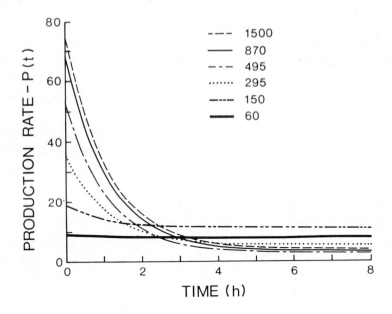

Fig. 6. Model output for the series of adaptive time course data in Fig. 3.

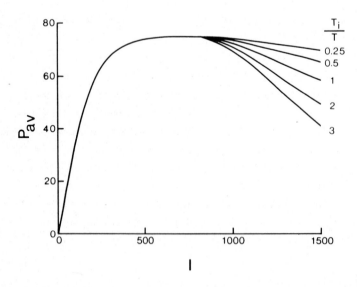

Fig. 7. Model simulation of P-I curves obtained from incubations with different values of the ratio of incubation time to P_s adaptive response time T_i/T.

The resulting P_{av}-I curves showed that photoinhibition at high light intensities in P-I curves obtained from incubations may be largely a function of length of incubation. We have simulated these results using (11) for a variety of different ratios of T_i/T (where we have used expressions (5) and (7)). The results plotted in Fig. 7 show an apparent photoinhibition effect that increases as the incubation time T_i becomes larger than T, the response time for adaptation of P_s. The onset of the apparent photoinhibition is equal to I_B, the threshold for the inhibiting ability of irradiance in (7). This result, that inhibition in P-I curves can occur in a model with no explicit inhibition, is not a general result. For a different species of phytoplankton whose long term adaptation was such that exposure to high light after low light conditioning resulted in upward adaptation of P_s, Lewis and Smith (1983) observed the greatest photoinhibition with the shortest incubation time. To obtain such a result with our model, we would have to include an explicit photoinhibition term that itself was capable of adaptation.

Diurnal cycles

The model should simulate the results of diurnal time course studies (e.g. Harris, 1973; Harris and Lott, 1973; Marra, 1978a) that documented the existence of photoadaptation both in laboratory cultures and in samples recently taken from lakes or the ocean. We should expect the model to simulate both the midday depression in production rate and the hysteresis in the P-I phase plots resulting from the response time required for adaptation. In Fig. 8 we show the results for model parameters similar to those used to reproduce the asymmetry

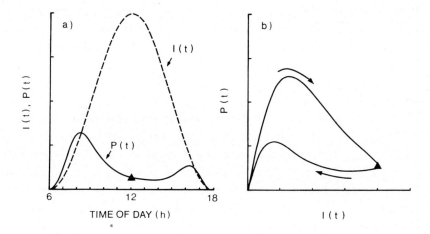

Fig. 8. Response of adaptive model to diurnal irradiance cycle. (a) Time course of irradiance I and production rate P. (b) Phase plot of P-I showing hysteresis resulting from midday and afternoon depression. Solid triangle denotes noon.

resulting from the incomplete recovery of the photosynthetic apparatus following the suppression caused by the high irradiances at midday.

Simulation of 'Greenhouse' data

Marra and Heinemann (1982) reported on a series of time course measurements of oxygen production by batch cultures of phytoplankton growing in a window greenhouse under natural cycles of irradiance. They found that the P-I phase plots were roughly linear on shady days with peak irradiances less than about 200 μE m^{-2} s^{-1} but that significant hysteresis occurred on days with peak irradiances greater than 400 μE m^{-2} s^{-1}. Marra, Heinemann and Landriau (1985) were able to simulate the low irradiance days from constant daily P-I curves, but only a statistical model with time varying coefficients could fit the data for days of variable irradiance (Neale and Marra, 1985).

One of the greenhouse experiments (Greenhouse IV) consisted of a 2 week time series with the same diatom species <u>Lauderia borealis</u> that was used in the 12 D: 12 L experiments that we used to develop our model. We ran our model on these data with the same adaptation time T = 1.1 h but with curves estimated to be the upper and lower envelopes about the data on a P-I phase plot (fig. 5 in Marra, 1980a; or fig. 1a in Neale and Marra, 1985). As before, only the asymptote P_s was allowed to adapt to the varying light history. Results are presented in Fig. 9 for 30-31 October 1979, partially sunny days with peak irradiances of about 950 μE m^{-2} s^{-1}. The main variations in instantaneous production rate are reproduced except for an overshoot in the early morning rise in production rate. Since the P-I curves were estimated indirectly from data, and the response time and model characteristics were determined from data on the same species but 3 years earlier under different conditions of growth, the extent of agreement is good. It is entirely possible that other coefficients (initial slope a or inhibition b) were also undergoing adaptation to the highly variable irradiance.

EFFECTS OF INTERNAL WAVES
No adaptation

In coastal regions, phytoplankton are often displaced vertically 10's of meters by passing internal waves (e.g. Denman and Herman, 1978; Haury et al., 1983). Because of the approximately exponential decay with depth of irradiance below the sea surface, those vertical displacements translate into fluctuations in the light intensity received by cells located on constant density surfaces. These fluctuations can be greater than a factor of 100 (Haury et al.); for example, with a vertical displacement of 10 m and an attenuance coefficient of 0.3 m^{-1}, the irradiance changes by a factor of about 20.

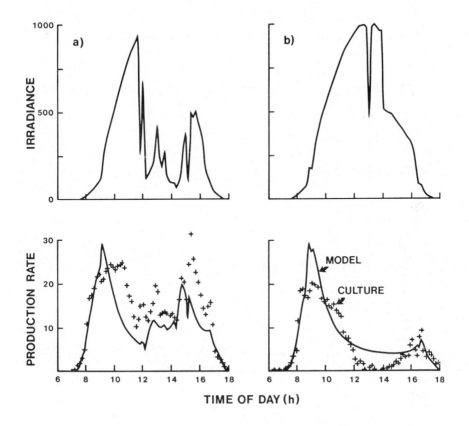

Fig. 9. Model results and data from cultures grown in a greenhouse (Marra and Heinemann, 1982) for (a) 30 and (b) 31 October, 1979. The model was run with functions (4) and (6) with P_s = 35, a = 0.3 and I_b = 200; units as before.

If the irradiance fluctuations experienced by cells were strictly sinusoidal, the negative curvature of the P-I curves would cause the average rate of photosynthetic production to be less than in the absence of internal waves. However, the irradiance cycle experienced by cells usually will not be sinusoidal because of nonlinearities in both the shape of the internal waves and the decay of irradiance with depth. Consider the irradiance I(Z) on a density surface being displaced vertically by internal waves instantaneously at a depth Z(t). For PAR radiation, a single attenuance coefficient c is a good approximation

$$I(Z(t)) = I_s e^{-c\,Z(t)} \qquad\qquad (12)$$

where I_s is the PAR irradiance just below the sea surface. Internal waves near the sea surface often have flattened crests (e.g. fig. 5.1, Phillips, 1977) similar to that plotted in Fig. 10a. Depending on the degree of that flattening

(a)

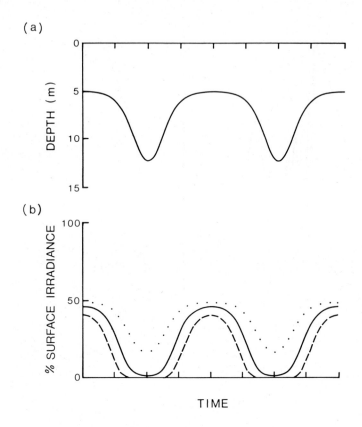

(b)

Fig. 10. (a) Idealized time series for the depth Z(t) of a phytoplankter located on an isopycnal surface being displaced vertically by a near-surface internal wave. (b) Possible forms for the irradiance fluctuations experienced by the cell in a light field decaying exponentially with depth.

and on the amplitude of the internal wave relative to the attenuance length c^{-1}, the irradiance fluctuations experienced by cells on a density surface at depth Z(t) may be roughly sinusoidal or skewed towards either the higher or lower values as depicted by the three curves in Fig. 10b.

For a purely sinusoidal internal wave, the exponential dependence of irradiance with depth (12) causes the average irradiance experienced by a phytoplankter to be greater than in the absence of the wave. Below some critical depth (where the linear portion of the P-I curve pertains and the nonlinearity in the irradiance profile dominates), the internal wave can then actually increase the average production (G. Holloway, pers. comm.). We have found examples of this effect but no general expression for the critical depth.

Adaptation present

Internal waves exist at periods from the local inertial period (17 h at 45°
lat) to the buoyancy period (of order 5 min in the upper thermocline). Because
the response time T for the type of photoadaptation examined here is 1-3 h, we
expect minimal adaptation to light fluctuations caused by internal waves with
periods much shorter than T and substantial adaptation to fluctuations at
periods much greater than T. To test this hypothesis with our model, we present
an idealized case for which we can obtain an analytic solution that illustrates
the nature of adaptation to fluctuations at internal wave frequencies.

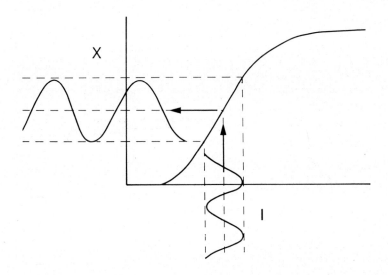

Fig. 11. Idealized case where sinusoidal irradiance I results in an adapting
capability X(I) also approximately sinusoidal in time.

Consider the case where the irradiance impinging on cells being displaced up
and down by internal waves is given by

$$I(t) = R_1 + R_2 \cos 2\pi ft, \tag{13}$$

i.e. we approximate the middle curve in Fig. 10b with a sinusoid. Now let the
inhibiting strength of that irradiance also be approximated by a sinusoid

$$X(t) = g_1 + g_2 \cos 2\pi ft, \tag{14}$$

i.e. the fluctuating irradiance is on the linear part of the X(I) curve as
depicted in Fig. 11. For our model with $P_s(t)$ adapting, the step response
function is (Jenkins and Watts, 1968)

$$Y(t) = g_1 + g_2 \frac{\cos \left[2\pi ft + \varphi(f)\right]}{[1 + (2\pi fT)^2]^{1/2}} \tag{15}$$

where $\varphi(f) = -\text{atan} (2\pi fT)$. As before we substitute (15) into (1) for $P_s(t)$ which is then substituted into (5) for the instantaneous rate of production $P(t)$.

To get the mean rate of production for any frequency f, we must average $P(t)$ over a complete period $1/f$. The solution consists of two parts: a non-adapting term P_n, and an adapting term P_a. The non-adapting term has the form

$$P_n = [P_D - (P_D - P_L) g_1] [1 - I_0(aR_2/P_D) e^{-aR_1/P_D}] \tag{16}$$

where $I_0(z) = \frac{1}{\pi} \int_0^\pi e^{\pm z \cos \theta} d\theta$ is a modified Bessel function of zero order (p. 376, Abramowitz and Stegun, 1965). $I_0(z)$ is a monotonically increasing function of z with $I_0(0) = 1$.

The adapting term, in which we are most interested, can also be expressed in terms of a modified Bessel function

$$I_1(z) = -I_1(-z) = \frac{1}{\pi} \int_0^\pi \cos \theta \, e^{z \cos \theta} \, d\theta \tag{17}$$

and has the form

$$P_a = g_2 (P_D - P_L) I_1(-a R_2/P_D) G(f) \cos \varphi(f) e^{-a R_1/P_D} \tag{18a}$$

where

$$G(f) \cos \varphi(f) = \frac{\cos 2\pi fT}{[1 + (2\pi fT)^2]^{1/2}}. \tag{18b}$$

For $P_D > P_L$, we get $P_a < 0$; that is, for our model parameters, the adaptation to internal waves reduces the average rate of photosynthetic production. Further-more, $P_a \to 0$ when $f \gg 1/T$, so the reduction only occurs for internal waves with periods long compared to T. We plot (18b) as a function of fT in Fig. 12. For internal wave periods just twice the adaptation response time T, P_a is reduced to about 0.1 of its value when $f \ll T^{-1}$. For $T \approx 1$ h, there will be significant adaptation to internal waves at the semidiurnal tidal frequency, but at frequencies approaching the buoyancy frequency (1–10 cph), the adaptation will be negligible.

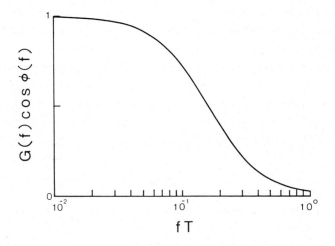

Fig. 12. Transfer function for adaptation to irradiance fluctuations induced by internal waves, plotted as a function of internal wave frequency f scaled by the adaptive response time T (= 1.1 h for the data of Marra, 1978b).

The reduction of average production in fluctuating light and the lack of adaptation at high frequencies are contrary to the results of Walsh and Legendre (1983) and others. They found the coefficients a and P_s to increase under fluctuating light in certain cases. However, their experiments were with fre quencies of light fluctuation of 1-10 Hz, at which other biophysical processes associated with light stimulation may be operating (e.g. Abbott, Richerson and Powell, 1982), rather than the type of adaptation addressed here.

FUTURE WORK

We have shown that a linear first order exponential response model of the adaptation of photosynthetic rate reproduces the essential features in data obtained from batch cultures of phytoplankton grown under natural and artificial lighting. The model did not include an explicit photoinhibition term; only one coefficient, the asymptotic high irradiance rate of production P_s was allowed to adapt. Small increases in P(t) at low irradiances after dark preconditioning were not reproduced.

Because the model is linear, it can be extended such that each coefficient can adapt separately and can have multiple time scales of adaptation. Short time incubation data (of the type presented in Lewis et al., 1984; and Lewis and Smith, 1983) giving the adaptive response separately for each coefficient in the P-I curves are required to extend the model in this manner. The functional form for X(I), the adapting capability of a given irradiance was chosen from scanty data; specific experiments must be designed to determine its form.

It takes little solar radiation to stabilize the upper ocean except under extreme conditions (Denman and Gargett, 1983; Woods and Barkmann, submitted). Thus, adaptation to vertical displacements by internal waves may be of more general importance than adaptation to vertical displacements caused by turbulent vertical mixing. To that end, we should extend the simulation of the adaptation of phytoplankton on internal waves over a range of combinations of wave shapes and irradiance profiles, and we should perform further laboratory experiments with programmed fluctuating lights.

ACKNOWLEDGEMENTS

We thank Paul Falkowski, Marlon Lewis, Pat Neale and Trevor Platt for helpful discussions. Howard Freeland, Greg Holloway, and Richard Thomson contributed to the internal wave analysis.

REFERENCES

Abbott, M.R., Richerson, P.J. and Powell, T.M., 1982. In situ response of phyto-plankton fluorescence to rapid variations in light. Limnol. Oceanogr., 27: 218–225.

Abramowitz, M. and Stegun, I.A., 1965. Handbook of Mathematical Functions. Dover, New York, 1046 pp.

Denman, K.L. and Gargett, A.E., 1983. Time and space scales of vertical mixing and advection of phytoplankton in the upper ocean. Limnol. Oceanogr., 28: 801–815.

Denman, K.L. and Herman, A.W., 1978. Space-time structure of a continental shelf ecosystem measured by a towed porpoising vehicle. J. Mar. Res., 36: 693–714.

Falkowski, P.G., 1980. Light-shade adaptation in marine phytoplankton. In: P.G. Falkowski (Editor), Primary Productivity in the Sea. Plenum, New York, pp. 99–119.

Falkowski, P.G., 1983. Light-shade adaptation and vertical mixing of marine phytoplankton: a comparative field study. J. Mar. Res., 41: 215–237.

Falkowski, P.G., Dubinsky, Z. and Wyman, K., 1985. Growth-irradiance relation-ships in phytoplankton. Limnol. Oceanogr., 30: 311–321.

Gallegos, C.L. and Platt, T., 1981. Photosynthesis measurements on natural populations of phytoplankton: numerical analysis. In: T. Platt (Editor), Physiological Bases of Phytoplankton Ecology, Can. Bull. Fish. Aquat. Sci., 210: 103–112.

Geider, R. and Platt, T., In prep. A mechanistic model of photoadaptation in microalgae.

Harris, G.P., 1973. Diel and annual cycles of net plankton photosynthesis in Lake Ontario. J. Fish. Res. Board Can., 30: 1779–1787.

Harris, G.P. and Lott, J.N., 1973. Light intensity and photosynthetic rates in phytoplankton. J. Fish. Res. Board Can., 30: 1771–1778.

Haury, L.R., Wiebe, P.H., Orr, M.H. and Briscoe, M.G., 1983. Tidally generated high-frequency internal wave packets and their effects on plankton in Massachusetts Bay. J. Mar. Res., 41: 65–112.

Jenkins, G.M. and Watts, D.G., 1968. Spectral Analysis and Its Applications. Holden-Day, San Francisco, 525 pp.

Lederman, T.C. and Tett, P., 1981. Problems in modelling the photosynthesis -light relationship. Bot. Mar. 24: 125–134.

Lewis, M.R., Cullen, J.J. and Platt, T., 1984. Relationships between vertical
    mixing and photoadaptation of phytoplankton: similarity criteria. Mar. Ecol.
    Prog. Ser., 15: 141-149.
Lewis, M.R. and Smith, J.C., 1983. A small volume, short-incubation-time method
    for measurement of photosynthesis as a function of incident irradiance. Mar.
    Ecol. Prog. Ser., 13: 99-102.
Marra, J., 1978a. Effect of short-term variations in light intensity on photo-
    synthesis of a marine phytoplankter: a laboratory simulation study. Mar.
    Biol., 46: 191-202.
Marra, J., 1978b. Phytoplankton photosynthetic response to vertical movement in
    a mixed layer. Mar. Biol., 46: 203-208.
Marra, J., 1980a. Vertical mixing and primary production. In: P.G. Falkowski
    (Editor), Primary Productivity in the Sea. Plenum, New York, pp. 121-137.
Marra, J., 1980b. Time course of light intensity adaptation in a marine diatom.
    Mar. Biol. Lett., 1: 175-183.
Marra, J. and Heinemann, K., 1982. Photosynthesis response by phytoplankton to
    sunlight variability. Limnol. Oceanogr., 27: 1141-1153.
Marra, J., Heinemann, K. and Landriau, Jr., G., 1985. Observed and predicted
    measurements of photosynthesis in a phytoplankton culture exposed to natural
    irradiance. Mar. Ecol. Prog. Ser., 24: 43-50.
Neale, P.J. and Marra, J., 1985. Short-term variation of Pmax under natural
    irradiance conditions: a model and its implications. Mar. Ecol. Prog. Ser.,
    in press.
Phillips, O.M., 1977. The Dynamics of the Upper Ocean. Cambridge University
    Press, Cambridge, 336 pp.
Platt, T., Gallegos, C.L. and Harrison, W.G., 1980. Photoinhibition of photo-
    synthesis in natural assemblages of marine phytoplankton. J. Mar. Res., 38:
    687-701.
Platt, T. and Gallegos, C.L., 1980. Modelling primary production. In: P.G.
    Falkowski (Editor), Primary Productivity in the Sea. Plenum, New York, pp.
    339-362.
Post, A.F., Dubinsky, Z., Wyman, K. and Falkowski, P.G., 1984. Kinetics of
    light-intensity adaptation in a marine planktonic diatom. Mar. Biol., 83:
    231-238.
Prézelin, B.B. and Matlick, H.A., 1980. Time-course of photoadaptation in the
    photosynthesis-irradiance relationship of a dinoflagellate exhibiting photo-
    synthetic periodicity. Mar. Biol., 58: 85-96.
Rivkin, R.B., Seliger, H.H., Swift, E. and Biggley, W.H., 1982. Light-shade
    adaptation by the oceanic dinoflagellates Pyrocystis noctiluca and P.
    fusiformis. Mar. Biol., 68: 181-191.
Slagstad, D., 1982. A model of phytoplankton growth - effects of vertical mixing
    and adaptation to light. Model. Ident. Control, 3: 111-130.
Walsh, P. and Legendre, L., 1983. Photosynthesis of natural phytoplankton under
    high frequency light fluctuations simulating those induced by sea surface
    waves. Limnol. Oceanogr., 28: 688-697.
Woods, J.D. and Barkmann, W., Submitted. The response of the upper ocean to
    solar heating. I: The mixed layer. Quart. J. R. Met. Soc.

THE EFFECTS OF THE BROAD SPECTRUM OF PHYSICAL ACTIVITY ON THE
BIOLOGICAL PROCESSES IN THE CHESAPEAKE BAY

A. BRANDT[1], C. C. SARABUN[1], H. H. SELIGER[2], M. A. TYLER[3]

[1]Johns Hopkins University, Applied Physics Laboratory, Laurel,
Maryland 20707, USA

[2]McCollum Pratt Institute and Dept. of Biology, Johns Hopkins
University, Baltimore, Maryland 21218, USA

[3]College of Marine Studies, University of Delaware, Newark,
Delaware, 19711 (currently on leave at National Science
Foundation, Washington, D.C.)

ABSTRACT

A multidisciplinary program to investigate the bio-physical processes and interactions in estuarine environments has been initiated.
Significant effects on the biological food chains can result from
physical activity on virtually all scales -- from seasonal circulation
and stratification patterns to the localized small-scale internal
wave and turbulent mixing activity. The focus of initial efforts
has been on the small-scale dynamic processes in the Chesapeake
Bay. Simultaneous biological, physical, and chemical measurements
have illustrated the nature of these processes in the Chesapeake
Bay and bring to light the key role they play in the total system
ecology.

INTRODUCTION

The Chesapeake Bay, one of the world's most productive estuaries,
has suffered a substantial decline in several of its noteworthy
biological species. In particular, the striped bass population has
declined to the point where the State of Maryland has prohibited
commercial and recreational fishing, and the oyster survival rate
has fallen to its lowest level in many years (Seliger, et al.,
1985). A recently completed U.S. Environmental Protection Agency
study (EPA, 1982) has documented the declining state of the Bay and
identified the general source of the problems: nutrient over-
enrichment due to urban and farm runoff. However, it is not
possible to identify the specific links in the food chain that were
detrimentally affected or to precisely evaluate the effects of the
increased volume of anoxic water that was evident from the historical
data. Thus, it is not possible to assess the potential effects of
proposed nutrient control techniques nor even to effectively
separate anthropogenic effects from natural effects.

The difficulty in assessing the specific nature of the Bay's problems results from the complexity of the estuarine processes. The relevant processes are clearly multidisciplinary in nature: physical, chemical, and biological (EPA, 1983). Moreover, these processes operate over a broad range of scales: from annual variations, in which differing physical and chemical properties of the water (e.g., temperature and dissolved oxygen levels) can affect annual spawning success rates to local subtidal* processes such as the turbulent mixing within the water column, which directly affects plankton photosynthesis and feeding.

To address estuarine processes in general, and the issues facing the Chesapeake Bay in particular, a multidisciplinary research team has been established. The long-term objectives of our team effort are to investigate the characteristics of the small-scale (subtidal) dynamic physical processes, to quantify the effects of the physical processes on biological activity and productivity, and to gain a fundamental understanding of the processes resulting in the development and evolution of anoxic water (Officer, et al., 1984; Seliger, et al., 1985).

The multidisciplinary team effort has focused on a series of experiments in the Chesapeake Bay in which simultaneous measurements of the physical, chemical, and biological processes were made at frequencies appropriate to the subtidal variability. These field experiments were conducted during May 1984 and May 1985.[†] May is a critical month in the Chesapeake Bay: because of the increased stratification resulting from the spring runoff and the rapidly increasing, spring temperature, the plankton biomass evolution is rapidly increasing and the rate of increase of anoxia is at a maximum. The test scenario for these studies was to anchor the research vessel at a location of interest so that frequent or continuous measurements could be made throughout a complete tidal cycle. The locations were selected on the basis of the local bathymetery, the historical phytoplankton distribution, and the anticipated dissolved oxygen levels. The locations are shown in Figure 1. A description of the instruments used for the multidisciplinary measurements at each station is presented in Table 1.

---

*In this report "subtidal" refers to processes having periods less than the tidal period.

[†]The 1985 test was conducted immediately following the Liège colloquium and, as the data are still being reduced, will not be extensively discussed herein.

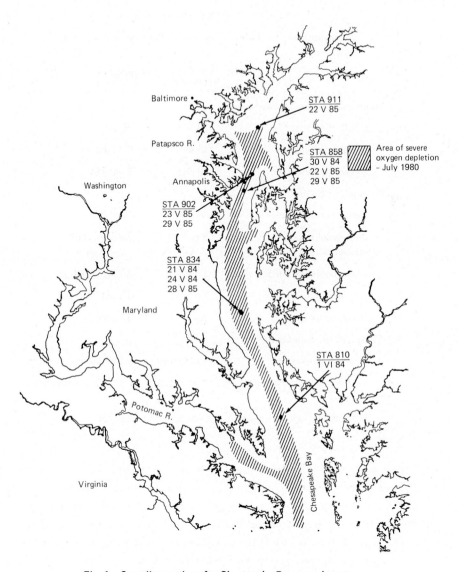

Fig. 1 Sampling stations for Chesapeake Bay experiments.

The purpose of this paper is to provide an overview of the broad range of physical processes acting on the biological material in this estuarine environment. Emphasis will be on the small-scale, subtidal processes and on phytoplankton ecology. These field studies provide a unique look at subtidal processes and will be used to illustrate the nature of these processes, rather than to

Table 1

Instrumentation for multidisciplinary Chesapeake Bay studies

	Sampling frequency	Resolution
Continuous vertical profiles		
C.T.D.	30 min	10 cm
Dissolved oxygen	30 min	0.1 ppm
In vivo fluorescence	30 min	1 μgl^{-1}CHL A
Discrete vertical profiles		
Current velocity, direction	30 min	1 cm s^{-1}
Phytoplankton species concentrations	2 h	Zero = $\leqslant$ 3 ml^{-1}
Chlorophyll a extraction	2 h	0.1 μgl^{-1}CHL A
Nutrient concentrations	2 h	0.5 μM
Dissolved oxygen	2 h	0.1 ppm
Time series measurements		
Acoustic echosounder	Continuous	0.1 m, particles $>$ 1 mm
Thermistor (10) – Conductivity (6) chain	Continuous	0.1°C; 0.1 mmho; 0.5 m
Microbiology optical imaging	Continuous	0.5 mm

provide a comprehensive picture of the Bay ecology. The following section will review the physical processes extant in the Chesapeake Bay and illustrate the ubiquity and surprising strength of the high-frequency internal wave field. The next section will illustrate, using results from our multidisciplinary studies, the major effects of the physical dynamics on key biological processes. The final section will discuss some wider implications of the subtidal processes and their variability.

PHYSICAL DYNAMICS OF THE CHESAPEAKE BAY ESTUARY

The Chesapeake Bay is classified as a partially mixed, salt-wedge estuary which is vertically stratified during the spring and summer and well mixed during the winter. Over the years, the Chesapeake Bay and its tributary rivers have been extensively studied, most notably from the physical point of view by Pritchard and co-workers (e.g.: Pritchard 1954, 1956). Scientific studies of the Bay's biological and physical processes and even anoxia date back to the early part of the century (Newcombe and Horne, 1938). The overriding emphasis of these studies has been on the mean, tidally-averaged processes, which are critical to the water circulation and to chemical and biological transport. It is only in recent years that the instrumentation necessary to precisely measure the subtidal processes has become available. With such measurements, the direct role of these small-scale, subtidal processes on the biological activity is beginning to come to light.

Processes and Scales

The physical processes affecting the stratification, or degree of vertical mixing, in an estuary can be classified on the basis of three time scales: (1) seasonal processes, (2) short-term processes, and (3) very short-period, small-scale mixing processes. Within the first class there are two principal forces working to affect the degree of stratification: solar heating and fresh water inflows. On the short-term, subseasonal time scales, the principal forces are wind, spring-neap tidal variations, the tidal variations themselves, long-period internal waves, cross-bay seiching, and diurnal variations in heat flux. Finally, in the third class, are the very short-period internal waves and convective and turbulent mixing processes. Each of these temporal classes of physical phenomena plays a direct role in developing, maintaining, or destroying the degree of vertical stratification.

In the Chesapeake Bay, the principal driving force on the seasonal time scale is the fresh water inflow, primarily from the Susquehanna River. This river influx establishes the gravitational estuarine circulation and, hence, the efficacy of the subsurface transport pathway (Tyler and Seliger, 1978), which is so important to the biological communities.

The degree of vertical mixing has important consequences for the biological communities in the Bay, especially with respect to the development and maintenance of anoxic conditions. When the halocline is strong enough to suppress vertical mixing, there is an effective isolation of the deeper layer from both the downward mixing of dissolved oxygen and the subsurface transport of more oxygenated water from the mouth of the Bay. Thus, on longer time-scales, factors that control the degree of vertical stratification are also important to the primary biological processes associated with anoxic conditions.

The dynamics of the mixing processes themselves, however, have important effects at much shorter time-scales. At time-scales of hours to weeks, there are a number of physical mechanisms that can impart significant energy to the processes that result in mixing. The largest energy source is, of course, tidal, but other mechanisms such as wind forcing, internal waves, and seiching can also be important. The actual dynamics of these mechanisms are only now being studied as other than a mean, averaged effect; see for example, Partch and Smith (1978). Such studies have shown that for a strongly two-layered system, significant vertical exchange at the interface may occur during limited portions of the tidal cycle and is often related to the breaking of internal waves.

Stratification in the Chesapeake Bay

Development of Tidally-Averaged Properties. The evolution of the tidally-averaged stratification in the Chesapeake Bay is illustrated by the sequence of Bay transect profiles shown in Figure 2. Here the increasing strength of the pycnocline during the spring months is quite evident. This strong layering results from the spring runoff, mostly input at the head of the Bay, and the counterbalancing, denser salt wedge flowing upstream along the bottom. This process typically culminates in early June with a strong salinity gradient and an associated oxycline, which separates hypoxic and eventually anoxic water in the lower depths from the oxygenated surface water, as shown in Figure 3.

In 1984 the river runoff was very strong, resulting in a rather high degree of stratification, an associated high degree of internal wave activity, and an extensive volume of anoxic water (Seliger, et al., 1985). In contrast, 1985 was a relatively weak runoff year and evidenced a somewhat weaker stratification and a lower level of internal wave and mixing activity, albeit still significant with regard to effects on the biological processes.

Subtidal Variability. Although the variations in tidal current during the course of a tidal cycle are well understood, the variations in the mean stratification during the course of a tidal cycle and the high frequency, internal wave activity have not often been considered. Prior to the present studies, only limited measurements of internal wave activity in the Chesapeake Bay or similar environments had been made (Clarke, et.al., 1983; Sarabun, 1980). In the current Chesapeake Bay investigation, considerable variations in the properties of the mean water column were evident. These

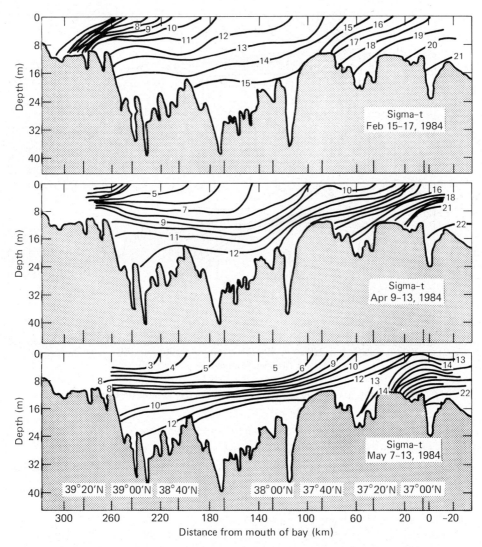

Fig. 2 Evolution of pycnocline stratification – Chesapeake Bay – Spring 1984.

variations are illustrated in Figure 4, which shows data obtained during a 26-hour period (2 tidal cycles) at a mid-Bay station.

The axial current contours shown in Figure 4a illustrate the typical ebb-flood cycle during the second tidal period; even this process may not always conform to the standard pattern, as illustrated by the pervasive flooding below ∿8 m depth during the entire first cycle. The salinity contours shown in Figure 4b illustrate the variability of the water column properties resulting from tidal forcing and internal wave activity.

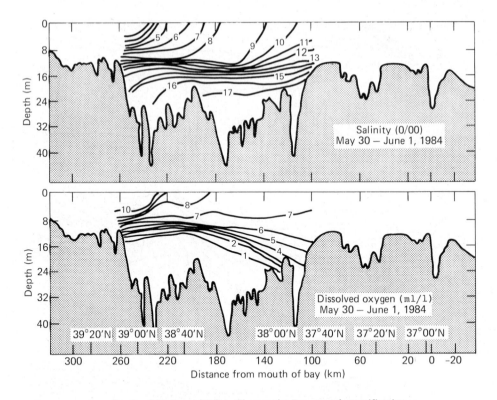

Fig. 3   Salinity and DO in Chesapeake Bay at peak stratification.

The Brunt-Vaisala (BV) frequency is a measure of the degree of stratification and, in a two-layered system, a measure of the strength of the interfacial layer.   The BV frequency is defined by

$$N \equiv - \left(\frac{g}{\rho} \frac{d\rho}{dz}\right)^{\frac{1}{2}},$$

(1)

where ρ is the fluid density, and z is the depth.   Figure 4c illustrates the BV frequency variation throughout the tidal cycle. The strong density gradient, characterized by the high BV values, forms an interfacial region that tends to inhibit the vertical transport of plankton and oxygen.

An additional illustration of the marked changes that occur during a tidal cycle is presented in Figure 5.   Here the contrast between the smooth gradient and step-like temperature profiles during the respective flood and ebb portions of the tidal cycle is quite evident.

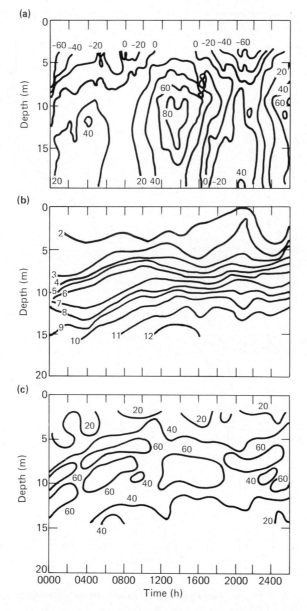

Fig. 4 Time series of vertical distributions of (a) axial tidal currents (in centimeters per second), (b) salinity (in parts per thousand) and (c) BV frequency for the 26-hour station occupied on May 30, 1984, just south of the Chesapeake Bay Bridge.

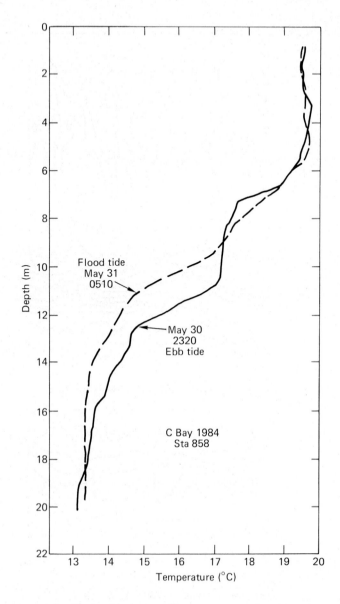

Fig. 5 Temperature profile variation during a tidal cycle.

High-Frequency, Subtidal Processes - Internal Waves. The
multidisciplinary field studies conducted in the spring of 1984 and
1985 were focused primarily on subtidal processes. During the
cruises, the research vessels occupied a series of stations at
fixed locations along the axis of the mid-Bay. Each station was
occupied for one to two tidal cycles (∿13 to 26 hours). The

physical oceanographic instruments deployed to measure the high-
frequency activity were a 200-kHz narrow-beam acoustic echosounder
and a 10-element, high-frequency-response thermistor chain array
(Table 1).

   The echosounder and thermistor chain data obtained during the
experiments illustrate the presence of strong, high-frequency
internal-wave and mixing activity in the central portion of the
water column. As an illustration of those observations, Figure 6

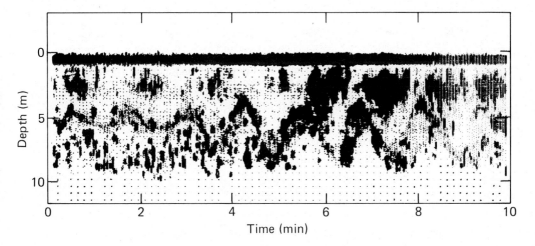

Fig. 6   Time series of gray-scale encoded acoustic backscatter intensity versus depth,
taken at a station just below the Chesapeake Bay Bridge on May 31, 1984.

shows a grey-scale encoded plot of acoustic backscatter intensity
for a short portion of the 26-hour period. A train of high-frequency
(1- to 2-minute period) internal waves is clearly evident, with the
largest waves attaining a peak-to-peak waveheights of 6 to 7 meters.
Between the crests of the largest waves are clouds of acoustic
scatterers that have apparently been collected by convergences in
the wave's velocity field. During this period, the acoustic
scatterers were small zooplankton (copepods), which accumulated in
the pycnocline along with the algae.

   Figure 7 is a thermistor chain record taken at about the same
time as the echosounder record in Figure 6. It also shows the
large amplitude internal waves. In these chain data, two wave
packets are evident as periodic, wave-like changes in temperature.
Associated with the 1- to 2-minute waves are bursts of high-frequency
turbulent fluctuations with periods on the order of seconds. The
upper portion of the thermistor chain was in an isothermal region

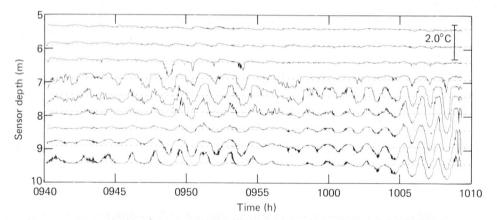

Fig. 7   Time series of temperature for the thermistor chain for May 31, 1984, taken just below the Chesapeake Bay Bridge. Thermistors are separated by 0.5 meter. Note the two wave packets between time periods 0945-0955 and 1002-1009.

during this period (see Figure 5) so that oscillations in the water column cannot be seen by the upper thermistors. These data were obtained during the same time period covered by the contour plots shown in Figure 4, where the existence of the high-frequency activity is masked by the sparseness of even the hourly samples.

In Figure 8, thermistor data from a different location in the Bay also show evidence of a broad range of physical activity, including long-period internal waves (20- to 30-minute periods), high-frequency internal waves having ∿1- to 2-minute periods, and

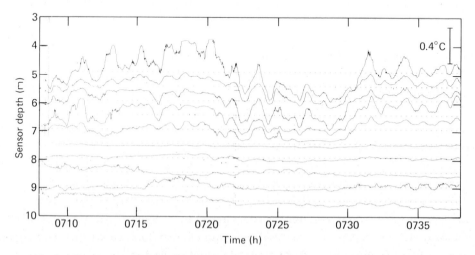

Fig. 8   Time series of temperature for the thermistor chain from May 24, 1984, taken in mid-Bay just south of Tilghman Island. Thermistors are separated by 0.5 meter.

turbulent motions with fluctuations measured in seconds. (Note that the temperature scale in Figure 8 is ∿4 times larger than that in Figure 7.)

The power spectra (power spectral density vs. frequency) for two individual thermistor records (30-minute segments) are shown in Figure 9. Most apparent is the dominant peak (evident on both traces) centered at ∿100 s, corresponding to the visually dominant oscillation on the chain records shown in Figures 7 and 8. The higher overall level of the spectral distribution at Station 858, Figure 9, reflects the high level of wave activity present at that location. For these data the maximum water column Brunt-Vaisala frequency is about 60 cyc/hr, or 0.016 Hz. The Brunt-Vaisala frequency determines the high frequency limit of wave activity. The spectral slope at lower frequencies is ∿1.8, which is in general agreement with the accepted value for moored measurements of internal-wave displacements (Garrett and Munk, 1979). The fall-off at higher frequencies is considerably higher ($\sim\omega^{-4}$), while the leveling off and rise at the highest frequencies, f > 0.05 Hz, is indicative of a high degree of turbulent activity, perhaps due to local wave overturning. (Note that for this experiment the thermistor response time was ∿7 ms, while the overall data resolution and actual digitization rate was 10 Hz.)

An attempt was made to correlate the level of internal-wave activity with the tidal phase to test the hypothesis that the internal waves were generated by the action of the ebb tidal current flowing over local variations in the bathymetery. Figure 10 shows a plot of the water-column-averaged wave intensity for 30-minute data segments centered at the indicated times, for two different Bay locations. The tidal phase corresponding to each data segment is indicated by E for ebb, F for flood, and no mark for tidal nodes. As no correlation is apparent, a more complex wave generation process may be at work; its resolution will have to await further, more comprehensive data.

These spring 1984 data graphically illustrate the energetic nature of high-frequency activity present in the Bay. While the mean stratification is an important seasonal controlling factor in the Bay, much of the vertical mixing results from phenomena with considerably shorter time-scales. It follows, therefore, that any realistic attempt to understand the Bay's biological and chemical processes must be carried out in close conjunction with the study of the physical transport and mixing processes.

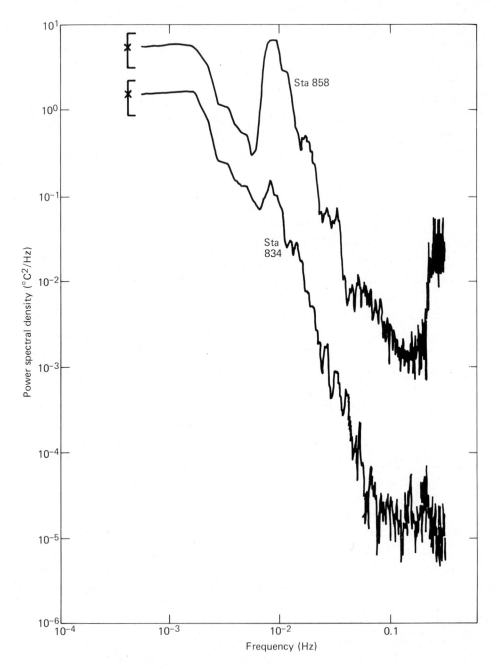

Fig. 9 Power spectral density of temperature fluctuations at two stations.

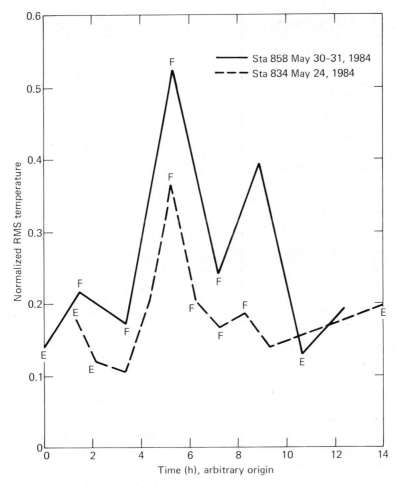

Fig. 10　Water column averaged wave intensity.

PHYSICAL EFFECTS ON BIOLOGICAL PROCESSES

The physical processes described above can have direct and often profound effects on the biological activity. In general, physical processes will have a direct effect on the biological processes that are of a comparable scale. An example is the effect of fluid turbulence on plankton: turbulent motions can result in patchy plankton distributions and increased motions of individual plankters leading to enhanced food availability. Physical processes of scales larger than the biological species will result in mean or integral effects. The transport of phytoplankton through the estuary is an example. The interactions at various scales are illustrated schematically in Figure 11. Examples of these processes

376

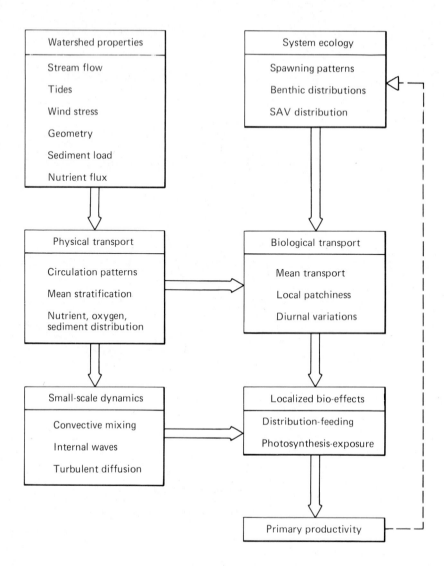

Fig. 11 Bio-physical interactions.

resulting from our studies in the Chesapeake Bay will be discussed in this section. Some of these effects are seen to be quite dramatic and heretofore underestimated, if not unanticipated.

Circulation and Transport

The tidally-averaged circulation patterns, driven by the Bay stratification (see Figures 2 and 3), affect the transport of the non-motile Bay species. This integral effect on the transport of phytoplankton, particularly Prorocentrum, has been studied

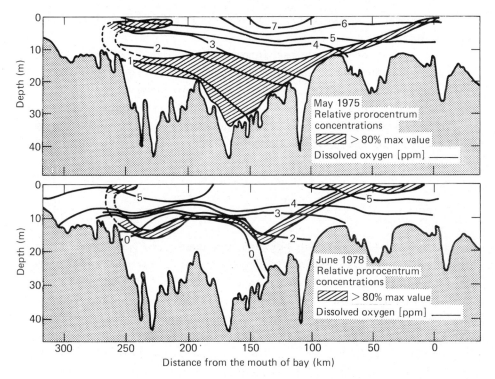

Fig. 12   Prorocentrum transport pathway in the Chesapeake Bay.

extensively by Tyler and Seliger (1978, 1981). Figure 12 illustrates
the up-Bay transport of phytoplankton by the net inflow of ocean
water. At the sharply shallowing up-Bay station, ~250 km from the
Bay mouth, the Prorocentrum are mixed into the surface water and
result in visible "red-tides". In 1984 the stratification was
sufficient to inhibit this vertical advection despite the intense
internal wave and turbulent mixing illustrated in Figures 6 and 7.

Internal Wave Effects

The high-frequency internal waves, shown in Figures 6 through 8,
also have significant direct effects on the plankton distributions.
The most evident and dramatic effect is the vertical oscillation of
the plankton layers induced by the high-frequency, internal waves.
This effect can be seen directly in Figure 6, as the backscattering
signal arises primarily from the zooplankton layer. In the 1984
experiment, internal-wave-induced oscillations of 5 to 7 meters
were observed frequently, while waves of 2 to 4 meters were a
common occurrence in the up-Bay stations. In 1985, because of the

differing mean stratification, the maximum wave heights observed
were ∿4 meters.

One direct result of these internal-wave oscillations on plankton
layers is the increased exposure to the higher light levels present
nearer the surface, which results in an increased potential for
photosynthetic activity. For example, if an exponential decrease
in light intensity with depth is assumed (Tyler and Preisendorfer,
1962), then the ratio of light intensity at two depths will be
given by

$$\frac{I_u}{I_\ell} = e^{-k\Delta z} \; ,$$

(2)

where Δz is the difference between the upper, u, and lower ℓ,
depths, and k is the extinction coefficient. For the relatively
turbid Bay water, a typical value of k is 1 m^{-1} (Kirk, 1983). This
represents a 1/e loss of light in 1 meter. Equation (2) then
indicates that the light intensity due to the internal wave activity,
centered at the mean level of the plankton layer*, would increase
by factors of 4.5, 12, and 33 for internal wave heights of 3, 5,
and 7 meters, respectively. Moreover, this variation can take
place over periods as short as 100 seconds!

The potential of such activity for increasing the net photosynthesis, especially of a nominally trapped "deep" layer is evident.
In addition, this level of wave activity should have a profound
effect on the local mixing and plankton-food exposure. This latter
effect is even more pronounced in the local convergence zones
between wave crests. The large backscatter regions (zooplankton
masses) visible on the right of Figure 6 result from the internal-
wave-induced velocity field in which the vertical wave oscillations
induce local horizontal currents and thus orbital fluid motions
that converge between wave crests. Suspended material (plankton
and sediment) in the wave field will be transported along these
orbital streamlines and will tend to accumulate at the velocity
nodes - convergence zones.

The Pycnocline Boundary

Evident throughout the present studies in the highly-stratified
environment is the observation that there is a confluence of
physical, biological, and chemical processes at the pycnocline.

------

*Using ½ of the peak-to-peak wave heights.

The strength of many estuarine pycnoclines in general, and that in the Chesapeake Bay in particular, is markedly greater than in the deep ocean, as illustrated in Figures 2 through 5. This strong gradient region is a region of intense internal wave and turbulent mixing activity. In addition, the stratification interface acts as a physical boundary, limiting biological and chemical transport.

The distribution and physical trapping of the phytoplankton mass is illustrated in Figure 13, which shows the distributions of salinity* and in vivo fluorescence (IVF) during one of the 26-hour stations. Vertical profiles illustrate this process even more vividly. Figure 14 shows that the dissolved oxygen (DO) is trapped below the maximum density gradient, σ_T, and that the biomass, IVF and P. minimum counts, are trapped below the density gradient and above the zero DO level.

Further evidence of this trapping effect is shown in Figure 15, from an earlier cruise. Here a vertical profile of organism distributions shows that bacteria, Prorocentrum, and the mysid shrimp Neomysis Americana have accumulated below the pycnocline and just above the anoxic layer. At this time tows at ∿15 meters produced larval anchovy, and reports by numerous fishing boats in the areas indicated that bluefish were being caught at ∿30 feet (9 meters), approximately the depth of the pycnocline.

These data illustrate the extreme degree to which life-supporting conditions can be confined to a very narrow, subpycnocline band during strong anoxic conditions.

FURTHER IMPLICATIONS OF HIGH FREQUENCY PROCESSES
Sampling Strategies

As indicated, the biological systems in the Chesapeake Bay can be strongly influenced by physical transport and mixing processes. In particular, the physical dynamics play a major role in determining the spatial and temporal distributions of biological activity. On a long-term mean basis, the estuarine circulation, coupled with the physiology of the organisms, gives rise to a strong layering in the vertical and intermittency in the horizontal and temporal dimensions. This interaction means that sampling strategies must take account of these factors in order to monitor properly the seasonal development and distribution of the biological organisms. For example, Figure 13

---

*This plot is a non-smoothed version of Figure 4b.

380

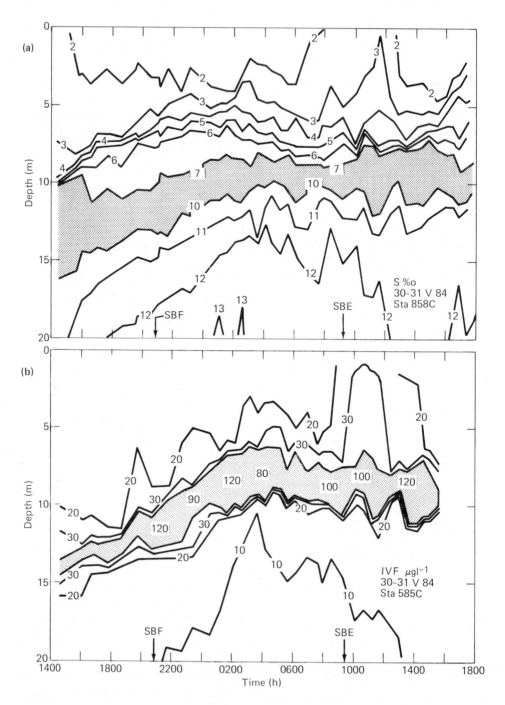

Fig. 13   Temporal evolution of (a) salinity and (b) in vivo fluorescence.

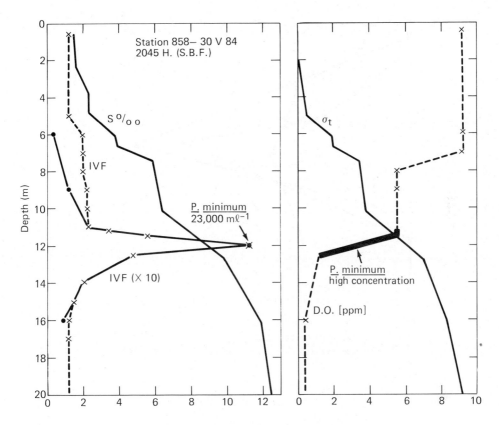

Fig. 14 Simultaneous vertical profiles of Bay parameters.

shows clearly how sampling at a limited number of pre-set depths might easily miss a rather thin layer of <u>Prorocentrum</u> at mid-depth. In the presence of short period internal waves as shown in Figures 6 and 7, with crest-to-trough heights of up to 25 percent of the total water column depth, a water sample drawn from a depth of 8 meters could be in the pycnocline, above it, or below it -- all within the space of less than 2 minutes.

In addition, there is the problem of when (or where horizontally) to sample. The high-frequency, strongly-nonlinear, internal waves have two further compounding effects apart from the periodic vertical displacement of layers. The periodic velocity field associated with the internal waves gives rise to convergence zones between the crests which, as seen in Figure 6, result in the concentration of scatterers between the crests. Thus, if one were to sample in a region clearly above the greatest vertical displacement of the pycnocline, one must still contend with the spatial/temporal

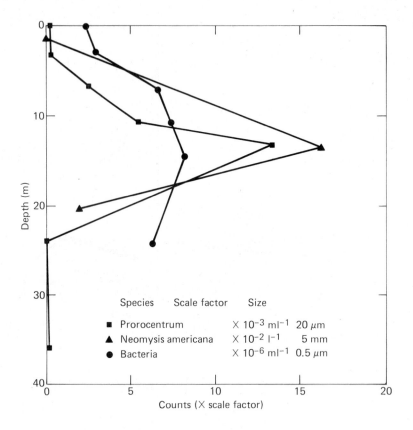

Fig. 15 Vertical distribution of acoustic scatterers in the lower Chesapeake Bay in late spring. The pycnocline is at 14 meters.

intermittency engendered by the internal-wave velocity field. Secondly, periodic breaking of the internal waves also occurs, resulting in actual physical exchange across the pycnocline.

Evolution of Anoxic Water

One of the principal reasons for concern with nutrient levels in the Bay (EPA, 1982), particularly nitrogen and phosphorus, is the strong historical correlation between an increase in these nutrient levels and a decrease in dissolved oxygen in the deeper areas of the Bay and the resulting effects on the Bay food chains (Seliger, et al., 1985). One mechanism postulated for this correlation is that the increase in algae growth due to the increase in nutrients ultimately results in the consumption of large amounts of dissolved oxygen when the algae die and decay (Officer, et al. 1984). The trapping of this oxygen depleted water below the pycnocline, as discussed above, can thus be a significant factor in the viability of critical stages of the Bay food chains. Thus, the decrease in dissolved oxygen levels below the pycnocline could be a determining factor for the observed decline in the harvest levels of several commercially important species.

SUMMARY

The present series of studies has brought to light several significant, multidisciplinary effects. In particular, we have found that large-amplitude (up to 25 percent of total water column depth), short-period internal waves are prevalent in the mid-Chesapeake Bay region. Indeed, it has been shown that significant physical activity exists at all scales in the Chesapeake Bay and that biological processes are affected at all scales. Particularly important are the direct effects of small-scale physical processes on the biological activity. These physical processes include: internal wave oscillations, local convergence zones, and the strong pycnocline interface.

Acknowledgements

J. R. Austin is gratefully acknowledged for the inspiration and continuing support for the multi-institute, multidisciplinary effort. We also express our thanks to our co-investigators J. A. Boggs, W. H. Biggley, R. B. Biggs, and to W. R. Drummond, J. E. Hopkins, G. D. Smith, and C. J. Vogt, who played key roles in our effort.

REFERENCES

Clarke, T. L., Crynock, V. F., and Proni, J. R., (1983), "Simultaneous Acoustic Doppler and Backscatter Observations of Estuarine Internal Waves," EOS Trans. Am. Geophys. Union, 64, 1022.
EPA, (1983) Fundamental Research on Estuaries: The Importance of an Interdisciplinary Approach, National Research Council, National Academy Press, Washington, D.C.
EPA, (1982) Chesapeake Bay Program Technical Studies: A Synthesis, U.S. Environmental Protection Agency, Washington, D.C.
Garrett, C. and Munk, W., (1979), "Internal Waves in the Ocean," Ann. Rev. Fluid Mech., Annual Reviews, Inc., 11, 339-69.
Kirk, J. T. O. (1983), Light and Photosynthesis in Aquatic Ecosystems, Cambridge University Press, Cambridge.
Newcombe, C.L., and Horne, W. A. (1938) "Oxygen-Poor Waters of the Chesapeake Bay," Science, 88, 80-81.
Officer, C. B., Biggs, R. B., Taft, J. L., Cronin, L. E., Tyler, M. A., and Boynton W. R., (1984) "Chesapeake Bay Anoxia: Origin, Development, and Significance," Science, 223, 22-27.
Partch, E. N. and Smith, J. D., (1978), "Time Dependent Mixing in a Salt Wedge Estuary," Est. Coast, Mar. Sci., 6, 3-19.
Pritchard, D. W., (1954), "A Study of the Salt Balance in a Coastal Plain Estuary," J. Marine Res., 13, 133-144.

384

Pritchard, D. W., (1956), "The Dynamic Structure of a Coastal Plain estuary," J. Marine Res., 15, 33-42.

Sarabun, C. C., (1980), "Structure and Formation of Delaware Bay Fronts," Ph.D. Thesis, College of Marine Studies, University of Delaware.

Seliger, H. H., Boggs, J. A., and Biggley, W. H., (1985), "Catastrophic Anoxia in the Chesapeake Bay in 1984," Science, 228, 70-73.

Tyler, J. E., and Preisendorfer, R. W. (1962) "Light," Ch. 8 in The Sea, Vol. 1, John Wiley and Sons, New York.

Tyler, M. A. and Seliger, H. H., (1978), "Annual Subsurface Transport of a Red Tide Dinoflagellate to its Bloom Area: Water Circulation Patterns and Organism Distributions in the Chesapeake Bay," Limnol. Oceanog., 23, 227-246.

Tyler, M. A. and Seliger, H. H., (1981), "Selection for a Red Tide Organism: Physiological Responses to the Physical Environment," Limnol. Oceanog., 26, 310-324.

ASPECTS OF THE NORTHERN BERING SEA ECOHYDRODYNAMICS

Jacques C.J. NIHOUL
GHER, University of Liege (Belgium)

INTRODUCTION

The Northern Bering Sea is a relatively shallow basin limited by the Bering Strait to the north and St Lawrence Island to the south (Fig. 1). The flow passing through the Bering Strait, from the Pacific Ocean to the Artic Ocean, penetrates the Northern Bering Sea through the Strait of Anadyr, to the west of St Lawrence Island, and by the Strait of Shpanberg, to the east. More than 60 % of the mean northward transport of water through the Bering Strait is derived from the "Anadyr Stream", a subsidiary of the Bering Slope Current which flows around the coasts of the Gulf of Anadyr, following the 60-70 isobaths, to the Anadyr Strait and the western part of the Shpanberg Strait (Coachman et al, 1975). The proportion of that stream which goes through the Strait of Anadyr or skirts St Lawrence Island, as well as the orientation, with respect to the Strait's axis, and seasonal variations of the entering flow, is likely to have a strong influence on the subsequent deployment of that flow in the Northern Bering Sea and in the Chukchi Sea.

Observations suggest that the Anadyr stream is the main source of nutrients and biological productivity in the Northern Bering Sea (Walsh et al, 1985).

In preliminary studies for the ISHTAR Research Project (e.g. Walsh et al, 1985), a series of numerical simulations were performed, with 2D barotropic and 3D baroclinic mathematical models, to test this hypothesis and determine if the residual circulation pattern in the Northern Bering Sea was indeed compatible with observed biological data (Walsh and Dieterle, 1986; Nihoul et al, 1986).

The results of these exploratory simulations confirm the general trend of the Anadyr Stream to spread to the east after passing the Anadyr Strait; the nutrient rich Anadyr waters deploying eastwards and progressively fostering biological productivity in the whole basin (Fig. 2, 3, 4).

Studies of the year-to-year variability of the flow pattern reveal however the existence of occasionally strong secondary flows in the form of eastwards propagating interleaving layers of frontal origin. These layers may contribute significantly to the cross-stream diffusion of nutrients and subsequent

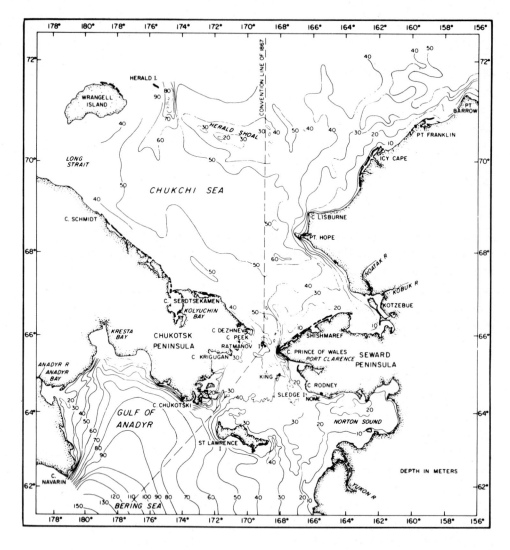

Fig. 1. The Northern Bering and Chukchi seas including Bering Strait.

biological activities.

It is shown in the following that the main features of these layers can be explained by a simple model of baroclinic instabilities predicting length scales and time scales in excellent agreement with the observations.

A scenario of the Northern Bering Sea Ecohydrodynamics, including the general circulation pattern and the local frontal secondary flows, is then presented as a working hypothesis to be tested by field surveys and mathematical models.

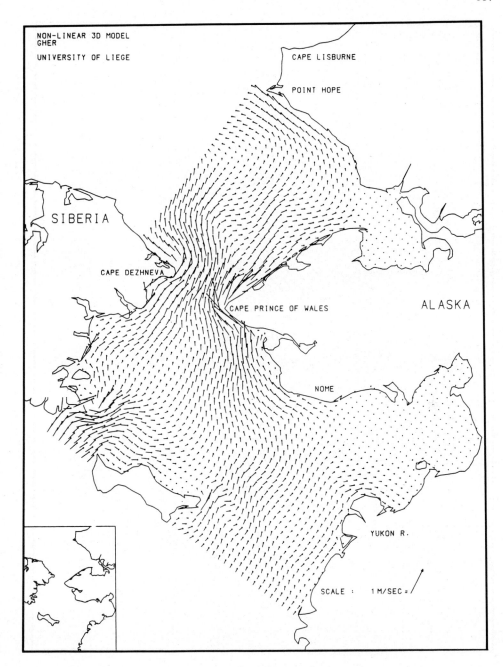

Fig. 2. Current pattern in the top layer of the Northern Bering Sea for a total flow through the Bering Strait of 1.8 $10^6 m^3 s^{-1}$.

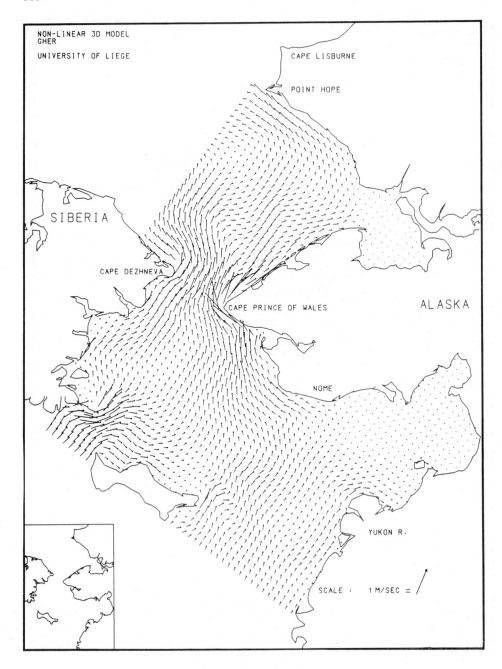

Fig. 3. Current pattern in the middle layer of the Northern Bering Sea for a total flow through the Bering Strait of $1.8 \ 10^6 m^3 s^{-1}$.

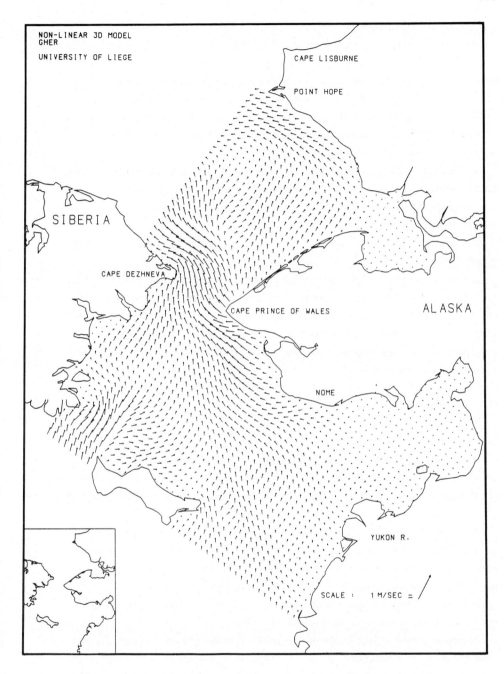

Fig. 4. Current pattern in the bottom layer of the Northern Bering Sea for a total flow through the Bering Strait of 1.8 $10^6 m^3 s^{-1}$.

The dateline ergocline

The marine system is characterized by fairly well-defined "spectral windows", i.e. domains of length-scales (inversely, wave-numbers) and time scales (inversely, frequencies) associated with identified phenomena. These windows may correspond to eigenmodes of the system (internal waves, inertial oscillations, Rossby waves, El Niño ...) or external forcing (annual or daily variations of insolation, tides, storms, atmosphere climate changes ...) (e.g. Monin et al 1977, Nihoul 1985).

In general, time scales and length scales are related and it is customary to associate high frequencies and high wave numbers, small frequencies and small wave numbers although the association may be different for eigenmodes and forced oscillations.

The transfer of energy between windows is effected by non-linear interactions.

Chemical and ecological interaction processes can also be characterized by specific time scales and the comparison between these time scales and those of hydrodynamic phenomena indicates which processes are actually in competition in the sea (Nihoul 1984, Denman and Powel 1984).

Obviously, at hydrodynamic scales much smaller than interaction scales, very little interaction takes place over time of significant hydrodynamic changes and basically the constituents are transported and dispersed passively by the sea. On the other hand, hydrodynamic processes with time scales much larger than interaction scales scarcely affect the dynamics of interactions over any time of interest.

Mesoscale, synopticscale and seasonalscale processes in the $10^{-4} - 10^{-7}$ s^{-1} range of frequencies form what one tends to call now the "weather of the sea" while longer time scale phenomena affect both the oceanic and the atmospheric climates.

Independently of climate problems, the year-to-year variability of the marine system, associated with globalscale processes, may be an important element in forecasting the reserves (population dynamics) and permissible hauls of the basic commercial fish (Nihoul 1985).

Evidence of year-to-year variability of the Bering Sea is given by Coachman et al (1975) and Aagaard et al (1985).

Although year-to-year variability is a climatic effect, one of its main consequences is a modification of the typical sea-weather patterns. This may result from changes in the general oceanic circulation and energy transports or (and) from similar changes in the atmosphere with differences in typical atmospheric-weather patterns over the marine area which is being investigated.

Thus, some exceptional years, rather important modifications will be observed in meso- and synoptic scale processes characteristic of a particular time
of the year and these will entail changes in the flow field and in many chemical and biological processes which depend on transport and dispersion.

A typical illustration of this phenomenon can perhaps be found in a sequence
of six remote sensing photographs taken on June 18, June 19, July 15, July 16,
August 1 and August 30, 1984. These photographs show a marked plume of cold
water, originating near Cape Chukotski on the Soviet Coast and spreading in
the Northern Bering Sea under the effect of the general Northward circulation
and lateral instabilities, eddies and extrusions progressing towards the East
(Fig. 5).

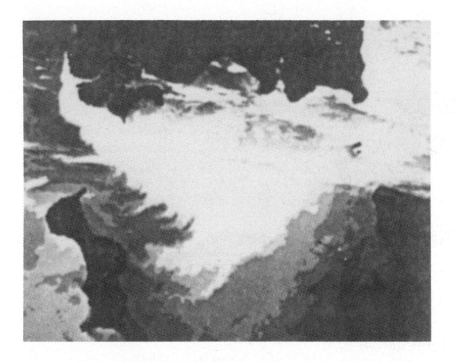

Fig. 5. Thermal image of the Northern Bering Sea for 16 July 1984.
        (Cold water in white)

Considering the climatological wind field for this time of the year, this
plume could be due to a rather intense upwelling bringing, to the surface,
cold bottom water with high nutrient concentrations.

The extruding layers which have typically a width of the order of 10 km, in the early stages of development, widen progressively as they flow eastwards, spreading the nutrient rich water over the Northern Bering Sea.

As shown in the next section, the formation of such layers can be explained by a baroclinic instability of the frontal edge of the cold plume.

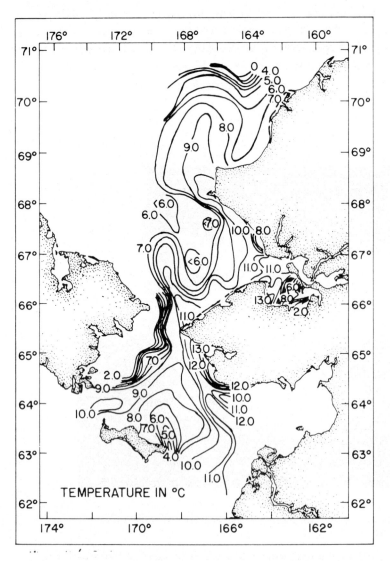

Fig. 6. Distribution of temperature °C at 5 m observed by "Brown Bear", 26 July–28 August 1960 (from Fleming and Heggarty, 1966).

Evidence of similar fronts, in the region and general direction of the dateline in the Northern Bering Sea can be found in several field surveys (e.g. Coachman et al 1975, fig. 6) but their intensity can be highly variable and, in this respect, the 1984 summer situation may have been exceptional[*].

Nevertheless the dateline front, when it occurs, constitutes an extremely efficient "ergocline" and the cross-front transport by the extruding layers is equivalent to a rather intense lateral mixing, extending the region of biological production and determining to a large extent the amount of organic matter which is ultimately transporter to the Chukchi Sea and further.

A simple model of baroclinic instability

The problem of baroclinic stability has been extensively studied and many models, with various degrees of sophistication and numerical skill, can be found in the literature (e.g. Eady 1949, Stone 1966, 1970, 1971, Tang 1971).

In a simple form, appropriate to the dateline ergocline in the Northern Bering Sea, the problem can be described as the determination of the conditions of instability of a depth-dependent horizontal current, flowing parallel to constant buoyancy surfaces in a region of significant buoyancy gradients (Fig. 6). The fastest growing mode of the linear stability problem gives rise to the observed extruding layers and its length-scale sets the width of these layers in the initial stages of development.

The basic equations applicable to this problem are the inviscid Boussinesq equations, viz.

$$\nabla \cdot \underline{v} = 0 \tag{1}$$

$$\frac{\partial \underline{v}}{\partial t} + \underline{v} \cdot \nabla \underline{v} + f \underline{e}_3 \wedge \underline{v} = -\nabla q + b \underline{e}_3 \tag{2}$$

$$\frac{\partial b}{\partial t} + \underline{v} \cdot \nabla b = 0 \tag{3}$$

where $\underline{v}$ is the velocity, f the Coriolis frequency, b the buoyancy, $b = -g \frac{\rho - \rho_0}{\rho_0}$, g the acceleration of gravity, ρ the density (ρ_0 its constant reference value), and where

$$q = \frac{p}{\rho_0} + g x_3 \tag{4}$$

($\underset{\sim}{e}_3$ is the unit vector along the vertical axis pointing upwards).

For the purpose of this study, taking into account that the length-scale of the perturbation is much smaller than the characteristic scale of variation of the basic flow, one may assume that the latter is horizontally uniform and extending to infinity. The buoyancy and velocity fields, in the unperturbed state, are then given by

$$\underset{\sim}{\nabla} b_o = - m^2 \underset{\sim}{e}_2 + n^2 \underset{\sim}{e}_3 \tag{5}$$

$$f \underset{\sim}{e}_3 \wedge \frac{d\underset{\sim}{v}_o}{dx_3} = - \underset{\sim}{\nabla} b_o \tag{6}$$

i.e.

$$\underset{\sim}{v}_o = (U + \frac{m^2}{f} x_3) \underset{\sim}{e}_1 \tag{7}$$

where n and m are, respectively, the "vertical" and "horizontal" Brunt-Väisälä frequencies, i.e.

$$n^2 = \frac{db_o}{dx_3} \quad , \quad m^2 = \frac{db_o}{dx_2} \tag{8), (9}$$

and where U is a constant of integration representing an eventual regional large scale flow directed along the front (i.e. with the approximation made in the Northern Bering Sea, parallel to the coast).

Assuming constant depth h and constant values of m and n, one can define a length-scale

$$\ell = m^2 h f^{-2} \tag{10}$$

and the following non-dimensional variables and parameters

$$x = x_1 \ell^{-1} \quad ; \quad y = x_2 \ell^{-1} \quad ; \quad z = x_3 \ell^{-1} \tag{11), (12), (13}$$

$$\tau = f t \quad ; \quad \pi = \hat{q} f m^{-2} h^{-1} U^1 \tag{14), (15}$$

$$a = \hat{b} f m^{-2} U^{-1} \quad ; \quad w = \hat{v}_3 m^2 f^{-2} U^{-1} \tag{16), (17}$$

$$u = \hat{v}_1 U^{-1} \quad ; \quad v = \hat{v}_2 U^{-1} \tag{18), (19}$$

$$r = f^2 n^2 m^{-4} \quad ; \quad s = U f m^{-2} h^{-1} \tag{20), (21}$$

where $\hat{}$ denotes a perturbation of the basic state.

The non-dimensional form of the Boussinesq equations for the perturbation may then be written, in the quasi-hydrostatic approximation,

$$\frac{\partial u}{\partial x} + \frac{\partial v}{\partial y} + \frac{\partial w}{\partial z} = 0 \qquad ; \qquad \frac{\partial \pi}{\partial z} = a \tag{22), (23}$$

$$\frac{\partial u}{\partial t} + (s + z)\frac{\partial u}{\partial x} + w - v = -\frac{\partial \pi}{\partial x} \tag{24}$$

$$\frac{\partial v}{\partial t} + (s + z)\frac{\partial v}{\partial x} + u = -\frac{\partial \pi}{\partial y} \tag{25}$$

$$\frac{\partial a}{\partial t} + (s + z)\frac{\partial a}{\partial x} - v + rw = 0 \tag{26}$$

The perturbation are assumed to be small and to have time and space dependences of the form $\phi(z)\exp[\,i\,(\alpha x + \beta y - \omega t)\,]$ where ω denotes a complex frequency (instability occurs when the imaginary part of ω is positive) and where ϕ is the appropriate amplitude, function of z.

Subisuting in eqs (22) – (26) and solving for $W(z)$ (the amplitude of w), one obtains

$$\zeta(\zeta^2 - 1)\,\ddot{W} + 2(1 - i\gamma\zeta)\,\dot{W} + W[\,r(1 + \gamma^2)\,\zeta + 2i\gamma\,] = 0 \tag{27}$$

where

$$\zeta = \alpha z - \omega + \alpha s \tag{28}$$

$$\gamma = \frac{\beta}{\alpha}$$

Eq. (27) belongs to the class of Heun's equations. It must be solved subject to the boundary conditions

$$W = 0 \quad \text{at} \quad z = 0 \tag{29}$$

$$W = 0 \quad \text{at} \quad z = 1 \tag{30}$$

The boundary value problem set by eqs. (28), (29) and (30) leads to a general complex dispersion relation of the form

$$R \ [\ (\omega_{r} - \alpha s), \ \omega_{i}, \ \alpha, \gamma, r] \ = \ 0 \tag{31}$$

where ω_{r} and ω_{i} are respectively the real and imaginery parts of ω.

Separating the real and imaginery parts of eq. (31) and eliminating $\omega_{r} - \alpha s$, one finds ω_{i} as a function of α, γ and r. The perturbation for which ω_{i} is maximum has the largest growth rate and generates the observed cross-ergocline secondary flows.

In the Northern Bering Sea, observations (e.g. Coachman et al 1975, Sambrotto 1984) indicate that

$$n^2 \sim 10^{-5} \quad ; \quad m^4 \sim 10^{-13} \quad ; \quad f^2 \sim 10^{-8}$$

hence

$$r = f^2 n^2 m^{-4} = \frac{\frac{db_0}{dx_3}}{\left\| \frac{dv_0}{dx_3} \right\|^2} \sim 1$$

For values of the Richardson number of that order, one can show (e.g. Stone 1966, 1970, Happel et al 1986) that the maximum growth rate is obtained for

$$\gamma_{max} \sim 0 \quad ; \quad \alpha_{max} \sim 1$$

The typical wave-length of the fastest growing perturbation is then given by

$$\lambda = \frac{2\pi \ell}{\alpha_{max}} = \frac{2\pi m^2 h}{\alpha_{max} f^2}$$

i.e., taking $h \sim 50$ m ,

$$\lambda \sim 10 \text{ km}$$

in agreement with the observations.

The secondary flow pattern of eastward extruding layers in the Northern Bering Sea may thus presumably be attributed to the baroclinic instability of the dateline ergocline which is formed, in well-defined environmental conditions, at the edge of a plume of cold upwelled water passing through the Anadyr Strait.

A scenario for the Northern Bering Sea Ecohydrodynamics

It is now believed that nutrients are essentially brought to the Northern
Bering Sea by the Anadyr Stream and that biological production in the area is
closely related to the deployment and residence time there of Anadyr Stream
waters, in different environmental conditions.

The occasional occurence of a marked upwelling plume swept along in the
Northern Bering Sea as an unstable frontal current and the subsequent development of extruding layers, flowing eastwards, contribute, in exactly the same
way, to the lateral diffusion of the nutrients and one may argue that the
productivity of the Northern Bering Sea depends on the intensity and the variability of both the primary and secondary flows.

It is illuminating, in this respect, to examine the results of the Second
Soviet-American Expedition in the Bering Sea, 27 June - 31 July 1984, a period
characterized, as pointed out before, by a well-marked ergocline event
(Sambrotto 1984).

The expedition had selected four polygons of observations ; three of which
were more or less distributed along the Anadyr Stream. While Polygon I upstream was characterized by relatively high surface temperatures (> 6°C) and
high nutrient concentrations, increasing with depth, the transect between
Polygons II and III showed a decrease in surface temperature down to 2°C,
a depletion of nutrients in surface waters due to active phytoplankton production, high concentrations of ammonia in the bottom layers indicating active
microbial degradation of organic matter (as confirmed by microbiological
studies) and high rates of organic sedimentation. The additional observation
of a larger sea birds' population at Polygon III suggest that the food chain
has completely developed along the Anadyr Stream before it penetrates the
Northern Bering Sea and the following ecohydrodynamic scenario appears quite
plausible.

Marine productivity in the Northwestern Bering and Chukchi Seas is achieved
in two successive phases : one downstream of the shelf-break and the other
downstream of St. Lawrence Island. The first involves intense primary production along the "Anadyr stream", with sedimentation of organic matter, both
south of Cape Chukotski and St. Lawrence Island, and, generally speaking, in
the outer-lagoon of the slowly revolving water in the Gulf of Anadyr's secondary gyre-flow (Fig. 7). The second phase develops north of the Anadyr Strait,
spreading to the eastern side of the Northern Bering Sea to a variable extent,
due to interannual changes in wind forcing. The eastward excursion of nutrient
enriched water is then a function, not only of the intensity and direction of
the inflowing current from the Gulf of Anadyr, but also of the intensity of the
coastal upwelling, the resulting plume development, and frontal instabilities.

398

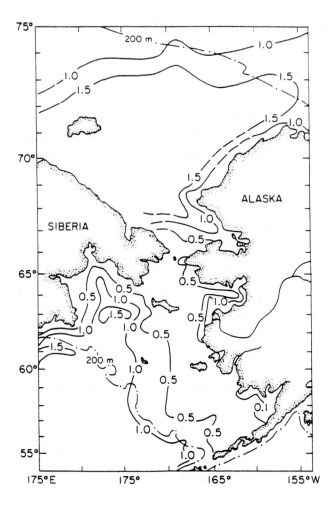

Fig. 7. The distribution of organic carbon (% dw) in the surface sediments of the Bering-Chukchi Seas (Walsh et al 1985).

At this stage, the scenario described above can be nothing more than daring hypotheses. A more thorough investigation requires more data (and in particular, more remote sensing data and meteorological information) and the development of mathematical models growing to full three-dimensional maturity.

This will hopefully be the program of the second phase of the ISHTAR project.

Acknowledgments

The author is indebted to the National Science Foundation for its support in the scope of the first phase of the ISHTAR project.

REFERENCES

Aagaard, K., Roach, A.T. and Schumacher, J.D., 1984. On the wind-driven variability of the flow through the Bering Strait. J. Geophys. Res., (in press).

Coachman, L.K., Aagaard, K. and Tripp, R.B., 1975. Bering Strait. The regional physical oceanography. Univ. of Washington Press, 172 pp.

Denman, K.L. and Powell, Th. M., 1984. Effects of physical processes on planktonic ecosystem in the coastal ocean. Oceanogr. Mar. Biol. Ann. Rev., 22 : 125-168.

Eady, E.T., 1949. Long waves and cyclone waves. Tellus, 1: 33-52.

Fleming, R.H. and Heggarty, D., 1966. Oceanography of the southeastern Chukchi Sea. In: Environment of the Cape Thompson Region, Alaska US Atomic Energy Comm., Div. of Tech. Information: 697-754.

Happel, J.J., Nihoul, J.C.J. and Deleersnijder, E., 1986. Some properties of the Heun equation with application to the study of baroclinic instability. To be published.

Heun, K., 1889. Math. Annalen, 33: 161-179.

Monim, A.S., Kamenkovich, V.M. and Kort, V.G., 1977. Variability of the oceans. Wiley - Interscience Publ., N.Y., 241 pp.

Nihoul, J.C.J., 1985. Perspective in marine modelling. JRC, Ispra, 404 pp.

Nihoul, J.C.J., Waleffe, F. and Djenidi, S., 1986. A 3D-numerical model of the Northern Bering Sea. Environmental Software, 1: 1-7.

Sambrotto, R.N., 1984. Cruise report of the second Soviet-American Expedition in the Bering Sea aboard the R/V Akademic Korolev, 27 June - 31 July 1984. Unpublished report.

Stone, P.H., 1966. On non-geostrophic baroclinic stability. J. Atmos. Sci., 23: 390-400.

Stone, P.H., 1970. On non-geostrophic baroclinic stability : Part II. J. Atmos. Sci., 27: 721-726.

Stone, P.H., 1971. Baroclinic stability under non-hydrostatic conditions. J. Fluid Mech., 45: 659-671.

Tang, Ch.M., 1971. The stability of continuous baroclinic models with planetary vorticity gradient. Tellus, 23: 285-294.

Walsh, J.J., Blackburn, T.H., Coachman, L.K., Goering, J.J., McRoy, C.P., Nihoul, J.C.J., Parker, P.L., Springer, A.M., Tripp, R.B., Whitledge, T.E. and Wirick, C.D., 1985. The role of the Bering Strait in carbon/nitrogen fluxes of polar marine ecosystems. Proceedings Fairbanks Conf. on Marine Living Systems of the Far North, May 1985. To be published.

Walsh, J.J. and Dieterle, J., 1986. Simulation analysis of plankton dynamics in the Northern Bering Sea. In: J.C.J. Nihoul (Editor), Marine Interfaces Ecohydrodynamics. (Elsevier Oceanography Series, 42) Elsevier, Amsterdam (this volume).

SIMULATION ANALYSIS OF PLANKTON DYNAMICS IN THE NORTHERN BERING SEA

J. J. WALSH and D. A. DIETERLE

Department of Marine Science, University of South Florida, St. Petersburg, Florida 33701 U.S.A.

INTRODUCTION

"She noticed a curious appearance in the air: it puzzled her very much at first, but after watching it a minute or two, she made it out to be a grin, and said to herself 'It's the Cheshire Cat: now I shall have somebody to talk to!' .... and she went on. 'Would you tell me please, which way I ought to go from here?' 'That depends a good deal on where you want to get to,' said the Cat. .... '-- so long as I get somewhere,' Alice added as an explanation. 'Oh, you're sure to do that' said the Cat, 'if you only walk long enough' .... and this time it vanished quite slowly, beginning with the end of the tail, and ending with a grin, which remained some time after the rest of it had gone."

(Carroll, 1865)

Reconstruction of a Cheshire Cat from just its grin, like the description of a marine ecosystem deduced from a few shipboard measurements, is an appropriate objective of simulation analysis. Simulation models, like isolated current meter data, nutrient measurements, or plankton tows only provide an accurate reconstruction, however, if the object of the study is already known. If one "walks long enough" in an iterative process of model confrontation with hopefully unaliased measurements (Walsh, 1972), a reasonable, if not unique, description of an unknown Cheshire Cat, i.e., natural system, emerges from this process. As part of a multidisciplinary study of the fate of dissolved carbon and nitrogen in the northern Bering Sea within the ISHTAR (Inner SHelf Transfer And Recycling) program, the present simulation analysis represents a first attempt to describe the plankton dynamics of this shelf ecosystem, similar to initial attempts to constrain the bounds of biological interactions within an upwelling ecosystem (Walsh and Dugdale, 1971).

After four cruises of the R/V Alpha Helix, Discoverer, and Akademik Korolov to the Bering/Chukchi Seas during July-August 1982-84, the vague outlines of the grin of this high-latitude Cheshire Cat became visible. During the 1983-84 survey cruises of the Alpha Helix and Akademik Korolev, for example, biological rate measurements had been made of ^{14}C and ^{15}N uptake

by phytoplankton, of the abundance and species distribution of macro-
zooplankton, of the grazing rates and abundance of micro-zooplankton, of the
abundance and production of bacterioplankton by ^{15}N dilution techniques, of
the oxygen consumption, ^{15}N and ^{35}S turnover by micro-benthos and meiobenthos, of macro-benthos abundance and their release of recycled nitrogen
compounds, all in relation to the distributions of temperature, salinity,
nutrients, chlorophyll, particulate carbon, and particulate nitrogen. These
initial data have been discussed by Walsh et al. (1986a) and are summarized
in the two carbon budgets of Figure 1. Alaska Coastal Water is a warmer,
less saline water type, derived from a mixture of Yukon River water and
Bering shelf water between Bering Strait and Shpanberg Strait, to the east of
St. Lawrence Island, while Anadyr Stream Water is the colder, more saline
shelf water between Bering Strait and Anadyr Strait, to the west of
St. Lawrence Island.

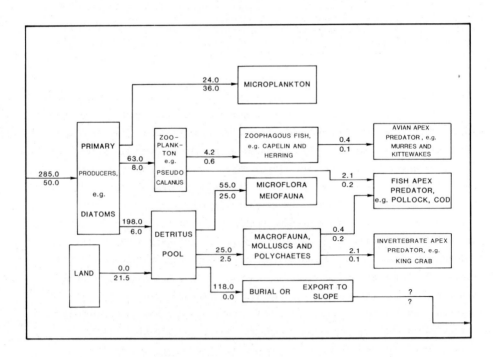

Figure 1.   The annual carbon flow (g C m^{-2} yr^{-1}) within food webs of the
Anadyr Stream Water (upper value) and of Alaska Coastal Water
(lower value).

Of the two sources of "new" nitrogen, i.e., nitrate, to the northern
Bering Sea from the Yukon River and from the shelf-break (Coachman and Walsh,
1981) south of St. Lawrence Island, the riverine input apparently leads to at
least five-fold less primary production in Alaska Coastal Water (Fig. 1). An
additional input of terrestrial detrital carbon from freshwater discharge is
also inferred in our budget for this part of the northern Bering Sea, in
contrast to the more productive communities of the Anadyr Stream Water.
Although the distances from the shelf-break of the northwestern Bering Sea to
the mouth of the Yukon River and to Anadyr Strait, on the eastern and western
sides of St. Lawrence Island, are about the same, nitrate is evidently not
stripped, en route from deep to shallow water on the western side of the
Bering Sea, within the Anadyr Stream Water.

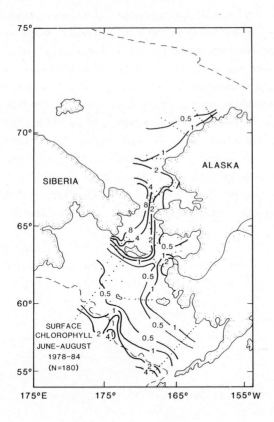

Figure 2. A chlorophyll composite ($\mu g \ \ell^{-1}$) of the surface
distribution of phytoplankton biomass within the
Bering/Chukchi Seas during June-August 1978-84.

Local upwelling in and north of Anadyr Strait may intensify the supply of nitrate in this shelf region as well, accounting for part of the higher primary production of Anadyr Stream Water (Fig. 1). Approximately 40% of the fixed carbon of algal photosynthesis was evidently not consumed by zooplankton, bacterioplankton, and benthos within Anadyr Stream Water, compared to complete utilization of marine and terrestrial carbon in Alaska Coastal Water. Consequently, a large signal of unconsumed algal biomass (Fig. 2) and productivity (Sambrotto et al., 1984) was found on the western side of Bering Strait, with the input of phyto-detritus to the downstream sediments of the Chukchi Sea (Walsh et al., 1985) presumably derived from this food web, not from that of Alaska Stream Water.

Based on these few, preliminary field measurements of the biological rates and previous estimates of the flow field (Coachman et al., 1975), we were able to assign, however, only mean spatially-averaged fluxes between the state variables of the annual carbon budgets in Figure 1. State equations could have been written for each variable, i.e., zooplankton, detritus pool, and microplankton, and solved numerically for their steady-state values at one point of a spatially homogeneous Bering Sea as a relatively trivial exercise. With the meager biological data set available to us from these and other surveys (50-100 rate measurements and <200 surface chlorophyll determinations over 6 years--see Fig. 2), we decided instead to construct a depth-averaged model of the flow, nitrate, and chlorophyll fields for exploration of the spatial consequences of both the carbon budgets (Fig. 1) and possible interannual changes of the physical habitat, i.e., a variation of 0.3-1.2 Sv northward transport through Bering Strait (Aagaard et al., 1985).

METHODS

Accordingly, we used a depth-integrated form of the Navier-Stokes and continuity equations, ignoring tidal forces, of

$$\frac{\partial U}{\partial t} = -gH \frac{\partial s}{\partial x} + F^x - B^x + fV \tag{1}$$

$$\frac{\partial V}{\partial t} = -gH \frac{\partial s}{\partial y} + F^y - B^y - fU \tag{2}$$

$$-\frac{\partial s}{\partial t} = \frac{\partial U}{\partial x} + \frac{\partial V}{\partial y} \tag{3}$$

where $U = \int_{z=-H}^{z=0} u \, dz$ and $V = \int_{z=-H}^{z=0} v \, dz$ are the horizontal transports in the x and y direction from the sea surface to the bottom, $-H$; where u and v are the

horizontal velocities; where g is the acceleration of gravity and s is the
sea surface elevation in the term for the pressure force of the slope of the
sea surface; where F and B are the wind and bottom stress components of the
terms for the frictional forces in the x and y directions; and f is the
Coriolis parameterization, assumed to be a constant of 1.32×10^{-4} at 65° N.
These are the linearized, barotropic equations of motion, since the terms for
the acceleration of momentum and for the density-driven buoyancy force of the
horizontal pressure gradient have been deleted.

The interior solutions of u and v were obtained by assuming $\frac{\partial u}{\partial z} = \frac{\partial v}{\partial z} = 0$,
i.e., dividing U and V by the local depth, –H, which was entered in the model
as a digitized form of the bottom topography (Fig. 3). At the upstream open

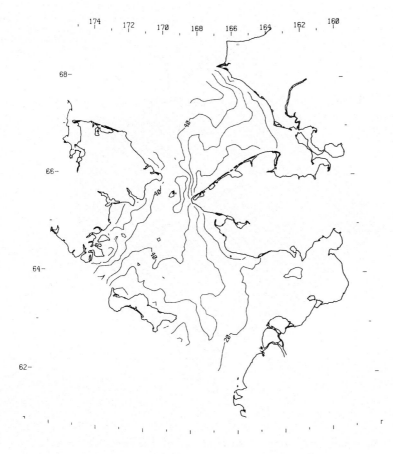

Figure 3.   Bottom topography (m) of a numerical model of the
            Bering/Chukchi Seas.

boundaries of the model (Fig. 4) across the Anadyr and Shpanberg Straits,
V was proscribed by assuming that of the total transport of the 3 cases (0.3,
0.6, and 1.2 Sv) through Bering Strait, 60% passes through Anadyr Strait and
40% through Shpanberg Strait, while s was then determined here from eq. (3).
At the downstream open boundary, U and s were functions of both adjacent,
interior solutions and the phase velocity, $\sqrt{gH}$, i.e., a radiative boundary
condition (Orlanski, 1976) in which only outward energy flux occurs, without
significant distortion of the solutions of eqs. (1)-(3). At the land boundaries, U = V = 0 such that there were 1825 active grid points (i,j) of 10 km
spacing in these simulations (Fig. 4).

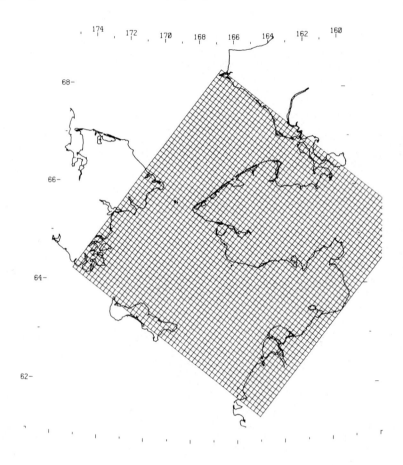

Figure 4. Eulerian grid, with 10 km spacing, of the numerical model.

At the bottom boundary, a mean, linearized bottom stress was used over two
adjacent grid points, i.e., instead of the usual $B^x = C\rho u^2$ and $B^y = C\rho v^2$, the
formulation was

$$B^x_{i,j} = (2\alpha C\rho)(H_{i,j} + H_{i-1,j})^{-1}(U_{i,j}) \tag{4}$$

$$B^y_{i,j} = (2\alpha C\rho)(H_{i,j} + H_{i,j-1})^{-1}(V_{i,j}) \tag{5}$$

where ρ is the density of sea water, C is the drag coefficient ($\sim 2 \times 10^{-3}$), and α is assumed to be a constant of 10 cm sec^{-1}. At the surface boundary, no wind stress was applied, i.e., $F^x = F^y = 0$, in the three cases of varying transport through the Bering Strait. One could think of the three circulation patterns of 0.3, 0.6, and 1.2 Sv, however, as the flow response under varying conditions of northerly or southerly wind forcing.

These equations were solved using a finite difference technique (Platzman, 1972) in which the state variables were discretized over space on a Richardson lattice in a staggered manner. The time integration of equations (1-3) was thus carried out by the following forward difference scheme:

$$U^{n+1}_{i,j} = U^n_{i,j} + 0.25 \, (\Delta t) \, (V^n_{i,j} + V^n_{i,j+1} + V^n_{i-1,j} + V^n_{i-1,j+1}) \, (f)$$

$$- \frac{g(\Delta t)}{2\Delta x} \, (H_{i,j} + H_{i-1,j})(s^n_{i,j} - s^n_{i-1,j}) - (\Delta t)(B^x_{i,j}) \tag{6}$$

$$V^{n+1}_{i,j} = V^n_{i,j} - 0.25 \, (\Delta t) \, (U^{n+1}_{i,j} + U^{n+1}_{i+1,j} + U^{n+1}_{i,j-1} + U^{n+1}_{i+1,j-1}) \, (f)$$

$$- \frac{g(\Delta t)}{2\Delta y} \, (H_{i,j} + H_{i,j-1})(s^n_{i,j} - s^n_{i,j-1}) - (\Delta t)(B^y_{i,j}) \tag{7}$$

$$s^{n+1}_{i,j} = s^n_{i,j} - \frac{\Delta t}{\Delta x} \, (U^{n+1}_{i+1,j} - U^{n+1}_{i,j}) - \frac{\Delta t}{\Delta y} \, (V^{n+1}_{i,j+1} - V^{n+1}_{i,j}) \tag{8}$$

At time n+1 in the numerical solution, eq. (6) is computed for the entire spatial grid (Fig. 4), then eq. (7) and then eq. (8); that is, only one time level is stored for any variable at each of the 1825 grid points. The time step, Δt, for this difference scheme of a two-dimensional grid must satisfy the requirement that $\Delta t < \Delta x / \sqrt{2gH}_{max}$ where H_{max} was set to 50 meters, south of St. Lawrence Island. Evaluation of this expression, with $\Delta x = 10$ km,

$g = 10 \text{ m sec}^{-2}$, and $H_{max} = 50 \text{ m}$, yields Δt must be less than 5 minutes; we used a 3-minute time step in the simulations.

Upon solution of eqs. (1)-(3), the barotropic velocities, u and v, were then entered in the non-linear state equations for nitrate, N, in units of $\mu g\text{-at } NO_3 \text{ } \ell^{-1}$, and phytoplankton, P, in units of μg chlorophyll ℓ^{-1},

$$\frac{\partial N}{\partial t} = \frac{\partial}{\partial x} (K_x \frac{\partial N}{\partial x}) + \frac{\partial}{\partial y} (K_y \frac{\partial N}{\partial y}) - \frac{\partial uN}{\partial x} - \frac{\partial vN}{\partial y} - aP \tag{9}$$

$$\frac{\partial P}{\partial t} = \frac{\partial}{\partial x} (K_x \frac{\partial P}{\partial x}) + \frac{\partial}{\partial y} (K_y \frac{\partial P}{\partial y}) - \frac{\partial uP}{\partial x} - \frac{\partial vP}{\partial y} + abP - w/H \text{ } P \tag{10}$$

where K_x and K_y are numerical diffusion coefficients (Hirt, 1968) derived from the finite-difference approximation of eqs. (9)-(10), and where a is the growth rate (hr^{-1}) of the phytoplankton expressed by

$$a = d(1.43 \sin 0.2618 \text{ } t)(N/q + N) \tag{11}$$

in which $d = 0.025 \text{ hr}^{-1}$ to allow a maximum daily growth rate of 0.43 day^{-1} over 12 hours of daylight, while the sine function is set to zero at night, i.e., a = 0 then as well, and (N/q + N) is a Michaelis-Menton expression for nitrate limitation of the algal growth, with the half-saturation constant, q, taken to be $1.5 \text{ } \mu g\text{-at } NO_3 \text{ } \ell^{-1}$ (Walsh, 1975). The other parameters of eq. (10) are b = 2, the assumed conversion ratio of chlorophyll/particulate nitrogen within the phytoplankton and w is their sinking velocity, 10 m day^{-1}. Since there is no term in eqs. (9)-(10) for light limitation of phytoplankton growth with depth, P and N are not really averages over the water column, but can be considered representative of nitrate and chlorophyll concentrations within the first attenuation depth of the water column, i.e., 1-10 m, where photosynthesis is maximal.

The algal sinking rate, w, of eq. (10) is divided by the local depth, H, such that the units of this loss term $(\mu g \text{ chl } \ell^{-1} \text{ hr}^{-1})$ are the same as those of the growth, diffusion, and advective terms. The implicit vertical balance of chlorophyll fluxes is thus taken to be the influx at the surface, which is zero, and the outflux at the bottom of the water column to the sediments, where P near the bottom is the same as P near the surface, i.e., a homogeneous distribution, $\partial P/\partial z = 0$. The selection of $w = 10 \text{ m day}^{-1}$, instead of sinking rates of $\sim 1 \text{ m day}^{-1}$ measured in the laboratory, will be discussed with respect to the simulation results.

The upstream boundary conditions of nitrate and chlorophyll across Anadyr and Shpanberg Straits were taken from previous cruises in 1969-84, e.g., Fig. 2, during which similar longitudinal gradients of both variables were found, decreasing from west to east. We assumed values of nitrate to be 10 µg-at NO_3 ℓ^{-1} in the middle of Anadyr Strait and 1 µg-at NO_3 ℓ^{-1} near the middle of Shpanberg Strait. At these boundaries, values of chlorophyll were 6 µg chl ℓ^{-1} in the middle of Anadyr Strait and 1 µg chl ℓ^{-1} in Shpanberg Strait. At the air and land boundaries, P and N were both taken to be zero, i.e., no influx of nutrients from either rainfall or the Yukon River. The initial conditions of nitrate and chlorophyll within the interior of the grid (Fig. 4) were N = P = 0 in simulation cases where growth and sinking were not considered (Figs. 8, 9, 17). The initial conditions were N = N + P/b from the previous solutions of the advective/dispersive cases, while P was set to 0.5 µg chl ℓ^{-1}, when other cases of growth and sinking were subsequently studied. Finally, at the downstream boundary conditions of N and P between Capes Dezhneva and Hope, an alternative method to the radiative boundary condition was used to solve eqs. (9)-(10), that of upstream finite differences (Molenkamp, 1968; Roache, 1972), which introduces the numerical diffusion estimated by K_x and K_y.

RESULTS

Currents

To specify the relative importance of the upstream boundary conditions, i.e., far field currents and previous biological interactions, on the solution of eqs. (1)-(3) and (9)-(10), a series of cases were run with different terms of these equations set to zero. With no forces acting on the interior flow field of the 0.6 Sv case, for example, there is no setup of the surface elevation, such that eq. (3) becomes

$$0 = \frac{\partial U}{\partial x} + \frac{\partial V}{\partial y} \qquad (12)$$

and eqs. (1)-(2) are not relevant, resulting in currents which neither follow the bottom topography, nor accelerate downstream, except within the constricted area of Bering Strait, with a maximum flow of 45.6 cm sec^{-1} (Fig. 5).

Inclusion of the Coriolis, pressure, and frictional forces of eqs. (1)-(2), acting on the boundary currents which enter the model's domain in the 0.6 Sv case, generated horizontal sea level gradients in both the x and y directions. At steady state, i.e., $\frac{\partial s}{\partial t} = \frac{\partial U}{\partial t} = \frac{\partial V}{\partial t} = 0$, the largest gradient, ∿20 cm/300 km, was between Anadyr Stream Water, off the Siberian peninsula, and Alaska Stream Water, off the mouth of the Yukon River (Fig. 6). Such a sea-surface slope of 0.07×10^{-5}, and $g = 10^3$ cm sec^{-2}, would lead to a fluid acceleration of ∿7 $\times 10^{-4}$ cm sec^{-2}, compared to ∿1 $\times$

410

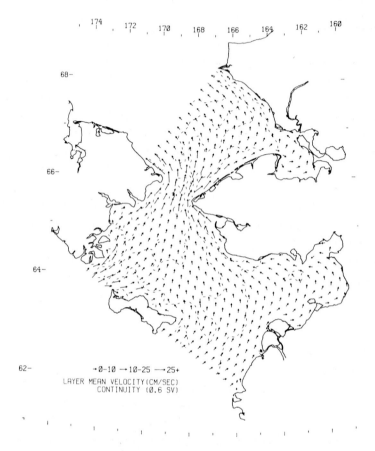

Figure 5.   Continuity solution of a depth-averaged flow field (cm sec^{-1}) with 0.6 Sv transport through Bering Strait.

10^{-4} cm sec^{-2} accelerations of the flow from either a 7 m sec^{-1} wind, exerting stress at the sea surface for one day (10^{5} sec), or from the usual tidal forces. Over one day, with no frictional losses at the bottom, a current of 70 cm sec^{-1} would develop from such an acceleration of the fluid in this 0.6 Sv case of the model.

Since eqs. (1)-(3) are linear, the current velocities for the 0.3 Sv and 1.2 Sv cases are half and double, respectively, of the 0.6 Sv case. Maximum speeds in Bering Strait were 36.5 cm sec^{-1} for the 0.3 Sv case, 72.9 cm sec^{-1} for the 0.6 Sv case (Fig. 7), compared to the continuity solution of 45.6 cm sec^{-1} for the same case (Fig. 5), and 142.7 cm sec^{-1} for the 1.2 Sv case (Fig. 16). A combination of the Coriolis correction for a rotating earth coordinate system, and the acceleration of the boundary currents induced by the horizontal pressure gradient, led to flow along the isobaths (Fig. 7), in

411

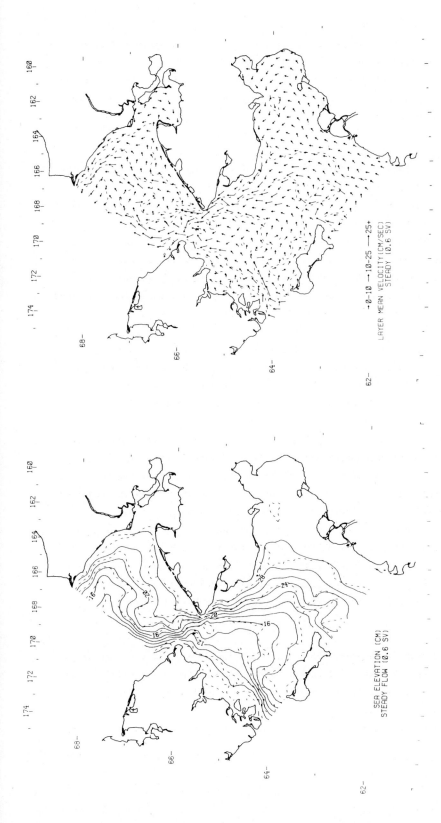

-0 -10 →10-25 →25+
LAYER MEAN VELOCITY (CM/SEC)
STEADY (0.6 SV)

Figure 7. The depth-averaged flow field (cm sec⁻¹) of
a circulation sub-model with horizontal
pressure, bottom friction, and Coriolis
forces after 240 hours in the 0.6 Sv case.

SEA ELEVATION (CM)
STEADY FLOW (0.6 SV)

Figure 6. The spatial distribution of sea level
elevation (cm) in a flow field with 0.6 Sv
transport through Bering Strait.

sharp contrast to the simple continuity case (Fig. 5). In this model, the
steady flow pattern can be simplistically described as the implied vorticity
balance of eqs. (1)–(3), in which the torque imparted to the water column by
the bottom stress is compensated by the cross–isobath movement of water to
conserve angular momentum.

Algal growth

Within the second flow field of the 0.6 Sv case (Fig. 7), deletion of the
growth and sinking terms of eqs. (9)–(10) represents an analysis of the
dispersion of nitrate (Fig. 8) and chlorophyll (Fig. 9) from their upstream
boundaries. In this situation, the distribution of chlorophyll north of
St. Lawrence Island is determined only by previous population growth south of
Anadyr and Shpanberg Straits and the currents flowing north to Bering Strait
(Fig. 7). This simple simulated chlorophyll field (Fig. 9) exhibited an
east–west gradient between Siberia and Alaska of the same magnitude displayed
by a composite of the surface chlorophyll measurements taken over 6 years
(Fig. 2). The northward termination of the 4 μg ℓ^{-1} isopleth of chlorophyll
in the model (Fig. 9) was also similar to the historical data base (Fig. 2)
for the Chukchi Sea (Hillman, 1984). With the algal population growth rates
of $\sim$0.6–1.0 day^{-1} observed during the 1983–84 ISHTAR cruises between
St. Lawrence Island and Bering Strait, however, an assumption of no in situ
growth is, of course, unrealistic.

Addition of a modest 0.43 day^{-1} growth term in eqs. (9)–(10) led to $\sim$50%
depletion of the nitrate stocks north of St. Lawrence Island (Fig. 10). In
situ growth of algal populations north of St. Lawrence Island, together with
phytoplankton biomass passing through Anadyr and Shpanberg Straits, resulted
in more than 12 μg chl ℓ^{-1} within the model, west of both Northeast Cape on
St. Lawrence Island and Cape Wales on the Alaskan mainland (Fig. 11). Such
an extensive accumulation of phytoplankton standing stocks was not observed
during short cruises in June–August, between 1978 to 1984 (Fig. 2), suggesting that either grazing losses (Fig. 1) or sinking losses must have a
significant impact on the primary production of the Bering–Chukchi seas.

Sinking loss

The large biomass (Stoker, 1981) and ingestion demands of the benthos, as
well as the possible export of carbon within Anadyr Stream Water (Fig. 1),
led us to parameterize all phytoplankton losses within the sinking term,
w/H P, of this simulation model (Walsh, 1983). Recent field evidence
suggests rapid sinking rates of 20–40 m day^{-1} for spring blooms within the

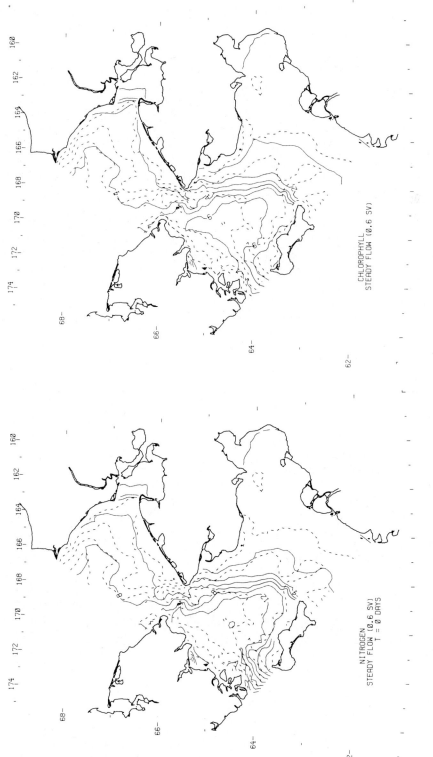

Figure 9. The steady-state spatial distribution of chlorophyll (mg m⁻³), with neither growth nor sinking, in the 0.6 Sv case.

Figure 8. The steady-state spatial distribution of nitrate (mg-at m⁻³), with neither growth nor sinking, in the 0.6 Sv case.

414

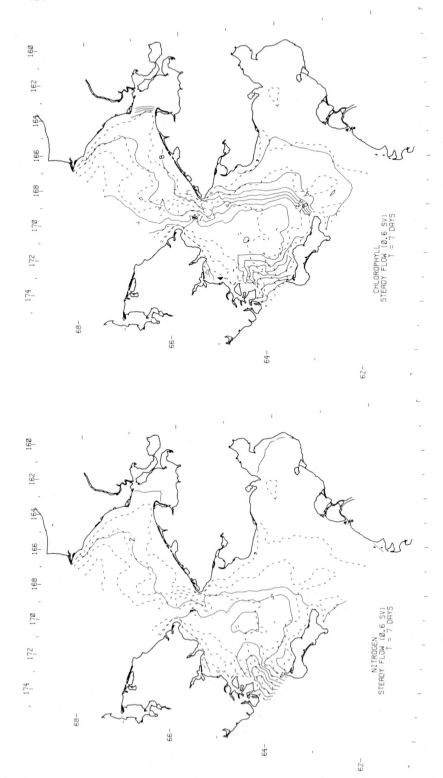

Figure 11. The spatial distribution of chlorophyll (mg m⁻³) after 7 days of growth, with no sinking, in the 0.6 Sv case.

Figure 10. The spatial distribution of nitrate (mg-at m⁻³) after 7 days of growth, with no sinking, in the 0.6 Sv case.

mid-Atlantic Bight (Walsh, et al., 1986b; Walsh et al., 1986c) and the Baltic Sea (Bodungen et al., 1981). A choice of a modest sinking rate of 10 m day^{-1} for w in eq. (10), however, allowed sufficient phytoplankton losses that the nitrate stocks formed a spatial pattern after 7 days (Fig. 12), which was similar to the case with no algal growth (Fig. 8). After 14 days of simulated time, however, the nitrate stocks were again depleted by ∿50% (Fig. 14) of those at 7 days (Fig. 12), but with a significant difference between this 14-day result and that spatial nitrate pattern derived from the growth, but no sinking case (Fig. 10)--halfway between St. Lawrence Island and Bering Strait a minimum in the nitrate field (Fig. 14) developed after 14 days of water movement (Fig. 7) from Anadyr Strait to Bering Strait.

At an average flow of 15-20 cm sec^{-1} within the 0.6 Sv case, a phytoplankton cell would take ∿10-14 days to be moved from Anadyr Strait to about 200 km downstream towards Bering Strait, where the nitrate minimum was found after 14 days of simulated time (Fig. 14). Within the first 4 days of such a phytoplankton trajectory, the chlorophyll biomass derived from primary production south of Anadyr and Shpanberg Straits (Fig. 9) would have sunk out of a 40 m water column at a sinking rate of 10 m day^{-1}. Removal of the previously grown 6-8 µg chl ℓ$^{-1}$ of algal biomass would deepen the euphotic zone, making available more light in the water column for initiation of primary production north of St. Lawrence Island.

After sinking losses were imposed for 7 days of simulated time, for example, ≤2 µg chl ℓ$^{-1}$ were found over most of the model's domain (Fig. 13), in sharp contrast to the growth only case of Figure 11. The algal population from the Anadyr Strait boundary condition has sunk out of the model, but t'⌐ local populations at the interior grid points have not had enough time to significantly increase their biomass within just one week. With a maximal population growth rate of only 0.43 day^{-1}, or a doubling time of ∿1.5 days, it would take at least 6-7.5 days, with no losses, for an exponential increase of the local populations to occur from the 0.5 µg chl ℓ$^{-1}$ initial condition to biomass levels of 8-16 µg chl ℓ$^{-1}$. These high levels of chlorophyll biomass were found after 14 days of simulated time (Fig. 15), however, in the region of the nitrate minimum (Fig. 14). These simulation results suggest that the accumulated algal biomass, grown south of St. Lawrence Island, sinks out between this island and Bering Strait, whereas that increment of algal biomass grown north of Anadyr and Shpanberg Straits is likely to sink out within the Chukchi Sea.

416

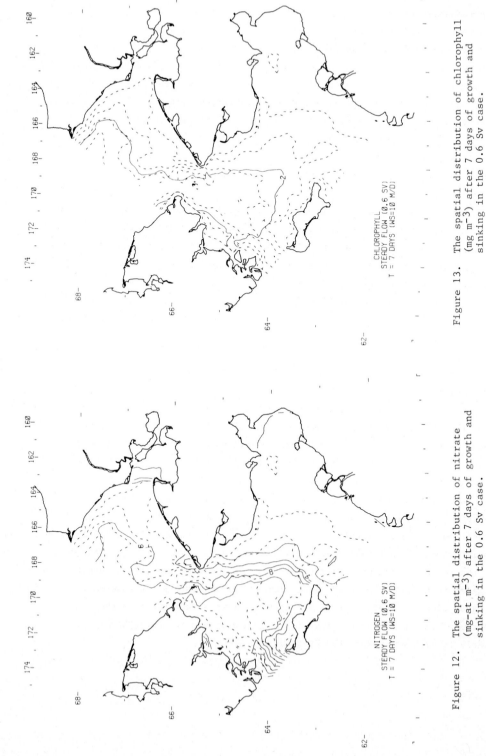

CHLOROPHYLL (0.6 SV)
STEADY FLOW (WS=10 M/D)
T = 7 DAYS

Figure 13. The spatial distribution of chlorophyll (mg m^{-3}) after 7 days of growth and sinking in the 0.6 Sv case.

NITROGEN (0.6 SV)
STEADY FLOW (WS=10 M/D)
T = 7 DAYS

Figure 12. The spatial distribution of nitrate (mg-at m^{-3}) after 7 days of growth and sinking in the 0.6 Sv case.

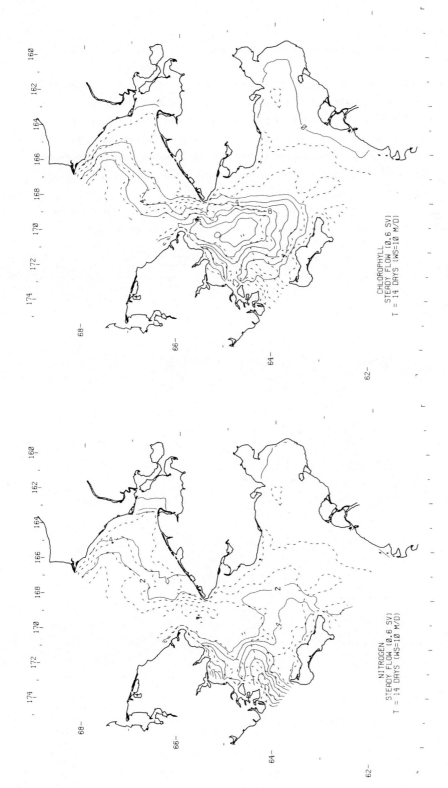

CHLOROPHYLL
STEADY FLOW (0.6 SV)
T = 14 DAYS (WS=10 M/D)

Figure 15. The spatial distribution of chlorophyll
(mg m^{-3}) after 14 days of growth and
sinking in the 0.6 Sv case.

NITROGEN
STEADY FLOW (0.6 SV)
T = 14 DAYS (WS=10 M/D)

Figure 14. The spatial distribution of nitrate
(mg-at m^{-3}) after 14 days of growth and
sinking in the 0.6 Sv case.

Habitat variability

A change in northward transport of water through the Bering Strait might alter the sites of the two depocenters of algal carbon suggested by our model. To explore the role of physical variability, interannual or seasonal, in determining the intensity of the sources for these depocenters, the southern one, fed from the boundary algal input, and the northern one, fed from in situ algal production downstream of St. Lawrence Island, we considered three flow regimes. In addition to the 0.6 Sv transport cases discussed above, we simulated biochemical interactions within both a weaker flow field of 0.3 Sv and a stronger one of 1.2 Sv.

Only the results of the chlorophyll fields within the 1.2 Sv flow regime (Fig. 16) are presented here. Under a situation of neither growth nor

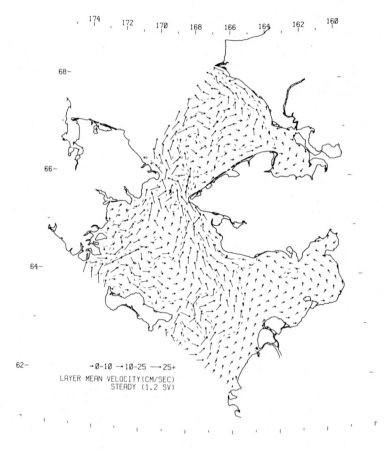

Figure 16.   The depth-averaged flow field (cm sec^{-1}) of a circulation sub-model with horizontal pressure, bottom friction, and Coriolis forces after 240 hours in the 1.2 Sv case.

sinking, the spatial pattern of chlorophyll in the 1.2 Sv case (Fig. 17) was
the same as the 0.6 Sv case (Fig. 9). Although the advective/diffusive gains
of chlorophyll were larger at each grid point in the 1.6 Sv case, the losses
were also larger, such that the net result was no change in the local
abundance of algal biomass. Of course, the nutrient patterns remained the
same as well.

In the situation of growth and dispersion, but no sinking, within the
faster flow field, the chlorophyll plume of Anadyr Stream Water extended
∿50 km farther north (Fig. 18) than before (Fig. 11). However, there was no
change in the lateral displacement of algal biomass to the east (Figs. 11,
18). The slower flow field of the 0.3 Sv case led to ∿50 km less penetration
north of the chlorophyll plume of the Anadyr Stream Water than that within
the 0.6 Sv regime (Fig. 11). Thus, we found about 100 km total displacement
along the Siberian coast of algal populations, derived from the boundary
condition, during possible interannual or seasonal variations of 0.3-1.2 Sv
flow north through the Bering Strait.

The results of our sinking loss experiment within the 1.2 Sv flow field
suggest, however, that the algal populations grown north of St. Lawrence
Island did not shift their locations in response to a change in the transport. After 7 (Fig. 19) and 14 (Fig. 20) days within the 1.2 Sv growth and
sinking case, the chlorophyll isopleths of the faster flow regime were still
∿50 km farther north off the Siberian coast than those within the 0.6 Sv flow
field (Figs. 13 and 15). Within the shelf region directly north of
St. Lawrence Island, however, there was little change in the strength of the
currents between the 0.6 Sv (Fig. 7) and 1.2 Sv (Fig. 16) cases.
Consequently, the 12 µg ℓ^{-1} isopleth of chlorophyll remained in the same
location after 14 days of both flow regimes (Figs. 15, 20). It continued to
be positioned between the 40 m isobaths on the Siberian and Alaskan sides of
the Chirikov Basin, north of St. Lawrence Island, i.e., where the smallest
daily sinking loss (0.25 day^{-1}) occurred in both flow regimes. The
implication of these simulation runs is that one might expect the algal
sources for the southern depocenter to migrate north or south, while the
sources for the northern depocenter might remain fixed, during interannual or
seasonal changes in the flow field linking the Bering and Chukchi Seas.

Verification

An example of a seasonal departure from the apparent "summer" steady-state
system, implied by the meager June-August data sets collected during 1-2 week
cruises in 1978-84 (Fig. 2), was obtained on the first leg (June 28-July 10)

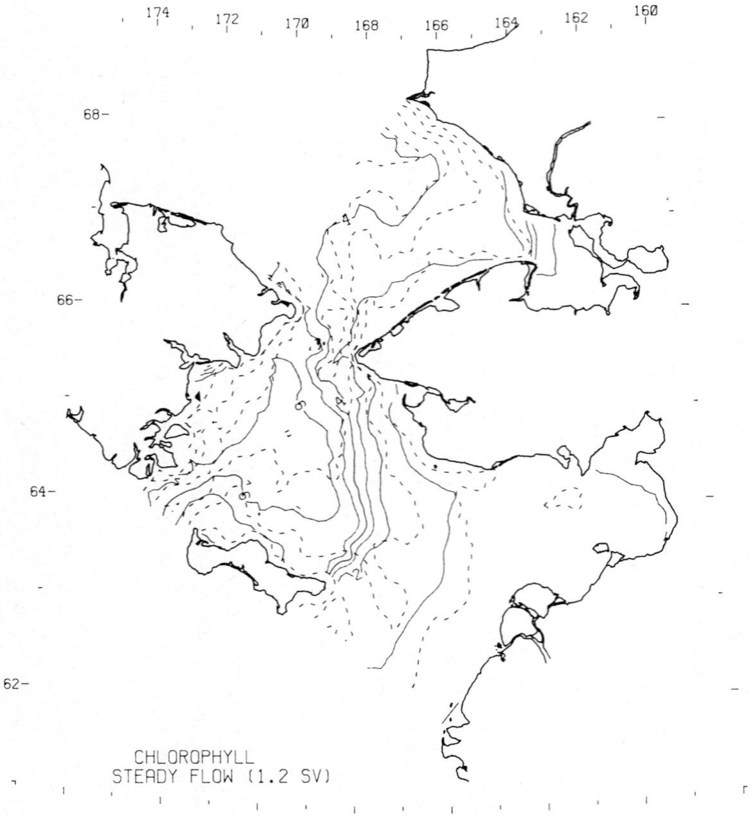

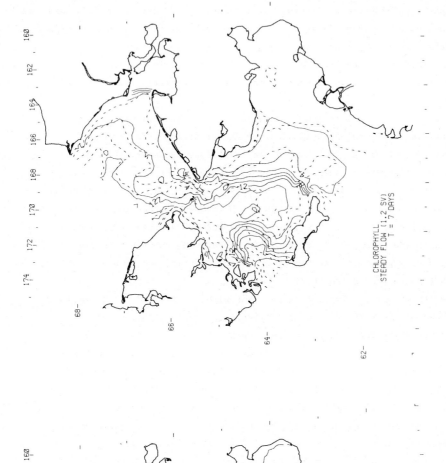

CHLOROPHYLL (1.2 SV)
STEADY FLOW
T = 7 DAYS

Figure 18.   The spatial distribution of chlorophyll
(mg m^{-3}) after 7 days of growth, with no
sinking, in the 1.2 Sv case.

CHLOROPHYLL
STEADY FLOW (1.2 SV)

Figure 17.   The steady-state spatial distribution of
chlorophyll (mg m^{-3}), with neither growth
nor sinking, in the 1.2 Sv case.

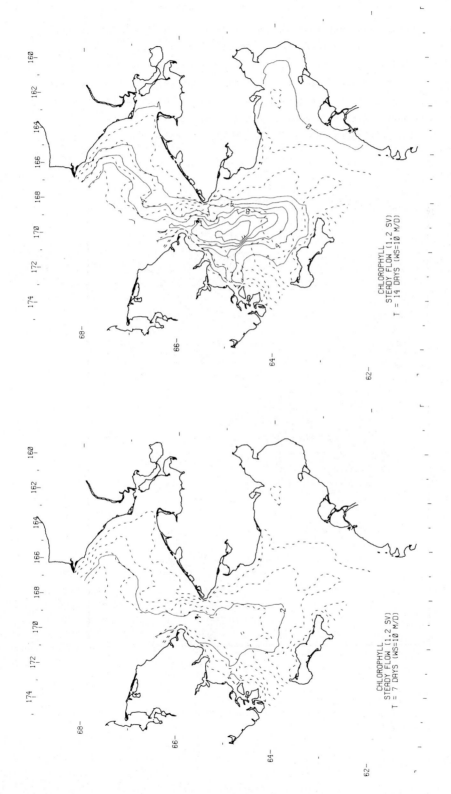

CHLOROPHYLL
STEADY FLOW (1.2 SV)
T = 14 DAYS (WS=10 M/D)

Figure 20. The spatial distribution of chlorophyll (mg m^{-3}) after 14 days of growth and sinking in the 1.2 Sv case.

CHLOROPHYLL
STEADY FLOW (1.2 SV)
T = 7 DAYS (WS=10 M/D)

Figure 19. The spatial distribution of chlorophyll (mg m^{-3}) after 7 days of growth and sinking in the 1.2 Sv case.

of the 1985 ISHTAR field experiment. As the result of a "late" spring, pack-ice was still so abundant between St. Lawrence Island and Norton Sound that Leg 1 of the 1985 field experiment had to be postponed for two weeks, thus representing a "spring snapshot" in July 1985 of this system rather than the usual "summer snapshots" of previous months of July. Consequently, surface (Fig. 21) and bottom (Fig. 22) maps of nitrate, measured at 54 stations (Fig. 21) on 4-9 July 1985, depict >20 μg-at NO_3 ℓ^{-1} throughout the column of

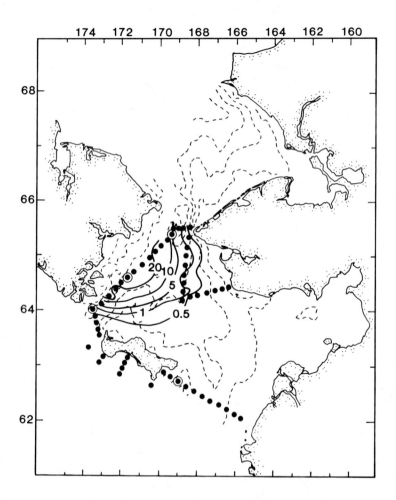

Figure 21.   The spatial distribution of nitrate (mg-at m^{-3}) within surface waters of the Bering Sea during 4-9 July 1985, with respect to the locations of the moored fluorometers (◉) and of hydrographic stations (•).

Anadyr Stream Water in a section taken along the International Dateline from
Anadyr Strait to Bering Strait. Less than 15 μg-at NO_3 ℓ^{-1} were found at the
surface in the same region on 12-17 July 1984 (Walsh et al., 1986a), similar
to previous observations during June 1969 (Husby and Hufford, 1969) and
simulated in the above model, e.g., Fig. 8.

As in previous years, however, <0.5 μg-at NO_3 ℓ^{-1} were found throughout
the water column of Alaska Coastal Water, with a monotonic west-east decline
of nitrate in bottom water (Fig. 22). The surface nitrate pattern contained
sharper horizontal gradients (Fig. 21) than the bottom nitrate map, with
<0.5 μg-at NO_3 ℓ^{-1} found west and north of St. Lawrence Island as well. The

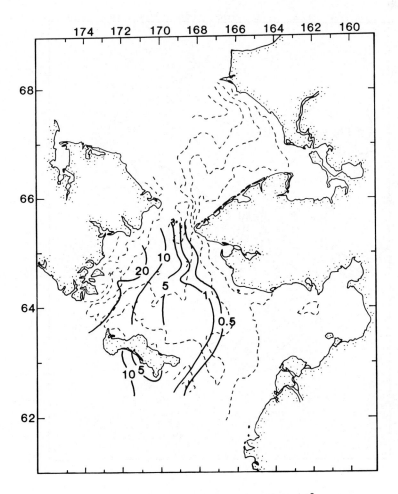

Figure 22.   The spatial distribution of nitrate (mg-at m^{-3}) within bottom
waters of the Bering Sea during 4-9 July 1985.

424

surface chlorophyll map (Fig. 23) was the mirror image of the surface nitrate pattern, with a downstream increase of algal biomass as the nitrate stocks declined, reminiscent of similar plankton dynamics of coastal upwelling areas off Peru, Baja California, and Northwest Africa (Walsh, 1983).

At this time of year in 1985, there was no surface signature of an input of phytoplankton from the boundaries of Anadyr and Shpanberg Straits (Fig. 23). The bottom chlorophyll map (Fig. 24) provided a striking contrast to the surface one, however, with a tenfold higher algal biomass found south of St. Lawrence Island and within the two straits. With low chlorophyll

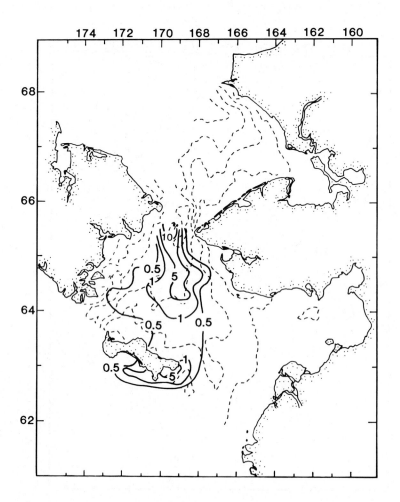

Figure 23. The spatial distribution of chlorophyll (mg m^{-3}) within surface waters of the Bering Sea during 4-9 July 1985.

content of the euphotic zone in these regions, the near-bottom algal biomass
must reflect either previous production events or an algal source farther to
the south and west of St. Lawrence Island. During July 1984, for example,
>40 µg chl ℓ^{-1} were found in the middle of the water column within Anadyr
Strait, yet 1-2 orders of magnitude less algal biomass were found here in
July 1985.

Farther north, between the 40 m isobaths and the Diomede Islands, 5-10 µg
chl ℓ^{-1} were found throughout the water column in July 1985 (Figs. 23-24),
indicative of intense _in situ_ production and similar to observations of

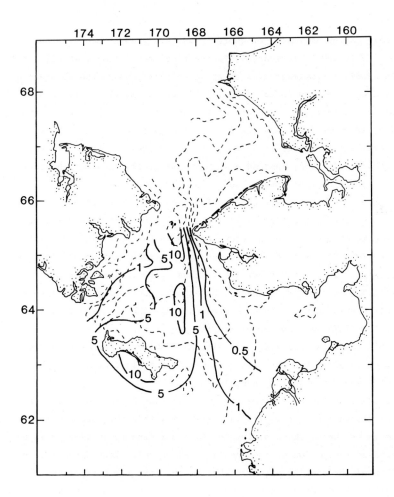

Figure 24. The spatial distribution of chlorophyll (mg m^{-3}) within bottom
waters of the Bering Sea during 4-9 July 1985.

previous years (Fig. 2). During the second leg of the 1985 ISHTAR field experiment in the Chukchi Sea, >40 μg chl ℓ^{-1} were found in the middle of the water column, just north of Bering Strait, implying that the source of carbon for the second depocenter had not shifted location between this and previous years. We do not yet have available the current meter data for June 1985, but our simple simulation model would suggest an "atypical" flow pattern had occurred within and south of Anadyr Strait in response to the late spring of 1985. This latest ISHTAR field experiment will, of course, provide additional verification data for the models and an opportunity to upgrade them.

CONCLUSIONS

Any theory can be improved, of course, and this present model is no exception. We intend to increase 1) the vertical resolution (Nihoul, 1984), 2) the bio-chemical complexity (Walsh et al., 1981), and 3) the accuracy of particle trajectories (Walsh and McRoy, 1985) of our models as the ISHTAR field experiments generate more data, taken at higher temporal and spatial frequencies. Figures 21-24 are examples of the data base that will eventually be available for additional simulation analyses. Taken over only 5 days during 4-9 July 1985, instead of 6 years (Fig. 2), these shipboard surface and bottom maps of nitrate and chlorophyll provide insight into the seasonal, interannual, and vertical complexities of this Arctic ecosystem.

Furthermore, during the first leg of the 1985 ISHTAR field experiment, nine current meter moorings were recovered from a collaborative SAI deployment the previous October 1984, while ten more current meter moorings were installed for retrieval in October 1985. Thus, year-long current meter records will be available from Anadyr, Shpanberg, and Bering Straits to update our circulation sub-models at the end of this ISHTAR field experiment. At four of the ISHTAR current meter moorings (θ of Fig. 21), seven internally recording fluorometers (Whitledge and Wirick, 1986) were deployed as well, with six located along the date-line and one in Shpanberg Strait as the result of the above simulation analysis.

Upon recovery of these fluorometers, we will also have time series of chlorophyll fluctuations at 15 minute intervals within the euphotic and aphotic zones of the northern Bering Sea for ∿3 months to justify the proposed increased resolution of the models, i.e., a sufficient data base for verification of the simulation output before the next field experiment (Walsh, 1972). As Alice noticed, however, "in another minute the whole head appeared .... The Cat seemed to think that there was enough of it now in sight, and no more of it appeared"; reconstruction of a natural system is thus a stepwise process. Our next simulation analysis will consist of a three-dimensional model of more biological state variables, when the reduced data set justifies the additional computation of a multi-layered resolution of the vertical structure of the water column. Considering the progress over the last decade of remote sensing, of in situ instrumentation, and of the speed and capacity of digital computers, we are confident that, in the next decade, reconstruction of a number of local marine ecosystems will occur, providing the basis for eventual networking of simulation models on a global scale.

ACKNOWLEDGEMENTS

This research was mainly funded by the Division of Polar Programs, National Science Foundation, as part of the ISHTAR program. Additional support for our computer facility was provided by funds from the Department of Energy and the National Aeronautics and Space Administration.

REFERENCES

Aagaard, K., Roach, A. T. and Shumacher, J. D., 1985. On the wind-driven variability of the flow through Bering Strait. J. Geophys. Res. (In press).
Bodungen, B., Brockel, K., Smetacek, V. and Zeitzschel, B., 1981. Growth and sedimentation of the phytoplankton spring bloom in the Bornholm Sea (Baltic Sea). Kiel. Meer. Son., 5: 460-490.
Carroll, L., 1865. Alice's Adventures in Wonderland. MacMillan, 164 pp.
Coachman, L. K., Aagaard, K. and Tripp, R. B., 1975. Bering Strait: The Regional Oceanography. University of Washington Press, Seattle, 172 pp.
Coachman, L. K. and Walsh, J. J., 1981. A diffusion model of cross-shelf exchange of nutrients in the Bering Sea. Deep-Sea Res., 28: 819-837.
Hillman, S. R., 1984. Near-shore chlorophyll concentrations in the Chukchi Sea. Polar Record, 22: 182-186.
Hirt, C. W., 1968. Heuristic stability theory for finite-difference equations. J. Computat. Phys., 2: 339-355.
Husby, D. M. and Hufford, G. L., 1969. Oceanographic investigation of the northern Bering Sea and Bering Strait, 8-21 June 1969. U. S. Coast Guard Oceanogr. Rep. No. 42, CG 373-42, Washington, D. C.
Molenkamp, C. R., 1968. Accuracy of finite-difference methods applied to the advection equation. J. Appl. Meteor., 7: 160-167.
Nihoul, J. J., 1984. A three-dimensional general marine circulation model in a remote sensing perspective. Ann. Geophys., 2: 433-442.
Orlanski, I., 1976. A simple boundary condition for unbounded hyperbolic flows. J. Computat. Phys., 21: 251-259.
Platzman, G. W., 1972. Two dimensional free oscillations in natural basins. J. Phys. Oceanogr., 2: 117-138.
Roache, P. J., 1972. On artificial viscosity. J. Computat. Phys., 10: 169-184.
Sambrotto, R. N., Goering, J. J. and McRoy, C. P., 1984. Large yearly production of phytoplankton in the western Bering Strait. Science, 225: 1147-1150.
Stoker, S., 1981. Benthic invertebrate macrofauna of the eastern Bering/Chukchi continental shelf. In: D. W. Hood and J. A. Calder (Editors), The Eastern Bering Sea Shelf: Oceanography and Resources. University of Washington Press, Seattle, pp. 1069-1091.

Walsh, J. J., 1972. Implications of a systems approach to oceanography. Science, 176: 969–975.

Walsh, J. J., 1975. A spatial simulation model of the Peru upwelling ecosystem. Deep-Sea Res., 22: 201–236.

Walsh, J. J., 1983. Death in the sea: enigmatic phytoplankton losses. Prog. Oceanogr., 12: 1–86.

Walsh, J. J. and Dugdale, R. C., 1971. A simulation model for the nitrogen flow in the Peruvian upwelling system. Inv. Pesq., 35: 309–330.

Walsh, J. J., Rowe, G. T., Iverson, R. L. and McRoy, C. P., 1981. Biological export of shelf carbon is a neglected sink of the global CO_2 cycle. Nature, 291: 196–201.

Walsh, J. J. and McRoy, C. P., 1985. Ecosystem analysis in the southeastern Bering Sea. Cont. Shelf Res. (In press).

Walsh, J. J., Premuzic, E. T., Gaffney, J. S., Rowe, G. T., Harbottle, G., Stoenner, R. W., Balsam, W. L., Betzer, P. R. and Macko, S. A., 1985. Organic storage of CO_2 on the continental slope off the mid-Atlantic Bight, the Southeastern Bering Sea, and the Peru coast. Deep-Sea Res., 32: 853–883.

Walsh, J. J., McRoy, C. P., Blackburn, T. H., Coachman, L. K., Goering, J. J., Nihoul, J. J., Parker, P. L., Springer, A. M., Tripp, R. B., Whitledge, T. E. and Wirick, C. D., 1986a. The role of Bering Strait in the carbon/nitrogen fluxes of polar marine ecosystems. In: L. Rey and V. Alexander (Editors), Marine Living Systems of the Far North. McMillan Co. (In press).

Walsh, J. J., Dieterle, D. A. and Esaias, W. E., 1986b. Satellite detection of export of the 1979 spring bloom from the mid-Atlantic Bight. Deep-Sea Res. (Revised).

Walsh, J. J., Wirick, C. D., Pietrefesa, L. J., Whitledge, T. E. and Hoge, F. E., 1986c. High frequency sampling of the 1984 spring bloom within the mid-Atlantic Bight: synoptic ship-board, aircraft, and in situ perspectives of the SEEP-I experiment. Con. Shelf Res. (Submitted).

Whitledge, T. E. and Wirick, C. D., 1986. Development of a moored in situ fluorometer for phytoplankton studies. M. J. Bowman, C. M. Yentsch and W. J. Petersen (Editors), Tidal Mixing and Plankton Dynamics. Springer-Verlag (In press).

THE TERRESTRIAL-MARINE INTERFACE: MODELLING NITROGEN TRANSFORMATIONS
DURING ITS TRANSFER THROUGH THE SCHELDT RIVER SYSTEM AND ITS
ESTUARINE ZONE

G. BILLEN, C. LANCELOT, E. DEBECKER and P. SERVAIS

Groupe de Microbiologie des Milieux Aquatiques
University of Brussels, 50 av, F. Roosevelt, B-1050 Bruxelles (Belgium)

ABSTRACT

A budget of nitrogen transformations during its transfer through the river system and the estuary of the Scheldt shows that as much as 70% of total N discharged into the river system is eliminated before reaching the sea. Denitrification and, indirectly, primary production appear as significant processes in this elimination. The factors controlling the intensity of these processes, both in the river system and in the estuary, are identified. This allows mathematical modelling of these processes and prediction of the impact of large scale waste water purification policies on the input of nitrogen into the Belgian coastal zone.

INTRODUCTION

For the needs of a rational management of the quality of coastal waters models relating human activities (agricultural, domestic and industrial) to the ecological working of coastal ecosytems must be established. In such models, not only the estuarine zones, but also the whole river system, across which most discharges from human activities flow before they reach the sea, can be viewed as an interface where intense biological and physico-chemical processes occur, resulting in partial elimination or immobilization of the substances discharges

It is with this concern in mind that the present study of nitrogen transfers through the river system and estuary of the Scheldt, was undertaken. It had the following purposes:

(i) To establish a budget of nitrogen transformation processes during its transit from agricultural soils of the Scheldt watershed to the Belgian coastal zone.

(ii) To understand the mechanisms controlling the intensity of the most important processes involved in this budget.

(iii) To build up a model of nitrogen (and carbon) behaviour in the river system and in the estuarine zone in order to relate discharge into the rivers and outputs into the sea. This model is intended to predict the effect of different waste water purification policies on the coastal areas.

The results of the first step of this study, i.e. the budget of nitrogen transformations in the Scheldt watershed and estuary, has been published elsewhere (Billen et al, 1986). It is summarized by Fig. 1 which allows identification of the main biological processes involved in nitrogen transformations.

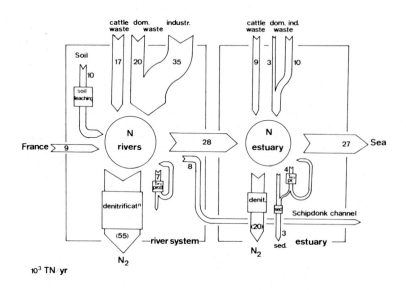

Fig. 1: Nitrogen budget for the Scheldt river system and estuary (from Billen et al, 1986). Fluxes are expressed in 10^3 TN/yr. Figures in brackets have been evaluated by difference.

Denitrification in the river system eliminates more than 60% of the total nitrogen load. Uptake of nitrogen by primary producers plays only a minor direct role in the budget, representing less than 8% of total inputs. Primary production is however of indirect importance for the nitrogen cycle, because of its role as a source of rapidly biodegradable organic carbon.This is of particular significance in the upper estuarine section of the Scheldt, where decaying fresh water phytoplankton contributes to sustain high heterotrophic activities which result in the establishment of anaerobic conditions favourable to denitrification in the water column. This process is responsible for the elimination of 40% of the nitrogen load entering the estuarine zone.

Thus, as a result of denitrification in river and in the upper estuary, only 28% of the total nitrogen discharged into the Scheldt river system does actually reach the sea. It is therefore likely that any modification of the factors affecting denitrification in the hydrographical network, e.g. as a result of a large scale program of waste water purification could deeply affect the input of nitrogen into the coastal areas.

In this paper, we try to identify the factors controlling the processes of denitrification and primary production, both in the river system and in the estuary. For each of these processes mathematical sub-models will be established and their coherency will be checked against in situ observations. Together, these sub-models allow to predict the effect of an extreme waste water purification scenario on the discharge of nitrogen to the Belgian coastal zone.

DENITRIFICATION IN RIVERS

As an anaerobic process, denitrification can occur either in the water column, when deprived of oxygen or within the sediments, below the oxic layer. Anoxic surface waters are not unusual in the Scheldt hydrographical network, receiving large amounts of untreated waste water. The extent of anaerobic zones is however highly variable, and difficult to evaluate at the scale of the entire river system. Benthic denitrification on the other hand, is more widespread and a semi-empirical study of its dependence on the organic matter content of the sediments will allow evaluation of its rate at the scale of the whole hydrographical network.

Control of bentic denitrification by organic matter content of sediments

According to the diagenetic model of Vanderborght and Billen (1975), the flux of nitrate across the sediment-water interface ($F_{NO_3}^0$) is given by:

$$F^0_{NO_3} = - Di \frac{kn/Di \ (zn^2/2 + \sqrt{Di/kd} \ . \ zn) - C^0_{NO_3}}{zn + \sqrt{Di/kd}} \qquad (1)$$

where kn   is the rate of nitrification in the oxic layer of the
             sediments;
       zn   is the depth of the oxic layer;
       kd   is the first order constant of denitrication
   $C^0_{NO_3}$ is the concentration of nitrate in the overlying water
       Di   is the diffusion coefficient in the pure water.

The depth of the oxic layer, zn, is given by

$$zn = \frac{2 \ . \ Di \ . \ C^0_{O_2}}{F^0_{O_2}} \qquad (2)$$

where $C^0_{O_2}$ is the oxygen concentration in the overlying water
       $F^0_{O_2}$ is the flux of oxygen across the sediment water interface

432

The flux of oxygen, $F^0_{O_2}$, can be expressed as

$$F^0_{O_2} = R_{O_2} \cdot zn \qquad (3)$$

where R_{O_2} is the zero order rate of oxygen consumption by the sediments

Combining (2) and (3) yields

$$F^0_{O_2} = \sqrt{2 \cdot Di \cdot C^0_{O_2} \cdot R_{O_2}} \qquad (4)$$

Direct measurements show that the rate of oxygen consumption by sediment samples can be related to organic matter content and temperature (Fig. 2).

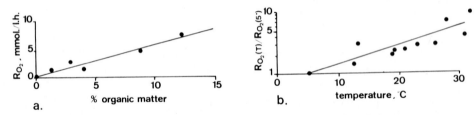

Fig. 2: a. Relationship observed between the rate of oxygen consumption by sediment sample (at 5°C) and organic matter content.

b. Temperature dependence of oxygen consumption rate.
R_{O_2} (mmole/m^3.h) = 608. organic matter %. $2^{\frac{T - 5°C}{10°C}}$

Combining the empirical relationship illustrated in Fig. 2 with relation (4) allows calculation of the flux of oxygen as a function of the organic matter content of the sediments. Calculated values compared well with experimental determinations performed in various Belgian rivers following the procedure described by Dessery et al (1982). Figure 3, provided Di is adjusted in the range 10^{-6} - 10^{-5} cm^2/sec.

The rate of denitrification, kd, as experimentaly determinated by short term kinetics of nitrate comsumption after nitrate addition to a sediment sample, was also found to be related to the organic matter content (Fig. 4).

The rate of nitrification, kn, was determined in a few river sediments according to the method described by Hansen et al (1981). Values in the range

0 - 300 mmoles/m^3.h were recorded. Such rates can be shown to have only a negligeable effect on the flux of nitrate as calculated by relation (1).

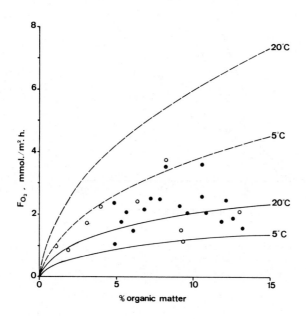

Fig. 3: Relationship between the flux of oxygen across the sediment–water interface and the organic matter content of the sediments, calculated for Di = 10^{-6}cm^{2}sec^{-1} (——) and 10^{-5}cm^{2}sec^{-1} (--), at 5 and 20°C.

Experimental data are from Edwards and Polly (1965) (●) and from our own measurements in Belgian rivers (○).

Relations (1), (2) and (4), along with the empirical relationships relating R and kd to organic matter content, allow calculation of the flux of nitrate consumed by river sediments. These calculations compare well with direct determination carried on in a range of Belgian rivers (fig. 5).

<u>Evaluation</u> <u>of</u> <u>benthic</u> <u>and</u> <u>water</u> <u>column</u> <u>denitrification</u> <u>at</u> <u>the</u> <u>scale</u> <u>of</u> <u>river</u> <u>system</u>

The semi-empirical model described above allows to evaluate benthic denitrification provided the total area of river bottom and the organic content of the sediments are known. The former information can be obtained by means of the geomorphological analysis of river sytems introduced by Horton (1945). In this analysis, each river is affected by an order.

Small rivers without tributaries are zero order, rivers having only tributaries of order (n - 1) are called order (n). On basis of the analysis of small representative sub-watersheds, it is then possible to deduce general relationships between order on the one hand, and the number of tributaries,

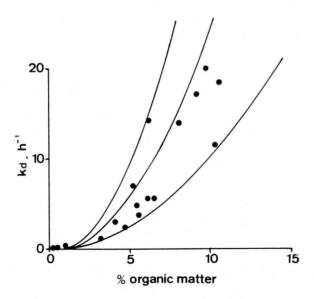

Fig. 4: Relationship observed between the first order constant of denitrification (kd) of river sediment samples and the organic matter content.

$$kd(h^{-1}) = \alpha \ . \ (\% \ OM - 0.5)^2 \ 2^{\frac{T - 20}{10}}$$

with α in the range $0.45 \stackrel{}{-} 0.1$ (mean 0.3).

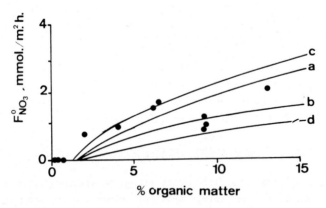

Fig. 5: Relationship between the flux of nitrate consumption by river sediments
and their organic matter content, as calculated by the model developed
in the text, for the following values of the parameters:

$$Di = 10^{-5} cm^2/sec$$
$$C^0_{O_2} = 350 \ \mu M \ - \ C^0_{NO_3} = 400 \ \mu M$$

Temperature = 20°C (a, c, d) or 5°C (b)
= 0.3 (a, b), 0.45 (c) or 0.1 (d)
Dots represents direct measurements in Belgian rivers

their mean length, the area of their watershed and their width, on the other hand.

Table 1 summarises the results of this analysis for the Scheldt river system. On basis of these data the total area of river bottom in the Belgian Scheldt watershed limited at Rupelmonde can be estimated at 55 km^2.

TABLE 1

Horton analysis of the Scheldt watershed

Order	Number of tributaries	Mean length (km)	Mean width (m)	Total area of river bottom (km^2)	Area of bottom/ total area of water- shed (km^2/km^2)
1	1000	1.8	0.9	1.6	73
2	340	4.5	1.8	2.8	128
3	110	11	4.1	5.0	229
4	32	26	9.2	7.7	353
5	10	60	17	10	472
6	3	140	41	17	794
7	1	330	260	86	–

The organic matter content of river sediments depends on the "density" of organic discharges, i.e. the ratio between the amount of organic matter (from domestic, industrial or agricultural origin) discharged into a river section and the bottom area of this section.

Available data relative to several Belgian districts are reported in Fig. 6. In spite of a rather large disperison it shows a trend of increasing mean organic matter content with increasing density of discharge .

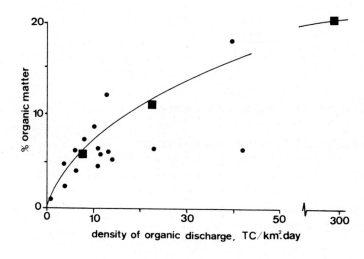

Fig.6: Observed rela-
tionship bet-
ween the mean
organic matter
content of ri-
ver sediments
and density of
organic dis-
charge for
several Belgi-
an districts.

On basis of these relationships, the rate of denitrification in the sediments has been evaluated, as shown in Table 2, for the hydrographical network of each district from the Belgian Scheldt watershed. A total value of $13-22 \cdot 10^3$ TN/yr is obtained. This represents 25-40% of total elimination of nitrogen in the river system as evaluated by difference in the nitrogen budget (Fig. 1). The estimation non accounted for by denitrification in the sediments must probably be explained by denitrification in the water column of anaerobic rivers.

TABLE 2

Evaluation of benthic denitrification in the hydrographical network of the districts from the Scheldt watershed

District	Mean O_2	Mean NO_3	river bottom	Mean % organic matter	Nitrate flux (TN/yr)	
	μm	μm	km^2		$20°$ C	$5°$ C
Malines	0	400	4.56	12	2636	1567
Turhout	0	400	2.82	16.5	2268	1348
Bruxelles	0	670	0.34	19	530	315
Leuven	0	400	2.42	6.1	681	405
Nivelles	sat	500	2.27	5.6	390	218
Halle-Vilvorde	0	670	1.96	17	2723	1619
Kortijk	sat	500	1.59	12	504	294
Roulers	sat	550	0.57	19	279	164
Tielt	sat	430	0.68	9	148	85
Aalst	0	670	0.98	17.5	1403	834
Audenarde	0	550	2.67	4	646	384
Termonde	0	400	14.75	5	3337	1984
St Niklaas(2/3)	0	400	8	5	1810	1076
Gent	0	400	2.26	20	2216	1317
Ath	sat	400	1.02	8	188	107
Mons	sat	400	1.22	15	365	213
Soignies	sat	400	1.08	7.5	189	107
Tournai	sat	400	2.45	8.8	487	279
Mouscron	sat	460	0.21	18	82	48
Hasselt	sat	400	1.89	16	593	347
Maaseik	sat	400	1.23	12.5	321	187
Total			55		22000	13000

PHYTOPLANKTON GROWTH IN THE RIVER SYSTEM AND THE ESTUARY

Dependence of phytoplankton growth on environmental factors

The variations of a phytoplankton community Phy in a river or estuarine system are the resultant of both biological and hydrodynamical processes i.e. growth and mortality on the one hand and dilution or dispersion linked to the hydrodynamic of the system on the other hand.
The evolution equation can be written:

$$\frac{d\ Phy}{dt} = \mathbf{X}\ Phy + (\mu - d)\ Phy \qquad (5)$$

where $\mathbf{X}$ is the hydrodynamical operator which will be discussed
below for the case of river and estuary respectively.

μ   is the growth rate

d   is a first order mortality rate including grazing by
herbivorous zooplankton, spontaneous lysis and
sedimentation.

The growth rate ,μ ,depends on both available light intensity and
temperature. Major nutrient concentrations are always saturating in Belgian
rivers and do not control phytolankton growth.

The classical Vollenweider relationship (Vollenweider, 1965) simplified
for non-photoinhibited communities was used to express the control by light
intensity on the growth rate. This in turn was estimated by the integration of
this equation on the variations of available light intensity with the time and
depth:

$$\mu = \mu max\ \frac{1}{24} \int_0^\lambda dt\ \frac{1}{h} \int_0^h dz\ \frac{(I/I_k)}{(1 + (I/I_k)^2)^{1/2}}$$

with $I = I_0\ e^{-\eta \cdot z}$

where I_0 is the incident photosynthetically available light
intensity (range 400-700 nm)

λ   is the photoperiod

h   is the depth

η   is the light attenuation coefficient

μmax   and I_K are physiological parameters which characterize
the light-photosynthesis curve.

Determinatiom of μmax , the growth rate at saturating ligh intensity,
yielded values of about 0.3 h^{-1} at $20^\circ C$ in rivers and about 0.5 h^{-1} in the
downstream part of the Scheldt estuary. The dependence on temperature can be
expressed by a Q_{10} of 2.

Temperature itself varies seasonally according to:

$$T = 15 - 9 \cos \omega\ (t - 30)$$

I_k the saturating light intensity level, was found to be physiologically
adjusted to the amount of light energy available for the water column (Fig.
7).

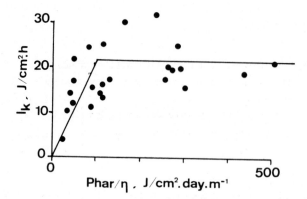

Fig. 7 : Relationship observed between I_k and the available light energy
for the water column, Phar/η

η   the light attenuation coefficient, depends both on the content of detrital
suspended matter and on the self-shading of phytoplankton itself:

$$\eta = \eta_{det} + \eta_{Phy}$$

The dependence of η_{det} on the suspended matter content of water (Fig.   8a)
was established   during periods without important phytoplankton biomass. From
this knowledge, self-shading by phytoplankton could be estimated for flowering
periods. Its control by phytoplanktonic biomass is given by figure 8b.

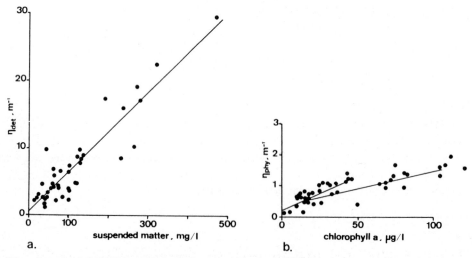

Fig. 8 :   a.   Relationship between η_{det} and suspended matter
          b.   Relationship between η_{Phy} and phytoplankton biomass

Incident photosynthetically available light intensities and photoperiods were calculated from integrated measurements of total incident solar radiation collected every 30 min at Uccle by the "Institut Royal Meteorologique de Belgique).

The mortality rate, d, is a composit parameter difficult to be experimentally determined.

Previous observations in other freshwater aquatic systems (Billen et al, 1983) give estimates around 0.005 h^{-1}.

Modelling phytoplankton growth in the river system

Smith et al (1984) have shown that in the case of a river the hydrodynamical operator $\mathbf{X}$ in equation (5) is reduced to a dilution factor depending on pluviosity, drainage conditions and watershed shape. It can be written

$$\mathbf{X} = \frac{1}{A} \frac{dQ}{dx} \tag{6}$$

where A is the wetted section of the river and
$\frac{dQ}{dx}$ is the longitudinal gradient of discharge,
owing to the inputs of water by the affluents,
supposed to be devoided of biomass.

It has been shown elsewhere (Billen et al, 1985; Debecker et al, in preparation) that in small rivers (with Horton order less than 0.6), the dilution factor is always higher than the net growth rate of phytoplankton. In these rivers, therefore only macrophytes are able to develop. The development of a significant phytoplankton population is restricted to rivers with order 6 or higher, i.e., in the case of the Scheldt estuary, in the Scheldt itself and the Dyle-Rupel.

For a river of given geometry, equation (5) allows calculation of phytoplankton biomass reached at a fixed observation station, after the travel of the water masses from the source, where phytoplankton is supposed to be present at a low but non zero concentration.

The dilution term is calculated according to relation (6). The wetted section is the product of mean width (W) and mean depth (H). The former is calculated from the following empirical relationship with watershed area, (SW), found for Belgian rivers.

$$\omega(m) = 0.8 \sqrt{SW} \tag{7}$$

The mean depth is calculated from Manning's formula:

$$H(m) = \left(\frac{Q \cdot n}{\omega \cdot \sqrt{S}} \right)^{3/5}$$

where   S   is the slope of the river

and n   the Manning coefficient adjusted to 0.07.

The term $\frac{dQ}{dx}$ in relation (6) is evaluated from the seasonal variations of discharge at the outlet of the river, assuming that discharge increases linearly with distance from the source.

Figure 9 presents the results of these calculations for the Scheldt river at St Amands (km 220). Comparison with observed data of phytoplankton biomass shows that the major trends of the seasonal variations of phytoplankton are satisfactorily simulated by this, yet oversimplified, model.

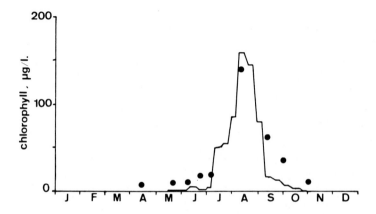

Fig. 9 : Observed and simulated seasonal variations of phytoplankton biomass in the Scheldt at St Amands (km 220).

Modelling phytoplankton growth in the estuary

Transport and dispersion processes in the estuary result both from a residual downwards circulation, linked to the freshwater discharge and from flood and ebb currents linked to the tides. It is possible to describe the resultant effect of these complex hydrodynamical processes by a simplified model taking into account advection by the residual discharge and dispersion by tide currents.

The operator $\mathbf{X}$ equation (5) is thus written:

$$\mathbf{X} = \frac{Q}{A(x)} \frac{\partial}{\partial t} + \frac{1}{A(x)} \frac{\partial}{\partial x} \left(A(x) D_s \frac{\partial}{\partial x} \right) \tag{7}$$

where   Q   is the freshwater dicharge

A(x)   is the wetted section which obeys the following relationship:

$$A(x) \ (m^2) = 151350.10^{-0.02 \ x} \ (km)$$

Ds   is an apparent mixing coefficient adjusted on basis of experimentally observed salinity profiles:

$$Ds(m^2/h) = 10^5 + 4500 \ . \ Q(m^3/sec) \qquad \text{for } x < 75 \text{ km}$$
$$= 10^5 + 22500 \ . \ Q(m^3/sec) \qquad \text{for } x > 80 \text{ km}$$

Phytoplankton in the estuary, consists in two distinct populations, adapted respectively to fresh and seawater. The dependence of their maximum growth rate on salinity (S) is experimentally shown to obey the following relationships:

$$\mu max(S) = \mu max \ . \ e^{-(\frac{S}{10})^2} \qquad \text{for the freshwater population}$$

$$\mu max(S) = \mu max \ . \ e^{-(\frac{S - 17}{5})^2} \qquad \text{for the sea water population}$$

Their mortality rate, d, was supposed to be increased by a factor 3 above 5 g Cl$^-$/l and below 10 g Cl$^-$/l respectively.

Equation (5) allows calculation of the dynamics of these two populations.

The seasonal variations of phytoplankton biomass at St Amands simulated by the river system model described above, are used as the upstream limit condition for freshwater population. The seawater population is supposed to be absent at this limit. Conversely, the seasonal variations of phytoplankton biomass observed in the Belgian coastal zone (Joiris et al, 1982) is used as the downstream limit condition, where freshwater phytoplankton is set at zero.

For setting the value of the extinction coefficient   , a complete model of the suspended matter in the estuary should be necessary. Instead, empirical relationships between suspended matter mes, deduced from the observation of a large number of longitudinal profiles, were used:

$$\begin{aligned}
&mes(mg/l) = 4000 \ . \ 1/Q(m^3/sec) \qquad &&\text{in the Antwerp zone}\\
&mes(mg/l) = 60 \qquad &&\text{for } Q \leqslant 50 \ m^3/sec\\
&\qquad = 0.6 \quad Q(m^3/sec) + 30 \quad &&\text{for } Q > 50 \ m^3/sec \ \text{in the mixing zone}\\
&mes(mg/l) = 60 \qquad &&\text{in the downstream zone}
\end{aligned}$$

442

Figure 10 shows some of longitudinal profiles of phytoplankton biomass simulated by this model for the year 1983. Available experimental observations are also plotted, and show that the simulation reproduces the major trends and levels of both the spatial and temporal phytoplankton variations within the estuary.

Figure 11 shows the simulation of the seasonal variations of biomass and primary production at Antwerp (km 80) and Doel (km 60).

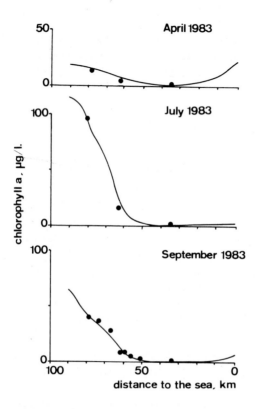

Fig. 10 : Simulated and observed longitudinal profiles of phytoplankton biomass in the Scheldt estuary in 1983.

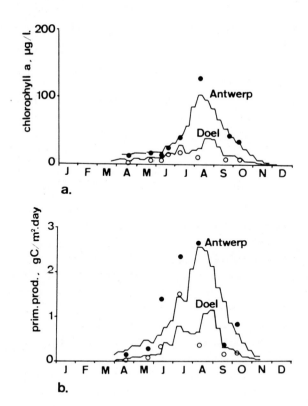

Fig. 11 : Observed and calculated seasonal variations of phytoplankton biomass
(a) and primary production (b) at Antwerp (•) and Doel (∘), in 1983.

NITROGEN TRANSFORMATIONS IN THE SCHELDT ESTUARY

Control of heterotrophic activity by organic matter.

Organic matter in natural water is mostly supplied under the form of
macromolecular biopolymers which have to be hydrolysed before they can be
taken up by microorganisms.This exoenzymatic hydrolysis is thought to
represent the limiting step in the microbial utilization of organic matter and
obeys a Michaelis-Menten kinetics (Somville and Billen, 1983; Billen, 1984;
Fontigny et al, in press).

It is customary to consider that organic matter in natural waters is
present under at least two classes with different biodegradabilities
(Westrich, 1983; Garber, 1984) i.e. with different succeptibilities of
exoenzymatic hydrolysis.

Because exoenzymes in natural waters are generally attached to the external envelopes of bacteria and are not subject to regulation (Vives-Rego et al,
1984; Fontigny et al, submitted), the concentration of exoenzymes is directly
proportional to the bacterial biomass.

The process of microbial organic matter degradation is thus represented by the following system of equations:

$$\frac{dH_1}{dt} = H_1 + P_1 - e_{1max} \frac{H_1}{H_1 + K_1H} \cdot B$$

$$\frac{dH_2}{dt} = H_2 + P_2 - e_{2max} \frac{H_2}{H_2 + K_2H} \cdot B$$

$$\frac{dB}{dt} = B + Y (e_{1max} \frac{H_1}{H_1 + K_1H} + e_{2max} \frac{H_2}{H_2 + K_2H}) B - dB$$

where H_1 and H_2 represent the concentration of each class of organic matter

P_1 and P_2 ,their rate of supply

e_{1max} and e_{2max} their maximum rate of exoenzymatic hydrolysis

K_1H and K_2H the half-saturation constant of their hydrolysis

B   is the bacterial biomass

Y   is the growth yield constant, i.e. the ratio of bacterial biomass formed to the total organic matter metabolized.

is generally close to 0.3 in nitrogen non limited environments (Lancelot and Billen, 1985; Servais and Billen, in press).

d   is the first order mortality constant of bacteria, which is close to 0.01 h (Servais et al, 1985).

The kinetic parameters of biopolymers hydrolysis were determined by adjustment on the results of a batch experiment in which the bacterial utilization of the organic matter from Scheldt water, untreated or concentrated by ultrafiltration, was followed for 30 days (Billen et al, 1983). The following values of the parameters at 20°C were obtained:

e_{1max} = 0.4 h^{-1}

K_1H = 1 mg C/l

e_{2max} = 0.06 h^{-1}

K_2H = 13 mg C/l

The dependence of e_{max} on temperature was supposed to be characterized by a Q_{10} of 2.

Organic matter is supplied by discharge of domestic, agricultural or industrial waste (Table 3) and by mortality of phytoplankton (calculated by the phytoplankton model).

It will be considered that organic matter from waste is made of 70% rapidly usable organic matter (H_1), and 30% more slowly degradable organic matter (H_2). For organic matter supplied by phytoplankton lysis, these proportions are 80 and 20% respectively.

TABLE 3
Domestic, agricultural and industrial discharge of organic matter and nitrogen in the Scheldt estuary (from data quoted in Billen et al, in press).

Domestic wastes

     total population: 1 130 000 inhab.
     existing purification capacity (1982): 980 500 inhab. equiv.

        organic C discharge: 34 TC/day
        total N Discharge: 9 TN/day

Industrial wastes

     Total discharge before purification: 190 TC/day
                                    45 TN/day

     % purification (1982)   80%

        organic C discharge: 30 TC/day
        total N discharge: 26 TN/day

Agricultural wastes
(cattle waste produced in excess of spreading capacity)
        organic C discharge: 84 TC/day
        total N discharge: 44 TN/day

Redox processes and N transformations

    For modelling the effect of heterotrophic activity on water quality (including ammonium and nitrate concentrations), the whole redox balance must be considered. The principles of this modelling has been described in details by Billen and Smitz, 1976.

    The redox state of the water is characterized by a function F defined as the weighted sum of all oxidants involved in microbial respiration:

$$F(meq/l) = 4[O_2] + 2 [MnO_2] + 8 [NO_3] + 1 [Fe(OH)_3] + 8 [SO_4]$$

    This variable is subject to hydrodynamical transport processes, to consumption by organotrophic activity (as calculated by the model described above), to production by photosynthesis (as calculated by the phytoplankton model, taking into account a photosynthetic quotient of 1.25 moles O_2 /mole C (Williams et al, 1979)) and to supply by reaeration (proportional to the oxygen saturation deficit, with a reaeration coefficient equal to 0.1 m h^{-1}/depth (m).):

$$\frac{\partial F}{\partial t} = F - \text{Org. Act (meq/l)} + \text{Prim Prod (meq/l)} + \text{Reaer.} \qquad (8)$$

The value of F is known at the upstream (see Fig. 12) and downstream (225.556 meq/l) limits of the model.

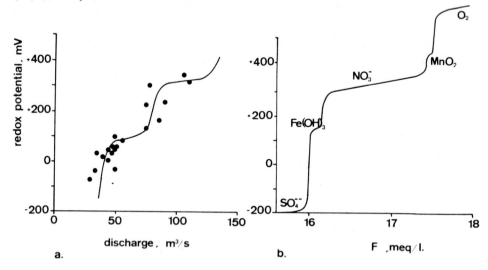

Fig. 12 : a. Empirical relationship observed at Rupelmonde between redox potential and discharge
b. Theoretical relationship between the F function (weighted sum of oxidants) and redox potential for the composition of water at Rupelmonde (Redox titration curve)

The link between F value and redox potential is represented by a filtration curve similar to that shown in figure 12b for the case of Rupelmonde. It depends on the total concentration of each redox couples involved, which must be calculated at each point of the longitudinal profile. For MnO_2/Mn^{++}, $Fe(OH)_3 / Fe^{++}$ and $SO_4^{=}/HS^{-}$. The total concentration is assumed to be conservative (possible losses by sedimentation of solid species are neglected). For mineral nitrogen, a complete balance is established at each point, taking into account:

- input of ammonium nitrogen by water discharges
- mineralization of organic nitrogen linked to organotrophic activity
- denitrification representing the total of organotrophic activity when nitrates are the best oxidants present in the water.

The relationship between F and Eh described in figure 12b results from an external thermodynamical equilibrium hypothesis. Owing to the slow development of nitrifying bacteria, this hypothesis can lead to overestimate the rate of nitrification. An empirical limit of the rate of nitrification is therefore introduced within the model. This limit is set to 1 mole/l.h on basis of experimental observations (Somville, 1978).

<u>Longitudinal ammonium and nitrate distribution and calculation of denitrification</u>

A few examples of observed longitudinal profiles of ammonium and nitrate concentration in the Scheldt estuary are shown in figure 13. Figure 14 shows the profiles simulated with the model described above for the year 1983.

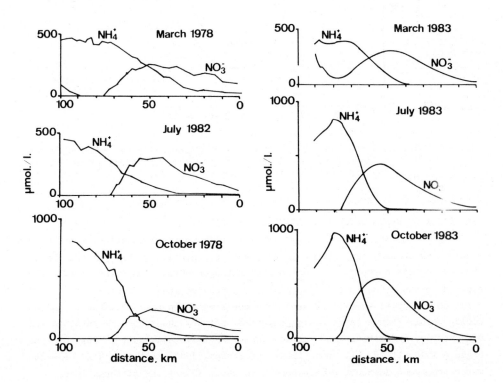

Fig. 13: Observed longitudinal profiles of ammonium and nitrate concentrations in the Scheldt estuary

Fig. 14: Calculated longitudinal profiles of ammonium and nitrate concentrations in the Scheldt estuary in 1983

Figure 15 presents the seasonal variations of nitrogen exportation into the sea and of nitrogen importation into the estuarine zone from the river system and waste discharges. The difference between both curves represents elimination by denitrification. It can be evaluated to 14 10^3 TN/yr in reasonable agreement with the estimation of 20 10^3 TN/yr based on the N budget of the estuary (see fig. 1).

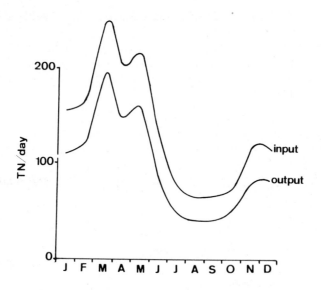

Fig. 15: Calculated seasonal variations of the nitrogen fluxes imported to the estuarine zone from the river system and the discharges in the Antwerp region and exported from the estuary to the sea.

SIMULATION OF WASTE WATER PURIFICATION SCENARIO

Owing to the dominant role of denitrification in the N budget of the Scheldt watershed, any change in the factors controlling this process can be thought to have a profound effect on the nitrogen output into the sea. The sub-models described in this paper allow to predict these effects.

As an exampple, we will present here the simulation of the situation which would result from a large scale program of primary and secondary purification of domestic, industrial and agricultural wastes, eliminating 90% of the organic load. It is well known (Bund and Straub, 1980) that this kind of waste water treatment does not retain more than about 30% of the nitrogen load.

Such a program would therefore result in a considerable improvement of the quality of the river system but only in a small reduction of nitrogen inputs. On the contrary, disappearance of anaerobic reaches of rivers and reduction of the organic matter content of the sediments could lead to a severe reduction of denitrification both in the river system and the estuary, thus resulting in increased inputs into the sea.

Using the model described in the first part of this paper, denitrification in river sediments in the purification scenario considered here, can be estimated to 2 - 3 10^3 TN/yr. Considering in addition that no more denitrification would occur in the water column of the river system, the input of nitrogen into the estuarine zone would be increased by at least a factor 2 (Compare fig. 1 and fig. 16).

The redox model of the Scheldt estuary, run for these new set of limit conditions, yields the results shown in figure 17. Denitrification in the estuary would becomes negligeable and the output into the sea would increased from 27 10^3 TN/yr in the present situation to 70 10^3 TN/yr (Fig. 16).

This simulation, however, does not take into account a possible change in the growth conditions of phytoplankton owing to reduction of suspended matter. In spite of this limitation, it can be concluded that generalized waste water purification without tertiary treatment should thus paradoxically result in an increase of nitrogen discharge into coastal water, increasing the risks of eutrophication.

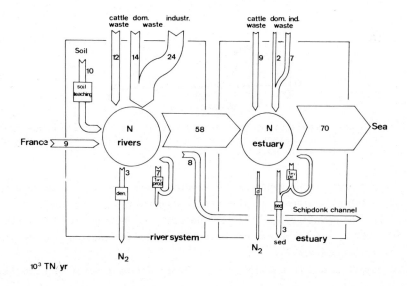

Fig. 16: Simulated nitrogen budget for the Scheldt river system and estuary with a 90% reduction of the organic load without tertiary treatment.

450

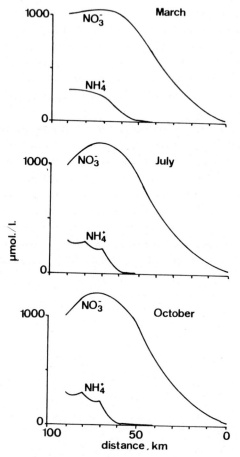

Fig. 17: Simulated longitudinal profiles of nitrate and ammonium concentrations
in the Scheldt estuary for the scenario of 90% reduction of the organic
load in the whole watershed without tertiary treatment.

ACKNOWLEDGMENT

   This paper present the synthesis of several years studies of the Scheldt
and its watershed. These studies were supported by the EEC (contract
ENV-522-B), the Ministry of Public Health (convention with the North Sea
Management Unit), the Ministry of Scientific Policy (Action de Recherche
Concertee en Oceanographie).

   G. Billen is Research Associate of the Fonds National Belge de la Recherche
Scientifique.

REFERENCES

Billen, G., Smitz, J., Somville, M and Wollast, R. 1976. Degradation de la matiere organique et processus d'oxydo-reduction dans l'estuaire de l'Escaut. In: Wollast, R. and Nihoul, J.C.J. (Editors). Modele Mathematique de la Mer du Nord. Rapport de Synthese. Ministere de la Politique Scientifique Bruxelles, Vol. 10, pp 102-152.

Billen, G. Dessery, S., Lancelot, C. Meybeck, M and Somville, M. 1983. Suivi et Modelisation de l'amelioration de la qualite de l'eau dans le bassin de storage de Mery s/Oise. 2d Rapport d'avancement. Compagnie Generale des Eaux, Paris, 80 pp.

Billen, G., Lancelot, C., Servais, P., Somville, M. and Vives Rego, J. 1983. Etablissement d'un modele predictif de la qualite de l'eau de l'estuaire de l'Escaut. Rapport final. Unite de Gestion du Modele Mathematique de la Mer du Nord. Ministere de la Sante Publique, Bruxelles, 67 pp.

Billen, G. 1984. Heterotrophic utilization and regeneration of nitrogen. In: Hobbie, J.E. and Williams, P.J. LeB, (Editors). Heterotrophic activity in the sea. Plenum Press. New York, pp 313-355.

Billen, G., Debecker, E., Lancelot, C., Mathot, S., Servais, P. and Stainier, E. 1985. Etude des processus de transfert, d'immobilisation et de transformation de l'azote dans son cheminement depuis les sols agricoles jusqu'a la mer. Rapport de synthese.Contrat ENV-522(B). Commission des Communautes Europeennes, Bruxelles, 101 pp.

Billen, G., Somville, M. Debecker, E. and Servais P. 1986. A nitrogen budget of the Scheldt hydrographical basin. Neth. J. Sea Res. in press.

Bond, R. G. and Straub, C. P. 1980. Handbook of environmental control. IV Wastewater treatment and disposal. CRC Press.

Dessery, S., Billen, G., Meybeck, M. and Cavelier, C. 1982. Evaluation et modelisation des echanges d'azote a travers l'interface eau-sediments dans le bassin de Mery-s/Oise. J. Franc. Hydrobiol., 13: 215-235.

Edwards, R. W. and Rolley, H. L. J. 1965. Oxygen consumption of river muds. J. Ecol. 53: 1-22.

Garber, J. M. 1984. Laboratory study of nitrogen and phosphorus remineralisation during the decomposition of coastal plankton and seston. Est. Coast. Shelf Sci., 18: 685-702.

Hansen, J. I. K., Henriksen, K. and Blackburn, T. H. 1981. Seasonal distribution of nitrifying bacteria and rates of nitrification in coastal marine sediments. Microb. Ecol. 7: 297-304.

Horton, R. E. 1945. Erosional development of streams and their drainage basins: hydrophysical approach to quantitative morphology. Geol. Soc. Am. Bull. 56: 275-370.

Joiris, C., Billen, G., Lancelot, C., Daro, M.H., Mommaerts, J.P., Bertels, A., Bossicart, M., Nijs, J. 1982. A budget of carbon cycling in the Belgian coastal zone: relative roles of zooplankton, bacterioplankton and benthos in te utilization of primary production.Neth. J. Sea Res. 16: 260-275.

Lancelot, C. and Billen, G. 1985. Carbon Nitrogen relationships in nutrient metabolism of coastal marine ecosystems. Adv. Aqu. Microbiol., 3, in press.

Servais, P., Billen, G. and Vives Rego, J. 1985. Rate of Bacterial mortality in aquatic environments. Appl. Env. Microbiol., 49: 1448-1454.

Servais, P., Billen, G. and Hascoet, M.C. 1986. Determination of the biodegradable fraction of dissolved organic matter in waters. Water Res., in pess.

Smitz, J., Descy, J.P., Everbecq, E., Servais, P. and Billen, G. 1985. Etude ecologique de la Haute Meuse et modelisation du fonctionnement de l'ecosysteme aquatique. Rapport final. Ministere de la Region Wallonne pour l'Eau, l'Environnement et la Vie Rurale. Namur, 250 pp.

Somville, M. 1978. A method for the measurement of nitrification rates in water. Water Res. 12: 843-848.

Vanderborght, J.P. and Billen, G. 1975. Vertical distribution of nitrate in interstitial water of marine sediments with nitrification and denitrification. Limnol. Oceanogr. 20: 953-961.

Vives Rego, J., Billen, G., Fontigny, A. and Somville, M. 1985. Free and attached proteolytic activity in water environments. Mar. Ecol. Progr. Ser. 21: 245-249.

Vollenweider, R. A. 1965. Calculation models of photosynthesis depth curves an some implications regarding day rate estimates in primary production measurements. In: Goldman, C. R. (Editor). Primary production in aquatic environments. University of California Press.

Westrich,J. T. 1983. The consequences and controls of bacterial sulphate reduction in marine sediments. Ph. D. Thesis., Yale University, USA.

Williams, P.J. LeB, Raine, R. C. and Bryan, J.R.. 1979. Agreement between the ^{14}C and oxygen methods of measuring phytoplankton production: reassessment of the photosynthetic quotient. Oceanol. Acta, 2: 411-416.

MOBILIZATION OF MAJOR AND TRACE ELEMENTS AT THE WATER-SEDIMENT INTERFACE IN THE BELGIAN COASTAL AREA AND THE SCHELDT ESTUARY

W. BAEYENS[1], G. GILLAIN[2], M. HOENIG[3] AND F. DEHAIRS[1]

[1]Analytical Chemistry, Vrije Universiteit Brussel, 1050 Brussel (Belgium)
[2]Laboratoire d'Océanologie, Université de Liège, 4000 Liège (Belgium)
[3]Institut de Recherches Chimiques, Ministère de l'Agriculture, 1980 Tervuren (Belgium)

ABSTRACT

In a shallow coastal area such as the Belgian coastal waters, a quite large amount of suspended matter reaches the bottom where a part of it is mobilized and released into the overlying water. Heterotrophic bacterial degradation of POM (Particulate Organic Matter), which is essentially limited to the first centimeters of the sediments, is the process which is responsible for that mobilization. The presence or absence of dissolved iron and manganese depends on the redox state, anoxic or oxic, respectively, of the sediment. The behaviour of strontium in the first centimeters seems to be almost independent of POM degradation as well as redox conditions.

The percentages of excess·Cu, Zn and Cd remobilized during the transition from suspended matter to the first centimeters of the bottom sediments are very high (100, 80 and 70% respectively) and independent of the type of sediment, suggesting POM is their carrier. Pb is more reluctant to mobilization and shows a higher mobilization percentage in anoxic sediments, suggesting that dissolution of Fe and Mn phases also releases a significant part of the solid Pb content.

Epibenthic fluxes of Cd, Cu, Pb and Zn have been estimated in two different ways : (1) by considering a similar behaviour between the metal and POM for which remobilisation rates are known and (2) by using vertical pore water data profiles. Both estimates agree fairly well for Cu, Zn and Cd, but they differ significantly for Pb.

1. INTRODUCTION

The Southern Bight of the North Sea is a productive ecosystem with a daily primary production from 10 to 280 mg N/m^2.day (an annual primary production of 25 g N/m^2.Y) in the coastal area, and from 10 to 154 mg N/m^2.day (annual 20.5 g N/m^2.Y) in the offshore area (Baeyens et al., 1983 ; Mommaerts et al., 1984 ; Baeyens et al., 1984). Near the Belgian coast only a small fraction of this organic matter is grazed directly by zooplankton : most of it is mineralized by bacteria. Since water depth in this area is only 10-30m, a significant part of the organic matter sedimentates in certain deposition areas and is mineralized in the sea floor.

Many mud areas, organically enriched, occur in the northeast of the Belgian

coastal area. These anoxic mud deposits, often polluted by heavy metals have ne-
matods as the only surviving metazoans (Heip et al., 1984). Coarse sands and
even gravels with a very low organic matter content and a very diverse infauna
occur more offshore.

The organic matter stock and mineralization rate in the sediments are essential elements to understand the fluxes of major and trace elements at the water/
sediment interface. In sandy sediments, at least in the upper layers only oxygen
is used as the terminal electron acceptor in the oxydation process. Elements
bound to POM (Particulate Organic Matter) may be transferred to the dissolved
phase and the breakdown of DOM (Dissolved Organic Matter) may change the specia-
tion of the dissolved species, but the geochemical behaviour is less complex
than in muddy sediments. In muds, a whole series of electron acceptors will suc-
ceed each other when the redox potential goes down: oxygen, manganese, nitrate,
iron and sulphate. Some of the reactions involved here will change the pH as
well. Many compounds will disappear, other will be produced when the Eh-pH conditions are modified: as an example, the reduction of sulphate to sulphide at
very low redox potentials will form hardly soluble Hg, Cd,...sulphides.

In this paper some effects of bacterial metabolism on the distribution of major and trace elements in different types of coastal sediments are discussed. In
addition epibenthic fluxes of trace elements have been estimated in the coastal
area and the Scheldt estuary.

2. SAMPLING AND ANALYSIS

The coring stations in the coastal zone are shown on Figure 1A. Bottom sediment properties at stations 33, 34 are quite different from these at stations
1149, 1150 and 1151 as is indicated on Figures 1B and 1C. At the latter stations
the fine sand fraction (< 74 μ m) ranges from 50 to 100% and the organic carbon
content from 8-16%, while at the stations 33 and 34 both parameters are much lower.

The coring stations in the Scheldt estuary (2,7,12,15,18,25) are shown in Figure 1D. Stations 2 and 7 (downstream) are characterized by a dominant sand
fraction. Stations 12 to 18 are situated in the sedimentation zone : the sediments are rich in organic mattter. At station 25 (the most upstream) the bottom
has a clay structure (Boom clay).

The sediment cores were sampled by divers inserting plexiglass tubes into the
sediments. After extrusion of the cores, they were sectioned into 2 cm intervals. Pore waters were expressed under nitrogen pressure using a teflon made
squeezing device of the Rheeburg type (Rheeburg, 1967). Suspended matter was
collected by filtration on Millipore 0.45 μ m filters and by centrifugation. Se-
dimenting particles were collected with sediment traps, similar in design to the
type described by Zeitzschel et al., (1978), but for single sample collection.

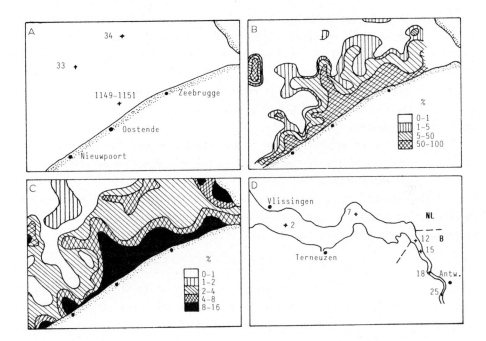

Figures : 1A-coring stations in the coastal zone ; 1B-distribution of fine sand fractions (< 74 μ m) ; 1C-distribution of organic carbon content of superficial sediment as measured by the weight loss at 550°C (Wollast, 1976) ; 1D-coring stations in the Scheldt estuary.

In solution Fe, Mn, Sr and P were analyzed by ICP (IL 2000), NO_3^- with an auto-analyzer (Technicon), Zn, Cd, Pb and Cu with DPASV (Bruker E 310). In the solid phase Fe, Mn, Sr, Al were analyzed by ICP (after fusion with $LiBO_2$), Cd, Pb, Cu and Zn by FAAS or GFAAS (after $HF/HCl/HNO_3$ mineralization), and N,C-inorganic and C-organic with a C, H, N-analyzer (Perkin-Elmer).

3. PARTICULATE ORGANIC MATTER (POM) DEGRADATION
Watercolumn

Biodegradation of high molecular weight organic material is a two-step process with rate limiting exoenzymatic hydrolysis intervening before bacterial uptake of amino acids. Billen et al., (1980) have demonstrated that the utilization rate of amino acids by heterotrophic bacteria during the spring bloom varied between 0.1 and 4% h^{-1} in the Belgian coastal area. Given the observed concentration of prominent amino acids in sea water (0.23-0.92 mmol/l), and their average N-content (17.6 mg N/mmol), the depth integrated concentration of nitrogen in the amino acids ranges from 60 to 240 mg N/m^2 (the average water depth is 15 m). The pelagic remineralization flux of nitrogen can thus vary between 1.4 and 230 mg N/m^2.day. However a high remineralization flux is observed only at the end of the spring bloom (3 g N/m^2.month) while it is almost insignificant at other periods (Baeyens et al., 1984).

The degradation of organic matter in the coastal waters has no influence on the redox potential of the system. Oxygen is the only electron acceptor involved and is always abundantly present. This is not always the case in the upstream part of the Scheldt estuary (stations 12 to 25). Oxygen is often exhausted and other electron acceptors such as mangenese and iron are then used. The reduction of these elements will also lead to their dissolution.

Sediments

A good estimate of protein degradation in the sediments is the ammonification rate. In the sediments of the Southern Bight of the North Sea, Billen (1976) has studied the ammonification rate as a function of organic matter content, depth, oxydants, etc... His main conclusions can be summarized as follows :

(1) The ammonification rate observed in the first 5 to 8 cm of North Sea sediments ranges from 0.2 to 3.10^{-6} μ moles N/cm^3.s. These rates are not substantially different for muddy and sandy sediments. Seasonal variations, if any, cannot be recognized in the broad spectrum of values created by local, spatial variations. The remineralization flux in the upper sediment layer, integrated over a month, thus ranges from 0.36 to 8.7 g N/m^2.month. These values are of the same magnitude as the maximal pelagic remineralization flux at the end of the spring bloom, but they are much higher than the pelagic fluxes in the other periods.

(2) In organic rich sediments, exhaustion of oxydants may severely reduce the utilization of hydrolysis products from organic macromolecules. These smaller metabolites (proteins and amino sugars) accumulate and limit their own production by inhibition of the exoenzymes. In anaerobic conditions (sulphate exhausted), ammonification is slower than 0.05×10^{-6} μ moles N/cm^3.s.

(3) In organic medium and poor sediments, exhaustion of organic matter will stop the production of dissolved nitrogen. However, in the deeper layers of most of these sediments, organic matter is not exhausted. It is thus surprising to see that the ammonification rate is there much lower than in the superficial layer. The most probable explanation is that the freshly deposited organic matter in the upper layer is much easily hydrolyzed by exoenzymes than the aged organic matter in the deeper layers, which is (1°) more depleted in easily

degradable nitrogen compounds such as proteins and (2°) more protected to exoenzymatic attack due to complexation with minerals.

(4) The turnover time of ammonia is short with respect to the period of variation of the environmental conditions. Therefore it is reasonable to consider the
ammonia profiles as stationary. The adjustment of the stationary equation describing the ammonia behaviour in the sediments (Billen, 1976) on the experimentally obtained vertical ammonia profiles yield a dispersion coefficient ranging
from 0.5 to 2.0×10^{-4} cm^2/s. These values are substantially higher than bioturbation and molecular diffusion coefficients, and suggest a turbulent diffusive
control as is likely ot occur in the shallow belgian coastal environment,
strongly affected by tides.

4. METALS MOBILIZATION

Major elements

Degradation of POM in the watercolumn does not influence the behaviour of
iron and manganese as long as aerobic conditions prevail. In the upstream part
of the Scheldt estuary these conditions are no longer fulfilled; both elements
serve as electron acceptors in the degradation process of POM and appear in the
dissolved phase. In the downstream area where oxygen reappears they precipitate
and are transported further seawards with the remainder of the suspended matter.

Considering the behaviour of Fe and Mn in the sediments, two sediment types
can be distinguished. In the top layers of the sandy cores (stations 33 and 34),
the ratio of Fe and Mn to Al is high relative to the soil reference value, indicating an excess situation. This ratio decreases with depth, ending in a no excess situation. No dissolved iron or manganese is observed in the upper layers.
In the deeper layers where oxygen is exhausted, a low dissolved concentration
may be found. In the muddy cores (M1149 to 1151) located more inshore, no excess
Fe or Mn is observed in the solid phase (Figure 2). They both appear rapidly in
the dissolved phase as oxygen is almost immediately exhausted. In fact, the dissolved and solid phase profiles appear to be complementary. Since these elements
are easily reoxydized when oxygen is present, their fluxes out of the sediment
may be only a small fraction of that which is produced. Sundby and Silverberg
(1984) found in their study that most of the dissolved manganese produced (71-
87%) does not escape into the watercolumn but precipitate within the sediment.

Strontium which is not directly involved in the oxydation process of POM
shows a near conservative behaviour; only some small dissolution and precipita-
tion occurs which seems to be correlated with the behaviour of the other elements.

458

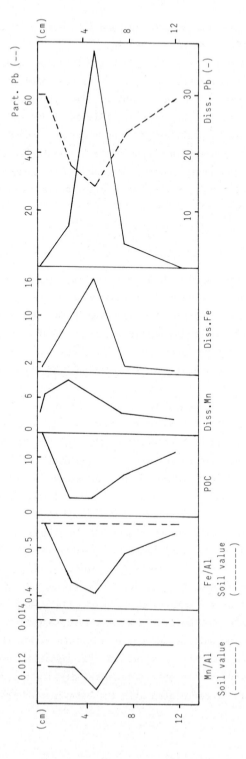

Figure 2 : Vertical profiles of Mn/Al, Fe/Al, POC (mg/g) and Pb (μ g/g) in the solid phase ; Mn (mg/l), Fe (mg/l), Pb (μ g/l) in the dissolved phase (Core M1149).

Trace elements

In oxic environments microbial degradation of POM is a major production process of dissolved trace metals. In anoxic environments dissolution of iron and
manganese oxydes or hydroxydes may also release adsorbed and incorporated trace
metals. Since the degradation rate and the residence time of POM in the sediments are much higher than in the watercolumn, much lower concentrations of POM
and associated metals (solid phase) can be expected there.

In Figure 3 we compare excess metal composition between suspended matter (total suspended matter and sediment trap material) and sediments. Particle size
effects, such as dilution by quartz grains, are corrected for by considering the
excess trace metal amounts relative to Al. This Al is assumed to be essentially
associated with the small-sized aluminosilicate fraction and to behave in a conservative way. From Figure 3 we see that for Cu, Zn and Cd more than 70% of the
relative contents in suspended matter have disappeared in the first centimeter
of the oxic sediments. For Pb this is only about 30%.

In anoxic sediments changes in redox potential or pH may also remobilize metals associated with redox sensitive particulate phases. While Cu, Zn and Cd are
mobilized to similar degrees, both in oxic and anoxic sediments, Pb is not. In
anoxic sediments Pb remobilization amounts to about 50% (of the amount in suspended matter) as compared to only 30% in oxic sediments. In the deeper layers
mobilization relative to suspended matter composition, increases only of a
further 7 to 11%, in agreement with the discussed lower rate of ammonification
in these layers. The complementarity between the dissolved and solid phase profiles of Pb in the muddy core M1149 (Figure 2) illustrates the effect of remobilization.

We also calculated changes in excess metal composition relative to aluminium
from suspended matter in a sediment trap to sediments. The mobilization percentages are similar to these of total suspended matter for Cu, slightly lower for
Zn and Pb and slightly higher for Cd.

5. EPIBENTHIC FLUXES OF METALS
Coastal area

Epibenthic fluxes have been estimated in two different ways.
Since the cycling of POM and the ratio metal to POM in the solid phase are
known, metals fluxes at the water/sediment interface can be deduced under the
condition that POM and metals behave similarly. This assumption is realistic: in
the previous paragraph we brought evidence that POM degradation in the first
centimeters of the sediments will liberate most of the excess metal in the solid
phase in sandy as well as muddy sediments. The sedimentary outflux of POM has
been based on the ammonification rate in the sediments and equal about 22 g N/m^2.Y or

Figure 3 : Change in excess metal composition relative to aluminium from suspended matter to sediments : the Belgian coastal zone.

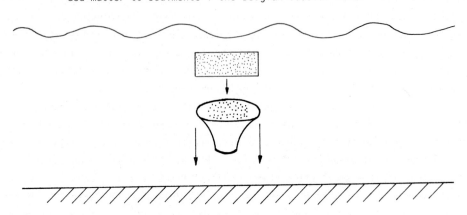

	OXIC SEDIMENTS			ANOXIC SEDIMENTS	
	At - 1 cm			At - 1 cm	
Elem.	Susp. Matter	Sed. Trap	Elem.	Susp. Matter	Sed. Trap
	(% remobilized)			(% remobilized)	
Zn	82	70	Zn	79	65
Cu	100	100	Cu	100	100
Cd	71	77	Cd	71	77
Pb	31	23	Pb	54	48

$$\frac{\text{OXIC}}{\text{ANOXIC}}$$

	At - 15 cm			At - 5 cm	
Zn	89	81	Zn	88	80
Cu	100	100	Cu	100	100
Pb	40	36	Pb	65	61

***

160 g C/m^2.Y (POM) . The ratio metal to POM in suspended matter in the coastal area is given in Table 1. The remobilization of POM in the sediments is almost 100 %. For the trace metals it varies from 100 (Cu) to 43 (Pb) as reported in Figure 3. The metal outfluxes shown in Table 1 are thus based on the POM outflux and the ratio of metal to POM in the outflux (this ratio equals the ratio in the suspended matter multiplied by the remobilization percentage in the sediments).

TABLE 1

Ratio of metal to POM in suspended matter (10^3R)

Hg	Zn	Cd	Pb	Cu
0.080	2.95	0.055	1.21	0.67

Remobilization percentage of metals
from the solid phase in the sediments

Hg	Zn	Cd	Pb	Cu
96	80	71	43	100

Epibenthic metals fluxes (tons/km^2.Y)

Hg	Zn	Cd	Pb	Cu
0.012	0.38	0.0062	0.082	0.11

The second approach is based on vertical pore water data profiles. As an example, such vertical profiles observed in a muddy core (51°18'20"N-2°58'40"E) are shown in Figure 4. From these profiles concentration gradients at the wa-ter/sediment interface can be inferred. These gradients are given in Table 2. These concentrations refer to the liquid phase and should be corrected for the porosity of the sediment. At the top of the sediment this porosity is 0.5 in the sandy cores (33 and 34) and 0.8 in the muddy cores (1149 to 1151). The diffusion coefficient (D_s), see part 3, has been estimated at 10^{-4} cm^2/s. The resulting outfluxes can then be calculated according to :

$$J = p \; D_s \; dC/dz \quad (1)$$

J = outflux (ng/cm^2.s) ; p = porosity ; D_s = diffusion coefficient (cm^2/s) ; C = concentration (ng/cm^3) ; z = depth (cm).

The Zn and Cd outfluxes based on POM remobilization seem to compare to those obtained from pore water data profiles. For Cu the agreement is within a factor of 3. However, for Pb there exists an order of magnitude difference, with the va-lue based on pore water profiles on the low side. The results obtained with the two different approaches suggest that Cd, Zn and Cu may have a sedimentation and mobilization behaviour which is closely related to that of organic matter. For

462

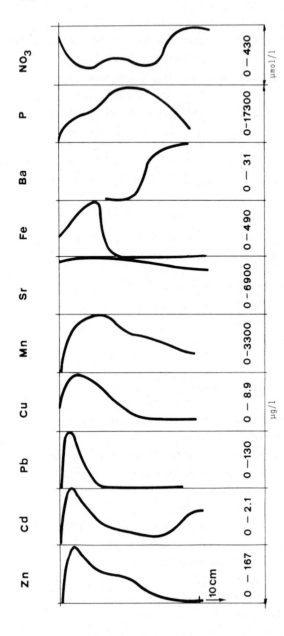

Figure 4 : Interstitial concentration profiles in the muddy core M1151

Pb the remobilization flux as based on Pb/POM ratios is 10 times larger than the calculated diffusive outflux. This indicates that Pb if bound to POM is more reluctant to mobilization or rather that Pb is associated mainly with other carriers than POM. Particulate Mn and Fe phases are likely involved as evidenced above by the increased remobilization of Pb in anoxic sediments. This is confirmed also by our recent data for North Sea suspended matter showing significant positive relationships between Pb and Fe, Mn and Al-silicate phases.

TABLE 2

Concentration gradients on the water/sediment interface (ng/cm^4)				
Core	Zn	Cd	Pb	Cu
33	9.6	0.68	0.22	5.7
1149	5.0	0.27	0.20	0.58
1150	21.0	0.13	0.43	0.52
1151	28.7	0.35	n.m.	1.37
Sedimentary metal outfluxes $(10^{-5} ng/cm^2.s)$				
Core	Zn	Cd	Pb	Cu
33	48	3.4	1.1	28
1149	40	2.1	1.6	4.7
1150	168	1.0	3.5	4.2
1151	230	2.8	-	11
Mean	122	2.3	2.1	12
Sedimentary metal outfluxes $(Tons/km^2.Y)$				
	0.44	0.0084	0.0077	0.044

n.m. = not measured

Scheldt estuary

Sedimentation fluxes in the estuary are very difficult to estimate due to the high tidal energy dissipated in the system; sediments are periodically deposited and eroded in most areas, particularly in the zone of the turbidity maximum (between 50 and 80 km from the mouth) (Baeyens et al., 1981). Upstream of the turbidity maximum the river is a narrow channel incised in the Boom clay and currents are very high so that accumulation of suspended matter at the bottom is almost prevented. The flux of suspended matter and hence particulate metals from the upstream zone into the area of turbidity maximum or sedimentation can thus be estimated in a fairly correct manner. The same can be done at the downstream boundary of the sedimentation zone where we find the particles which escaped from the trapping zone, mostly the finer material. The sedimentation rate of particulate metals in the sedimentation zone can thus be estimated based on their fluxes at the up- and downstream boundaries (Table 3). In addition the re- mobilization percentages of metals from the solid phase can be deduced from their concentrations in suspended and bottom sediments (Table 3). These frac- tions are not corrected for the soil component and perhaps therefore lower remo- bilization percentages are obtained in the estuary than in the coastal zone, ex- cept for Pb which shows the remobilization percentage for muddy sediments in the sea. The redoxpotential in the Scheldt estuarine sediments is also very low, so that a significant lead release may be due to the dissolution of iron and manga- nese oxides and hydroxides. Using the sedimentation rate and the remobilization percentage an epibenthic flux can be calculated. These outfluxes are in the Scheldt estuary from 5 to 50 times higher than in the coastal zone (Table 3).

TABLE 3

Sedimentation rates in the sedimentation zone of the Scheldt estuary (Tons/km^2.Y)				
Zn	Cd	Pb	Cu	Hg
2.5	0.092	0.74	0.96	0.0166
Remobilization percentage from the solid phase in the sediments				
Zn	Cd	Pb	Cu	Hg
63	65	57	77	-
Epibenthic metals fluxes (Tons/km^2.Y)				
Zn	Cd	Pb	Cu	Hg
1.6	0.06	0.42	0.74	-

Based on the pore water data profiles, we can calculate the sedimentary out- fluxes of metals in the same way as described in the previous section. For lack of a better estimate of the turbulent diffusion coefficient in the area we adop- ted the coefficient estimated for the coastal area ($D_s = 10^{-4}$cm^2/s). This value is, however, probably too low. Applying equation (1) to the Scheldt pore water data yield the epibenthic fluxes given in Table 4. These fluxes are all compara- ble within an order of magnitude with those derived from the sedimentation ra- tes, except for Pb which is thirteen times lower.

TABLE 4

Concentration gradients at the water/sediment interface in the Scheldt estuary (ng/cm^4)				
Core	Zn	Cd	Pb	Cu
12	56.9	0.30	2.3	10.0
15	17.7	0.13	-0.24	3.8
15-1	n.m.	0.02	0.20	1.8
18	n.m.	1.2	2.0	2.9
18-1	2.7	0.12	0.49	7.5
18-2	8.7	0.03	0.40	6.6
25	35.1	n.m.	2.6	3.4
Mean	24.2	0.30	1.11	5.1
Epibenthic metals fluxes				
	Zn	Cd	Pb	Cu
$10^{-5}ng/cm^2 s$	194	2.4	8.88	40.8
$Tons/km^2.Y$	0.71	0.0088	0.032	0.15

n.m. = not measured

CONCLUSIONS

The heterotrophic bacterial degradation of POM takes essentially place in the first centimeters of the bottom sediments, zone wherein the major fraction of the excess trace metal amount relative to Al is also liberated from the solid phase. The similarity between the Cu, Zn and Cd sedimentary outfluxes based on POM remobilization and those obtained from pore water data profiles on the one hand, and the fact that different redox potential conditions seem not to have a major effect on the mobilization of the three metals on the other hand, suggest that these metals are associated with POM. For Pb, other carriers are also involved : most likely Mn and Fe phases as is evidenced by the increased remobilisation of Pb in anoxic sediments.

ACKNOWLEDGEMENTS

This research was mainly sponsored by the Ministry of Scientific Policy as a part of Geconcerteerde Acties Oceanologie. Additional support was provided by the EEC (contract ENV-766-B).

REFERENCES

Baeyens W., Adam Y., Mommaerts J.P. and Pichot G., 1981. Numerical simulations of salinity, turbidity and sediment accummulation in the Scheldt estuary. In: J.C.J. Nihoul (Editor), Ecohydrodynamics, Elsevier, Amsterdam, 319-332.

Baeyens W., Goeyens L., Dehairs F. and Decadt G., 1983. Nitrogen cycles in a coastal and an open sea zone off the Belgian coast. Trans. Am. Geophys. Un., EOS, 64, 52, 1024 (Abstract).

Baeyens W., Mommaerts J.P., Goeyens L., Dehairs F., Dedeurwaerder H. and Decadt G., 1984. Dynamic patterns of dissolved nitrogen in the Southern Bight of the North Sea. Estuarine, Coastal and Shelf Science, 18, 499-510.

Billen G., 1976. Etude écologique des transformations de l'azote dans les sédiments marins. Ph. D. Thesis, Université Libre de Bruxellles, Brussels (Belgium), 266 pp.

Billen G., Joiris C., Wijnant J. and Gillain G., 1980. Concentration and microbiological utilization of small organic molecules in the Scheldt estuary, the Belgian coastal zone of the North Sea and the English Channel. Estuarine, Coastal and Shelf Science, 2, 279-294.

Heip C., Herman R. and Vinkx M., 1984. Variability and productivity of meiobenthos in the Southern Bight of the North Sea. Rapp. Pv.; Réun. Cons. Int. Explor. Mer, 183, 51-56.

Mommaerts J.P., Pichot G., Ozer J., Adam Y. and Baeyens W., 1984. Nitrogen cycling and budget in Belgian coastal waters : North Sea areas with and without river inputs. Rapp. P.-v. Réun. Cons. Int. Explor. Mer, 183, 57-69.

Rheeburg W.S., 1967. An improved interstitial water sampler. Limnol. Oceanogr., 12, 163-170.

Sundby B. and Silverberg N., 1985. Manganese fluxes in the benthic boundary layer. Limnol. Oceanogr., 30 (2), 372-381.

Wollast R., 1976. Propriétés physico-chimiques des sédiments et des suspensions de la Mer du Nord. In: J.C.J. Nihoul et F. Gullentops (Editors), Sédimentologie, Vol. 4 of the Final Report of the Project Sea. Ministry of Scientific Policy, Brussels.

Zeitschel B., Diekmann P. and Uhlmann L., 1978. A new multi sample sediment trap Mar. Biol., 45, 285-288.

***

SEASONAL NUTRIENT SUPPLY TO COASTAL WATERS

L. RYDBERG and J. SUNDBERG

Department of Oceanography, University of Gothenburg, Box 4038, S-400 40 Gothenburg (Sweden)

ABSTRACT

Monthly nutrient and salinity observations have been undertaken during the years 1982 - 1985 in the southeastern part of the Kattegat on the Swedish west coast. Two minor embayments in that area, the Laholm bay and the Skälderviken receive large amounts of inorganic nitrogen from a couple of small rivers. Oxygen deficit occurs as a frequent feature in the deep water outside the bays. The observations have been used to calculate bimonthly mean concentrations of total nitrogen, total phophorus, inorganic nitrogen and phosphate within three different water masses; one defined as local surface water within the Laholm bay, one as Kattegat surface water, outside the bay and the third as Kattegat deep water.

Using the observed landbased supply of nutrients, the deep water supply, calculated from entrainment theory and the measured nutrient gradients, we have determined the exchange of water and nutrients between the Laholm bay surface water and the Kattegat surface water. We have also calculated the "net assimilation" of inorganic nitrogen and phosphate within the Laholm bay. In an earlier report, we did similar calculations on an annual mean basis. Here, we have done a separation between a winter period from November to February when the assimilation is low, and the rest of the year, when most of the primary production occurs. From March to October, approximately 3/4 of the inorganic nitrogen supply to the local water is of land based origin while the phosphate supply is dominated by entrainment from the deep water. 2/3 of the total inorganic nitrogen supply is assimilated by the primary production within the local water. The rest may be found as a loss from the local water to the Kattegat, which occurs mainly during the winter, but also during the spring, when the supply of phosphate is frequently too small for the primary production. The supply of phosphate to the bay seems to be low for the rest of the season as well, and the IN/IP ratio based on the external nutrient supply is well above the Redfield ratio. Still, however, the surface water concentrations of inorganic nutrients points towards nitrogen as the limiting nutrient for the main part of the productive season, indicating a more effective internal regeneration of phophorus and/or denitrification at the bottom.

468

Fig 1. Map over the Kattegat. Stations are marked with black dots.

INTRODUCTION

Intensive algal blooms and the destruction of bottom fishery due
to oxygen deficit have been seen occasionally in the southeastern
Kattegat (fig 1) over the last fifteen years (Rosenberg, 1985). Severe
oxygen deficit and bottom death was observed both in 1980 and
1981. Since then there has been a gradual recovery of the deep
bottoms and the fishery. Oxygen measurements have also indicated
somewhat better conditions (Fig. 2).

The waters in this area are characterized by a strong vertical
salinity stratification, which is caused by the outflow of low
salinity water from the Baltic Sea and an inflow of high salinity
water of oceanic origin at deeper levels. The surface water has a
salinity between 12 - 25 , and a thickness of approximately 15 m.
The deep water salinity is normally between 32 - 34, and has its
origin in the Skagerrak.

The halocline depth is strongly influenced by the outflow from
the Baltic, but it varies also due to local winds which redistribute the surface waters and induce mixing, mainly as upward
entrainment. In this part of the Kattegat, the depth is generally
less than 30 m. The varying halocline depth thus implies that the
deep water volume sometimes becomes small or even disappears, at
least in the Laholm bay where the depth does not exceed 20 m.

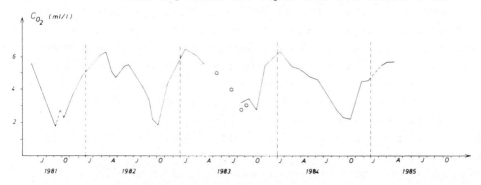

Fig 2. Oxygen concentrations within the deep water of the southeastern Kattegat from 1981 to 1985. The curve shows a mean value
for the stations K5, K7 and K9, 1 m above the bottom (mean depth,
27 m, mean salinity, 33.1). The characteristic minimum value for
each station individually has been 1 ml/l, while the inner
stations have had even lower concentrations when deep water has
existed at those stations.

The phytoplankton production in the Kattegat begins with a springbloom, usually in early March (1-2 gC/m^2 day), followed by a rather stable summer production of the order of 0.4 gC/m^2 day. The production is normally low after the end of October. The annual primary production seems to be just above 100 gC/m^2 year (Aertebjerg et al., 1981). The production in the outer part of the Laholm bay has been measured since 1981 by Edler (pers.comm) who has found a value of the order 150 gC/m^2 year, thus a little higher than in the open Kattegat.

The Department of Oceanography at the University of Gothenburg started an intensive field program in the southeastern Kattegat in February 1982, with emphasis on the situation in the Laholm bay. The bay recieves a large nutrient load, especially inorganic nitrogen through the rivers Lagan and Nissan (Fig. 3c). The purpose was to determine to what extent the local supply of nutrients influenced the severe oxygen conditions within the area, and if so, whether a decrease in the nitrogen supply from land (which origins from fertilizers used in the agricultural districts) could contribute to better oxygen conditions.

Since the beginning, we have carried out approximately one cruise per month to the area. The field program, which includes hydrography and nutrient chemistry at 15 stations will be continued until the end of 1985. Parallell to our programme, the local government is running a monthly follow-up of the landbased input of nutrients to the area (Fleischer et al, 1985), and biologists from the universities of Lund and Gothenburg are studying, among other things, the development of the phytoplankton and zooplankton production and the various species within the area. In 1984 we made a first rough estimate of the relative importance of the local landbased nitrogen supply, by doing a comparison with the supply of deep water nitrogen. That approach was based on yearly mean values, determined from 22 surveys (Rydberg and Sundberg, 1985). Today, we have carried out another year of monthly observations and will now make a next step forward by using bimonthly mean nutrient concentrations to determine the seasonal (winter/summer) nutrient fluxes and the phytoplankton nutrient assimilation within the bay.

In the future, we shall put large efforts into studying the

relation between nutrient supply and oxygen consumption. This implies that we must follow the variability of the parameters involved on a monthly time scale, or even shorter. Whether this will be sucessful or not is still an open question and also whether one should try a "real time description" or whether one should use a "mean quantity description" as we do here.

OBSERVATIONS AND METHODS

The monthly programme included CTD-profiling (Neil Brown, MK III) and discrete sampling with a rosette water sampler (Gen. Oceanics). At each of approximately 15 stations (of which the results from 11 stations are used here, see Fig. 1) between 2 and 6 water samples were taken for analyses of salinity, oxygen, nitrate, nitrite, ammonium (IN = Σ(nitrite + nitrate + ammonium), total nitrogen (TN), phosphate (IP) and total phophorus (TP). For the calculations made here, we shall make use of data from 29 expeditions during the period from February 1982 to December 1984. The time spacing appears from Fig. 3a. More details concerning observations and methods are given in Rydberg (1985).

Data treatment

To give a general view of the time and space variability of the observed parameters, we have done as follows: Spatial mean concentrations of IN (IP) and TN (TP) within the surface water of the Laholm bay and the Skälderviken have been determined for each cruise separately (Figs. 3a, b and 4 a, b). These are based on the eight inner stations, i.e. 11-16, 4, 6 and 7. The surface water is defined as the water between 0-10 m, or when the halocline is shallower than 10 m (has happened once) as the water with salinities S<30. These "local" mean values can be compared with the corresponding mean values for the Kattegat surface water (formed by addition of the values at the stations K5-K9), which are also shown in the same figures. To get a detailed picture of the spatial variations between the various stations, we have also determined the annual mean surface water concentrations of IN, TN, IP and TP for each station individually (Figs. 3c, d and 4c, d).

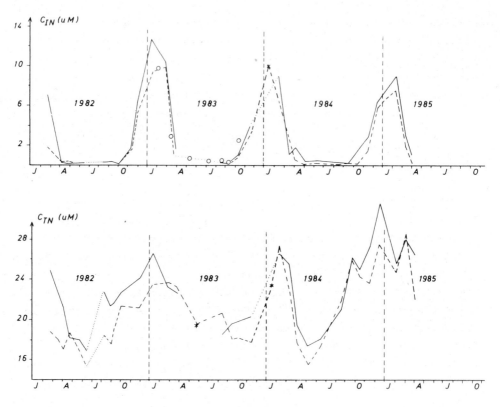

Fig 3a, b. Mean surface water concentrations of IN and TN, within the Laholm bay and the Skälderviken(_____) and within the Kattegat (-----) . The mean values are determined for the eight inner stations (11-16, 4-7) and for the three outer stations K5 - K9. Surface water has been defined as the water in the interval 0-10 m. The dots (o,*) refers to observations performed by other institutes.

We have furthermore calculated bimonthly surface water mean concentrations of IN, TN, IP and TP for the Laholm bay (stns 11 - 16, with half weight for the stns 11 and 16) and for the Kattegat stations K7-K9 (Figs. 5 a - d, and Table 1). Finally, we have determined the bimonthly mean concentrations for the same para- meters 1m above the bottom at the stations K5 - K9, which are the only stations where high salinity deep water has always been present (Figs. 5 a - d)

The monthly observations of nutrient concentrations (IN, TN, IP and TP) in the rivers, the freshwater supply and the municipal se- wage, measured by the local government, were used to calculate the

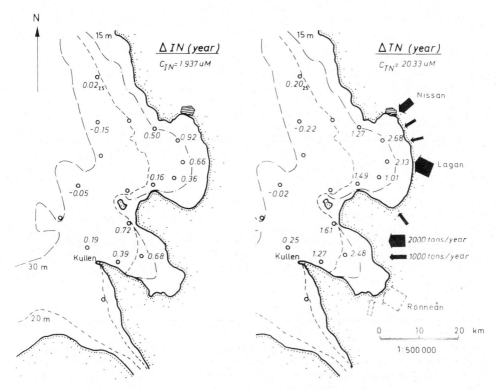

Fig 3c, d. Annual mean deviation, ΔC of IN and TN from a spatial mean concentration, C within the Kattegat surface water. This mean value (CIN = 1.937 uM, CTN = 20.33 uM) is determined from the observations at the stations K5 - K9. The results are based on 29 expeditions from Feb 1982 - Nov 1984. The local land based supply of TN is indicated by arrows (the supply of IN is approximately 50 % of the TN supply).

bimonthly land based supply of nutrients (and freshwater) to the Laholm bay. These are seen from Figs. 6 a - c and also from Table 1.

In earlier reports we have done a comparison between the land based supply of nutrients and the deep water supply, where the later was calculated from the observed nutrient concentrations multiplied by a "theoretically" determined entrainment trans-port. In Fleischer et al., 1985, for example, we made use of the observed bimonthly mean depth of the halocline (depth to isohaline 25),h_1 , the observed mean salinity difference, ΔS between the surface water (see above) and the deep water and observations of the long time mean wind at the Danish weather station Christiansö.

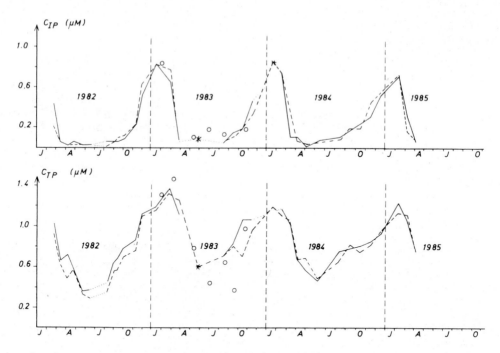

Fig 4a, b. Mean surface water concentrations of IP and TP, within the Laholm bay and the Skälderviken(_____) and within the Kattegat (-----) . The mean values are determined for the eight inner stations (11-16, 4-7) and for the three outer stations K5 - K9. Surface water has been defined as the water in the interval 0-10 m. The dots (o,*) refers to observations performed by other institutes.

The entrainment velocity, w, was then calculated according to the formulas;

$$w = (2\ m_0\ u_*^3)/(g\ \beta\ \Delta S\ h_1)\quad \text{and}\quad u_*^2 = ((c\ \varrho_A)/\varrho)\ W_*^2$$

where u_* is the friction velocity, g is the gravity constant and W_* the wind velocity. $\beta = 8 * 10^{-4}$ relates the salinity to the sea water density, while the expression $(c\ \varrho_A)/\varrho = (1.25 * 10^{-3})^2$. Stigebrandt, 1983, determined a mean value for the constant $m_0 = 1$, within the Kattegat, implying a mean entraiment velocity of approximately 25 cm/day. w is lower in the southern Kattegat due to a stronger vertical salinity gradient.

In this report, the bimonthly entrainment transport, $q_0 = w * A$ (where A is the halocline area within the bay) was corrected, with just marginal changes, for true windspeed at Spodsbjerg, Denmark.

TABLE 1.
Bimonthly mean concentrations (C, (uM)) of TN, IN, TP and IP
within the Laholm bay surface water (L), the Kattegat surface
water (1) and the Kattegat deep water (0), and bimonthly supply
of nutrients (R, (g/s)) and freshwater (q_f, (m^3/s)) to the
Laholm bay (1982-84).

	Jan – – Feb	Mar – – Apr	May– – Jun	Jul – – Aug	Sep – – Oct	Nov – Dec
CTNL	25.77	20.73	17.69	21.57	22.34	25.61
CTN1	23.92	19.23	16.26	20.08	20.61	21.41
CTNO	23.42	21.97	22.62	24.43	24.32	24.50
CINL	10.59	1.41	0.34	0.20	0.76	4.50
CIN1	8.16	0.55	0.22	0.22	0.31	3.77
CINO	10.72	11.12	12.97	10.27	9.90	8.86
CTPL	1.17	0.76	0.45	0.64	0.82	0.94
CTP1	1.13	0.76	0.48	0.59	0.70	0.93
CTPO	1.35	1.49	1.53	1.33	1.67	1.42
CIPL	0.71	0.07	0.04	0.08	0.16	0.42
CIP1	0.71	0.07	0.03	0.06	0.13	0.48
CIPO	1.00	0.91	1.07	1.02	1.12	1.06
RTN	234.1	197.9	110.7	58.6	106.6	205.9
RIN	133.2	108.1	49.5	22.1	43.9	103.2
RTP	5.01	5.35	3.20	1.82	3.14	4.19
RIP	2.00	2.14	1.28	0.73	1.26	1.68
q_f	206.0	174.0	88.0	45.0	94.0	165.0

These observations were kindly given to us by Stigebrandt and
refer to the same period (1982 – 1984) as the other quantities.
The result is shown in Fig 6 c. We like to stress that the
calculation of the entrainment transport is one uncertain point in
this report. There are large variabilities in the halocline depth
and in the salinities as well. Sometimes there is no deep water at
all in the Laholm bay, implying zero deep water transport , and it
has not been possible for us to cover these variabilities in a
sufficient way. In the report mentioned before (Rydberg and
Sundberg, 1985), we did an alternative approach, where we treated
the entrainment transport as an unknown, which was determined by
using one more equation than we shall do here.

Comments

   The surface water concentrations of IN and IP in the Kattegat
(Figs. 5a, c) follow an annual cycle with maximum values of 8.2 uM
(0.7 uM for IP) during January/February, a rapid decrease after
the springbloom in March and thereafter very low concentrations

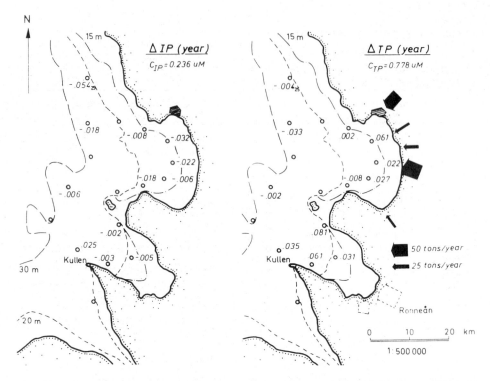

Fig 4c, d. Annual mean deviation, ΔC of IP and TP from a spatial mean concentration, C within the Kattegat surface water. This mean value (CIP = 0.236 uM, CTP = =.778 uM) is determined from the observations at the stations K5 - K9. The results are based on 29 expeditions from Feb 1982 - Nov 1984. The local land based supply of TP is indicated by arrows.

during the summer. There is weak increase in the IP concentrations during the summer, which can be found both in our measurements and in those by the Danish Agency of Environmental Protection during the years 1975-1978 (see also Rydberg and Sundberg, 1984). A rapid increase in IP and IN occurs from October when the primary production decreases. The concentrations of IN are higher within the Laholm bay than outside, especially during the winter months when we also expect the assimilation to be small. The difference indicates the large local supply of IN. During the summer period, the concentrations of IN are the same as in the Kattegat, thus indicating a rapid assimilation of the local supply (in fact, even the stations nearest to the river mouths, which are not shown here,

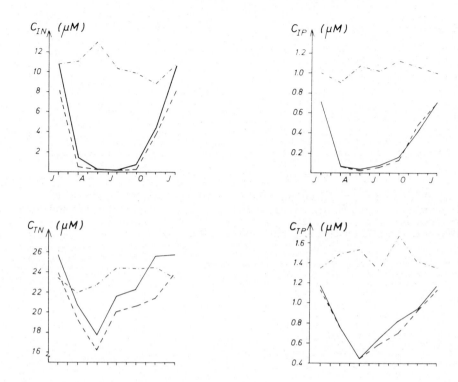

Fig 5 a - d. Bimonthly mean concentrations of IN, TN, IP, and TP within the surface water of the Laholm bay (the stations 11-16 with half weight for 11 and 16, _____), within the surface water of the Kattegat (the stations K7 and K9, -----), and in the deep water of the Kattegat (the stations K5, K7 and K9, -·-·-·-).

show very low IN concentrations). The landbased IP supply, on the other hand, is small. Consequently, the corresponding gradients in IP are also small.

The importance of the local nutrient supply is easily seen also from the surface water concentrations of TN and TP, shown in Figs. 5b and 5 d. There is a horizontal gradient in TN, but in this case as a year round feature, which is exepected as TN is nearly (see discussion) conservative. The gradient in TP, on the other hand, is relatively weak. The decrease in TN and TP concentrations during the spring follows the decrease in IN and IP and is of the same order, but goes slower. The phase lag is of the order of 1 month which seems to indicate that the spring bloom sinks to the bottom within that time. If that interpretation is correct, the phyto-plankton settling velocity is of the order of 15 m / month. The

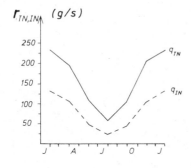
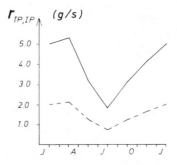

Fig 6a, b. Bimonthly, landbased supply of IN and TN, IP and TP, respectively, to the Laholm bay. The supply follows the freshwater input (compare Fig. 6c), and the calculations are based on monthly observations during the years 1982 - 1984.

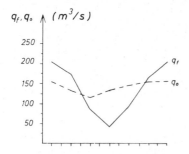

Fig 6c. Bimonthly supply of freshwater, q_f (measured) and deep water, q_0 (calculated, see text) to the Laholm bay surface water.

later increase in both TN and TP concentrations may indicate an ecological succession, where the external nutrient supply is conserved in the surface water by the living organisms.

The Figs. 5 a - d also show the deep water nutrient concentrations. These are influenced by the water exchange with the rest of the Kattegat deep water but also by the local biochemical processes. There are no large variations in these components during the year, and except for the increased concentrations of IN after the spring bloom, it is difficult to observe even the slightest coupling to what occurs in the surface water. A comparison between the deep water oxygen concentrations (Fig. 2) and the deep water IN and IP concentrations also shows that we should not expect to find a simple consumption/production ratio <u>on the time scales</u> <u>studied here</u>. The strong decrease in oxygen concentrations during the summer is actually accompanied by a weak <u>decrease</u> in IN concentrations and nearly constant IP concentrations. The decrease may indicate that denitrification at the bottom is an important sink for nitrogen. Still however, the concentrations of IN and IP are rather high in this area (11 (1.0 for IP)uM), compared to the mean values for Kattegat deep water (6 (0.5 for IP)uM).

The landbased supplies of nutrients are also subject to strong seasonal variability (Figs. 6a, b). These are actually dominated by the variations in the river supply (Fig. 6c), while the concentrations in the rivers are nearly constant throughout the year. The entrainment transport, q_0, also shown in Fig. 6c, is almost constant, although the windstress is much larger during the winter. A larger windstress means a rapidly increasing entrainment velocity, but at the same time a deepening halocline and, for the Laholm bay, a decreased halocline area and thus a suppressed entrainment transport.

A comparison between the Figs. 5a and 5c indicate a (weak) phosphorus limitation within the bay (but not within the Kattegat) during the early spring, and a clear nitrogen limitation from June/July and towards the autumn. This is in accordance with the size of the surface water nutrient pool before the spring bloom (IN/IP is nearly 16 (by atoms) within the bay, but a large extra supply of IN is added during the bloom; IN/IP is 12 in the Kattegat, also with an extra addition of IN - but relatively smaller). The gradually increasing concentrations of IP (Fig. 5c) during the later part of the season is due to a decreasing IN/IP ratio in the deep water (more effective P regeneration ?) which can also be seen from the Figs. 5a and 5c.

A SIMPLE BOX MODEL

We will now make use of the observations in a simple box model consisting of three water masses; Laholm bay surface water (indexed L), identified by the spatial mean surface water concentrations at the stations 11 - 16 (with half weight for stns 11 and

480

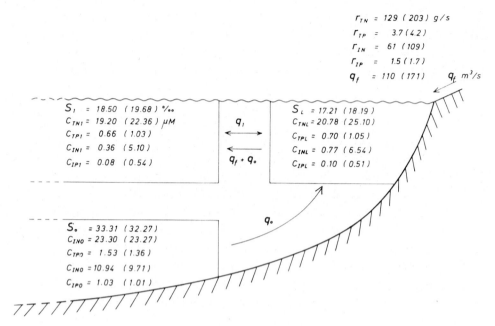

Fig 7. A box model for the Laholm bay and the southeastern Kattegat, indicating the observed summer and (winter) mean concentrations and fluxes within and between each of three water masses
defined in the text. There are no physical borders involved in
this model.

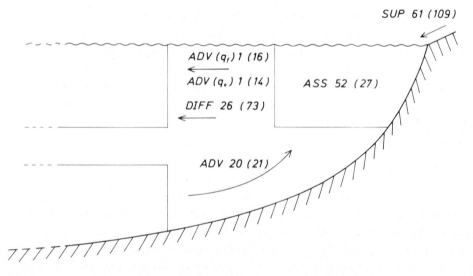

Fig 8. Observed and calculated fluxes and assimilation of IN (g/s)
according to the model in fig 7.

16), Kattegat surface water (1), identified by the mean surface water concentrations at the stations K7 and K9 and Kattegat deep water (O), finally, by the mean concentrations 1 m above the bottom at the stations K5 - K9. The model will be used to determine the exchange of water between the surface waters of the Laholm bay and the Kattegat and the assimilation of nutrients within the bay.

The model is shown in Fig. 7, where we have inserted the winter (November - February) and the summer (March - October) mean concentrations of IN, TN, IP, TP and S, for each of the three boxes. The model also indicates the size of the landbased nutrient fluxes (RTN, RIN,. .....) and the volume fluxes (q_1,....) between the boxes, which are defined as follows;

q_1 diffusive volume flow between box L) and 1) - doubly directed
q_f fresh-water supply, advective volume flow from box L)
q_0 net upward entrainment flow from box O) to box L)

We implicitly assume that the Kattegat boxes are infinite and include an effective horizontal mixing. This implies that the fluxes of properties from the Laholm bay does not influence the properties in the other boxes.

We may now write down the following equations for the conservation of TN and IN within the bay;

$$q_1 (CTNL-CTN1) + q_f CTNL - q_0 (CTNO-CTNL) - RTN = 0 \qquad (1a)$$
$$q_1 (CINL-CIN1) + q_f CINL - q_0 (CINO-CINL) - RIN - PIN = 0 \qquad (1b)$$

where CTNL (RTN) is the concentration (landbased supply) of TN within (to) the Laholm bay surface water, and so on. PIN finally is meant to be the planktonic assimilation of IN within the bay, but may as well be a global term, which contains all other sources and sinks for IN . We thus assume that TN is conservative, which is not self-evident. Unfortunately, the outflow of low salinity Baltic water through the Sound causes larger salinity gradients perpendicular to the coast (due to the earth's rotation) than does the local freshwater supply. This implies that there are problems to use the salinity as a tracer. Similar problems occur with the

phosphorus components. These various difficulties will be brought
up in the discussion.

By inserting summer and winter mean values, taken from Table 1,
(weighted for the number of observations during each two month
period) it is now possible to calculate the assimilation of IN,
PIN and the water exchange, q_1. The results are seen from Fig. 8.

DISCUSSION

During the period from March to October, the supply of IN to
the bay is dominated by the transport from land, which is 61 g/s,
while the supply from the deep water is 20 g/s (Fig. 8). The
diffusive outflow to the Kattegat is 29 g/s and the net assimilation 52 g/s (corresponding to a mean net assimilation of 1 mM/m^2
day, assuming an area of the bay of 300 km^2. We did not discuss
any physical borders for the bay water earlier). The flux of IN to
the Kattegat waters, mainly occurs in March and April, when the
landbased supply is large, and there is still a gradient between
the bay and the Kattegat waters (see Fig. 6).

The supply of IN from land is larger during the winter than
during the summer, while the deep water supply is approximately
similar. The assimilation is expected to be low during these
months, however, but it is seen from Fig. 8 that the calculations
indicate a surprisingly large net uptake of 26 g/s. This seems too
high. Contrary, IN (like TN) should behave almost like a
conservative substance during these months. If we assume that this
is the case, we may use Eq. 1b (with the term PIN excluded) to
determine the water exchange, q_1, during the winter. q_1 then
becomes 5000 m^3/s, which is larger than the value determined from
eq. 1a (3600 m^3/s) and also marginally larger than the summer
exchange, 4600 m^3/s. It seems more convenient to have a larger
water exchange during the winter, and thus we believe that the
value for q_1, determined from the TN equation is too small. One
reason could be that the observations of TN are too few to give a
correct horizontal gradient during the winter; we do have some
problems with the observations of TN. Approximately every
twentieth value seems wrong (which means too high), and for

unexplainable reasons we have accepted only values below 40 uM. Still however, just very few incorrect values at the stns K7 and K9 may be enough to give a change in q_1 of the order 1000 m^3/s . On the other hand there are other sources of errors, which will be taken up in the forthcoming discussion.

One possibility is that the supply of IN to the surface water of the bay is underestimated, due to a mineralization of some of the externally supplied TN. The mineralization of the landbased organic nitrogen supply is probably a slow process, compared to the water exchange, which occurs within a week. A qualified guess is that no more than 10 % of the organic nitrogen supply (approximately 50 % of the external TN supply) is mineralized within a week, which implies a contribution of the order <6 g/s during the summer. The atmospheric supply of IN was also left outside the equations. The term is small but not quite negligible (<10 g/s IN). The net supply of IN may thus be underestimated with <15 g/s (and thus, the net assimilation as well).

A net loss of TN from the surface water occurs due to a sinking primary production, but also due to sinking organic matter derived from land. The sedimentation must be most important during the spring bloom, and carpets of dead phytoplankton and organic matter have also been found on the bottom . A maximum TN content of nearly 1000 tons was observed in these areas in April 1984 (Håkansson and Floderus, 1985). The net uptake of IN during the spring-bloom is of the order of 500 tons (based on the surface water nutrient pool including a limited external supply during the springbloom period). It does not seem unreasonable to assume that the springbloom, as a whole, represents a loss of TN from the bay waters. Only 50 % of the bloom will go to the bottom locally,however, as a result of the low sinking rate (a phytoplankton organism or cell reaches the bottom of the bay in 2 weeks) compared to the retention time for the surface water (1 week). If we put this assumption into eq. 1a, it represents a sink for TN with mostly 12 g/s (and a 15 % decrease in q_1). This is not much compared to the landbased supply, which is above 100 g/s, but the loss could be a bit higher due to sedimentation of organic nitrogen during the rest of the season. The sedimentation has been the subject of many discussions and observations in the bay, with changing success however.

In the calculations based on yearly mean values (Rydberg and
Sundberg, 1985) we determined an assimilation ratio IN/IP (by
atoms) of 14. This result was obtained using an equation for IP,
equivalent to the one for IN (eq. 1b). Since then we have added
observations from another year, and also made a division into a
summer and a winter period. These changes influenced the phosphate
gradients in such a way, that if we do the same calculations today, we get a very small net assimilation of only 2.8 g/s during
the summer. There is also a loss of IP from the local waters during the summer of 2.8 g/s, which seems curious. One would expect
the supply of phosphate to the surface water to be very effectively assimilated, due to the large supply and the rapid uptake
of IN. The IN/IP supply ratio is approximately 30 during the summer and nearly 100 for the landbased supply alone. For the moment,
the use of IP (and TP as well) in the budget calculations seems as
hazardous as using salinity. The "signal" from land is simply too
weak, while the more important supply from the deep water cannot
be accurately measured. We may note however, that it seems reasonable to have substantial contribution to the local waters from
mineralization of organic phosphorus from land (the TP supply is
approximately 4 g/s, and organic phosphorus seems more readily
mineralized than does organic nitrogen). This contribution implies
a substantial increase to the assimilation value given above.

To summarize, we see that even though we have really put large
efforts into observations of the nutrient conditions, the results
are still not good enough to do "marine modelling". On the other
hand, we have got a good insight in the distribution of the local
nutrient supply and its importance relative to the contributions
from the "sea". The equations give us a possibility to calculate
changes in the nitrogen conditions due to a local decrease in
nitrogen supply (which could be done, as the origin of this
nitrogen is the heavy farming districts on land). Such calculations are of limited value, however as long as we cannot present a
direct coupling to the deep water oxygen consumption, which in
turn requires a model based on the monthly variations (at least).
It probably also requires that internal processes such as remineralization within the surface, deep and bottom waters
(occuring on timescales which are short compared to those consi-

dered here) and denitrification be considered. Here we have only investigated the external inputs. To manage these coming problems, we have carried out a series of intensive observations during shorter periods where the aim is to observe processes like oxygen consumption, nutrient assimilation and remineralization etc., on the timescales of the processes themselves, i.e. from days to hours.

REFERENCES

Aertebjerg, G., Jacobsen, T., Gargas, E. and Buch, E., 1981. The Belt Project. Evaluation of the physical, chemical and biological measurements. The National Agency for Environmental Protection, Copenhagen, Denmark, 122 pp.

Fleischer, S., Rydberg, L., Stibe, L. and Sundberg, J., 1985. Temporal variations in the nutrient transport to the Laholm bay (in swedish with english abstract). Vatten, 41:29-35

Håkansson, L. and Floderus, S., 1985. Bottom lenses and nutrient dynamics in the Laholm bay (in swedish with english abstract). Vatten, 41: 20-28

Miljöstyrelsen, 1984. Iltsvind i de danske farvand (with english summary) The National Agency for Environmental Protection, Copenhagen, Denmark.

Rosenberg, R., 1985. Eutrophication - the future marine coastal nuisance?. Mar. Poll. Bull., 16(6): 227-231.

Rydberg, L. and Sundberg, J., 1984. On the supply of nutrients to the Kattegat. Rep. no 44, Oceanografiska institutionen, Göteborgs universitet, Box 4038, S-400 40, Gothenburg, Sweden. 17pp

Rydberg, L. and Sundberg, J., 1985. External nutrient supply to coastal waters. A comparison between different sources. Rep. Journ. Mar. Res. Inst., Reykjavik, Island (in print).

Rydberg, L., 1985. Some observations of nutrient fluxes through the coastal zone. I.C.E.S. C.M. no 62, 1984 (to appear in Rapp.P.-v. Reun. Cons. int. Explor. Mer.)

LARVAL SETTLEMENT OF SOFT-SEDIMENT INVERTEBRATES: SOME PREDICTIONS BASED ON AN
ANALYSIS OF NEAR-BOTTOM VELOCITY PROFILES*

CHERYL ANN BUTMAN**

Ocean Engineering Department, Woods Hole Oceanographic Institution, Woods Hole,
Massachusetts 02543 (USA)

ABSTRACT
    During settlement, planktonic larvae may actively select habitats, they may
be passively deposited onto the seabed, or both processes may apply, but for
different spatial or temporal scales or for different flow regimes. Proposing
realistic settlement scenerios involving both passive deposition and active
habitat selection can profit from a priori analyses of near-bed flow
characteristics relative to known aspects of larval biology (i.e., swim speeds
and fall velocities). Toward this end, smooth-turbulent velocity profiles were
calculated for everyday tidal flows at a shallow subtidal study site, where
continuous near-bed flow measurements were available. Velocity profiles were
constructed for a realistic range of flow conditions. Rough-turbulent flow
profiles also were calculated, assuming storm waves periodically are sufficient
to resuspend sediments and make a rippled seabed. Under most flow conditions
analyzed, mean flow speeds exceed maximum larval swim speeds, even to within
tenths of millimeters from the bed. In the smooth-turbulent flows, larvae
generally would encounter no opposed velocity if they swam vertically in the
viscous sublayer, to heights of about 0.25-cm above the bed. In rough-turbulent
flows, eddies regularly penetrate to within tenths of millimeters of the bed, so
larvae would experience eddy velocities with components in all directions very
close to the bed. It is concluded that, at least at this study site, larvae
probably do not search for preferred habitats by horizontal swimming. Larvae
may swim vertically down to test the substrate and then swim vertically up to be
advected downstream. However, it also is noted that measured larval swim speeds
and fall velocities are about the same order-of-magnitude, so at best, larvae
may only be able to maintain position when swimming vertically.

INTRODUCTION

    In temperate latitudes, most infaunal organisms have planktonic larvae that
eventually settle onto the seabed and become benthic adults. Larval settlement
sites may be actively selected by larvae, larvae may be passively deposited onto
the seabed, or both processes may operate but on different temporal or spatial
scales. There is support in the literature for both active selection and
passive deposition; however, hydrodynamical conditions in the field that may
permit either process have not been determined. In the present study, some
realistic bottom boundary-layer flow profiles are constructed, based on physical
measurements from a specific field study site. Characteristics of the flow

*Contribution number 6046 from Woods Hole Oceanographic Institution
**Previously published as Cheryl Ann Hannan

very close to (i.e., < millimeters of) the seabed are analyzed relative to
pertinent aspects of larval biology (e.g., measured swim speeds and fall
velocities). Based on these results, insight can be gained regarding flow
profiles that would permit active selection and flow profiles where larvae would
be advected and deposited like passive particles. In addition, results of the
near-bed flow analysis indicate how the larvae may actually move between habitats
in the field, thereby suggesting reasonable selection mechanisms for future study.

Active habitat selection by a variety of soft-sediment invertebrate larvae
and meiofauna has been demonstrated at very small spatial scales (millimeters to
centimeters) in still-water laboratory experiments (e.g., see reviews by Meadows
and Campbell, 1972; Scheltema, 1974; Strathmann, 1978). Active selection also
is strongly suggested from results of field experiments (e.g., Oliver, 1979;
Williams, 1980; Gallagher et al., 1983) conducted at larger spatial scales (tens
of centimeters to 20 meters). Experiments performed in controlled laboratory
flow regimes that mimic specific field environments are required, however, to
determine hydrodynamic conditions that would permit active selection in the
field and to specify the spatial scales involved.

Specific mechanisms whereby larvae perceive information about available
habitats and then select a particular location for settlement are poorly
understood and are primarily speculative for soft-substrate invertebrates (but
see Crisp's [1974] and Burke's [1983] reviews of the hard-substrate literature
on this topic). Observations of some larval species during settlement in still
water indicate that the organisms must contact a surface to perceive a specific
cue (e.g., Wilson, 1968; Caldwell, 1972; Cameron and Hinegardner, 1974;
Eckelbarger, 1977) and, more recently, Suer and Phillips (1983) demonstrated
that the chemical factor promoting metamorphosis of their soft-substrate study
organism was effective only if it was absorbed onto a solid surface. Thus, the
"tactile chemical sense," coined by Crisp and Meadows (1963) to describe the
process of chemoreception in barnacle cyprids, also may apply to the settlement
of soft-substrate larvae. Information on the way larvae may move between
potential habitats (i.e., by swimming, hopping, crawling, or by being passively
distributed) during selection in moving fluid is scant, being limited to some
early observations of settling polychaete larvae (Whitlegge, 1890, cited in
Gray, 1974; Wilson, 1948, 1958; but see the quantitative work on barnacle
cyprids by Crisp, 1955; Crisp and Meadows, 1962).

Until recently, only a handful of researchers (including Pratt, 1953;
Baggerman, 1953; Fager, 1964; Moore, 1975; Tyler and Banner, 1977) considered
passive deposition of larvae as a realistic alternative hypothesis to active
selection. In recent experiments on the role of physical processes in sinking,
settlement and recruitment of larval infauna or meiofauna, hydrodynamic null

hypotheses generally could not be rejected. These studies showed that, from fluid-dynamical considerations, it is possible to account for patterns of certain organism distributions by passive accumulation (Eckman, 1979, 1983; Hogue and Miller, 1981), passive sinking (Hannan, 1984a, b) and passive resuspension and transport (Palmer and Gust, 1985; but see also Grant, 1981). The results stipulate that near-bed flow processes must be added to the list of potential factors controlling the population dynamics of soft-sediment organisms.

Active habitat selection and passive deposition need not be mutually exclusive alternative hypotheses to account for patterns of larval settlement. For example, hydrodynamical processes may sort and distribute larvae over relatively large areas (meters to tens of kilometers) of the seabed, just as sediments are sorted and distributed. Then, once larvae have been initially deposited in a particular sedimentary environment, they may redistribute at smaller spatial scales (millimeters to centimeters) by actively choosing a preferred microhabitat. A variety of other scenerios are possible where passive deposition and active selection operate at different spatial or temporal scales. Considerable insight into the plausibility of each scenerio can be obtained through an analysis of velocity profiles that are likely to occur close to the seabed in habitats where larvae settle. The mean flow speed at a given height above the bed sets, for example, the required swim speed for a larva to effectively maneuver in a plane parallel to the mean flow and also sets the horizontal distance a larva would be advected if only passive sinking occurred.

In the present study, near-bed velocity profiles are calculated for a specific soft-sediment environment, where experiments with settling larvae have been conducted since 1980 (see Hannan, 1984a, b). Sufficient data on near-bed flows at this site are available to permit profile calculations for a realistic range of flow conditions. The resulting profiles are constrained by the assumptions underlying the calculations (see PROFILE CALCULATIONS AND RESULTS), and thus, they may or may not commonly occur at the study site. However, the profiles shown here are meant only to be illustrative. They represent a first attempt at gaining quantitative insight regarding the order-of-magnitude of flow speeds potentially encountered by a larva as it gets closer and closer to the seabed. In addition, these profiles can be modeled in a laboratory flume, allowing experimental tests of the hypotheses generated from this study.

STUDY SITE AND FLOW MEASUREMENTS

Study site description and surface circulation

The field study site (Fig. 1), Station 35 (from Sanders et al., 1980), is located in Buzzards Bay, Massachusetts (USA) in 15 m of water. Bottom sediments primarily are medium sand (250-500µm), periodically overlain with a mud veneer.

490

Fig. 1. Map of Southeastern Massachusetts (from Sanders et al., 1980) showing location of Buzzards Bay, on the western border of Cape Cod. The location of Station 35 is indicated by an asterisk.

Sanders et al. (1980) characterized the sediments as "moderately well to poorly sorted," based on monthly samples of the top 4-cm of sediment for one year; sediment composition was 0.5-6.7 percent gravel (> 2 mm), 59-90 percent sand (63 µm - 2 mm), and 10-37 percent mud (silt + clay, < 63 µm) during this time.

Previous descriptions of the surface circulation of Buzzards Bay have presumed that currents were primarily tidal (e.g., Redfield, 1953), but until recently (see below), few flow measurements were made. Because the main axis of the bay is oriented northeast/southwest (see Fig. 1), tidal currents generally are oriented along this axis. In some areas of the bay, however, there is a slight tendency for a counterclockwise gyre in the surface circulation of Buzzards Bay. Surface tidal currents generally are weak, rarely exceeding 50 cm/sec, and are slightly stronger and of longer duration during the flood than during the ebb tide.

During the summer, when larvae are settling, the prevailing winds are from the southwest as a result of the Bermuda high-pressure system lying to the southeast of Cape Cod. Winds are strongest in the afternoon, when local seabreezes augment the prevailing southwesterly winds. At Station 35, winds from the southwest experience the longest fetch, so local seas at the study site can reach heights of 1-1.2 m in 2-3 hrs. However, under these non-storm conditions, locally generated wind waves in the bay are fetch-limited to ~ 4 sec and rarely penetrate to the bottom at the study site. The entire bay generally

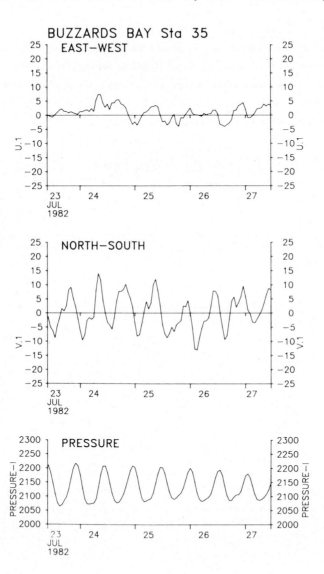

Fig. 2. Plots of the east-west and north-south components to the near-bottom currents (0.5-m above the bed) and near-bottom pressure at Station 35 during a larvae experiment (see Hannan, 1984a, b) from 7/23/82 through 7/27/82. The values plotted are edited one-hour averages during the interval.

is vertically stratified during the summer (Rosenfeld et al., 1984) due to surface heating. Because of variations in bottom topography, relatively cold water can persist at depth in the south and southwestern portions of the bay; this cross-bay temperature gradient may result in weak density-driven flows during the summer (W.D. Grant, personal communication).

Near-bottom flow measurements

During larvae experiments by the author in the summer of 1982 (see Hannan, 1984a, b), Dr. Bradford Butman (U.S. Geological Survey, Woods Hole) deployed a bottom-moored tripod instrument system to continuously measure near-bottom flows. The tripod system (described in Butman and Folger, 1979) has instruments for measuring current speed and direction, pressure, light transmission

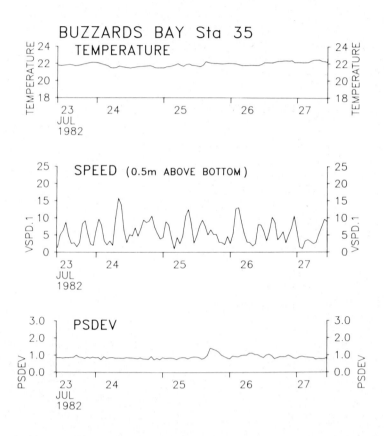

Fig. 3. Plots of near-bottom water temperate (°C), current speed (0.5-m above the bed) and pressure standard deviation ("PSDEV") at Station 35 during a larvae experiment (see Hannan, 1984a, b) from 7/23/82 through 7/27/82. The values plotted are edited one-hour averages during the interval.

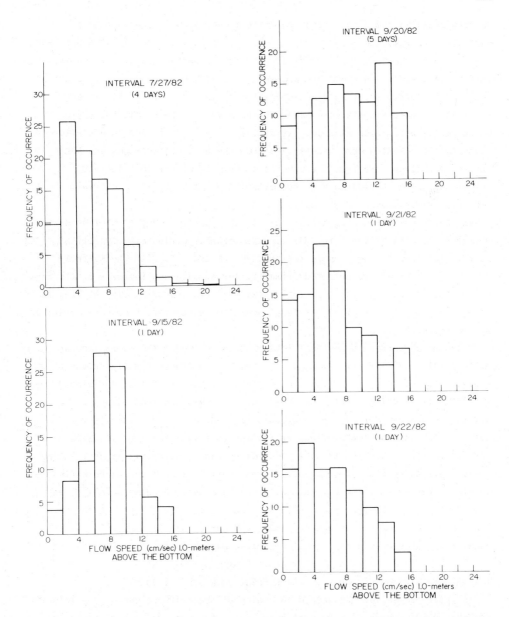

Fig. 4. Flow speed-frequency histograms for currents measured 1.0-m above the bottom at Station 35 during five different intervals when larvae experiments were conducted (see Hannan, 1984a, b). Intervals are identified on the graphs by the date that they ended; interval duration, in days, also is shown. Average values for the edited "burst" measurements are plotted for all intervals except 7/27/82, where measurements taken at the midpoint of the 3.75-min intervals are plotted.

and temperature, and also is equipped with a camera that takes bottom
photographs. Savonius rotors for measuring current speed are located 0.5- and
1.-m above the seabed; small vanes are mounted below each rotor for sensing
current direction. Currents and pressure were sampled in two ways (see Butman
and Folger, 1979); an average measurement was made over a 3.75-min interval and
a "burst" of measurements were taken in the middle of this interval (24 burst
measurements were taken at 2-sec intervals). The current speed and pressure
measurements reported here usually are from the 3.75-min averages and the
current directions are from the burst samples. Light transmission and
temperature were sampled only at the midpoint of each 3.75-min interval. Bottom
photographs were taken every hour.

The near-bed flow measurements indicate that bottom flows at Station 35 are
primarily tidally driven (Fig. 2). The semidiurnal periodicity typical of tides
at this latitude can be seen in the pressure record. As with the surface
currents, the flows are oriented primarily north-south and there is little flow
east-west, indicating that the tidal flows traverse approximately the long axis
of Buzzards Bay (see Fig. 1), at least near the coast where flows are polarized
by the shore.

Near-bottom current speed oscillates between approximately a minimum and
maximum value twice daily (Fig. 3), as expected for these tidally driven flows.
However, because other physical phenomena (e.g., density-driven and wind-driven
currents) also contribute to the flows, current speeds do not always go to zero
and the curves are not smooth. Periodically, surface storm activity was
detected in the near-bottom flows at Station 35 (e.g., see peak in PSDEV on
7/25/82 in Fig. 3); such strong surface winds cause the regularly oscillating
tidal flows to deviate substantially. Near-bottom water temperature varied
little on the short-term, but gradually cooled about 5°C between 7/27/82 and
9/22/84 (Hannan, 1984b). Flow speed 1.0-m above the bed varied between zero and
a maximum of 22 cm/sec during the summer and early fall of 1982 (Fig. 4);
however, usually only a maximum of 16 cm/sec was reached.

GENERAL DESCRIPTION OF BOUNDARY-LAYER FLOWS OVER SOFT SEDIMENTS

As water flows over the seabed, a region of shear (the slope of the velocity
profile, du/dz, where u = the horizontal velocity component and z = the
perpendicular distance from the surface; refer to Fig. 5) develops as a result
of the retarding effect of the boundary on the flow. This region of shear near
the bed is referred to as a "boundary layer". The mean velocity profile is
constrained by conditions at each end of the boundary layer: u = 0 at z = 0
(the "no-slip" condition at the boundary) and u = U (the free-stream velocity at
z = δ) (the boundary-layer thickness) (see Fig. 5). The shape of the velocity

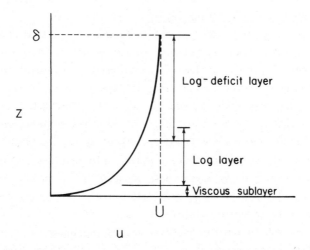

Fig. 5. Diagram of a turbulent bottom boundary layer plotted on a linear scale for both axes, showing the relative positions of the viscous sublayer, the log layer, and the log-deficit layer.

profile varies depending on flow properties (e.g., the flow Reynolds number, the background turbulence, and accelerations), fluid properties (e.g., stratification induced by temperature, salinity, and suspended sediments) and boundary characteristics (e.g., the bed roughness and the cohesiveness of sediments). The boundary-layer thickness depends on the boundary shear stress and inversely on the forcing frequency for the flow, $\kappa u_*/\sigma$, where u_* is the bottom shear velocity ($\sqrt{\tau}/\rho$, where τ is the bottom shear stress and ρ is the fluid density), κ is von Karman's constant of 0.4, and σ is $2\pi/P$ (where P is the period of the flow).

Because larval settlement takes place inside the bottom boundary layer, it is instructive to briefly review relevant characteristics of boundary layers that may form over soft sediments under simple, steady-flow conditions (see also the recent discussion by Nowell and Jumars, 1984). Velocity profiles which may occur at the study site then are calculated (see PROFILE CALCULATIONS AND RESULTS), based on both field data and assumptions about profile characteristics . The following discussion is somewhat idealized, for the sake of posing logical predictions concerning the role of hydrodynamical processes in larval settlement; for this modest goal, the idealization does not significantly affect the outcome of this study. For recent reviews of the state-of-the-art in geophysical boundary-layer flows, see Nowell (1983) and Grant and Madsen (1986).

General characteristics of boundary layers

This discussion considers steady, uniform flow over a bottom which is uniform over large horizontal distances, relative to the height off the bed. In theory, the bottom boundary layer may be laminar or turbulent, depending on the relative importance of viscous versus inertial forces in the flow, as characterized by the flow Reynolds number, $Re_f = LU/\nu$ (where L = the characteristic length scale for the flow, U = the characteristic reference velocity of the flow, and ν = the kinematic viscosity of the fluid). Laminar boundary layers occur at low Re_f where turbulent fluctuations are relatively unimportant. Laminar boundary layers have pronounced stream-wise stability; any disturbance to the layer will be quickly dissipated by viscosity downstream, restoring the profile to the predisturbance case. Thus, only horizontal velocities are present in laminar boundary layers in steady, uniform flows. Turbulent boundary layers occur at high Re_f. Here, velocities have both a mean and a fluctuating component; fluctuations are due to turbulent eddies, which can have velocity components in all directions. Transfer of mass and momentum within the layer occurs due to products of coherent velocity fluctuations associated with these eddies. Near the bottom, the energetic eddies scale with height above the bed. The turbulence is produced by the product of vertical shear and Reynolds stress due to the presence of the boundary.

The Re_f is a good predictor of laminar or turbulent boundary layers for flows over smooth flat plates, but other factors are important to this prediction in ocean flows traveling over sediments or bumpy seabeds. Turbulence may be generated in the flow by a source away from the bed (e.g., wave breaking) or turbulence may be "tripped" at the seabed by a relatively large flow disturbance on the bottom. For turbulent flows, the roughness Reynolds number, $Re_* = u_* k_b/\nu$ (where k_b = the hydrodynamic bed roughness scale), is a better predictor of bottom boundary layer characteristics. However, turbulence is such a pervasive feature of ocean flows that even if local Re_* are in the laminar range, the flows often are turbulent (see Yaglom, 1979). In essence, laminar boundary layers are rare in the ocean.

Turbulent boundary layers

Turbulent flows are classified as smooth, rough, or transitional (e.g., see Schlichting, 1979), depending on Re_* of the flow. In the immediate vicinity of the boundary, viscous forces dominate the flow. A pronounced viscous sublayer (see Fig. 5) may develop in the case of flow over hydrodynamically smooth bottoms (e.g., see Eckelmann, 1974) occurring at relatively low Re_*. The viscous sublayer has characteristics of laminar boundary layers. Over hydrodynamically rough bottoms, viscosity still acts at the boundary, but no distinct well-behaved sublayer forms comparable to the smooth case and eddies may penetrate to within tenths of millimeters of the bed; thus, in rough-turbulent flow the velocity structure close to the bed is complicated (e.g., see Yaglom, 1979) and not well-known. For intermediate Re_*, transitional flow occurs, with characteristics intermediate between smooth- and rough-turbulent. For pipes, flows are shown to be smooth-turbulent for $Re_* < 5$ and rough-turbulent for $Re_* > 70$ (see Schlichting, 1979); for open-channel or geophysical flows, these values may be more like 3.5 and 100, respectively (e.g., see review of Nowell and Jumars, 1984).

Based on empirical studies and scaling arguments (see Clauser, 1956), turbulent boundary layers in the laboratory can be divided into three regions (refer to Fig. 5). Adjacent to the boundary, in the viscous sublayer (for smooth-turbulent flows), velocity varies linearly with distance from the boundary according to $u/u_* = u_* z/\nu$, the scaling parameters for this flow region. The outer region of flow is called the log-deficit layer because the deficit velocity, $(u-U)/u_*$, is logarithmically related to z/δ. Between these two layers (and overlapping with the lower portion of the log-deficit layer) is the log layer, a _major_ feature of steady, uniform flows. The velocity profile in the log layer is described by:

$$\frac{u}{u_*} = \frac{1}{\kappa} \ln \frac{z}{k_b} + B \tag{1}$$

(where B = the empirically defined constant of integration). The velocity scale of eddies (i.e., the root-mean-square of the velocity fluctuations) in the log layer is about 10 percent of the free-stream velocity, U (see Hinze, 1975). For smooth-turbulent flows, the shape of the velocity profile in the log layer depends on u_* and ν. For fully rough-turbulent flows, the velocity profile depends on u_*, ν and bed geometry. From empirical studies of smooth-turbulent pipe flows (see Schlichting, 1979), the lower limit of the log layer is approximated by $11.6\nu/u_*$ and the upper limit of the viscous sublayer by $5.0\nu/u_*$. Between these heights, there is a complicated wake layer that cannot be described simply. In channel flows and geophysical boundary layers, the wake region may be larger (see reviews of Nowell, 1983; Grant and Madsen, 1986).

Ocean bottom boundary layers

Typical oceanic bottom boundary layers vary between smooth-turbulent and fully rough-turbulent. For example, the detailed velocity profiles measured in a laboratory flume by Grant et al. (1982) over an area of uniform intertidal sands taken from Barnstable Harbor, Massachusetts, typified a classical smooth-turbulent boundary layer. Other examples include the profiles measured in the laboratory flume studies of Nowell and Church (1979), Nowell et al. (1981),

Eckman et al. (1981), Eckman (1983) and see also the review of Jumars and Nowell (1984). In the ocean, smooth-turbulent profiles were measured by Chriss and Caldwell (1982), transitional by Grant et al. (1985), and rough-turbulent profiles by Gross and Nowell (1983), Grant et al. (1984); many other examples exist. Note that, at a given study site, a flow can be smooth-turbulent under one flow condition and rough-turbulent under another condition, for example, due to changes in z_0 (a parameterization of the bed roughness length scale, k_b) or in other sediment properties caused by bioturbation, sediment transport or bedform development (see Grant and Madsen, 1979, 1982).

PROFILE CALCULATIONS AND RESULTS

Profiles of current speed within the log layer at a site can be calculated, given the following assumptions. (1) There is quasi-steady, uniform, neutrally stratified flow over the bed. (2) The bed is uniform over large horizontal distances, relative to the height above the bed of the calculated velocities. (3) Bottom roughness is small, compared to the boundary-layer thickness. In addition, information must be available on velocities occurring at some height above the bed within the log layer and on bottom roughness characteristics. These assumptions periodically are met at Station 35; for example, during flood or ebb tide when near-bed flows are only tidally driven and there are no complications from wind-driven circulation, density-driven circulation or surface waves. Thus, the profiles calculated here accurately represent near-bed flow conditions only a certain percentage of the time. The rest of the time, the velocity profiles resulting from unsteady or non-uniform flows are imposed on the steady-flow case (e.g., the log layer profile), so composite profiles of flows that would be measured over the bed are difficult to predict (for a discussion of these features, see Grant and Madsen, 1986). Some of these complicated boundary-layer flows have been modeled (e.g., Smith and McLean, 1977; Grant and Madsen, 1979, 1982), but such calculations are not necessary for the first-order approach of this paper (see GENERAL DESCRIPTION OF BOUNDARY-LAYER FLOWS...).

Smooth-turbulent profiles

Smooth-turbulent velocity profiles were calculated for everyday flow conditions at the study site. Smooth-turbulent profiles were indicated by the estimated range in ke_* (see Table 1) for the range of measured near-bed flows (Fig. 4), by the observed seabed roughness and because of the preliminary results of detailed velocity measurements near the bed, made by Dr. William D. Grant (WHOI). Current speed and direction were measured over a 6-hr period, during non-storm conditions at Station 35 in October 1982, using four vertically

TABLE 1

Parameter values for velocity profiles shown in Figs. 6, 7 and 8.

u_{50} [a] (cm/sec)	u_* (cm/sec)	C_D [b] $(\times 10^{-3})$	z_0 $(cm \times 10^{-3})$	Re_* [c]	$(0.1)(\delta)$ (cm)
	Smooth-Turbulent				
A. 15.3	0.60	1.53	1.8	1.2	165
B. 9.8	0.40	1.66	2.8	0.8	111
C. 4.6	0.20	1.89	5.6	0.3	55
	Rough-Turbulent				
D. 15.3	0.98	4.14	100	294	270
E. 15.3	0.82	2.90	30	74	226

[a]u_{50} = u at z = 50 cm.
[b]For z = 50 cm.
[c]For smooth-turbulent flow, k_b = 200 μm and for rough-turbulent flow,
 k_b = (30)(z_0).

stacked acoustic-time-travel current meters (described in Grant et al., 1984)
mounted at distances of approximately 30-, 50-, 100- and 200-cm above the bed.
From these direct flow measurements, if the velocity profile is logarithmic, it
is possible to estimate u_* from equation (1) using the profile technique
(Grant et al., 1984), since u_*/κ is given by the slope of the velocity
profile. Thus, it is possible to calculate Re_* to determine if flows are
smooth-turbulent, rough-turbulent, or transitional. The preliminary results
indicate that, during non-storm conditions, the flow is smooth-turbulent to
transitional (W.D. Grant, personal communication).

For smooth-turbulent laboratory pipe flows, empirical results show that the
general log-layer equation given in (1) has the specific form of:

$$\frac{u}{u_*} = \frac{1}{\kappa} \ln \frac{z \, u_*}{\nu} + 5.5 \qquad (2)$$

(see Schlichting, 1979); note that the constant differs slightly for channel
flows and geophysical flows. To calculate a profile from this relationship
requires estimates of ν, u(z) measured inside the log layer, and u_*; also,
some iteration is necessary. For all calculations, ν = 0.01 cm^2/sec was used.

To choose u(z) requires an estimate of the thickness of the log layer. This
thickness can be approximated by (0.10)(δ) (Clauser, 1956; Grant and Madsen,
1986), where δ = the boundary-layer thickness. For a tidal flow, $\delta = \kappa u_*/\sigma$,
where σ = the tidal frequency ($2\pi/P$, where P is the tidal period of ~ 12 hr, in
this case). For log-layer thicknesses estimated here (see Table 1), velocities
measured at 0.5-m above the bed will always be in the log layer; for slower
flows at the site, measurements at z = 1.0 m may be above the log layer. To be

conservative, u_{50} (i.e., u at z = 50 cm) was used in calculations here. The difference in mean velocities measured at z = 0.5 m and z = 1.0 m was consistently between 1 and 2 cm/sec (B. Butman, personal communication), and thus, for z = 50 cm, the range of velocities (4.6 to 15.3 cm/sec, see Table 1) used to calculate profiles here seems reasonable based on the flow measurements at z = 1.0 shown in Fig. 4.

The choice of values for u_* needed to calculate smooth-turbulent profiles was constrained by the requirement that C_D, the bottom drag coefficient ($\sqrt{C_D} = u_*/u_{50}$), must be about 1×10^{-3} to 2×10^{-3} (typical values measured for smooth-turbulent flows). Some iteration was required to obtain the values

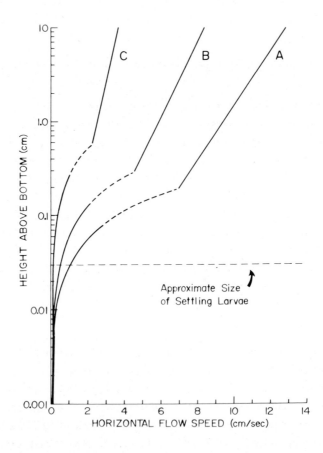

Fig. 6. Smooth-turbulent velocity profiles on a semi-log plot, calculated for a range of near-bottom flow speeds measured at Station 35. Parameter values are listed in Table 1. The log layer is the straight-line portion of each curve. Below this, the dashed curves indicate approximately the region of the wake layer, where velocities are difficult to estimate (see GENERAL DESCRIPTION OF BOUNDARY-LAYER FLOWS ...).

listed in Table 1.

   To determine Re_* for the profiles, k_b = 200 μm was used for the smooth-
turbulent case (see Table 1). This value was chosen because, while sediments at
the study site are primarily sands (250-500 μm, see STUDY SITE AND FLOW
MEASUREMENTS), surface sediments are heavily pelletized by the dominant infaunal
organism, <u>Mediomastus ambiseta</u> (a small polychaete worm). This organism occurs
in abundances of up to 2×10^5 per square meter (Sanders et al., 1980); it feeds
below the sediment surface and deposits discrete cylindrical fecal pellets
(~ 80 μm x 200 μm) on the sediment surface. Note, however, that k_b as large
as 830 μm still would result in $Re_* < 5.0$ for even the largest value of u_*
listed in Table 1 for smooth-turbulent flows.

   Smooth-turbulent velocity profiles are shown in Fig. 6 for z between 0.001
and 10 cm. Also shown by a horizontal dashed line on the figure, is the
approximate size (300 μm) of a settling polychaete larva; however, in temperate
latitudes, settling larvae can vary in size by approximately an order-of-
magnitude (from 100 to 1000 μm). Below this height, a 300 μm larva would not
have room to maneuver in a flow by horizontal swimming. It would be sitting on
the bottom or crawling along the bed and, at most, the flows would cause it to
roll. For a more detailed look at velocities very near the bed (i.e., at

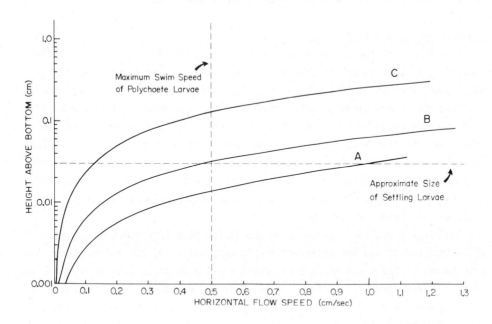

Fig. 7. Smooth-turbulent velocity profiles in the viscous sublayer on a
semi-log plot, calculated for a range of near-bottom flow speeds measured at
Station 35. Parameter values are listed in Table 1.

distances relevant to settling larvae), the same smooth-turbulent profiles (see Table 1) are plotted for z between 0.001 and 0.2 cm in Fig. 7. In addition to the approximate size of settling larvae (horizontal dashed line), the maximum measured swim speed of a polychaete larva (from the review of Chia et al., 1984) is shown as a vertical dashed line on the figure. Flow speeds to the right of this line would advect larvae; larvae may effectively maneuver by horizontal swimming in flows to the left of this line. Thus, larvae would be expected to effectively maneuver by horizontal swimming only for flows occurring in the upper left-hand quadrant of the figure.

The results indicate that larvae can horizontally swim only in the slowest profile plotted (see C in Fig. 7) and only to a height of about 0.1-cm above the bed. Above this height in profile C and for velocities at $z > 300$ μm in profiles A and B, larvae essentially would be advected by the flow. The velocities plotted in Fig. 7 all lie within the viscous sublayer (see Fig. 6), and thus, mean flow components occur only in the stream-wise (horizontal) direction. While they are being advected horizontally, larvae could still swim vertically to heights of at least 0.1-cm above the bed and face no opposed velocity. However, even in smooth-turbulent flows, the viscous sublayer periodically is subject to turbulent eddy penetration so vertical velocities, of-the-order $(0.1)(U)$, could be present from time-to-time.

Rough-turbulent profiles

In addition to the smooth-turbulent case for everyday flows, it is possible to construct rough-turbulent profiles at the site for conditions following a major storm with sufficient bottom stress to move sediments. It is observed at the site that storm winds oriented down the long axis of the bay (see Fig. 1) generate sufficient bottom stress to cause ripples to form on the seabed. After the storm, the ripples persist until they are obliterated by benthic biological processes. Because ripples set a much larger bottom roughness scale than grain roughness or fecal pellet roughness, rough-turbulent flows can result for the same range of everyday forcing conditions that produced smooth-turbulent profiles for the non-rippled bed.

Two rough-turbulent profiles were calculated here (see Table 1), using $u_{50} = 15.3$ cm/sec and two different ripple heights, h, of 0.5 and 0.15 cm. In both cases, a ripple steepness (h/1, where 1 = the distance between ripple crests) of 0.2 was used. This corresponds to the maximum ripple steepness observed under waves (see Grant and Madsen, 1982). For a rippled bed, the bottom roughness parameter, z_0, can be estimated by $z_0 = h$ (h/1) (Grant et al., 1984), so $z_0 = 0.1$ cm and 0.03 cm for the 0.5-cm and 0.15-cm tall ripples, respectively.

For rough-turbulent flow, empirical results show that the general log-layer equation given in (1) has the form:

$$\frac{u}{u_\star} = \frac{1}{\kappa} \ln \frac{z}{z_0} \quad .$$

(3)

Again, $u_\star$ can be calculated once z_0 and a reference value for $u(z)$ are known. For the same $u(z)$ used in the smooth-turbulent case, C_D is expected to be higher for rough-turbulent flows, i.e., $C_D > 2 \times 10^{-3}$ (see Table 1).

The two rough-turbulent profiles are plotted, along with a smooth-turbulent profile, for the same u_{50}, in Fig. 8. The slopes of the curves for the log

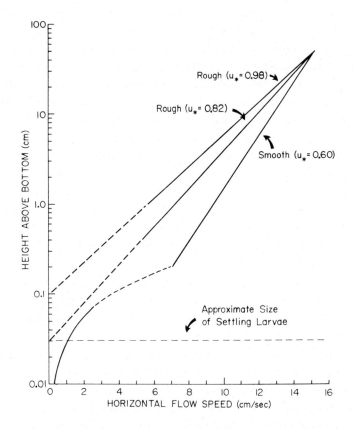

Fig. 8. Two rough-turbulent velocity profiles and a smooth-turbulent profile, all having the same u_{50}, but different values of $u_\star$ and z_0 (see Table 1). A log layer is known to accurately describe a rough-turbulent profile at distances > $(100)(z_0)$ from the bed, and is a reasonable predictor for distances between $(10)(z_0)$ and $(100)(z_0)$. Thus, on the figure, the rough-turbulent profile curves are dashed below $(10)(z_0)$, indicating that velocities in this region may be described by some other wake function. The dashed portion of the smooth-turbulent profile represents the wake layer (see Fig. 6).

layer are smaller in the rough-turbulent cases, than in the smooth-turbulent case. Thus, at a given height above the bed (below $z = 50$ cm), velocities are lower in the rough-turbulent flows. This simply reflects the fact that, in rough-turbulent flows, eddies close to the bed are mixing low-momentum fluid near the bed with higher-momentum fluid away from the bed so that near-bed mean velocities are lower, relative to the smooth-turbulent case.

The rough-turbulent profiles intercept the ordinate at z_0 and, in these cases, z_0 is greater than or equal to the approximate size of settling larvae (see Fig. 8). Thus, it appears that in rough-turbulent flows larvae may have a lot of vertical distance to maneuver by horizontal swimming before flow speeds reach a value that the organisms cannot swim against. However, it is important to realize that z_0 is a roughness parameter, reflecting where the flow effectively goes to zero. The boundary layer actually can attach anywhere between the trough and the crest of the ripples; in fact, internal boundary layers with different profile characteristics form in this complicated flow region close to the seabed (see caption to Fig. 8). Depending on where the larva is situated relative to the roughness elements, the animal could experience relatively high or low velocities. For example, very low flows generally would be expected in the lee of a ripple crest in a steady flow, but eddies also can be shed from these crests.

Even though, for a given z, mean horizontal velocities are lower in the rough-turbulent flows than in the smooth-turbulent flow plotted in Fig. 8, eddies regularly reach to within tenths of millimeters of the seabed in rough-turbulent flows (see GENERAL DESCRIPTION OF BOUNDARY-LAYER FLOWS ...). As previously mentioned, the most energetic eddies in the flow have velocities of about 10 percent of the free-stream velocity, U. For example, if U was 15.3 cm/sec (at $z = 50$ cm) for the flows in Fig. 8, then eddy velocities are a maximum of 1.5 cm/sec at this level above the bed, easily exceeding values required to prohibit effective swimming by a larva in any direction (see Fig. 6); eddy velocities closer to the bed are smaller and less energetic.

In summary, under most flow conditions analyzed here, horizontal flow speeds exceed maximum larval swim speeds even to within one body diameter of the organism from the seabed. If larvae actively maneuver in a flow, then vertical swimming to get into higher or lower horizontal flows seems most likely. These kinds of behaviors often have been proposed for planktonic organisms in the water column, for example, to account for vertical migrations of copepods (see reviews by Longhurst, 1976; Pearre, 1979; and also the recent collection of papers in Angel and O'Brien, 1984). Mileikovsky (1973) also proposed that "high" vertical swim speeds of soft-sediment invertebrate larvae may account for their retention in near-shore and estuarine waters; retention of, especially, crustacean and bivalve larvae in estuaries by active vertical movements of the organisms has a burgeoning literature (e.g., see symposium on this subject in Kennedy, 1982). However, previous to the present study, the relative effectiveness of horizontal versus vertical swimming for organisms in flows very close to the seabed has never been investigated quantitatively.

Other calculations

As with other particles, larvae have mass so they always are sinking through the water at a speed specific to their size, shape and density. Fall velocities of anesthetized polychaete larvae were measured directly by the author (see Hannan 1984a, b) and span about an order-of-magnitude, from 0.01 to 0.3 cm/sec,

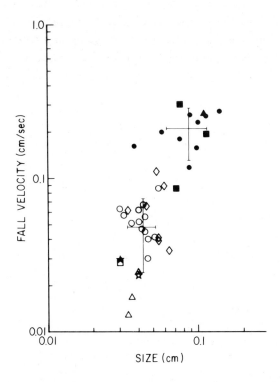

Fig. 9. Relationship between fall velocity and size for polychaete larvae tested in the study of Hannan (1984a, b). The length of the organisms, after they were anesthetized, is plotted against their measured fall velocity. Fall velocities were measured in two different settling chambers for two different groups of larvae, indicated by the closed versus the open symbols. The different symbols represent different anesthetizing treatments, but there were no significant differences in fall velocity that could be attributed to treatment (see Hannan, 1984b). The crosses represent mean fall velocity (± SD) and mean length (±SD) for the two groups of organisms tested. Details of the methods are given in Hannan (1984b).

roughly increasing with increasing body size (Fig. 9). It is interesting that this range overlaps the range of measured swim speeds for polychaete larvae (0.05 to 0.52 cm/sec, see review of Chia et al., 1984). Thus, even when larvae are swimming vertically it is possible that they are capable only of standing still!

The previous analyses have focused on how very near-bottom flow velocities may limit or allow active larval movements near the bed. It also is fruitful to look at the other extreme. Assuming that larvae only sink toward the bed like passive particles (see Hannan, 1984a, b), I have calculated the horizontal distance they would be carried by specific flows before reaching the bottom, given various starting heights above the bed (see Fig. 10). These results are useful, for example, in predicting distances between habitats where larvae would be able to test the substrate. Once deposited, if larvae are carried to a certain distance above the bed (i.e., by resuspension or by vertical swimming), then profile characteristics determine the horizontal advection distance (i.e., where the next test location would be) from that height. For the flows

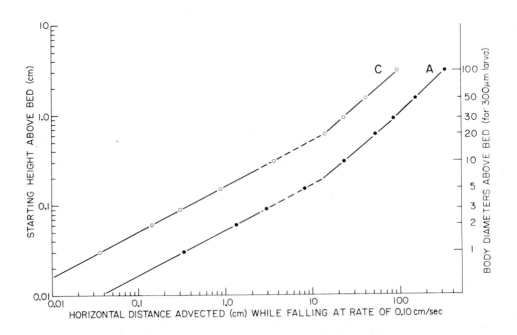

Fig. 10. Horizontal distances that passively sinking larvae would be advected given various starting heights above the bed. Larval fall velocities were taken as 0.1 cm/sec (see Fig. 9) and horizontal velocities for various starting heights above the bed were taken from smooth-turbulent profiles A and C in Fig. 6.

considered, larvae may be carried from tenths of millimeters to meters, for starting heights up to 3 cm (or about 100 body diameters of a 300 μm larva) above the bed. If this were the mechanism of habitat selection by larvae, then these spatial scales apply.

DISCUSSION AND CONCLUSIONS

Biologists have observed and quantified swimming in planktonic organisms for over half a century (see review of Chia et al., 1984). With the exception of vertical migrations, it is usually assumed that the organisms have little control of their position in the water column through swimming, since horizontal flow speeds greatly exceed their swim speeds (Mileikovsky, 1973). In fact, technically this distinguishes between plankton, "the drifters", and nekton, "the swimmers" (see Hardy, 1965). Likewise, larval dispersal in the plankton usually is assumed to be physically controlled (e.g., see Scheltema, 1971; Boicourt, 1982; Levin, 1983). However, until recently, settlement of larvae onto the seabed was assumed to be biologically controlled, through active habitat selection by the animals (see reviews cited in INTRODUCTION). An underlying assumption to this tenet is that the organisms can exert some control over their position close to the seabed, in order to select habitats. The precise mechanisms involved (e.g., horizontal or vertical swimming, hopping or crawling) have never been clear, but for an organism to choose a habitat, it seems necessary for the animal to be able to peruse the available sites (but see also the "threshold stimulus" hypothesis of Doyle, 1975, 1976).

It seems reasonable to expect that there will be some limiting height above the seabed, below which flow speeds would be low enough to allow organisms to maneuver effectively. This follows from the "no-slip condition" in fluid dynamics; flow speed must go to zero at the boundary. The rate at which velocity decreases with distance from the bed (i.e., the shape of the velocity profile) determines this "limiting maneuvering height". If larvae are choosing habitats by swimming around near the bed, it should be possible to constrain the limiting maneuvering height, for a given mean-stream flow. This was a goal of the present study.

The smooth-turbulent velocity profiles constructed here for everyday flow conditions at Station 35 in Buzzards Bay, Massachusetts (see Fig. 1) indicate that only the slowest flow modeled (profile C in Figs. 6 and 7) would permit effective horizontal swimming by larvae near the bed. In this case, the limiting maneuvering height is about 0.1 cm (or about three body diameters of a 300-μm larva), for a maximum swim speed of 0.5 cm/sec. This flow profile was constructed for a measured flow speed of 4.6 cm/sec at z = 50 cm. During five larvae experiments in the summer and early fall of 1982, flows < 6.0 cm/sec at z = 100 cm occurred from 24.4 to 56.8 (mean = 43.2) percent of the time (see Fig. 4). These are the flows for which larvae would be expected to effectively maneuver, at least to heights of 0.1-cm above the bed, because flows measured at z = 100 cm were 1 to 2 cm/sec faster than those measured at z = 50 cm (B. Butman, personal communication).

These results suggest that during about 40 percent of the tidal cycle at this site, it is physically realistic for larvae to swim around near the bed, exploring available habitats for settlement. However, for about half of these flows (those between 4 and 6 cm/sec, see Fig. 4), larval maneuvering would be confined to distances of only 0.1-cm (or three body diameters of a 300-μm larva) above the bed. Obviously, the larger the settling larva, the smaller the maneuvering height. As mentioned earlier (see PROFILE CALCULATIONS AND RESULTS), larvae would encounter no opposed velocity if they swam vertically to heights of about 0.25 cm above the bed, within the viscous sublayer, in all of the smooth-turbulent profiles plotted (see Fig. 6). However, turbulent eddies, with velocities on-the-order-of 1 cm/sec or less and with components in all directions, are expected above the sublayer and periodically even inside the sublayer, making larval maneuvering in any direction difficult.

Rough-turbulent flows are expected, a priori, to have relatively lower mean velocities close to the bed than smooth-turbulent flows with the same mean-stream velocity, due to turbulent mixing near the bed. This was demonstrated in the cases modeled here (see Fig. 8). The advantages incurred by lower mean velocities near the bed may not outweigh the disadvantages of increased eddy penetration to within larval-body diameters of the bed, however, since larvae would constantly experience fluctuating eddy velocities in all flow directions. On-the-other-hand, larvae may find some refuge in the microtopography in slow-flow regions behind flow obstacles (e.g., see Eckman 1979, 1983).

The profiles calculated here are for quasi-steady (i.e., current) boundary layers, without consideration of possible effects of surface waves. Although wave-generated velocities do not reach the bottom at the study site discussed here, wave effects on the bed are prevalent in many common coastal habitats where larvae settle. Where wave effects extend to the seabed, wave boundary layers can form, in addition to current boundary layers. The combined effects of wave and current boundary layers on near-bed velocity profiles and sediment transport are discussed in Grant and Madsen (1979, 1982). In general, wave boundary layers are thinner than current boundary layers and higher stresses occur closer to the seabed in the wave boundary layer. These higher near-bed velocities have obvious implications to larval settlement.

A conclusion of this study is that, at least at the study site modeled, larvae probably do not search for preferred habitats by active horizontal swimming near the bed, since the bulk of the flows modeled would not permit such searches even for maximum measured larval swim speeds. The larvae may swim vertically in smooth-turbulent flows, going down to test the substrate and up to be advected to another site downstream. It is curious that measured swim speeds and fall velocities of polychaete larvae are the same order-of-magnitude, suggesting that larvae may only be able to maintain position in the water column while swimming up; measurements of swim speeds and fall velocities for the same individual are required to test this hypothesis.

An estimate was made here of advection distances between substrate tests by a larva that used, for example, the "balloonist technique" (coined by P.A. Jumars, personal communication) where an organism swims or is lifted up off the bottom, is advected with the flow and then passively sinks to a new site downstream. Using this technique, larvae could test substrates separated by scales of millimeters to meters, depending on their starting height above the bed (see Fig. 10).

This first attempt to determine near-bed flow velocities relative to aspects of larval settlement biology has suggested some realistic flow-regime dependent settlement mechanisms meriting further study. The calculations were necessarily idealized, in some cases, but the idealizations do not significantly affect the outcome of the study. The modeled field profiles represent first-order-type solutions for the purposes of hypothesis development; direct measurements are needed to test the ideas presented here. The analyses suggest many important areas for future research on larval biology (i.e., quantifying swim speeds and directions, fall velocities, and excursion heights above the bed during searches) and on larval ecology during settlement (i.e., quantifying habitat selection for a realistic range of field flows modeled in a laboratory flume).

ACKNOWLEDGEMENTS

I thank B. Butman for making the physical measurements in the field and P. Shoukimas for processing these data. I am very grateful for the invaluable tutoring and advice given to me by W.D. Grant on boundary-layer flows. G. McManamin skillfully typed the manuscript on very short notice, for which I should be shot. The writing was supported by a grant from the PEW Memorial Trust and NSF OCE-85000875; ongoing field studies on aspects of near-bottom flows, sediment transport and benthic biology in Buzzards Bay are funded by the W.H.O.I. Sea Grant Program (R/P-21) NOAA Contract No. NA84AA-D-00033 and the Coastal Research Center at W.H.O.I.

510

REFERENCES

Angel, M.V. and O'Brien, J.J. (Editors), 1984. The diel migrations and distributions within meospelagic communities in the north east Atlantic. Prog. Oceanogr., 13: 245-511.

Baggerman, B., 1953. Spatfall and transport of Cardium edule L.. Arch. Neerl. Zool., 10: 315-342.

Boicourt, W.C., 1982. Estuarine larval retention mechanisms on two scales. In: V.S. Kennedy (Editor), Estuarine Comparisons. Academic Press, pp. 445-457.

Burke, R.J., 1983. The induction of metamorphosis of marine invertebrate larvae: stimulus and response. Can. J. Zool., 61: 1701-1719.

Butman, B. and Folger, D.W., 1979. An instrument system for long-term sediment transport studies on the continental shelf. J. Geophys. Res., 84: 1215-1220.

Caldwell, J.W., 1972. Development, metamorphosis, and substrate selection of the larvae of the sand dollar, Mellita quinquesperforata (Leske, 1978). Masters Thesis, University of Florida, 63 pp.

Cameron, R.A. and Hinegardner, R.T., 1974. Initiation of metamorphosis in laboratory cultured sea urchins. Biol. Bull., 146: 335-342.

Chia, F.-S., Buckland-Nicks, J., and Young, C.M., 1984. Locomotion of marine invertebrate larvae: a review. Can. J. Zool., 62: 1205-1222.

Chriss, T.M. and Caldwell, D.R., 1982. Evidence for the influence of form drag on bottom boundary layer flow. J. Geophys. Res., 87: 4148-4154.

Clauser, F.H., 1956. The turbulent boundary layer. Advances in applied mathematics, IV: 1-51.

Crisp, D.J., 1955. The behaviour of barnacle cyprids in relation to water movement over a surface. J. Exp. Biol., 32: 569-590.

Crisp, D.J., 1974. Factors influencing the settlement of marine invertebrate larvae. In: P.T. Grant and A.M. Mackie (Editors), Chemoreception in marine organisms. Academic Press, N.Y., pp. 177-265.

Crisp, D.J. and Meadows, P.S., 1962. The chemical basis of gregariousness in cirripedes. Proc. Roy. Soc. Lond. B., 156: 500-520.

Crisp, D.J. and Meadows, P.S., 1963. Absorbed layers: the stimulus to settlement in barnacles. Proc. Roy. Soc. Lond. B., 158: 364-387.

Doyle, R.W., 1975. Settlement of planktonic larvae: A theory of habitat selection in varying environments. Amer. Natur., 109: 113-126.

Doyle, R.W., 1976. Analysis of habitat loyalty and habitat preference in the settlement behavior of planktonic marine larvae. Amer. Natur., 110: 719-730.

Eckelbarger, K.J., 1977. Metamorphosis and settlement in the Sabellaridae. In: F.-S. Chia and M.E. Rice (Editors), Settlement and metamorphosis of marine invertebrate larvae. Elsevier, N.Y., pp.145-164.

Eckelmann, H., 1974. The structure of the viscous sublayer and the adjacent wall region in a turbulent channel flow. J. Fluid Mech., 65: 439-459.

Eckman, J.E., 1979. Small-scale patterns and processes in a soft-substratum, intertidal community. J. Mar. Res., 37: 437-457.

Eckman, J.E., 1983. Hydrodynamic processes affecting benthic recruitment. Limnol. Oceanogr., 28: 241-257.

Eckman, J.E., Nowell, A.R.M. and Jumars, P.A., 1981. Sediment destabilization by animal tubes. J. Mar. Res., 39: 361-374.

Fager, E.W., 1964. Marine sediments: Effects of a tube-building polychaete. Science, 143: 356-359.

Gallagher, E.D., Jumars, P.A. and Trueblood, D.D., 1983. Facilitation of soft-bottom benthic succession by tube builders. Ecology, 64: 1200-1216.

Grant, J., 1981. Sediment transport and disturbance on an intertidal sandflat: Infaunal distribution and recolonization. Mar. Ecol. Prog. Ser., 6: 249-255.

Grant, W.D. and Madsen, O.S., 1979. Combined wave and current interaction with a rough bottom. J. Geophys. Res., 84: 1797-1808.

Grant, W.D. and Madsen, O.S., 1982. Moveable bed roughness in unsteady oscillatory flow. J. Geophys. Res., 87: 469-481.

Grant, W.D. and Madsen, O.S., 1986. The continental shelf bottom boundary layer. Ann. Rev. Fluid. Mech., 18: 265-305.

Grant, W.D., Boyer, L.F. and Sanford, L.P., 1982. The effects of bioturbation on the initiation of motion of intertidal sands. J. Mar. Res., 40: 659-677.

Grant, W.D., Williams, A.J., 3rd and Glenn, S.M., 1984. Bottom stress estimates and their prediction on the Northern California Continental Shelf during CODE-1: The importance of wave-current interaction. J. Phys. Oceanogr., 14: 506-527.

Grant, W.D., Williams, A.J., 3rd and Gross, T.F., 1985. A description of the bottom boundary layer at the HEBBLE site: Low frequency forcing, bottom stress and temperature structure. J. Phys. Oceanogr., 66.

Gray, J.S., 1974. Animal-sediment relationships. Oceanogr. Mar. Biol. Ann Rev., 12: 223-261.

Gross, T.F. and Nowell, A.R.M., 1983. Mean flow and turbulence scaling in a tidal boundary layer. Cont. Shelf Res., 2: 109-126.

Hannan, C.A., 1984a. Planktonic larvae may act like passive particles in turbulent near-bottom flows. Limnol. Oceanogr., 29: 1108-1116.

Hannan, C.A., 1984b. Initial settlement of marine invertebrate larvae: The role of passive sinking in a near-bottom turbulent flow environment. Doctoral Dissertation, WHOI/MIT Joint Program, 534 pp.

Hardy, A., Sir, 1965. The open sea: It's natural history. Houghton Mifflin, Boston, 322 pp.

Hinze, J.O., 1975. Turbulence. 2nd edition, McGraw-Hill, Auckland, 790 pp.

512

Hogue, E.W. and Miller, C.B., 1981. Effects of sediment microtopography on small-scale spatial distributions of meiobenthic nematodes. J. Exp. Mar. Biol. Ecol., 53: 181-191.

Jumars, P.A. and Nowell, A.R.M., 1984. Fluid and sediment dynamic effects on marine benthic community structure. Amer. Zool., 24: 45-55.

Kennedy, V.S. (Editor), 1982. Estuarine comparisons. Academic Press, N.Y., 709 pp.

Levin, L.A., 1983. Drift tube studies of bay-ocean water exchange and implications for larval dispersal. Estuaries, 6: 364-371.

Longhurst, A.R., 1976. Vertical migration. In: D.H. Cushing and J.J. Walsh (Editors), The ecology of the seas. Blackwell Scientific Publications, Oxford, pp. 116-137.

Meadows, P.S. and Campbell, J.I., 1972. Habitat selection by aquatic invertebrates. Adv. Mar. Biol., 10: 271-382.

Mileikovsky, S.A., 1973. Speed of active movement of pelagic larvae of marine bottom invertebrates and their ability to regulate their vertical position. Mar. Biol., 23: 11-17.

Moore, P.G., 1975. The role of habitat selection in determining the local distribution of animals in the sea. Mar. Behav. Physiol., 3: 97-100.

Nowell, A.R.M., 1983. The benthic boundary layer and sediment transport. Rev. Geophys. Space Physics, 21: 1181-1192.

Nowell, A.R.M and Church, M.A., 1979. Turbulent flow in a depth-limited boundary layer. J. Geophys. Res., 84: 4816-4824.

Nowell, A.R.J. and Jumars, P.A., 1984. Flow environments of aquatic benthos. Ann. Rev. Ecol. Syst., 15: 303-328.

Nowell, A.R.M., Jumars, P.A. and Eckman, J.E., 1981. Effects of biological activity on the entrainment of marine sediments. Mar. Geol., 42: 133-153.

Oliver, J.S., 1979. Processes affecting the organization of marine soft-bottom communities in Monterey Bay, California and McMurdo Sound, Antarctica. Doctoral dissertation, Univ. of Calif., San Diego, 300 pp.

Palmer, M.A. and Gust, G., 1985. Dispersal of meiofauna in a turbulent tidal creek. J. Mar. Res., 43: 179-210.

Pearre, S., Jr., 1979. Problems of detection and interpretation of vertical migration. J. Plank. Res., 1: 29-44.

Pratt, D.M., 1953. Abundance and growth of Venus mercenaria and Callocardia morrhuana in relation to the character of bottom sediments. J. Mar. Res., 12: 60-74.

Redfield, A.C., 1953. Interference phenomena in the tides of the Woods Hole region. J. Mar. Res., 12: 121-140.

Rosenfeld, L.K., Signell, R.P. and Gawarkiewicz, G.G., 1984. Hydrodynamic study of Buzzards Bay, 1982-1983. Woods Hole Oceanographic Institution Tech. Rept., WHOI-84-5 (CRC-84-01), 134 pp.

Sanders, H.L., Grassle, J.F., Hampson. G.R., Morse, L.S., Garner-Price, S. and Jones, C.C., 1980. Anatomy of an oil spill: long-term effects from the grounding of the barge Florida off West Falmouth, Massachusetts. J. Mar. Res., 38: 265-380.

Scheltema, R.S., 1971. Larval dispersal as a means of genetic exchange between geographically separated populations of shallow-water benthic marine gastropods. Biol. Bull., 140: 284-322.

Scheltema, R.S., 1974. Biological interactions determining larval settlement of marine Invertebrates. Thal. Jugosl., 10: 263-296.

Schlichting, H., 1979. Boundary-layer theory. 7th edition, McGraw-Hill, N.Y., 817 pp.

Smith, J.D. and McLean, S.R., 1977. Spatially averaged flow over a wavy surface. J. Geophys. Res., 82: 1735-1746.

Strathmann, R.R., 1978. Larval settlement in echinoderms. In: F.-S. Chia and M.E. Rice (Editors), Settlement and metamorphosis of marine invertebrate larvae. Elsevier, N.Y., pp. 235-246.

Suer, A.L. and Phillips, D.W., 1983. Rapid, gregarious settlement of the larvae of the marine echiuran Urechis caupo Fisher and MacGinitie 1928. J. Exp. Mar. Biol. Ecol., 67: 243-259.

Tyler, P.A. and Banner, F.T., 1977. The effect of coastal hydrodynamics on the echinoderm distribution in the sublittoral of Oxwich Bay, Bristol Channel. Est. Coast. Mar. Sci., 5: 293-308.

Williams, J.G., 1980. The influence of adults on the settlement of spat of the clam, Tapes japonica. J. Mar. Res., 38: 729-741.

Wilson, D.P., 1948. The relation of the substratum to the metamorphosis of Ophelia larvae. J. Mar. Biol. Assoc. U.K., 27: 723-760.

Wilson, D.P., 1958. Some problems in larval ecology related to the localized distribution of bottom animals. In: A.A. Buzzati-Traverso (Editor), Perspectives in marine biology. Univ. Calif. Press, Berkeley, pp. 87-103.

Wilson, D.P., 1968. The settlement behavior or the larvae of Sabellaria alveolata (L.). J. Mar. Biol. Assoc. U.K., 48: 387-435.

Yaglom, A.M., 1979. Similarity laws for constant-pressure and constant-gradient turbulent wall flows. Ann. Rev. Fluid Mech., 11: 505-540.

TURBIDITY AND COHESIVE SEDIMENT DYNAMICS

Emmanuel Partheniades, Professor

Department of Engineering Sciences, University of Florida, Gainesville, Florida, U.S.A.

ABSTRACT

This review paper summarizes the most important recent advances on cohesive sediment flocculation, deposition and resuspension and ways of application in turbidity control. Equations for the degree and rates of deposition and resuspension are presented together with a simplified model of cohesive sediment-flow interaction which explains the observed phenomena. Factors affecting deposition and resuspension are also discussed.

INTRODUCTION

Turbidity in natural water bodies amounts to a partial or complete blockage of light transmission with serious environmental impacts the most direct and important of which is the interference with the photosynthetic process of plankton production. In coastal and estuarine waters turbidity is caused primarily by fine sediment (silt and clay) in suspension. These sediments may also cause severe pollution of the benthic layers and marine feeding grounds by adsorption of dissolved pollutants, such as heavy minerals, on the surfaces of suspended particles and subsequent flocculation and deposition. The muddy layer, thus formed, may contain a pollutant concentration much higher than that in the water. In addition to the environmental pollution, fine sediments constitute the primary source of shoaling in estuarine waterways and certain harbors as a result of which expensive dredging maintenance is required. The annual cost of such maintenance in the United States only approaches the half-a-billion dollar mark (Partheniades, 1979). Fine sediments in estuaries may be of marine or alluvial origin or may be introduced during dredging and filling operations. They constitute the main bed material in drowned river valley-type estuaries formed after the ice melt (Pritchard, 1967).

The need for a rational control of shoaling and turbidity motivated extensive research during the last thirty years on the fundamental and applied aspects of cohesive sediment dynamics as well as on the engineering and rheologic properties of estuarine sediments. The to-date acquired knowledge, although not yet complete, contributed to a substancial understanding of the processes of flocculation, transport, deposition and resuspension of fine cohesive sediments. This knowledge can already be incorporated in mathematical or hybrid models for the prediction of shoaling zones and shoaling rates. Of particular interest to the prediction and control of turbidity are the initiation, degree and rates of deposition of suspended sediment and the initiation, degree and rates of resuspension of deposited cohesive sediment beds. The most recent advances in these areas are herewith summarized and discussed.

THE PROCESS OF FLOCCULATION

Fine sediments are composed of particles small enough and of specific area high enough for the surface physico-chemical forces to become dominant. The nature of these forces and the factors affecting them have been summarized and discussed elsewhere (Corps of Engineers 1960, Partheniades 1962, 1965, 1973, 1979, and Partheniades and Paaswell 1970). Individual particles have a thin plate or needle shape with dimensions ranging from a fraction of one micron to a few microns. When dispersed in water the finer particles, specifically the ones with an average diameter of less than 2 microns, can be maintained indefinitely in supension by the Brownian motion of water particles, while even a relatively very small degree of agitation is sufficient to keep in suspension the larger range of the entire fine sediment population. A homogeneous dispersion of particles small enough that no measurable deposition takes place within a long time period is defined by van Olphen (Van Olphen, 1963) as "coloidal solution" or "sol"; otherwise it is defined as a "suspension". Most commonly, however, both cases are referred to as "clay suspensions".

Modern electron microscopy and X-ray diffraction techniques have revealed that clays are composed essentially of one or more members of a small group of clay minerals with a predominently crystaline structure whereby the atoms are arranged in definite geometric patterns. Argillaceous (clayey) material can be considered as made up of a number of these clay minerals stacked on each other in the form of a sheet of layered structure so that clay particles are shaped either as a book or sheet-like unit or as a bundle of needles, tubes and fibers.

Clay particles dispersed in water may be brought sufficiently close together, either by Brownian motion or by ambient turbulence, for the physico-chemical forces to interact. After collision they either separate or cling to each other. Under certain conditions and specifically in the presence of dissolved salts, the net effect of the interparticle physico-chemical forces is attraction. Colliding particles then stick together forming agglomerates which grow quickly to sizes sufficiently large for deposition. The phenomenon is known as "flocculation", the agglomerates are defined as "flocs" and the sol or suspension is called "flocculated" or unstable (Van Olphen, 1963). In a flocculated suspension the floc rather than the individual clay particle becomes the basic settling unit. The floc size distribution and, therefore, the deposition rates depend on both the interparticle physico-chemical forces and the flow conditions. The same physico-chemical forces provide the main resistance to erosion of deposited cohesive beds. These forces are functions of the clay mineralogy, the ions dissolved in the water and the stress history. In addition, the following factors may affect floculation, deposition and resuspension: (1) suspended sediment consentration as it affects the frequency of collision of suspended particles and flocs; (2) pH, which has been found to affect the electronegativity of the soil colloids; (3) temperature, for which there is limited evidence that it enhances flocculation and increases deposition rates; (4) dissolved organic matter, which promotes flocculation on the one hand by providing additional links between inorganic particles while on the other hand retards settling by reducing the overall density and the equivalent diameter of the flocs and (5) dissolved minerals when they affect the double-layer thickness or enter in some way into a physico-chemical reaction. These factors are discussed in references (Corps of Engineers, 1960; Partheniades, 1979) and summarized in (Partheniades, 1984).

To relate turbidity with cohesive sediment dynamics one has to consider the various stages flocculation goes through and the dominant factors and forces which control the growth of the particle agglomerates.

Assuming that the proper physico-chemical conditions for flocculation exist, colliding particles will form first the basic agglomerate unit previously defined as floc. In stationary or quasi-stationary waters, particle collision is caused by the Brownian motion only while in flowing waters the turbulent velocities far outweigh the Brownian motion as a flocculation factor. Settling flocs may collide among themselves or with other particles due to either differential settling, or turbulent velocities or both, thus forming second, third and higher order aggregates. For simplicity we shall consider only three classes of particle aggregates: the primary floc, which is the basic aggregate with the highest density; the floc aggregates, consisting of a number of flocs; and the aggregate network, consisting of a combination of floc aggregates. Networks are normally encountered in recent cohesive sediment deposits and at the lower depth of clay suspensions settling in quiescent waters. Krone in his studies of the rhelogic properties of estuarine sediments (Krone, 1963) estimated the floc density for seven clayey muds between 1.44 and 1.64 gr/cm^3. This estimate was based on a rheologic model for the viscosity of cohesive sediment suspensions and his experiments were conducted with a sediment concentration higher than 20000 ppm ($0.02 \ gr/cm^3$) and a water density of $\rho = 1.025 \ gr/cm^3$. This floc density range corresponds to a volume fraction of solid particles between 0.085 and 0.253 and to a water volume of 75% to 92% of the total floc volume. These values are comparable to those of consolidated recent cohesive sediment deposits in estuarine shoals. Figure 1 shows a schematic diagram of aggregate networks which is representative of the loosest state of recently deposited fine sediments (Partheniades, 1962, 1965).

The continuous growth of flocs and floc aggregates is counteracted by the stresses developed on the surfaces and within the settling units due to the drag forces exerted by the ambient water, to the collision with other settling flocs and aggregates and to the high near-bed velocity gradients, known also as "shear rates". In quasi-stationary waters the first two factors are essentially the only significant ones, while in moving waters the third factor, as recent research by the writer and co-workers have shown, (Mehta and Partheniades, 1973, 1975, 1979, 1982 and Partheniades, 1979), is the one which limits the maximum size of the settling aggregates. Equally important to the flow and/or settling-induced shear stresses is the strength of the flocs and of the higher order agglomerates. The agglomerate strength is determined by the magnitude of the interparticle physico-chemical bonds, which in turn are functions of the clay mineralogy and the chemistry of the water. The strength of flocs of montmorillonite mud from the San Francisco Bay was first studied by Krone (1962) in a laminar constant shear flow field generated between two mutually rotating cylinders. Considering the shear force developed during particle collision as the main factor limiting the floc size, he derived the relationship

$$\tau r = \frac{16 \, \Delta r}{3 \pi} \tau_{max} \tag{1}$$

where $\tau = \mu \, du/dy$ is the uniform shear stress in the laminar flow field, r is the radius of the floc or aggregate, Δr is the radius of the contact area during collision and τ_{max} is the shear strength of the floc or aggregate. From magnified photographs of settling flocs and from the measured total shear force at the inner cylinder, τ_{max} was estimated to be about 0.27 N/m^2. This value is within the eroding range of shear stress for a dense bed with a macroscopic shear strength of 0.250 N/m^2 in the author's early research (Partheniades, 1962, 1965).

For constant shear strength, τ_{max}, the floc size is related to the field shear stress τ through Eq. 1, i.e.

$$r = \frac{16}{3\pi} \Delta r \frac{\tau_{max}}{\tau} \tag{2}$$

This increase of r in inverse proportion to τ has been experimentally verified by Krone (1962) down to $\tau = 0.06$ $dynes/cm^2$. Below that shear stress a rapid and erratic increase of r was observed, apparently due to the formation of floc aggregates.

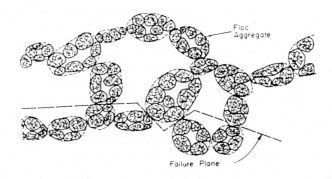

Fig. 1. Schematic picture of a flocculated cohesive bed at its loosest state
(Partheniades, 1962, 1965, 1973).

The interparticle cohesive bonds are expected to vary statistically within a wide
range. The law of distribution of these bonds is not known but it can reasonably be
assumed normal, (Partheniades, 1962, 1965). It follows that the strength of the
flocs and of the aggregates is expected to vary randomly. If the controlling near-
bed shear stress is constant, the maximum size, r, each aggregate may reach will be
also random, as determined by Eq. 2. This aspect will be further discussed in the
process of deposition and resuspension. Flocs and floc aggregates will eventually
approach the near-bed zone where the shear stress τ attains its maximum value. If

$$\tau_{max} > \frac{3\pi\tau r}{16\Delta r} \qquad\qquad (3)$$

the floc or aggregate will stick to the bed developing permanent physico-chemical
bonds with it; otherwise it will be broken up and reentrained to the main flow by
turbulence.

THE STRUCTURE AND PROPERTIES OF THE FLOCS

The microstructure of clay fabric has been under investigation for more than half
a century. Terzaghi (1925) and Casagrande (1932) presented a mechanistic model for
the clay fabric according to which numerous single clay particles are held together
by cohesion to form units or cells in the form of a honeycombed structure. A
discussion of the various theories and models on the microstructure of clay fabric
is beyond the purpose of this paper. A very good and concise summary was presented
by Quinn (1980). Until recently the prevailing concept was that the fundamental

unit in any clay fabric is the single particle. In fresh water deposits the dominant fabric was believed to be more open and of the cardhouse type with a prevailing edge to face attachment of particles, while in salt water the clay structure was assumed to be more random with edge to edge, face to face and edge to face attachments. More recent investigators challenged these ideas and claimed that the basic unit of clay fabric is the "clay packet", that is a domain of several clay particles in parallel array in the form of a book. This arrangement is referred to as "turbostratic structure" or "multiple aggregate particle fabric" and was recently investigated by Quinn under the supervision of Eades and the author. A scanning electron microscope was used to examine kaolinite flocs formed in quiescent water through a freeze-dried process (Quinn, 1980). The kaolinite particles ranged from 1 to 2 microns in average diameter. Regardless of clay concentration or electrolyte content, no "single particle behavior" was observed; instead, even in extremely dilute suspensions, the fundamental units were multiple particle books while the number of single particles was relatively small. The books appear in a form of steps; they are the product of feldspar weathering and, although they may weather further, this does not necessarily mean dispersion into individual platelets. Flocs formed in fresh water were found to become larger and structurally more complex as the clay concentration in suspension increased. A tendency was, moreover, detected for fresh water flocs to form elongated networks made up of platelets and packets joined together in edge-to-face and face-to-face stair-step configurations. The differences between fresh and salt water flocculation were not as drastic as expected. Apart from a somewhat more open structure in fresh water flocs, there is both edge-to-face and face-to-face flocculation of particles in either case. It could only be concluded that fresh water clays display a tendency to form open bookhouse-type randomly structured flocs with edge-to-face and edge-to-edge packet junctions while in salt water a greater tendency for more subparallel face-to-face packet junction was detected. Figures 2 and 3 give a good example of salt and fresh water flocs formed at the lowest suspended clay concentration (100 ppm). The clay packets are clearly displayed in both figures; however a single clay platelet can be observed in the fresh water floc which tends to produce a more open structure approaching the traditional picture of "cardhouse" structure. It was observed that the lower the concentration of suspended clay, the greater percentage there was of packets with less than five individual platelets per packet while the number of individual platelets per packet increases with increased concentration of suspended clay. This is clearly demonstrated in Figs. 4 and 5. In Fig. 4 an increase of face-to-face-oriented particles per packet is observed while the packets themselves are predominently oriented face-to-face. The packet in the left center is interpreted by Quinn as an example of a booklet that remained intact after initial formation via weathering of a feldspar. In the fresh water floc of Fig. 5, an increase of the number of particle platelets per packet is also observed; however, the latter follow more an open edge-to-edge and edge-to-face arrangement than those in the salt water floc of Fig. 4. Moreover, a single particle can be readily observed in the lower right contributing to the openness of the structure.

Figure 6, shows an example of a typical joint connecting flocs in floc aggregates and aggregate networks (Fig. 1). Figure 7 shows another floc aggregate formed in fresh water while Fig. 8 shows a similar floc aggregate formed in salt water under the same concentration of suspended clay as the previous two examples. The greater degree of randomness in comparison to the fresh water flocs is clearly observed. Finally, the same study indicated an increase of the average floc size with suspended sediment concentration (Fig. 9). This is, of course, to be expected, since the number of particles per unit volume and, therefore, the frequency of interparticle collisions are expected to be proportional to the suspended sediment concentration. It follows that the number of particles in a floc and, therefore, the floc volume at a particular time should also be approximately proportional to the suspended sediment concentration. This was indeed the case for $C = 0.1$ gr/ℓ, 0.2 gr/ℓ and 0.3 gr/ℓ. However, for concentrations higher than 0.3 gr/ℓ the floc volume was found to increase much more rapidly than C. There is no obvious explanation for this peculiar phenomenon at this time; it appears, however, that the answer should be sought in the initial floc-forming stage rather than in the collision frequency during settling.

The effect of ambient turbulence and shear stresses on the size, density and settling velocity of flocs was recently investigated by Kusuda et al (1981). It was found that both the average and maximum floc size decrease with increasing mixing intensity. The latter was quantitatively defined as

$$G = \frac{\sqrt{\varepsilon_0}}{\mu} \qquad (4)$$

where ε_0 is the rate of energy decay per unit volume of fluid, and μ is the dynamic viscosity of the fluid. G is indicative of the average rate of energy dissipation in the mixing chamber with rotating blades and can readily be computed. This mixing process generates a highly non-uniform flow field with zones of high velocity gradients and small scale eddies, such as between the blades, and zones of low velocity gradients and large scale eddies. The question arises as to which zone dominates the development of the flocs, since, due to the complexity of the flow field in the mixing chamber, the effect of each zone cannot be separated. Therefore, the conclusions reached in the study in question cannot very well be extrapolted to open channels. It was found next that when the clay suspension was first subjected to a high agitation rate, G_h, followed by a lower rate, G_ℓ, the flocs were less dense than the ones formed under the lower agitation rate G_ℓ only and that the floc density reduced with increasing ratio G_h/G_ℓ Apparently under constant agitation rate, G_ℓ, only primary flocs are formed with density and sizes as dictated by that particular rate, while in the case of a high mixing rate followed by a lower one, the final agglomerates are essentially floc aggregates consisting of groups of primary flocs formed during the first stage of high mixing rates. Moreover, even for a constant mixing rate the dry weight of flocs was found to be proportional to the 2.5 power of the floc diameter rather than the third power, which suggests that the average floc density decreases with increasing floc size. Expressions for the settling velocity and the floc density were given from curve fitting; however, for the aforementioned reasons, these expressions cannot be generalized.

Fig. 2. Salt water kaolinite floc at C = 0.1 gr/ℓ (Quinn, 1980)

Fig. 3. Fresh water kaolinite floc at C = 0.1 gr/ℓ (Quinn, 1980)

Fig. 4. Salt water kaolinite floc at C = 0.4 gr/ℓ (Quinn, 1980)

Fig. 5 Fresh water kaolinite floc at C = 0.4 gr/ℓ (Quinn, 1980)

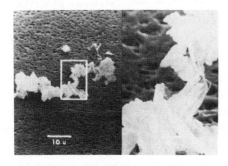

Fig. 6. Fresh water floc aggregate
at C = 0.5 gr/ℓ (Quinn, 1980)

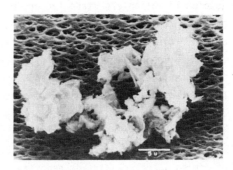

Fig. 7. Fresh water floc
aggregate at C = 0.5 gr/ℓ
(Quinn, 1980)

Fig. 8. Salt water floc aggregate
at C = 0.5 gr/ℓ (Quinn, 1980)

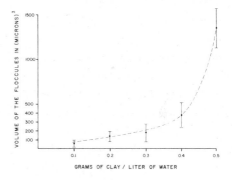

Fig. 9 Floc size - clay concentration relationship (Quinn, 1980)

DEPOSITION OF FLOCCULATED SUSPENSIONS

Deposition starts as soon as flocs grow large enough to obtain measurable settling velocities. In the hypothetical case of flocculation starting and developing
in a perfectly quiescent water environment, the Brownian motion constitutes the only
cause of particle collision. This would result in relatively slow flocculation
rates since the molecular diffusivity is of the order of kinematic viscosity, i.e.
$10^{-3} cm^2/sec = 10^{-1} mm^2/sec$. In contrast, in a typical case of an open channel with
an average velocity of 1m/sec, a Manning's friction coefficient n = 0.025 and a
hydraulic radius R_h = 10m the average eddy diffusivity is of the order of
$300 \ cm^2/sec$. In reality, however, natural water bodies very seldom remain in an

absolutely motionless state for prolonged time periods. Even in lakes, tideless bays and reservoirs, water inflow, winds, density currents and Coriolis acceleration are sufficient to generate a large scale motion and mixing. Moreover, the chances are that fine sediments are sufficiently flocculated for deposition to start before they enter the more stagnant zones of estuaries. Since deposition constitutes the basic mechanism for elimination or reduction of sediment-induced turbidity, the question arises as to what hydraulic parameters and physico-chemical properties of the sediment-water system control the initiation, degree and rates of deposition of suspended fine sediments as well as the initiation, degree and rates of resuspension of similar sediment already deposited. These questions have been the subject of intensive research by the author and his co-workers for the past twenty five years and led to a fundamental framework for cohesive sediment dynamics (Metha and Partheniades, 1973, 1975, 1979, 1982; Partheniades, 1979, 1984; Partheniades and Kennedy, 1966; Partheniades and Paaswell, 1970). The most important results of the work pertaining particularly to deposition are herewith summarized.

It should first be pointed out that deposition experiments in quiescent water may reveal some important aspects, such as the effect of salinity and other physico-chemical factors on flocculation and deposition, but bear little resemblance to the situation in flowing water. In the first case the only restriction to floc growth are the settling-generated shear stresses on the flocs and, possibly the stresses due to floc collision. All aggregates will eventually deposit and a settling inter-face is formed separating turbid water from almost clear water. This interface marks the lower limit of settling velocity of the flocs. In flowing waters the turbulent velocity fluctuations constitute the dominant mechanism of particle and floc collision. The floc and aggregates thus formed are expected to have sizes and strengths ranging between an upper and a lower limit. All of them will eventually approach the bed; however, only the ones with size and strength satisfying Eqs. 1 and 2 will reach the bed and will become part of it. Otherwise, they will be broken up and reentrained back into the main flow. It appears, therefore, that, although the flocs and floc aggregates are formed in the far-bed flow region, where the shear rates are minimal and the turbulence structure nearly homogeneous, the degree and rates of deposition will be determined in a near-bed zone where the shear rate attains its highest value. Upon contact with the bed, flocs and floc aggregates will develop bonds with the latter so that a lift force considerably higher than the submerged weight of the settled unit will be required for resuspension. Moreover, it has been shown (Blinco and Partheniades, 1971) that the turbulent velocity fluctuations and the turbulent diffusivity in the far-bed region are functions of the bed friction and the near bed flow conditions. It follows that the initiation, degree and rates of deposition are expected to be functions of the near bed flow conditions. Unlike in quiescent settling, in flowing sediment suspensions no interface between turbid and clear water is formed. Due to the low settling velocities, the concentration gradients of suspended sediment are very low, provided that other factors, such as density stratification, are not present. The recent research on the subject has been in determining the relationships linking the initiation, degree and rates of deposition to the near-bed flow conditions.

From deposition experiments in quiescent water with a silty-clay sediment from the San Franscisco Bay, Krone (1962) found that, for constant initial sediment concentration, substancial flocculation takes place for a salinity of 1000 ppm. The median settling velocity increases with increasing salinity attaining a maximum for a salinity ranging from 5000 ppm (for C = 120 ppm) to 15000 ppm (for C = 1000 ppm) where C is the concentration of suspended sediment. The median settling velocities corresponding to the optimum salinity were found to follow the law

$$v_s = \alpha\, C^{4/3} \tag{5}$$

where α is a constant. Krone explained the above relationship on the basis of Kruyt's law for the kinetics of flocculation (Krone, 1962) which reads

$$N_n = \frac{N_0\,(t_f/t_c)^{n-1}}{(1 + t_f/t_c)^{n-1}} \tag{6}$$

where N_0 is the number of primary particles in the suspension, N_n is the number of flocs with an average number of n primary particles per floc; t_f is the time of flocculation; and t_c is the average time between collision or the reciprocal of the collision probability. For $t \gg t_c$ Eq. 6 can be approximated by

$$\frac{N_0}{N_n} = (\frac{t_f}{t_c})^2 \tag{7}$$

N_0/N_n is equal to the number of particles n per floc. If r is the floc diameter then N_0/N_n is proportional to r^3 and r is proportional to $(N_0/N_n)^{1/3}$ or to $(t_f/f_c)^{2/3}$. The settling velocity v_s then being proportional to r^2, is also proportional to $(t_f/t_c)^{4/3}$. The collision probability $1/t_c$ is expected to be proportional to C; therefore, for constant flocculation time t_f, v_s should be proportional to $C^{4/3}$, which indeed was the case. Krone attempted next to find a sediment deposition law in terms of the bed shear stress for the San Francisco Bay mud, which consists of about equal proportions of silt and clay. The clay portion is predominently composed of montmorillonite with some illite and traces of some fine sand and organic matter. Krone first gave a critical shear stress for deposition equal to 0.6 dynes/cm^2 and then he derived the following experimental concentration-time laws:

For suspended sediment concentration less than 0.3 gr/ℓ

$$C = C_o \exp \left[- \frac{v_s p_r t}{h} \right] \qquad (8)$$

where C_o is the initial mass concentration of sediment; t is the time; h is the depth of flow, $p_r = 1 - \tau_b/\tau_\ell$; τ_b is the bed shear stress and τ_ℓ is the "threshold of τ_b" for deposition being equal to 0.6 dynes/cm². Krone claimed that for $\tau_b > 0.6$ dynes/cm² no sediment deposits. The term P_r was interpreted as the "probability of a particle sticking to the bed".

For C > 0.3 gr/ℓ he gave the law

$$\log C = - k \log t + \text{constant} \qquad (9)$$

where k is a constant depending on τ_b/τ_ℓ. For concentrations 0.3 gr/ℓ < C < 10 gr/ℓ he found a value for τ_ℓ equal to 0.8 dynes/cm² rather than 0.6 dynes/cm². The reason for the existence of two different laws was attributed to the hinderance during settlings. The limit of 0.3 gr/ℓ, however, is too low to account for any drastic change in the flocculation and deposition process. From Eq. 8 v_s was found equal to 6.6×10^{-4} cm/sec which was about 20 percent of the settling velocity in quiescent water.

Equation 8 can be explained by the sediment continuity equation:

$$\frac{dC}{dt} = - \frac{mN \, v_s P_r}{h} \qquad (10)$$

where N is the number of particles or flocs per unit volume and m is the average mass of each particle or floc. Obviously mN = C which leads to Eq. 8 provided that v_s does not change with time. Equation 9 does not seem to follow any sediment continuity model.

The author (Partheniades, 1962, 1965, 1979, 1984; Partheniades and Paaswell, 1970) investigated first the erosion mechanism of dense and flocculated cohesive sediment beds and presented a model for the erosion process giving the erosion rates as a function of the bed shear stress and the physico-chemical properties of the soil-water system. His first deposition experiments suggested that no interchange of suspended and bed particles takes place. This conclusion was confirmed in the following twenty years of research on deposition and resuspension.

The observed independent occurence of erosion and deposition of cohesive sediments led to separate investigations for each process notwithstanding the fact that they both constitute two phases of the same phenomenon. The more recent work started with the study of the depositional behavior of fine sediment suspensions, first at M.I.T. in 1963 (Partheniades and Kennedy, 1966, Partheniades, Cross and Ayora, 1968) and later on at the University of Florida from 1968 to

1983 (Mehta and Partheniades, 1973, 1975, 1979, 1982; Partheniades, 1973, 1979, 1984). The experiments were conducted in a special apparatus developed first at M.I.T. and subsequently enlarged and improved at the University of Florida. The last one is pictured is Fig. 10 and outlined schematically in Fig. 11. Its main component consists of an annular channel 60 in. (1.525 m) in mean diameter, 4 in. (10.2 cm) wide and with a maximum depth of 18 in. (45.8 cm) containing the water-sediment mixture, and a vertically movable annular ring positioned within the channel and in contact with the water surface. A simultaneous rotation of the two components in opposite directions generates a turbulent uniform flow field free of floc disrupting elements, such as pump blades, return pipes and diffusers, which have to be present in conventional laboratory flumes. The effect of the rotation-induced secondary currents on the deposition has been eliminated by a proper selection of the speeds of the channel and of the ring so that the sediment deposits uniformly accross the channel. At these operational speeds the bed shear stress across the channel measured by a Preston tube was found to be uniform (Mehta and Partheniades, 1973, 1975). The apparatus is properly instrumented for a direct and precise measurement of the shear stresses at the ring and at the channel bottom.

Figure 12 shows four typical velocity profiles in the middle portion of the channel displaying two wall layers of high shear rates and a core segment of near constant velocity and near homogeneous turbulence. A plot of the near bed velocity profiles suggested a logarithmic velocity distribution law of the form:

$$\frac{u}{u_*} = \frac{1}{\kappa} \log \frac{y}{y_0} \tag{11}$$

where u_* is the friction velocity at the bed; $\kappa = 0.40$, as in pipes and open channels; and y_0 is the distance at which $u = 0$ found to be equal to about 0.5 mm.

Figure 13 shows a typical example of suspended sediment concentration-time curves in the outlined experimental system where C is the concentration at time t and C_0 is the value of C at the start of deposition. It is observed that, after a short transient period, the concentration reaches a constant value herein defined as "equilibrium concentration" symbolized by C_{eq} which decreases with decreasing bed shear stress, τ_b. C_{eq} becomes zero for τ_b below a critical value, $\tau_{b\ min}$.

Figure 14 shows a concentration-time plot for kaolinite clay in distilled water for constant flow conditions but variable initial concentration, C_0 (Partheniades, 1973, 1984 and Partheniades and Kennedy, 1966). It is observed that the relative equilibrium concentration, $C_{eq}^* = C_{eq}/C_0$, remains constant and independent of C_0. This means that a given flow can maintain in suspension a constant

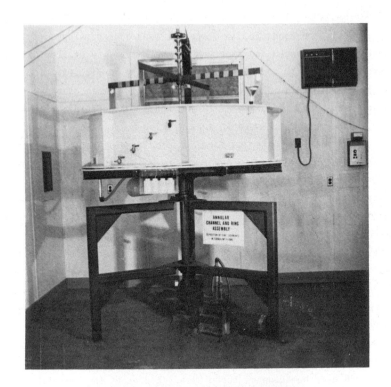

Fig. 10. Annular channel and ring assembly (Mehta and Partheniades, 1973, 1975).

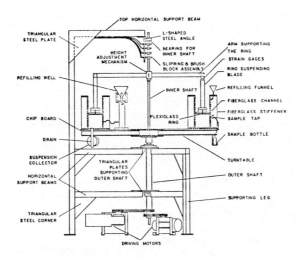

Fig. 11. Schematic outline of annular channel and ring assembly (Mehta and Partheniades, 1973, 1975).

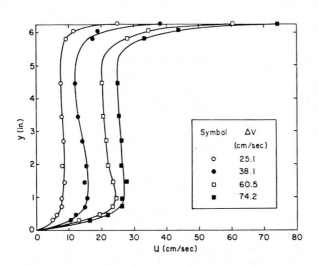

Fig. 12. Typical velocity profiles in the annular channel (Mehta and Partheniades, 1973).

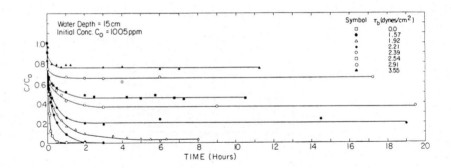

Fig. 13. Typical suspended sediment concentration-time curves for cohesive sediments (Mehta and Partheniades, 1973).

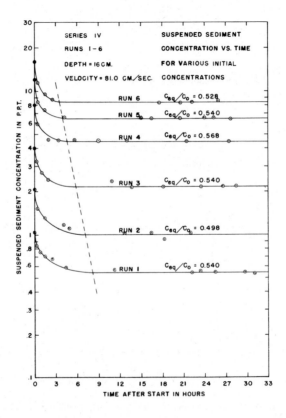

Fig. 14. Variation of suspended sediment with time (Partheniades, 1973, 1984 and Partheniades and Kennedy, 1966).

fraction of a particular sediment regardless of the absolute value of the concentration. This conclusion was verified for a variety of clay types and water
chemistry. It can be concluded that C_{eq}^{*} represents that part of suspended
sediment which can never form flocs with sufficiently strong bonds to resist the
near-bed disruptive shear stresses and join the bed. Likewise, the ratio
$C_{eq}^{**} = C_0 - C_{eq}/C_0 = 1 - C_{eq}^{*}$ represents that portion of the suspended sediment
that is able to form flocs with sufficiently strong bonds to reach the bed and
become part of it. The existence of the equilibrium concentration precludes any
exchange between bed and suspended particles or flocs during deposition and the
material in suspension at equilibrium is composed of the same particles. This
conclusion has been directly verified by gradually replacing the water-sediment
mixture in the rotating channel at equiblirium with clean water (Partheniades,
Cross and Ayora, 1968).

Under steady-state conditions, i.e. when equilibrium concentration is attained, there is a balance between the downward sediment mass flux due to gravitational settling and the upward diffussive transport, according to the equation

$$-v_s C = \varepsilon_y \frac{dC}{dy} \tag{12}$$

where ε_y is the turbulent diffusivity in the y direction normal to the bed. For
the zone far from the bed, where the velocity is nearly constant and the turbulence appears to be nearly homogeneous, ε_y is expected to be almost constant and
independent of y. Indicating by ε that constant value, integration of Eq. 12 for
constant v_s yields

$$\frac{C}{C_a} = \exp\left[-\frac{v_s}{\varepsilon} (y - a) \right] \tag{13}$$

where C_a is the concentration at the reference point y = a. Figure 15, shows a
concentration distribution plot for τ_b = 1.85 dynes/cm^2 which satisfies quite
closely Eq. 12 with v_s/ε = 0.026 in.$^{-1}$.

For τ_b = 2.95 dynes/cm^2, v_s/ε was equal to 0.015 in.$^{-1}$ (5.12 x 10^{-3}cm^{-1}).
Early velocity distribution measurements in an open flume with suspended sediment
by the writer with the help of a specially designed Prandtl tube, indicated that
the velocity profiles were not measurably affected by the suspended sediment even
at concentrations near 10,000 ppm (Partheniades, 1962). This suggests that the
turbulent eddy diffusivity is unaffected by the suspended sediment at least for
concentrations up to that limit. Indeed, experiments with initial concentration
with kaolinite C_0 equal to 7,680 ppm and 16,900 ppm and at bed shear stresses of
1.87 dynes/cm^2 and 1.85 dynes/cm^2 respectively gave values of v_s/ε equal to
0.022 in.$^{-1}$ (8.66 x 10^{-3}cm^{-1}) and 0.026 in.$^{-1}$ (10.24 x 10^{-3}cm^{-1}) respectively,

i.e. a deviation of the order of the experimental error (Mehta and Partheniades, 1973). These conclusions, however, should be considered as tentative. Moreover, the eddy diffusivity as well as the overall turbulence structure is expected to be affected by the suspended sediment for concentration above a certain limit. This phase has not been studied yet. The presented results indicate that in the uniform velocity region ε is a function of the bed shear stress. From Eq. 12 the actual settling velocity v_s can be obtained if ε can be measured by a hot film anenometer.

Figure 16 shows a plot of C_{eq}^* vs the bed shear stress, τ_b, for various depths and initial concentrations. An extrapolation of the plot intersects the abscissa at τ_{bmin}^* which is the limit for complete deposition. It can be concluded that C_{eq}^* depends only on τ_b and that there is no such a thing as a "critical shear for deposition" τ_ℓ, as implied in earlier investigations (Krone, 1962); a measurable degree of deposition can take place for τ_b as high as ten times the value of τ_{bmin}.

The specific law for the degree of deposition, as represented by C_{eq}^*, was next sought. Figure 17 shows a normal-logarithmic plot of C_{eq}^* versus the "non-dimensionalized excess bed shear stress" $\tau_b^* - 1$, defined as the "bed shear stress parameter", where $\tau_b^* = \tau_b/\tau_{bmin}$, for kaolinite clay in distilled water. This plot leads to the relationship

$$C_{eq}^* = \frac{1}{\sqrt{2\pi}} \int_{-\infty}^{Y} \exp\left(-\frac{\omega^2}{2}\right) d\omega \tag{14}$$

where

$$Y = \frac{\log\left[(\tau_b^* - 1)/(\tau_b^* - 1)_{50}\right]}{\sigma_y} \tag{15}$$

Here σ_y is the geometric standard deviation; $(\tau_b^* - 1)_{50}$ is the geometric mean, i.e. the value of the bed shear stress parameter for which 50 percent of the total sediment deposits; and ω is a dummy variable. The data of Fig. 17 give $\sigma_y = 0.049$ and $(\tau_b^* - 1)_{50} = 0.84$.

The unique dependance of C_{eq}^* on τ_b confirms the earlier stated notion that deposition of cohesive sediments is controlled at the bed while the turbulence in the far-bed region contributes only to the formation of the flocs. The same conclusion was reached for the erosion of cohesive sediment beds.

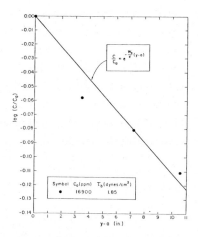

Fig. 15. Depth-concentration profile at equilibrium (Mehta and Partheniades, 1973).

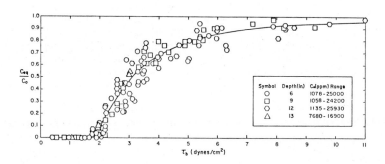

Fig. 16. Relative equilibrium concentration versus bed shear stress (Mehta and Partheniades, 1973).

Equations 14 and 15 were subsequently extended to other clay types and water qualities. In Fig. 18 two sets of data have been plotted, one for a mixture of equal parts of kaolinite and San Francisco Bay mud in salt water at ocean salinity (series C) and another for San Francisco Bay mud only also at ocean salinity (series D), together with the average dotted line from Fig. 17. The same data are plotted in Fig. 19 together with the data of series A and B for kaolinite clay in salt water at ocean salinity for depths of 6 in. and 9 in., respectively versus the parameter $(\tau_b^* - 1)/(\tau_b^* - 1)_{50}$. The data by the author et al from experiments at M.I.T. (Partheniades and Kennedy, 1966; Partheniades, Cross and Ayora, 1968), the only data by the author from his aforementioned original experiments in an open flume with San Francisco Bay mud (Partheniades, 1965, 1962) and the data by Rosillon and Volkenborn from experiments in an open flume with water at ocean salinity and sediment from the Maracaibo Bay in Venezuela (Rosillon and Volkenkborn, 1964) have also been plotted. The agreement for sediment-water systems so drastically different is indeed remarkable. It can be concluded that all the physico-chemical properties of the sediment-water system, which determine the distribution of the size and the strength of the flocs, can be represented by two readily determinable parameters: τ_{bmin} and $(\tau_b^* - 1)_{50}$.

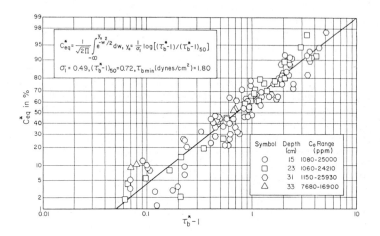

Fig. 17. Relative equilibrium concentration versus bed shear stress parameter $\tau_b^* - 1$ (Mehta and Partheniades, 1973, 1975).

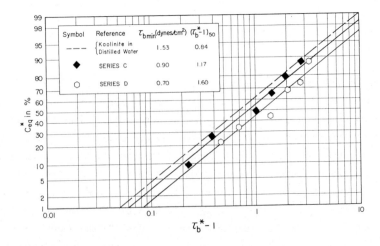

Fig. 18. Relative equilibrium concentrations versus $\tau_b^* - 1$ for various sediment suspensions (Mehta and Partheniades, 1973, 1975).

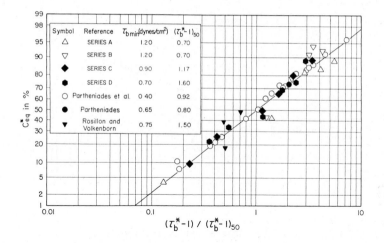

Fig. 19. Relative equilibrium consentrations versus $(\tau_b^* - 1)/(\tau_b^* - 1)_{50}$ for various sediment suspensions (Mehta and Partheniades, 1973, 1975).

For four different sediment-water systems $(\tau_{bmin}^* - 1)_{50}$ appears to be closely related to τ_{bmin} through the equation

$$(\tau_b^* - 1)_{50} = 4 \exp(-1.27 \ \tau_{bmin}) \tag{16}$$

with τ_{bmin} in dynes/cm^2 (Mehta and Partheniades, 1973). It appears, therefore, that τ_{bmin} is really the only parameter representing the effect of the physico-chemical properties of the sediment-water system on the degree of deposition. This conclusion has to be confirmed for more cases. There appears to be a corre-lation of both τ_{bmin} and $(\tau_b^* - 1)_{50}$ with the cation exchange capacity (CEC) of the sediment. However, CEC is a property of the sediment only and not of the sediment-water system. Its use, therefore, for the estimate of the degree of deposition is questionable.

The time rates of deposition of the depositable part of the sediment were in-vestigated next. In a closed system, like the experimental set-up used, the depositable part of the sediment can be represented by the concentration differ-ence $C_0 - C_{eq}$. Therefore, the ratio $C^* = (C_0 - C)/(C_0 - C_{eq})$ where C, the suspended sediment concentration at time t, represents the fraction of the depositable sediment deposited during time t after the beginning of deposition. Figures 20a and 20b show two logarithmic-normal plots of C^* versus the non-dimensionalized time, t/t_{50}, for low and high initial concentrations, where t_{50} is the time at which $C^* = 0.50$.

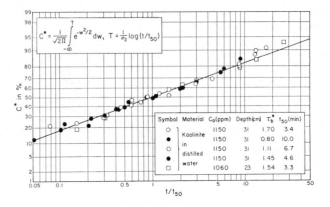

a. Low initial concentration.

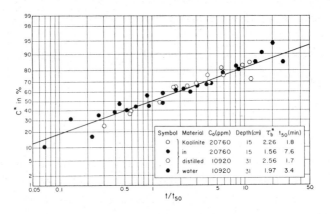

b. High initial concentration.

Fig. 20. Deposition rates (Mehta and Partheniades, 1973, 1975).

All points fall very closely on straight lines described by the equations

$$c^* = \frac{1}{\sqrt{2\pi}} \int_{-\infty}^{T} \exp\left(-\frac{\omega^2}{2}\right) d\omega \tag{17}$$

where

$$T = \frac{1}{\sigma_2} \log\left(t/t_{50}\right) \tag{18}$$

where σ_2 is the geometric standard deviation. Therefore, the deposition rates can be expressed by one and the same law, at least up to the limit of 20,000 ppm rather than by the three different laws (Eqs. 7, 8 and 9) given by Krone (1962). For initial sediment concentrations above 25,000 ppm the deposition rates start deviating from Eqs. 16 and 17 (Mehta and Partheniades, 1973). Similar deviations and somewhat erratic behavior were also observed in some cases in which τ_b was substantilaly lower than τ_{bmin}. It appears that for low values of τ_b, the deposition rates depend more and more on the size of aggregates formed in the main flow and that the random process of the formation of these aggregates is reflected in the erratic nature of the deposition. Deposition of kaolinite suspensions in salt water and of Maracaibo sediments (Rosillon and Volkenborn, 1964) also followed the laws given by Eqs. 16 and 17. The same is true for deposition rates of San Francisco Bay mud obtained by Krone (1962) and the writer (Partheniades, 1962, 1965) in flume experiments; however, in the two last cases t_{50} was much higher than in the experiments in the annular channel. The obvious explanation for this difference lies in the intensity and structure of turbulence in the far-bed region. In the rotating channel-ring system the turbulence structure in that region is defined, as already explained, exclusively by the near bed flow conditions and specifically by the bed shear stress. In the flume experiments the turbulence level was significantly influenced by the conditions in the return pipe, where the boundary stresses were much higher than those in the flume, and by the turbulence generated by the pump blades and the diffusers.

Figure 21 shows an example of correlation of the basic deposition rate parameters σ_2 and t_{50} with τ_b^* and initial sediment concentration, C_0, from which the following general conclusions have been reached:

1. The mean deposition time, t_{50}, increases first with τ_b^* reaching its peak for values of τ_b^* between 1 and 1.5. From then on it decreases with increasing τ_b^*.

2. For the same τ_b^*, t_{50} seems to increase with increasing depth although there is considerable overlapping.

3. The standard deviation, σ_2, increases initially with increasing τ_b^* reaching its peak, like t_{50}, for values of τ_b^* between 1 and 1.5. From then on, it appears to remain constant or to slightly decrease with increasing τ_b^*.

4. σ_2 seems to be affected very little by the depth of flow but the nature of that effect is not clear yet.

These conclusions can be explained as follows. We recall first that σ_2 is a measure of the spread of the settling time of the floc population, and t_{50} is the mean settling time. Indicating by h the channel depth and by U_{sf} the settling velocity of a floc from the part of the entire floc population which eventually reaches the bed, then the time t_s given by the ratio

$$t_s = \frac{h}{U_{sf}}$$

(19)

could be considered as a measure of settling time for that particular floc.

Figure 22, shows a schematic probability density function, $f(\log t_s)$ for the logarithm of the settling time, t_s. For near zero velocities the flocs are expected to reach their maximum size and t_{s50} its minimum value, since the floc disrupting stresses are then negligible and the floc size would then depend solely on the frequency of particle and floc collision as they settle to the bed. Moreover and for the same reason, the smaller flocs joint readily together to form larger units something that would narrow down the spread of the distribution and would decrease σ_2. With increasing shear stress more and more flocs are eliminated from the permanently depositable population starting from the smallest floc diameter and increasing gradually. Therefore, the depositable population is composed of larger and larger flocs as the bed shear stress increases. At the same time the turbulence intensity and the coefficient of vertical eddy diffusivity increase thus retarding the eventual deposition of a depositable floc to the bed. In fact both U_{sf} and t_s in Eq. 19 are apparent values in a turbulent flow field. An increase of the apparent settling time is reflected in a shifting of $\log t_{s50}$ to the right in Fig. 22. Thus the settling time and, therefore, the rates of deposition, are subjected with increasing τ_b to two opposing effects: an increase of the sizes of flocs which eventually settle and an increase of the turbulent diffusivity. According to Fig. 21, the diffusion effect appears to be initially dominant. However, the settling velocity increases in proportion to the third power of the floc diameter. An optimum value of τ_b is, therefore, expected to exist beyond which the weight of the flocs becomes dominant. At that point t_{50} reaches its highest value while beyond it the increase of U_{sf} will overcompensate the increase of eddy diffusivity. The net result is a decrease of the overall settling time and a shifting of $\log t_{s50}$ to the left.

Regarding now the standard deviation, σ_2, we first observe that an increase of τ_b breaks down the weaker flocs and floc aggregates thus increasing the spread of the distribution of the floc sizes. As long as $\tau_b \lesssim \tau_{bmin}$ eventually all flocs deposit; however, the spread of the floc size distribution is expected to increase with τ_b; this is indeed the case in Fig. 21. When the equilibrium concentration, C_{eq}, becomes substantial, the spread starts narrowing down and eventually it either levels off or even decreases. Indeed, Fig. 21, as well as other experiments not shown in this paper, indicate a sharp increase of σ_2 to a value of τ_b^* between 1 and 1.5 while for values of τ_b^* above the limit σ_2 either remains constant or decreases but at a rate much smaller than the rate of increase for

$\tau_b^* < 1$ to 1.5 (Partheniades, 1984). It should be noted that the peak of the log t_{50} – τ_b^* curves occurs within the same range of τ_b^*.

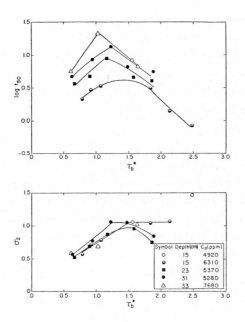

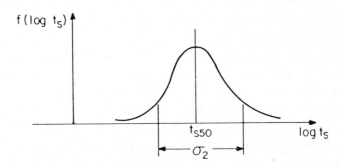

Fig. 21. Example of variation of t_{50} and σ_2 with τ_b^* (Mehta and Partheniades, 1973, 1975)

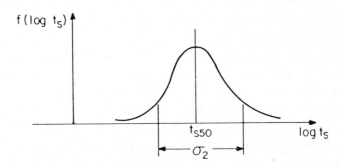

Fig. 22. Schematic probability density function of log t_s.

From Eqs. 16 and 17 the following expression can be derived for the time-rate of deposition:

$$\frac{dC^*}{dt} = \frac{0.434}{\sqrt{2\pi}\sigma_2 t} \exp\left(-\frac{T^2}{2}\right) \tag{20}$$

It should be kept in mind, however, that this equation is valid only for a non-dispersive system, i.e. for a flow system without longitudinal concentration gradients.

The depositional parameters appearing in Eqs. 14, 15, 17 and 18 change as the sediment disperses and as it goes through the deposition-resuspension cycles. These changes have been recently investigated by Mehta, the author and others and they reflect changes in the size and strength of flocs (Dixit, Mehta and Partheniades, 1982; Mehta, Partheniades and McAnnaly, 1982). A brief summary of this investigation is given by the writer (1984).

A MODEL OF INTERACTION BETWEEN FLOW, SUSPENDED SEDIMENT AND BED

The summarized erosional and depositional behavior of cohesive sediments appear at first glance to contradict the hydrodynamic behavior of coarse sediments while they resemble in certain aspects the wash load (Einstein, 1950). A theoretical model explaining the observed behavior of cohesive sediments and linking the wash load and bed material load has recently been developed by the author (1977). An outline of the essentials of that model is herewith presented.

According to Einstein's original approach, a near bed sediment grain in suspension will deposit if the instantaneous hydrodynamic lift force, L, exerted by the flow on the particle, is less than the submerged weight of the particle, W_b; that is if $W_b/L > 1$. Likewise, a bed particle will be entrained if L exceeds the summation of the erosion-resisting forces. For coarse grains Einstein considered the submerged weight of the grain, W_b, as the only resisting force. In general, however, the entrainment will also be resisted by the interparticle or interfloc friction and interlocking forces and, in the case of cohesive sediments, by the physico-chemical bonds developed between the flocs or particles and the bed. Friction and interlocking can be incorporated in W_b through a dimensionless coefficient, β, larger than unity. The physico-chemical bonds can be represented by a net attractive force, F_a, between the particle or floc and the bed (Fig. 23). Thus, the total resisting force is $\beta W_b + F_a$ and the condition for erosion becomes:

$$\frac{\beta W_b + F_a}{L} \leqslant 1 \tag{21}$$

Next, Einstein, on the basis of his experiments with El Sammi (Einstein and El Samni, 1949), assumed the following stochastic form for L:

$$L = \overline{L} (1 + n) \tag{22}$$

in which $\bar{L}$ is the mean value of L and n is a dimensionless random variable with mean zero and standard deviation n_0 found to be equal to $\frac{1}{2.75}$ for semicircular spheres. The probability of erosion, P_e, then becomes:

$$P_e = P_r \left(\frac{\beta W_b + F_a}{\bar{L}} - 1 \leqslant n \right) = 1 - P_r \left(n \leqslant \frac{\beta W_b + F_a}{\bar{L}} - 1 \right) \qquad (23)$$

while the probability for deposition, P_d, takes the form:

$$P_d = P_r \left(n \leqslant \frac{W_b}{\bar{L}} - 1 \right) \qquad (24)$$

It becomes obvious from the above equations that:

(1) if $W_b < L < \beta W_b + F_a$ neither erosion nor deposition can possibly occur;
(2) if $L < W_b$ only deposition occurs; and (3) if $L > \beta W_b + F_a$ only erosion takes place.

In the original Einstein's model L was considered taking values from $-\infty$ to $+\infty$ while following the normal distribution law (Einstein, 1950; Einstein and El Samni, 1949). It would be more realistic, however, to assume that L is limited by an upper bound, L_u, and a lower bound L_ℓ. The relative magnitude of the intervals $L_e - L_d$ and $L_u - L_\ell$ and the relative location of the points L_e, l_d, L_u, L_ℓ determine the kind of the sedimentation process.

We consider first the situation where $L_e - L_d > L_u - L_\ell$ which specifically applies to cohesive sediments, since for the latter F_a becomes much larger than W_b and where β attains a maximum value due to the irregular shape of the clay particles and flocs. The interval then $L_e - L_d$ cannot possibly lie between L_u and L_ℓ, which means that L cannot reach values smaller than W_b and larger than $\beta W_b + F_a$. It follows that in this case simultaneous erosion and deposition is not possible. The following three special cases may occur:

1. $L_u > L_e$. In this case erosion may occur but no deposition since L can attain values larger than L_e but never smaller than L_d. This corresponds to the case of relatively high velocities.

2. $L_d < L_\ell < L_u < L_e$. In this case, as already explained, neither erosion nor deposition can possibly occur. The sediment is simply transported through the channel, like a wash load, leaving no traces on the bed.

3. $L_\ell < L_d$ in which case $L_u < L_e$. That means that deposition can occur since the lift forces are sufficiently low for deposition but never high enough for erosion. This case corresponds to relatively low flows and low bed shear stresses.

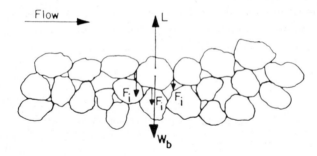

Fig. 23. Forces due to friction, interlocking, and physico-chemical attraction (Partheniades, 1977).

It follows from the above observations that for simultaneous erosion and deposition of sediment particles to take place and for the existence of a bed load function (Einstein, 1950) the interval $L_e - L_d$ should lie entirely within the interval $L_u - L_\ell$.

It should be noted that as the sediment becomes coarser, the forces F_a decrease and tend to zero. Likewise, with increasing grain size and as the sediment particles approach the spherical state, the coefficient β tends to unity. This is the reason why sediments which display the behavior of bed material load and for which a bed load function exists, are unexceptionally coarse (Einstein, Anderson, and Johnson, 1940).

RESUSPENSION OF DEPOSITED FINE SEDIMENT

Resuspension of deposited sediment is part of the normal deposition-resuspension cycles in tidal estuaries but it can also be the result of extreme events such as storms, waves and wind driven currents. An understanding of the resuspension process of naturally deposited flocs as well as equations for the initiation, degree and rates of resuspension are needed in order to be able to formulate an appropriate sediment source function in mathematical models for sediment transport and to predict turbidity levels.

The process of resuspensions has been recently investigated by Mehta and the author (Metha and Partheniades, 1979, 1982). A summary of this work is given in (Partheniades, 1984). The discussion of this research phase is here limited to the structure and engineering properties of the deposited bed since the latter is directly related to the structure of the floc aggregates which constitute the main theme of this paper.

It was recognized since the author's early experiments (Partheniades, 1962,

1965) that flocculated cohesive sediment beds formed by deposition display an increasing resistance to erosion with depth (Fig. 24). In contrast, artifically prepared cohesive beds of uniform consistency get eroded at constant rates (Fig. 25).

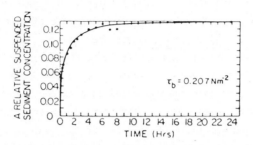

Fig. 24. Relative suspended sediment concentration versus time for a stratified kaolinite bed deposited in distilled water and eroded under τ_b = 0.207 N/m^2 (Mehta and Partheniades, 1979, 1982).

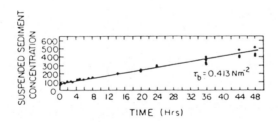

Fig. 25. Suspended sediment concentration versus time for a uniform kaolinite bed in distilled water eroded under τ_b = 0.413 N/m^2 (Mehta and Partheniades, 1979, 1982).

Since there is no deposition and no exchange between bed flocs and suspended flocs, the gradual decrease of the resuspension rates of naturally deposited sediment and, under certain circumstances, the eventual termination of scouring can be explained only by an increase of the interparticle cohesive bonds with depth. The latter cannot be attributed to the overburden consolidation pressure first because that pressure is extremely low and second, because if it were important it would have shown up in the original uniform bed of Fig. 25 where the average density was of the order of the density of a deposited bed. The bond increase is simply inherent in the internal floc structure as flocs segregate during deposition with larger and denser flocs depositing first thus forming a stratified bed of increasing density and strength with depth. Figures 26 and 27 show the variation of the density and the shear strength of the deposited bed respectively. The first was determined by a specially designed 2.5 cm diameter metal tube in which bed core samples were frozen by an alcohol-dry ice mixture and by subsequent slicing of the frozen samples. The shear strength τ_s was taken as equal to the applied shear stress τ_b when the rate of concentration increase of suspended sediment, $\frac{dC}{dt}$, reaches a near zero value.

The increase of the overall bed density with depth reflects an increase of the floc density. The latter can be attributed, to both a denser arrangement of particles and packets in each aggregate and to a larger number of particles per packet, as observed by Quinn (1980). The increase of shear strength of the bed reflects the resistance of the particles and of the various aggregates against removal and reentrainment by the flow-induced drag and lift forces (Partheniades, 1977). Indeed, both the submerged weight of an aggregate and the number of the physico-chemical bonds holding it to the bed are expected to increase with increasing size, density and number of units (particles and packets) per aggregate.

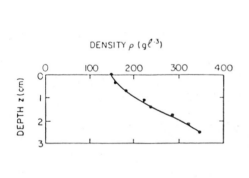 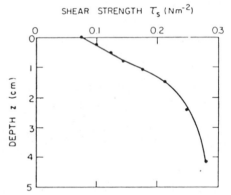

Fig. 26. Variation of bed density with depth (Mehta and Partheniades, 1982).

Figure 27. Variation of bed shear strength with depth (Mehta and Partheniades, 1982).

For a given bed shear stress, τ_b, resuspension will proceed at reducing rates until the total depth of erosion Δz will be the one corresponding to $\tau_s = \tau_b$ in Fig. 27. The instantaneous resuspension rates E have been found to be functions of the excess bed shear stress $\tau_b - \tau_s$ according to the equation:

$$\frac{E}{E_0} = \exp\left(\alpha \frac{\tau_b - \tau_s}{\tau_s} \right) \tag{25}$$

where E_0 and α are empirical coefficients. The concentration of suspended sediment during resuspension can be described by the law

$$C = C_s (1 - e^{-\beta t})\qquad\qquad(26)$$

where β is a coefficient ranging between 0.013 hr^{-1} and 0.038 hr^{-1}; t is the time in hours and C_s is the steady-state concentration when any further resuspension has ceased, i.e. when erosion has proceeded to a depth for which $\tau_s = \tau_b$. The distribution of density ρ_s of the deposited bed follows the law

$$\frac{\rho_s}{\overline{\rho}_s} = \zeta \left(\frac{H - z}{H} \right)^{-\xi}\qquad\qquad(27)$$

where $\overline{\rho}_s$ is the average bed density; H is the bed thickness and ζ and ξ are dimensionless coefficients equal to 0.794 and 0.288 respectively (Mehta and Partheniades, 1982). Of significant importance to resuspension is the consolidation time, T_{dc}, i.e. the time between the completion of the deposition process and the beginning of resuspension. Figure 28 shows the distribution of the erosive shear strength $\tau_c = \tau_s$ for the indicated four values of T_{dc}. It is obvious that τ_s increases significantly with time particularly for the first few hours after deposition but this increase may continue for a very long time and it appears to be due to a thixotropic rearrangement of particles within the flocculated network of the deposited bed. It is reflected in an increase of β and F_a in Eqs. 21 to 24.

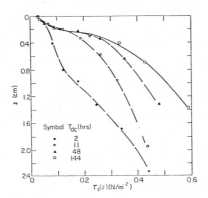

Fig. 28. Variation of critical erosion stress with depth for kaolinitic bed in tap water for various consolidation times, T_{dc} (Mehta and Partheniades, 1982).

SUMMARY AND CONCLUSIONS

Turbidity due to fine sediment suspensions in non-stratified natural water bodies is linked to the flow-induced bed shear stresses and to the physico-chemical properties of the sediment-water system. Considering deposition, the entire flow field can be divided into a near-bed region of high shear rates and a far-bed region of low shear rates and of nearly homogeneous turbulence. The former determines the degree of deposition, that is that proportion of the entire sediment which forms flocs with sufficiently high bonds to settle on the bed, while the latter contributes only to the formation of flocs and floc aggregates. Equations for the degree and rates of deposition have been developed in terms of the bed shear stress and other parameters representing the physicochemical properties of the sediment-water system. In natural flow systems these parameters are expected to vary as the sediment disperses during every deposition-resuspension cycle.

Deposited sediment beds display a strong degree of stratification with an increasing density and shear strength with depth. This is due to a natural process of floc seggregation with respect to size, density and strength during the process of deposition. Indeed, as recent electron microscope studies have shown, flocs are composed of packets of individual clay particles in book-type arrangements. The denser and larger flocs contain larger packets thus possessing higher shear strength than smaller flocs. As a result, the erosion rates diminish rapidly as the erosion proceeds and they may stop altogether at a depth where the flow-induced shear stresses become equal to the erosive shear strength of the bed. Experimental equations have been derived for the distribution of bed density with depth, the instantaneous erosion rates and the time variation of suspended sediment concentration during the process of resuspension. The consolidation time, that is the time between the completion of deposition and the beginning of resuspension, has a significant effect on the erosive shear strength of the bed and on the erosion rates.

A mathematical model of sediment-flow interaction has been developed which explains the above phenomena and links the behavior of the bed material load and of the wash load.

REFERENCES

Blinco, P. H. and Partheniades, E., "Turbulence Characetristics in Free Surface Flows over Smooth and Rough Boundaries", Journal of Hydr. Res., I.A.H.R., Vol. 9, No. 1, pp. 44-71, 1971.

Casagrande, A., "The Structure of Clay and its Importance in Foundation Engineering", Contrictions to Soil Mechanics, Boston Soc. of Civil Engrs, 1940, p. 72 and Journal, Boston Soc. of Civil Engrs., Vol. 19, April, 1932.

Committee on Tidal Hydraulics, Corps of Engineers, U. S. Army, "Soil as a Factor in Shoaling Processes, a Literature Review", Tech. Bull. No. 4, 1960, 47 pages.

Dixit, J. G., Mehta, A. J. and Partheniades, E., "Redepositional Properties of Cohesive Sediments Deposited in a Long Flume", Report No. UFL/COEL-82/002, Dept. of Coastal and Oceanographic Engrg., August, 1982.

Einstein, H. A. "The Bed-Load Function for Sediment Transportation in Open Channel Flows", Technical Bulletin No. 1026, U. S. Department of Agriculture, Washington, D. C., 1950.

Einstein, H. A., Anderson, A. G., and Johnson J. W., "A Distinction between Bed Load and Suspended Load in Natural Streams", Transactions, American Geophysical Union, Vol. 21, Part 2, 1940, pp. 628-633.

Einstein, H. A., and El Samni, S. A., "Hydrodynamic Forces on a Rough Wall", Review of Modern Physics, Vol. 21, 1949, pp. 520-524.

Krone, R. B., "Flume Studies of the Transport of Sediment in Estuarial Shoaling Processes". Final Report, Hydr. Engrg. and Sanitary Engrg Res. Lab., Univ. of California, Berkeley, Calif., 1962.

Krone, R. B., "A Study of Rheologic Properties of Estuarial Sediments". Final Report No. 63-8 Hydr. Engrg Lab and Sanitary Engrg Res. Lab., University of California, Berkeley, Calif., 1963.

Kusuda, T., Koga, K., Yorozu, H., and Hwaya, Y., "Density and Settling Velocity of Flocs", Memoirs of the Faculty of Engrg., Kyushu Univ., Vol. 41, No. 3, 1981, pp. 269-280.

Mehta, A. J., and Partheniades, E., "Depositional Behavior of Cohesive Sediments", Technical Report No. 16, Coastal and Oceanographic Engrg Lab., University of Florida, Gainesville, Fla., March 1973.

Mehta, A. J., and Partheniades, E., "An Investigation of the Deposition Properties of Flocculated Fine Sediments", Journal of Hydraulic Research, Vol. 12, No. 4, 1975, pp. 361-381.

Mehta, A. J. and Partheniades, E., "Kaolinite Resuspension Properties", Tech. Note, Journal of the Hydraulics Div., ASCE, Vol. 104, No. HY4, Proc. Paper 14477, April, 1979, pp. 409-416.

Mehta, A. J., and Partheniades, E., "Resuspension of Deposited Cohesive Sediment Beds", Proceedings, 18th Coastal Engineering Conference, Vol.2, Cape Town, So. Africa, Nov. 14-19, 1982, pp. 1569-1588.

Mehta, A. J., Partheniades, E. and McAnally, W., "Properties of Deposited Kaolinite in a Long Flume", Proceedings, Hydr. Div. Conf. on Applied Research to Hydraulic Practice, ASCE, Jackson, Miss., Aug., 1982.

Partheniades, E., "A study of Erosion and Deposition of Cohesive Soils in Salt Water", thesis presented to the University of California, at Berkeley, Calif. in 1962, in partial fulfillment of the requirements for the degree of Doctor of Philosophy.

Partheniades, E., "Erosion and Deposition of Cohesive Soils", Journal of the Hydraulics Division, ASCE, Vol. 91, No. HY1, Proc. Paper 4204, Jan., 1965, pp. 105-139.

Partheniades, E., "Unified View of Wash Load and Bed Material Load", Journal of the Hydraulics Division, ASCE, Vol. 103, No. HY9, Proc. Paper 13215, September, 1977, pp. 1037-1057.

Partheniades, E., "Engineering Properties of Estuarine Sediments", Lecture No. 16, North Atlantic Treaty Organization, Advanced Study Institute of Estuary Dynamics, Lisbon, Portugal, June 1973.

Partheniades, E., "Cohesive Sediment Transport Mechanics and Estuarine Sedimentation", Lecture Notes, Internat. Course on Sediment Transport in Estuarine and Coastal Environment', Poona, India, Nov. 26-Dec. 15, 1979, p. 164.

Partheniades, E., "A Fundamental Framework for Cohesive Sediment Dynamics", Proceedings, Cohesive Sediment Dynamics Workshop, Tampa, Florida, Nov. 12-14, 1984 (In print).

Partheniades, E., and Kennedy, J. F., "Depositional Behavior of Fine Sediment in a Turbulent Fluid Motion", Proceedings, 10th Conference on Coastal

Engineering, Tokyo, Japan, Vol. II, Chapt. 41, Sept., 1966, pp. 707-729.

Partheniades, E., Cross, R. H., and Ayora, A., "Further Results on the Deposition of Cohesive Sediments", Proceedings, 11th Conference on Coastal Engineering, London, England, Vol. II, Chapt. 47, Sept., 1968, pp. 723-742.

Partheniades, E., and Paaswell, R. E., "Erodibility of Channels with Cohesive Boundary", Journal of the Hydraulics Division, ASCE, Vol. 96, No. HY 3, Proc. Paper 7156, March 1970, pp. 755-771.

Pritchard, D. W., "What is an Estuary: Physical Viewpoint", Estuaries, Edited by G. H. Lauff, Publ. No. 83, Am. Assoc. Adv. of Science, 1967.

Quinn, M. J., "A Scanning Electron Microscope Study of the Microstructure of Dispersed and Flocculated Kaolinite Clay taken out of Suspension", A thesis presented to the University of Florida in partial fulfillment of the Requirement for the Degree of Master of Science, 1980.

Rosillon, R., and Volkenborn, C., "Sedimentación de Material Cohesiva en Agua Salada", Diploma Thesis, Dept. of Civil Engrg., Univ. of Zulia, Maracaibo, Venezuela, 1964.

Terzaghi, K., "Erdbaumechanik", F. Deuticke, Vienna, Austria, 1925.

Van Olphen, H., "An Introduction to Clay Colloid Chemistry", Intersicence Publication, 1963.

APPENDIX

C	= instantaneous suspended sediment concentration
C_a	= reference concentration at $y = a$
C^*	= $C_o - C/C_o - C_{eq}$
CEC	= cation exchange capacity
C_{eq}	= equilibrium concentration of suspended sediment
C_{eq}^*	= C_{eq}/C_o
C_{eq}^{**}	= $1 - C_{eq}^*$
C_o	= initial concentration of suspended sediment
C_s	= steady state concentration when resuspension ceases
E	= instantaneous resuspension rate
E_o	= empirical reference resuspension rate
$f(\ )$	= function of $(\ )$
F_a	= a net attractive physico-chemical force between a particle or floc and the bed
G	= mixing intensity
h	= depth of flow
H	= bed thickness
k	= proportionality constant
L	= lift force on sediment grain or floc
$\bar{L}$	= average value of L
L_d	= lift force at the threshold for deposition
L_e	= lift force at the threshold for erosion
L_ℓ	= lower bound of L
L_u	= upper bound of L

m	= average mass of a floc
n	= Manning's friction coefficient
N	= number of particles or flocs per unit volume
N_n	= number of flocs with a number of n primary particles per floc
N_0	= number of primary particles in the suspension
Pr	= $1 - \tau_b/\tau_\ell$
P_e	= probability for erosion
P_d	= probability for deposition
P_r	= probability of
r	= radius of a floc
t	= time
t_c	= average time between floc collision
t_f	= time of flocculation
t_{50}	= mean value of deposition time
t_s	= settling time
t_{s50}	= mean value of t_s
T	= $\dfrac{1}{\sigma_2} \log(t/t_{50})$
T_{dc}	= consolidation time
u	= local velocity
u_*	= friction velocity
U_{sf}	= settling velocity of a floc
v_s	= median settling velocity of a sediment suspension
W_b	= submerged weight of sediment particle or floc
y	= distance from theoretical bed
y_0	= distance from bed at which u = 0
Y	= $[\log (\tau_b^* - 1)/(\tau_b^* - 1)_{50}]/\sigma_y$
z	= distance below bed surface
z'	= H - z
α	= dimensionless coefficient
β	= dimensionless coefficient accounting for interparticle friction and interlocking; also exponent
Δr	= radius of contact area between two flocs
ε	= uniform turbulent diffusivity
ε_0	= rate of energy decay per unit volume of fluid
ε_y	= turbulent diffusivity at a distance y from the bed
ζ	= dimensionless coefficient
η	= normally distributed dimensionless lift coefficient
η_0	= standard deviation of η
κ	= Karman's universal constant
μ	= dynamic viscosity
ξ	= dimensionless exponent
ρ	= water density
ρ_s	= bed density
$\overline{\rho}_s$	= average value of ρ_s
σ_2	= standard deviation of C^*

550

σ_3 = dimensionless exponent

σ_{max} = maximum interparticle tensile force

σ_y = standard deviation of distribution of C_{eq}^*

τ = shear stress in general

τ_b = average bed shear stress

τ_{bmin} = minimum bed shear stress below which all sediment deposits

τ_ℓ = threshold bed shear stress according to Krone's definition

τ_b^* = τ_b/τ_{bmin}

τ_{max} = maximum shear stress

$(\tau_b^* - 1)_{50}$ = mean of $\tau_b^* - 1$

τ_c = threshold value of τ_o for the initiation of erosion

τ_s = erosive strength of deposited beds

ω = dummy variable

ADRIA 84
A JOINT REMOTE SENSING EXPERIMENT

P. SCHLITTENHARDT
Commission of the European Communities, Joint Research Centre, Ispra Establishment, 21020 Ispra (Va), (Italy).

ABSTRACT
    A joint experiment has been carried out in the Northern Adriatic Sea to calibrate and evaluate remotely sensed data and to contribute to the integration of in situ measurements, remote sensing and hydrodynamic modelling. After a summary of the experiment examples are shown of the airborne sensor data, seatruth measurements and satellite images.

INTRODUCTION
    Within the last few years, the application of remote sensors for studies of ocean-and-coastal zones have demonstrated important capabilities for oceanographic research. The Joint Research Centre (JRC) of Ispra of the European Communities started in 1977 a research activity on the use of Coastal Zone Color Scanner (CZCS) data for quantitative determination of chlorophyll and suspended matter concentration. Different algorithms have been developed for the atmospheric correction and interpretation of CZCS data and series of bio-optical and atmospheric measurements data sets were gathered within this research activity.

    As a result of these studies in the past, to extend the application of remote sensing techniques and to promote the collaboration of the involved institutes and organizations, the ADRIA 84 experiment was organized.

THE EXPERIMENT
    The ADRIA 84 experiment has been conceived to attain several objectives :
-   application objectives include the assessment of remote sensing techniques together with in situ measurements for a quantitative investigation of dynamical processes ;
-   measurement method objectives include a study of an airborne Lidar system and its comparison with other remotely sensed data and in situ measurements ;
-   oceanographic objectives include test and evaluation of existing hydrodynamic models.

This large-scale experiment should contribute to the integration of in situ measurements, remote sensing and hydrodynamic modelling as a new overall method for oceanographic research and sea protection.

REMOTE SENSING
- Ocean Colour (CZCS)
- temperature (NOAA-7 IR)

Field measurement
- ATM/METEO data
- vertical profiles
- non-visible parameters

Modelling
- calibration
- initiation
- operation

The experiment has been carried out between August 23 and September 7. A large number of scientific teams from national laboratories have participated. The experiment was mainly sponsored by the EEC (flights and measurements), CNR (oceanographic vessels and platform), Regione Emilia-Romagna (oceanographic vessel), Regione Friuli-Venezia-Giulia (flight), and DFVLR (Lidar flight).

The selected test site, the Northern part of the Adriatic Sea, is the shallowest area of the Mediterranean. Large river run-off, extended lagoon systems and the seasonally variable heat-exchange at the surface give rise to relatively large levels and gradients of phytoplankton. The lagoon systems and the rivers, as well as water drained from heavily cultivated surrounding land, are responsible for the pollutants in these regions. For protection and surveillance a deeper understanding is necessary of the mechanisms of evolution, propagation and the impact of the pollutants.

The main activities are summarized in Table I, the involved institutes are listed in Table II.

TABLE I - Summary of the main activities

Spacecraft (radiometers)		Ship and platform measurements
NIMBUS 7 : CZCS		EOS radiometer
NOAA 7/8 : AVHRR		PRT 5 radiometer
LANDSAT : TM		CTD profiles
METEOSAT		Water sampling
		Chlor. fluorescence (cont.)
Airborne instruments		Alga luminiscence
IR scanner	CESNA	Phytoplankton analyses
LL television camera	CESNA	Particle size distribution
MSS scanner		Trace elements
OCR/SCR	DO - 28	Mini-Lidar
PRT 5		Spectral radiometer
Lidar system	DO - 28	

TABLE II - Involved institutes

Istituto di Automazioni Navali del CNR (Viale Causa 18, I-16141 Genova,
  Mr. Siccardi)
Istituto di Biologia del Mare, CNR (Riva 7 Martini 1364/A I-30122 Venezia,
  Mr. Franco)
DFVLR Oberpfaffenhofen (D-8031 Wessling, V. Amman, Chr. Werner)
GKSS Forschungszentrum (D-2054 Geestacht, R. Doerffer)
Institut für Meereskunde (Düsternbrooker Weg 21 D-2300 Kiel, A. Schmitz-Pfeiffer)
Istituto di Ricerca sulle Onde Elettromagnetiche (Via Panciatichi 64 I-50127
  Firenze, L.P. Pantani)
Istituto per lo Studio della Dinamica delle Grandi Masse (Ca' Papadopoli, 1364
  San Paolo, I-30125 Venezia, L. Alberotanza)
Regione Emilia Romagna, Assessorato Ambiente e Difesa del Suolo (Via dei Mille
  21, I-40121 Bologna, G. Nespoli, L. Montanari)
Osservatorio Geofisico Sperimentale (PO Box 2011, I-34016 Trieste, A. Michelato)
Università di Firenze, Istituto Botanico (Via Michelli 1, I-50127 Firenze,
  M. Innamorati)
Università di Bologna, Istituto CNR (Via Pichat Bert C.R., I-40127 Bologna,
  R. Guzzi)
Universität Oldenburg (Ammerländer Heerstrasse 67, D-2900 Oldenburg, R. Reuter)
Regione Friuli-Venezia-Giulia, Direzione Regionale Lavori Pubblici, Centro
  Rilevamento Idrometeorologico (Riva Nazario Sauro 8, I-34124 Trieste,
  Mr. Verri)
Universität Regensburg, FB Physik (Universitätstr. 31, D-8400 Regensburg,
  H. Krause)
Istituto di Fisica dell'Atmosfera, CNR (Piazzale L. Sturzo 31, I-00144 Roma,
  G. Dalu)
Scientific Data (Via Bolis 10, I-35100 Padova, A. Ongaro)

SHIPS
R/V Bannock (CNR) - Trieste/Ancona
R/V Umberto d'Ancona (CNR) - Venezia - S. Giorgio
R/V Litus (CNR) - Venezia - S. Giorgio
R/V Daphne (Reg. Emilia-Romagna) - Cesenatico

AIRBASE
Aeroporto Venezia/Lido LIPV, AEROCLUB G. Aniellotto, S. Nicolà, Venezia-Lido
Aeroporto Venezia/Tessera, Servizio Meteorologico TESSERA

EXAMPLES OF MEASUREMENTS

The data collected during the ADRIA 84 campaign will be edited in a joint
Data Catalogue. The examples of airborne sensor data and seatruth data is a
first indication of the measurements, while the satellite maps - from 1982
and 1983 - represent the typical summer situation in the Northern Adriatic Sea.

Airborne sensors

For the airborne measurements three aircrafts have been involved : two DO 28
from DFVLR and one CESNA Skywagon 28 from the Regione Friuli-Venezia-Giulia.
Overflights of about 2 hour duration have been executed between 11.00 a.m. and
1.00 p.m. local time, on several days during the experiment. The general flight
line is shown in Fig. 5 together with the shipstation grid.

Fig. 1 shows the characteristics of the passive sensors on board of one airplane, while Fig. 2 presents the characteristics and principal way of operation of the involved Lidar system.

To provide an overview of the spatial colour and SST variations, three representative signals were selected from the OCR, SCR and PRT data, and plotted along the flight lines. These signals are :
- GBR the Green-to-Blue Radiance Ratio of the OCR channels 3 (551 nm) and 1 (446 nm), indicative of changes in water colour ;
- SST the sea surface temperature from the PRT-readings ;
- FLH the relative Fluorescence-Line-Height of the chlorophyll fluorescence peak (682 nm) from the SCR channels.

Fig. 3 gives an example of this "pattern quicklook" for flight N°. 3 (30.8. 1984). On the presented in-shore track the river Po plume is marked clearly by the GBR signal and the FLH signal. Fig. 4 shows the same signals for flight N°.4 (31.8.1984). The different pattern of these two days illustrates the short-term changes in ocean colour in the Po area over one day. In Fig. 4 the signals of this passive sensor are also compared with the Lidar signals of the same day and time. The Lidar system is working with two pulse lasers which make it possible to use two excitation wavelengths simultaneously : one in the uv at 380 nm, another in the visible at 451 nm. Seven optical bands are available for signal detection between near uv and the red part of the spectrum. The average bandwidth is about 10 nm.

In Fig. 4, the following signals are shown :
- excitation 451 nm/detection 684 nm - signal proportional to the concentration of chlorophyll in the water ;
- excitation 308 nm/detection 533 nm - signal proportional to the concentration of Gelbstoff in the water ;
- excitation 451 nm/detection 533 nm - signal inversely proportional to the light attenuation coefficient at the corresponding wavelength ;
- excitation 308 nm/detection 344 nm - signal inversely proportional to the light attenuation coefficient at the corresponding wavelenght.

Ship board measurements

Surface parameters can be correlated with remote sensing data, while vertical hydrologic measurements such as salinity and temperature are necessary for hydrodynamic studies. Therefore, the ship-board measurements have been carried out using continuous sampling methods at the sea surface as well as vertical sampling and profiles at the stations. The standard grid of the stations is shown in Fig. 5. The four involved ships have covered different but overlapping areas. Fig. 6 shows - as an example - the vertical distribution of oxygen,

SENSOR FIELD OF VIEW

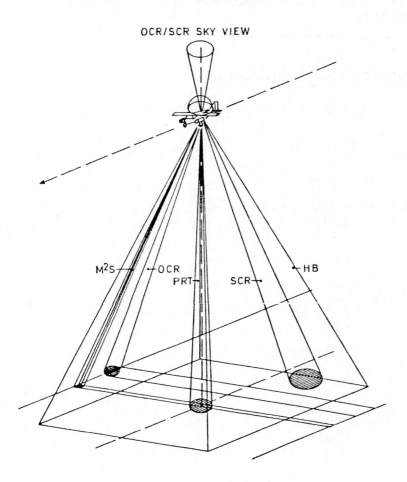

Fig. 1.a. Passive airborne sensors.

```
┌─────────────────────────────────────┐
│  DFVLR - OBERPFAFFENHOFFEN           │
│  V. AMANN                            │
└─────────────────────────────────────┘
```

PASSIVE SENSOR CHARACTERISTICS

MULTISPECTRAL LINESCANNER (M²S, BENDIX)

F.O.V	± 50°
J.F.O.V	2,5 mrad
Spectral Bands	8 VIS ($\Delta\lambda$ 40-60nm), 2 NIR ($\Delta\lambda$ 90nm)
	1TIR (8-14 µm)
SNR	14-235

OCEAN COLOR SCANRADIOMETER (OCR, DFVLR)

F.O.V	± 50°
Angular Resol.	3,6° (60 mrad)
Spectral Bands	5 VIS ($\Delta\lambda$ 11-13 nm), 1 NIR ($\Delta\lambda$ 13 nm)
SNR	det. by 8Bit Encoding

SIX-CHANNEL RADIOMETER (SCR, DFVLR)

F.O.V	5°
Spectral Bands	5 RED (10-13nm), 1 NIR ($\Delta\lambda$ 13nm)
SNR	$> 10^3$

PRECISION RADIATION THERMOMETER (PRT5, BARNES)

F.O.V	2°
Spectral Band	9,5 - 11.5µm
NEΔT	0,1K

HASSELBLAD CAMERA (HB)

Focal length	40 mm
Film	Agfa Color 50S, Kodak IR 2443

CALIBRATION SOURCES

Optical range	Ulbricht Sphere,
	Radiance Standard
Thermal range	NBS Blackbodies

Fig. 1.b. Passive airborne sensors.

LIDAR CHARACTERISTICS

lasers	excimer	dye	dye
emission/nm	308	450	533
pulse energy/mJ	70	5	5
pulse length/ns	12	8	8
repetition rate	max 20 Hz		
receiver	Cassegrain f/10, f = 4 m		
detection/nm	344/380/500/533/650/685		
bandwidth/nm	10		
photomultipliers	EMI 9812, 9818		
transient recorder	500 MHz, 6 bit		

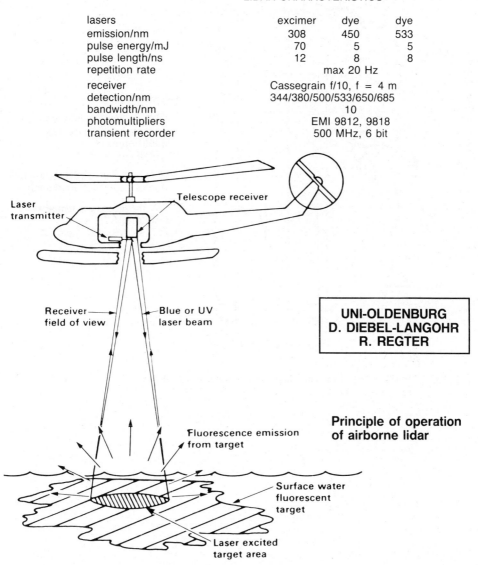

Laser transmitter

Telescope receiver

Receiver field of view

Blue or UV laser beam

UNI-OLDENBURG
D. DIEBEL-LANGOHR
R. REGTER

Fluorescence emission from target

Principle of operation of airborne lidar

Surface water fluorescent target

Laser excited target area

Fig. 2. Lidar characteristics.

558

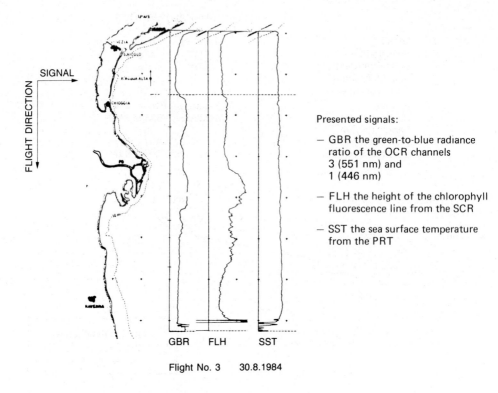

Presented signals:

— GBR the green-to-blue radiance
 ratio of the OCR channels
 3 (551 nm) and
 1 (446 nm)

— FLH the height of the chlorophyll
 fluorescence line from the SCR

— SST the sea surface temperature
 from the PRT

Flight No. 3   30.8.1984

Fig. 3. Signals from airborne passive sensors - Flight N°3 -

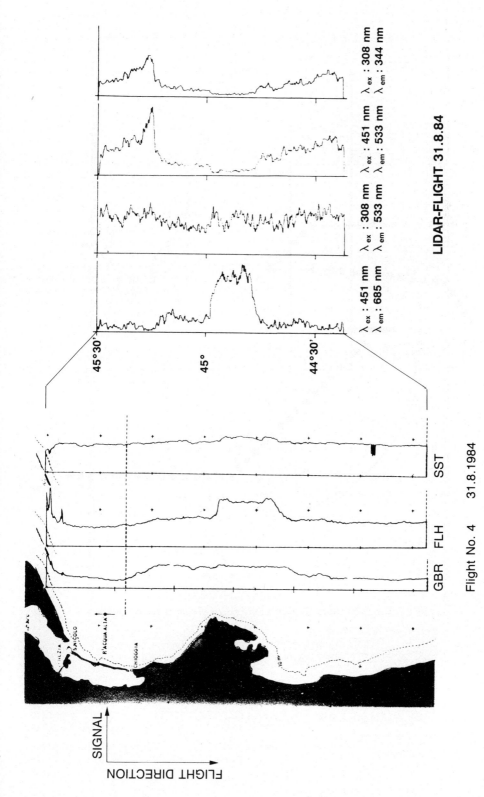

SIGNAL

FLIGHT DIRECTION

GBR   FLH   SST

Flight No. 4   31.8.1984

λ_{ex} : 451 nm λ_{ex} : 308 nm λ_{ex} : 451 nm λ_{ex} : 308 nm
λ_{em} : 685 nm λ_{em} : 533 nm λ_{em} : 533 nm λ_{em} : 344 nm

LIDAR-FLIGHT 31.8.84

45°30'

45°

44°30'

Fig. 4. Comparison of passive sensor signals and lidar signals for Flight N°4.

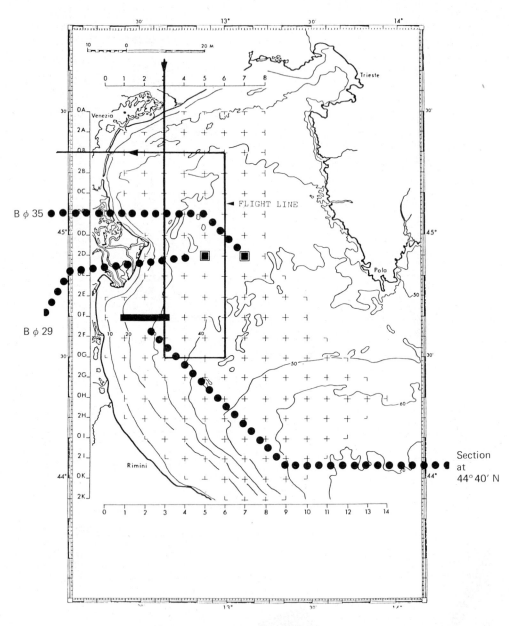

GRID OF SHIP STATIONS FROM P. FRANCO
ISTITUTO DI BIOLOGIA DI MARE, VENEZIA

Fig. 5. ADRIA 84. - Flight Track - Ship Stations.

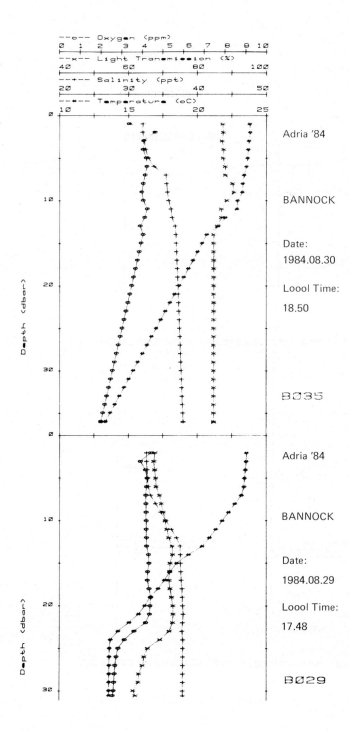

Fig. 6. Two vertical CTD profiles (kms).

562

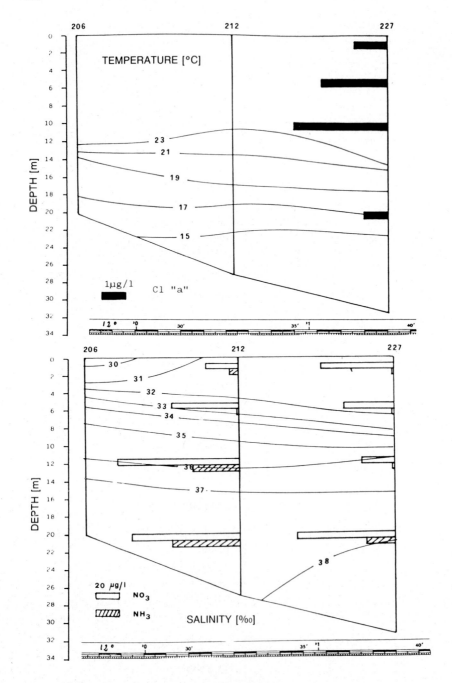

REGIONE EMILIA ROMAGNA, G. MONTANARI

Fig. 7. Distribution of temperature and salinity at 44°40' North.

light transmission, salinity and temperature for the two stations B035 and
B029 (Bannock). Fig. 7 presents the distribution of temperature and salinity
as measured in a section at 44°40'N. The salinity distribution derived from
CTD profiles is illustrated in Fig. 8.

Satellite images

Different satellite data have been acquired within the experiment. A listing
of the scenes received from the Coastal Zone Color Scanner (CZCS) on NIMBUS-7
and the Advanced Very High Resolution Radiometer (AVHRR) on NOAA-satellite is
given in Table III. The archived data will be elaborated within the next few
months.

TABLE III   -   Acquired CZCS and NOAA-7 scenes

| | CZCS | | NOAA-7 |
	Orbit number	Time GMT	Time GMT
840817	29359	1011	-
18	29373	1034	-
19	29387	1047	-
840821	29414	0941	-
22	29428	0959	-
23	29442	1017	-
24	29456	1035	-
840827	29497	0947	0309/1439
28	-	-	- /1427
29	29525	1023	0244/ -
30	-	-	0413/1401
31	-	-	0401/1349
840901	-	-	0348/1336
02	-	-	0335/1324/1506
03	-	-	0323/1453
840905	29622	1036	-
840908	29663	0958	-
840910	29692	1036	-

Three satellite images from 1982 and 1983 are presented in Figs. 8, 9a and
9b. These images illustrate the typical summer situation of the Northern
Adriatic Sea dominated by the Pô river outflow - a similar situation to the
period of the campaign. Fig. 8 is an example of sea surface temperature in °C
derived from AVHRR-data (NOAA-7) of 7 July 1983. The correction of the measured
temperature for the atmospheric absorption was obtained by an improved split
window technique. Figs. 9a and 9b are axamples of chlorophyll concentration
maps in mg/m^3 derived from CZCS-data (NIMBUS-7). The images have been corrected
for atmospheric effects and are processed to geographically rectified maps.

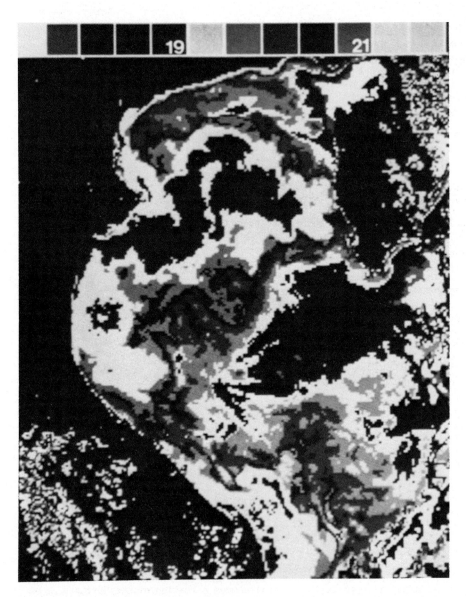

Fig. 8. Example of sea surface temperature in °C derived from AVHRR-data on NOAA-7 satellite. Day : 7.6.1983.

Fig. 9a. Chlorophyll concentration in mg/m^3 derived from CZCS on NIMBUS-7 satellite. a) Orbit 18121 ; Day 27.05.1982 ; GMT 10.34.

566

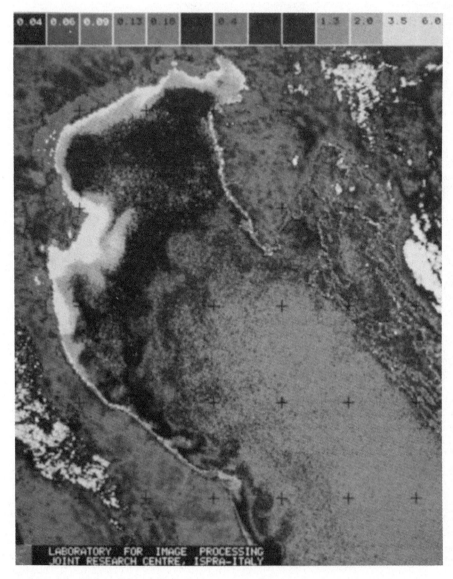

Fig. 9b. Chlorophyll concentration in mg/m^3 derived from CZCS on NIMBUS-7 satellite. b) Orbit 18176 ; Day 31.05.1982 ; GMT 9.55.

The atmospheric correction procedure and the pigment algorithm were elaborated especially for the rather turbid near coastal water of the Northern Adriatic Sea. In Fig. 9a -Orbit No.18121, Day 27.05.1982, GMT 10.34 - the largest part of the area is characterized by low chlorophyll concentration ($< 0.1 \, mg/m^3$) but near the Italian coast the chlorophyll concentration increases to even more than 10 mg/m^3. Fig. 9b - Orbit No.18176, Day 31.05.82, GMT 9.55 - is of very high quality nearly without any haze. It gives very good details of the Pô river outflow and the horizontal distribution of the river water.

CONCLUSIONS

From the first evaluation of the experimental data, it is believed that the main goals of the experiment could be achieved.
- The in situ measurements of physical, biochemical and optical water parameters can be used for a descriptive analysis of the hydrographic situation as well as a comparison with the remotely sensed data.
- The different flight and satellite data will contribute to the evaluation of passive and active sensors.
- The complete data will be significant for the calibration and validation of the hydrodynamic model.

The further elaboration and interpretation of the collected data will contribute to the integration of in situ measurements, remote sensing and hydrodynamic modelling.

REFERENCES

Camagni, P. et al., 1983. Marine remote sensing activities of JRC Ispra. Proc. EARSeL/ESA Symp. Remote Sensing Applications for Environment Studies, Brussels, 26-29 April 1983, 35-49.

Ferrari, E. et al., 1984. Remote monitoring of sediments and chlorophyll as tracer of pollutant movements in a mediterranean coastal area. IGARSS'84 Symp., Strasbourg, 27-30 August 1984, 701-707.

Nykjaer, L. and Schlittenhardt, P. and Sturm, B., 1984. Qualitative and quantitative interpretation of ocean color - NIMBUS-7 CZCS imagery of the Northern Adriatic Sea from May to September 1982, JRC technical note SA/1.04.E2.84.05.

Schlittenhardt, P.M. (editor), 1984. Workshop on Remote Sensing of Coastal Transport in the Northern Adriatic Sea, 11-12 October 1983, S.A. 1.05.E2.85. 03, JRC-Ispra, Italy.

Schlittenhardt, P.M., 1984. ADRIA 84 - a joint remote sensing experiment, Tech. Note No.1.05.E2.84.91, JRC-Ispra, Italy.

Schlittenhardt, P.M., 1985. ADRIA 84 - a joint remote sensing experiment, 2nd Report : A short summary of the experiment, Tech. Note No.1.05.E2.85.20, JRC-Ispra, Italy.

IDENTIFICATION OF HYDROGRAPHIC FRONTS BY AIRBORNE LIDAR MEASUREMENTS OF GELBSTOFF DISTRIBUTIONS

D. DIEBEL-LANGOHR, T. HENGSTERMANN and R. REUTER

Universität Oldenburg, Fachbereich Physik, P.O.Box 2503, 2900 Oldenburg

(Federal Republic of Germany)

ABSTRACT

Laser remote sensing (lidar) allows the measurement of various hydrographic parameters of interest for water quality monitoring and oceanographic research. Operated from aircraft a nearly synoptic investigation of extended areas of the sea is achieved which is particularly important if hydrographic conditions are changing rapidly in time, e.g. due to tides. Among the substances that are detectable with lidar, dissolved organic matter (Gelbstoff) gives rise to very dominant fluorescence signals and is thus sensitively measured in coastal waters where the concentration is generally high. Due to its good stability Gelbstoff can be used as a natural tracer for the study of transport and mixing, and for the identification of characteristic water masses and frontal systems. Results obtained during flights over the North Sea and the northern Adriatic with the Oceanographic Lidar System (OLS) developed at the University of Oldenburg are presented.

INTRODUCTION

In the past decade the use of remote sensing methods has become an important tool in oceanographic research. The reason for this lies in the complex dynamics of the ocean where the characteristic length scales of many relevant processes range between 10 and 1000 km. The occurrence of eddies as a general feature of the oceanic circulation has been verified with remote sensing. Fronts in coastal regions produced by river plumes and thermal processes in the upper water layer with high gradients of the physical and biological parameters have been observed which are changing rapidly in time due to tides. Classical methods of experimental oceanographic research, in particular the collection of data by use of shipboard instruments, do not always lead to an information representative for larger areas of the sea. An interpretation of these phenomena requires additional measurements with sensor systems, from which synoptic data with high horizontal resolution can be deduced.

The use of satellites for measurements of the colour of the sea by means of scanning optical radiometers has led to remarkable results over the open ocean. These spaceborne systems allow a sensitive and quantitative delineation of the biological productivity in the surface layer since the water colour can be related to the presence of phytoplankton (Gordon et al., 1983). However, problems arise for the interpretation of data obtained over coastal waters where other substances such as suspended minerals and dissolved organic matter (Gelbstoff) optically compete with phytoplankton pigments and also influence the water colour. Due to this, and because of the masking effect of the atmosphere which is responsible for about 90 % of the detected signal (Fischer, 1984), a quantitative evaluation of these different substances has not been achieved so far.

Another method of remote sensing makes use of lidar systems (light detection and ranging) installed in aircraft. A closer distance between the instrument and the water surface - the flight height being typically 100-500 m - reduces atmospheric effects on the detector signal significantly and allows the application of an artificial monochromatic light source, the laser. It will be shown below that laser spectroscopic methods result in a specific measurement of certain optically detectable water column parameters. Some of these quantities can be detected as vertical profiles down to water depths corresponding to about 6 optical attenuation lengths, if laser pulses of a few nanoseconds length and a fast high resolving signal receiver system are utilized. This depth profiling capability is discussed in a separate paper given in this volume.

At an operational level, data on the spectral light turbidity and on the concentration of phytoplankton and Gelbstoff can be obtained. In particular the latter which is mainly brought into the sea by river run-off is sensitively measured due to its specific fluorescence and possesses a notable stability with respect to biological and chemical degradation. It thus respresents a conservative property of seawater and can be used as a natural tracer for the description of characteristic water masses and for the identification of hydrographic fronts in coastal waters where the concentration is generally high.

The oceanographic lidar is thus a sensor allowing
- an extension of shipboard measurements with limited horizontal resolution by airborne data of the surface layer obtained over larger areas,
- a nearly synoptic calibration of large-scale satellite data along the aircraft flight track in terms of the different water constituents determining the water colour measured by the satellite,
- a specific measurement of hydrographic parameters characterizing frontal systems in coastal waters; this can be done within time periods that are short compared with the time scales of dominant circulation processes, e.g. tidal currents.

SPECTROSCOPIC PROPERTIES OF SEAWATER

Irradiation of seawater with monochromatic light leads to an emission of secondary light due to elastic and inelastic optical interactions with the water itself and with particulate and dissolved matter. An emission spectrum obtained

in vitro from a water sample taken from the German Bight, Fig. 1, displays
structures that are related to elastic scattering at water molecules and hydrosols, to inelastic water Raman scattering, and to the fluorescence of Gelbstoff
and chlorophyll a (the latter, not given in the figure, peaking at 685 nm with
a half width of 20 nm). Since these structures are spectrally well separated
and are constant in shape in a first approximation they can be specifically
measured by an appropriate choice of a few detection wavelengths. The partial
interference of Gelbstoff fluorescence with water Raman scattering and chlorophyll a fluorescence has to be taken into account by additional detection
channels nearby these signals for an identification of their true baseline.

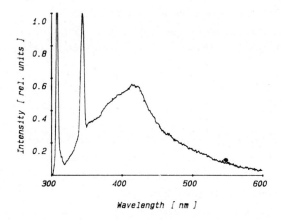

Fig. 1.

Emission spectrum of a natural
water sample taken from the
German Bight. Excitation wavelength is 308 nm. The peaks at
308 and 344 nm are due to
elastic scattering and water
Raman scattering, respectively.
The broad fluorescence band
centred at 420 nm is due to
Gelbstoff. The curve is corrected for the spectral response
of the instrument and normalized
to the Raman scatter intensity.

Among these spectral data, water Raman scattering represents an important
information for the lidar measuring process. The corresponding peak shown in
Fig. 1 is due to the stretching vibration of the water molecule being shifted
by a wavenumber of 3400 cm^{-1} with respect to the excitation line. The water
Raman scatter efficiency follows a $1/\lambda^4$ law with varying excitation wavelength,
with a value of the cross section of 4.5 ($\pm$ .3) $\bullet 10^{-29}$ cm^2/molecule $\bullet$ sr at
λ_{ex} = 488 nm (Slusher and Derr, 1975). Except for a weak temperature dependence
(Walrafen, 1967), the efficiency is constant with respect to other thermodynamic
quantities.

Besides instrumental factors as the laser intensity, the detector sensitivity
and the flight height, water Raman backscatter measured with lidar is thus given
by the penetration depth of the laser beam into the water column and the attenuation of the Raman scattered light on its way back to the water surface. As a
result of the theory the signal intensity is proportional to $c_{ex} + c_R$, the sum
of the light attenuation coefficients at the laser excitation and the Raman
scattering wavelength (Kung and Itzkan, 1976). In case of highly turbid waters

where multiple scattering at hydrosols becomes dominant, the signal measured
has a tendency to approach the diffuse attenuation coefficients $k_{ex} + k_R$
(Gordon, 1982).

In addition to turbidity measurements in the surface layer, water Raman
scattering serves as a calibration signal for normalizing the fluorescence data
of Gelbstoff and chlorophyll a to a constant measuring volume. The normalized
fluorescence is then proportional to the concentration of these substances;
concerning chlorophyll a, the dependence of the in vivo fluorescence efficiency
on the ambient light field must also be considered (Günther, 1985). The importance of this normalization has been pointed out by several authors, e.g. Hoge
and Swift (1982). Errors in the determination of Gelbstoff concentrations resulting from the necessity of measuring its fluorescence and water Raman
scattering at different detection wavelengths where the light attenuation coefficients and hence the water volume under consideration are not identical will
be discussed in the following section.

GELBSTOFF AS AN OPTICAL TRACER SUBSTANCE
Absorption

The existence of dissolved organic matter as a common feature of most coastal
areas has been well known for a long time. Because of its strong light absorption at blue wavelengths which results in a yellow colour of these waters,
Kalle (1937) characterized this substance by the term Gelbstoff. Chemically,
Gelbstoff is classified as humic substance which describes organic macromolecules with highly varying composition that are produced during the decay of
plants. The major part of Gelbstoff is of continental origin and is brought into
the sea by river discharge. An in situ production due to plankton blooms and
bacterial growth may be of relevance in the open ocean, in coastal waters such
contributions seem to be negligible (Højerslev, 1980).

Quantitative measurements of Gelbstoff in absolute concentration units are
hampered by the difficulty of establishing appropriate analytical methods which
can be used on a routine basis. The currently utilized procedure relates Gelbstoff to its light absorption at blue wavelengths after removing particulates
by filtration with a nominal pore width of 0.5 μm. Based on ultrafiltration of
Baltic water and weighing of the dry organic residual, Nyquist (1979) finds an
absorption coefficient $a = 0.212 \ m^{-1}$ at $\lambda = 450$ nm for a concentration of
1 mg/l.

However, the deduction of Gelbstoff concentrations from absorption data involves several assumptions: (i) filtration with the given pore size efficiently
separates dissolved and suspended absorbing material. This concerns the problem
that bacteria and mineral particles, both slightly absorbing, are not completely
retained and that the larger organic macromolecules show a tendency to form

colloids with increasing salinity and/or pH (Haekel, 1982). (ii) the composition of Gelbstoff is supposed to be constant in different areas of the sea including brackish and freshwater and possesses an invariant specific absorption spectrum. It will be shown that the absorption spectra of river and open sea water show distinct differences which might again be attributed to changes of chemical parameters.

As a result of investigations made by various working groups (e.g. Lundgren, 1976; Nyquist, 1979; Bricaud et al., 1979) and performed in different areas of the ocean, the absorption coefficient a_y of Gelbstoff - measured with respect to purified water as the reference medium - increases exponentially with decreasing wavelength in the near UV and blue portion of the spectrum according to

$$a_y \sim \exp(-b\lambda) \tag{1}$$

with $b = 0.014 \pm 0.0025$ nm^{-1}. Combining this with Nyquist's relation, it follows for the specific attenuation coefficient

$$s_y = .212 \exp[-b(\lambda-450)] = a_y/n \tag{2}$$

with n the Gelbstoff concentration.

Data obtained with highly resolving spectrophotometers having an optical path of 1 m length reveal that the spectrum is composed of several absorption bands covering the whole visible, their origin being actually unknown. Two intense overlapping bands with unidentified maxima in the UV yield an approximately exponential shape at blue wavelengths. This finding was first reported by Diehl and Haardt (1980) for Baltic Sea water and is also characteristic of the North Sea and the Adriatic (Fig. 2).

In contrast to this, the absorption spectrum of river water which has been examined mainly for the Adriatic up to now, is almost monotonic, whereby the exponential characteristic in the blue is about the same as with open sea waters (Fig. 3). The occurrence of distinct absorption bands is less obvious here.

Fluorescence

Another method for measuring Gelbstoff, which can be applied continuously in situ or by use of airborne lidar, takes advantage of its fluorescence (Fig. 1). The simultaneous registration of water Raman scattering allows an absolute calibration of the data with respect to the sensitivity of the instrument. A selection of the fluorescence detection wavelength sufficiently close to the Raman scatter wavelength reduces systematic errors efficiently resulting from the generally different attenuation coefficients at the detection wavelengths. As a result of the theory of the lidar measuring process discussed in the following section, the water Raman scattering and the Gelbstoff fluorescence signal P_R and

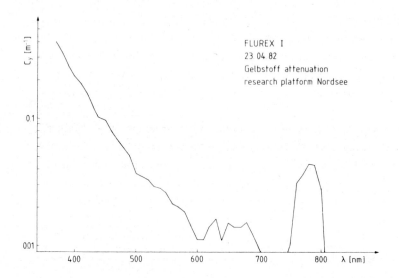

Fig. 2. Spectral attenuation coefficient of a water sample taken from the German Bight.

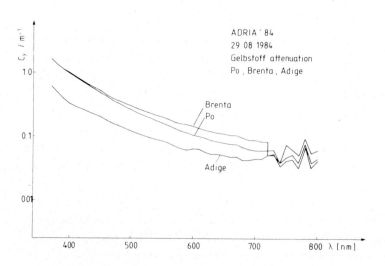

Fig. 3. Spectral attenuation coefficient of water samples taken from north Italian rivers.

P_y received from a homogeneous water column can be written as

$$P_R(\lambda_{ex}, \lambda_R) \sim \eta_R / (c_w + a_y + c_p) \tag{3}$$

$$P_y(\lambda_{ex}, \lambda_y) \sim n \cdot \eta_y / (c_w + a_y + c_p) \ ,$$

where η_R and η_y are the specific efficiencies of water Raman scattering and Gelbstoff fluorescence, n the Gelbstoff concentration, c_w and c_p the sum of the light attenuation coefficients at the laser excitation and the detection wavelengths of clear water and particulate matter, and a_y the sum of the respective absorption coefficients of Gelbstoff. With $a_y = n \cdot s_y$, s_y being the specific Gelbstoff absorption coefficient, the relative error for determining the Gelbstoff concentration with lidar is then

$$\frac{P_y / P_R}{n} \sim \frac{c_w(\lambda_{ex}, \lambda_R) + n \cdot s_y(\lambda_{ex}, \lambda_R) + c_p(\lambda_{ex}, \lambda_R)}{c_w(\lambda_{ex}, \lambda_y) + n \cdot s_y(\lambda_{ex}, \lambda_y) + c_r(\lambda_{ex}, \lambda_y)} \tag{4}$$

In Fig. 4 results are plotted as a function of the Gelbstoff concentration n with the particle attenuation coefficient at λ = 344 nm as the parameter. The wavelength dependence of c_p is assumed to be proportional to $1/\lambda$ which holds for most oceanic hydrosols. $s_y(\lambda)$ follows an exponential law according to (2). Data of c_w are taken from tables of the diffuse clear water attenuation coefficie t reported by Smith and Baker (1981).

Correlation between absorption and fluorescence

Fluorescence as an alternative technique for the examination of Gelbstoff necessitates that both methods lead to consistent results. This holds for the Adriatic (Russo, 1983) and the North Sea with oceanic values of salinity where a correlation coefficient of absorption and fluorescence of about 0.9 is found. However, the correlation drops down to about 0.5-0.7 in coastal areas with freshwater influence and for river water. It is surprising that Højerslev (1980) finds no correlation to be present in the Baltic. Near the Weser estuary in the German Bight drastic deviations of the fluorescence spectrum compared to the open sea are observed (Fig. 5) which must be due to a varying composition of Gelbstoff as a result of changing chemical conditions of the water.

Compared to absorption measurements performed at blue wavelengths and using a 1 m path length instrument, the fluorescence technique with near UV excitation shows a much better lower limit of sensitivity. In terms of absolute concentration units given above these limits are 0.1 and 0.01 mg/l, respectively.

Stability

Concerning the stability of Gelbstoff as a prior condition for its use as a

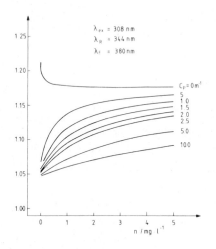
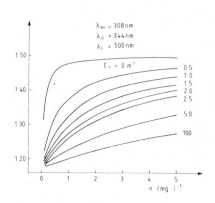

Fig. 4. Relative error for the lidar measurement of Gelbstoff, eq. (4), as a function of the concentration with the particle attenuation coefficient as the parameter. Excitation wavelength is 308 nm, water Raman scattering wavelength is 344 nm. The Gelbstoff fluorescence is detected at 380 and 500 nm, respectively. The importance of a fluorescence detection close to the excitation and water Raman scatter wavelengths is obvious.

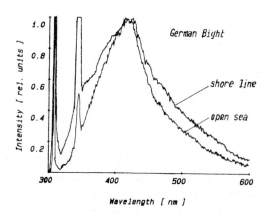

Fig. 5. Gelbstoff fluorescence spectra of water samples taken from the open sea and near the shore line (Weser estuary) of the German Bight. The curves are normalized to identical fluorescence maxima to enable a comparison of the shape of the spectra. The predominance of fluorescence at higher wavelengths for the shore line sample might be due to the higher concentration of large organic molecules.

natural tracer substance it is generally stated that a degradation due to biological processes does not exist. Haekel (1982) finds an invariant concentration
in water samples stored in the dark. The lifetime of Gelbstoff with respect to
photochemical decay has been examined in first experiments by the authors in
collaboration with Dr. R. Doerffer, GKSS Forschungszentrum Geesthacht. A decrease of the concentration by 50% over a period of 125 days measured with
fluorescence and absorption meters is identically found for 3 water samples exposed to an illumination which corresponds to daylight conditions, compared to
samples stored in the dark. This results in a good stability within time scales
characteristic for transport in the German Bight and most other coastal areas,
if the effective exposure of Gelbstoff in the water column to sunlight is taken
into account.

To summarize the characteristics of Gelbstoff, it can be stated that this
material is sensitively identified with optical methods, in particular with
fluorescence spectroscopy, in coastal waters due to its high concentration and
fluorescence yield. According to the present knowledge its stability is sufficiently high to provide a natural tracer for the study of characteristic
water masses. Since its concentration is only influenced by mixing processes
Gelbstoff turns out to be a better conservative parameter than temperature in
shallow coastal waters. However, a number of questions have to be examined in
detail:

- the changes in the composition and the optical properties of Gelbstoff during
  transport by rivers to the open sea, to be taken into account by appropriate
  algorithms applied to fluorescence and absorption data,
- the correlation of absorption and fluorescence as a function of varying
  salinity and pH,
- the role of a possible in situ production,
- the photochemical decay of Gelbstoff,
- a quantitative determination of the specific fluorescence efficiency as a
  function of the excitation and emission wavelength.

THE OCEANOGRAPHIC LIDAR

Theory of the measuring process

The oceanographic lidar is an active sensor by which an airborne probing of
optically detectable parameters in the upper water layers is achieved. In its
basic concept the lidar consists of a laser emitting at near UV or visible
wavelengths where water shows a good light transmission and of a telescope for
the detection of laser-induced radiation from the water column. The intensity
of the received light is measured at detection wavelengths which are characteristic for scattering and fluorescence of the substances under investigation

(Fig. 1). To overcome signal contributions due to sunlight, high power pulse laser systems are generally utilized whereby the background light can be eliminated by appropriate signal gating techniques.

A quantitative formulation of the measuring process is derived from the lidar equation (Browell, 1977) giving the laser-induced signal dP received from a depth interval dz at depth z:

$$dP = A \eta \frac{\exp \left(-\int_0^z c \, dz'\right)}{(z+mH)^2} \, dz \qquad (5)$$

with $z = 0$ at the water surface and z positive downwards, and where

$\eta = \eta(z, \lambda_{ex}, \lambda_{em})$ quantum efficiency of fluorescence or water Raman scattering,

$c = c_{ex}(z, \lambda_{ex}) + c_{em}(z, \lambda_{em})$ sum of light attenuation coefficients at the excitation and the emission wavelength,

H aircraft flight height,

m refractive index of water,

A includes instrumental factors, signal losses in the atmosphere, and effects of the rough water surface on the beam propagation.

The signal originating from a water layer at depth $z_1 \leq z \leq z_2$ reads:

$$P_{z_1 \to z_2} = A \exp \left(- \int_0^{z_1} c \, dz\right) \cdot \int_{z_1}^{z_2} \eta \, \frac{\exp \left(-\int_{z_1}^z c \, dz'\right)}{(z+mH)^2} \, dz \qquad (6)$$

$$= A \exp \left(- \int_0^{z_1} c \, dz\right) \cdot \phi \quad .$$

With $w = (z+mH)/mH$ (Browell, 1979) and assuming η and c to be constant at $z_1 \leq z \leq z_2$, the integral over this layer is written as

$$\phi = \eta \, \frac{\exp \left(mHcw_1\right)}{nH} \int_{w_1}^{w_2} \frac{\exp \left(-mHcw\right)}{w^2} \, dw \quad . \qquad (7)$$

With the further substitution

$$dw = - \frac{1}{mHc + \frac{2}{w}} \cdot \frac{w^2}{\exp \left(-mHcw\right)} \, d \, \frac{\exp \left(-mHcw\right)}{w^2} \quad ,$$

and approximating $mHc \gg 2/w$ which is equivalent to $c \gg 2/(z+mH)$ and will be valid in all practical cases, the integration of (7) is straightforward:

$$\Phi = \frac{\eta}{c} \frac{1}{(mH)^2} \left(\frac{1}{w_1^2} - \frac{\exp(-mHc(w_2-w_1))}{w_2^2} \right) .$$

The general solution of the lidar equation (6) for a water column with arbitrary stratification then reads

$$P_{z_1 \to z_2} = A \exp\left(-\int_0^{z_1} c\, dz\right) \frac{\eta}{c} \left(\frac{1}{(z_1+mH)^2} - \frac{\exp(-c(z_2-z_1))}{(z_2+mH)^2} \right) \tag{8}$$

and the total signal is obtained by the summation of all contributions (8) between $z=0$ and $z \gg 1/c$. For a homogeneous water column it follows

$$P_R = \frac{A}{(mH)^2} \frac{\eta_R}{c_R} \tag{9}$$

for water Raman scattering where η_R is constant, and

$$P_F = \frac{A}{(mH)^2} \frac{\eta_F}{c_F} \tag{10}$$

for fluorescence of Gelbstoff or chlorophyll a. The normalization of fluorescence signals to the water Raman scattering yields data on the concentration n_F of the fluorescing matter,

$$\frac{P_F}{P_R} = \frac{\eta_F}{\eta_R} \bigg/ \frac{c_R}{c_F} \sim n_F \tag{11}$$

as outlined in the preceding section, eq. (3) and (4).

Instrument description

The Oceanographic Lidar System (OLS) developed at the University of Oldenburg has been designed for installation in Do 28 or Do 228 research aircraft. Lasers and the detector system are mounted on an optical table to obtain a rigid alignment of the optical setup (Fig. 6). System operation and data processing is done in-flight by one operator. The total system weight is 500 kg; the payload

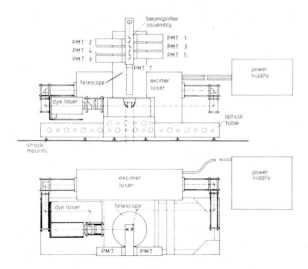

Fig. 6. Optical part of the Oceanographic Lidar System. The position of the telescope is above a bottom batch of the aircraft for free field of view towards the water surface. The ray path of the laser beams are shown as broken lines.

capacity of the Do 28 allows a flight time of 3 hours, that of the Do 228 of 6 hours.

A Lambda Physik EMG 101 excimer laser serves as the main light source with a peak power of 10 MW and a pulse length of 12 ns at a wavelength of 308 nm. Front and rear output of the laser are utilized as the lidar beam or as the pumping beam for a Lambda Physik FL 2001 dye laser with a peak power of 1 MW and a pulse length of 6 ns, respectively. The dye laser output wavelength is set at 450 nm. The maximum pulse repetition rate is 20 Hz.

The signal receiver is a Schmidt-Cassegrain f/10 telescope with a 40 cm aperture diameter. The field of view is set to 5 mrad, which corresponds to the beam divergence of the excimer laser. At a flight height of typically 200 m this results in a footprint diameter of 1 m. Dichroic beamsplitters deflect selected spectral ranges of the received light to interference and edge filters with typical bandwidths of 10 nm, and 10 dynode fast photomultipliers EMI 9812 and 9818. The multipliers are activated by a gating circuit for time periods which can be set between 100 ns and 3 μs following each laser shot to avoid a non-linear response due to sunlight-induced background.

The detection wavelengths are chosen at 344, 366, 380, 500, 533, 650 and 685 nm. The channels at 344 and 533 nm correspond to the wavelengths of water Raman scattering with 308 and 450 nm excitation, respectively. Chlorophyll *a* fluorescence is observed at 685 nm. The 366, 380, 500 and 650 nm channels are used for the measurement of the Gelbstoff fluorescence spectrum and for the identification of the baseline of water Raman scattering and chlorophyll a fluorescence (Fig. 1).

A calibration of the detection channels with respect to their relative spectral sensitivities is performed on the ground with a 1 m² plate of white teflon irradiated by the excimer laser. This material has the advantage of possessing a fluorescence spectrum covering the whole visible and near UV with an efficiency which corresponds well to the typical intensities of the water column return. Moreover, teflon is a stable material with good resistance against contamination, making it very suitable for use on an airfield.

Signal digitization is done with a Biomation Model 6500 transient recorder at a sampling rate of 500 MHz and a resolution of 6 bit. Since the digitizer is a one-channel instrument, 3 photomultipliers selected by the operator are sequentially combined on one signal line and fed to the transient recorder.

The system is controlled by a LSI 11/02 microcomputer, by which laser selection and triggering, selection of different detection modes, quick look data output and data storage on floppy discs or magnetic tapes is achieved. A schematic of the signal flow is shown in Fig. 7.

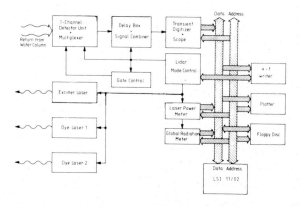

Fig. 7. Schematic of the signal flow of the Oceanographic Lidar System.

EXPERIMENTAL RESULTS

North Sea

Since 1984 the Institute for Marine Research, Bremerhaven, the German Hydrographic Office and the Universities of Hamburg and Oldenburg have conducted a joint research programme aiming at a better understanding of frontal systems and characteristic water masses of the German Bight. The scope of this project and results of biological and physical in situ observations are described by Krause et al. (1985) in a comprehensive paper given in this volume.

As part of this project, airborne experiments were done in parallel with shipboard investigations carried out by these institutions. A combination of these data and an integral interpretation are still in progress. This concerns also a calibration of the lidar measurements in absolute units which at present necessitates additional ground truth information. Results are thus shown in relative units, and only estimates of the substance concentrations are given which are derived from a few in situ data. New methods for an absolute calibration of OLS data without the need of ground truth are now under investigation.

Here we report some results of OLS flights which demonstrate the capability of lidar for large scale monitoring, in particular concerning the characterization of water masses and the identification of fronts at the water surface on the basis of Gelbstoff measurements.

Fig. 8 displays profiles obtained on a track from the island of Helgoland to Wilhelmshaven on October 25, 1983. In this flight which was the first airborne operation of OLS, the dye laser was emitting at a wavelength of 450 nm with a repetition rate of 2 Hz. Light attenuation is derived from the water Raman scattering signal at 533 nm according to equ. (10) after subtraction of background fluorescence; the data correspond thus to $c_{(450)} + c_{(533)}$. Gelbstoff concentration is calculated from the fluorescence measured at 500 nm normalized to the water Raman scattering, equ. (11). The same applies to chlorophyll $\underline{a}$ fluorescence detected at 685 nm yielding the respective concentration.

A close correlation of the light attenuation coefficient and of the Gelbstoff and chl $\underline{a}$ concentration is obvious from the data. In the open German Bight these parameters are horizontally homogeneous. The sharp peak at km 32 with a width of about 2 km is found at the Old Elbe Valley and is probably attributed to Elbe river water. The enhanced values at the end of the track, km 40 - 45, with a strong patchiness present at small scales, might be due to the influence of the Weser river plume.

Results obtained during a flight on March 25, 1985 along a west-east track

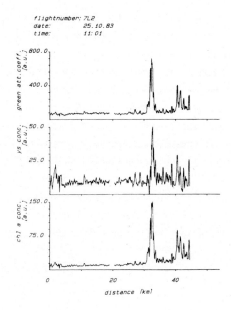

Fig. 8. Results of lidar measurements performed on a track from Helgoland to Wilhelmshaven, showing the light attenuation at blue/green wavelengths, and the Gelbstoff (ys) and chl a concentration. Flight height was 200 m, signal repetition rate 2 Hz.

on latitude 54°00' N, starting at 7°34' E north of the island of Langeoog and ending on the shore of the Elbe estuary, are shown in Fig. 9. In this experiment, the excimer laser was operated at a repetition rate of 10 Hz, its emission wavelength being 308 nm.

The UV attenuation coefficient is derived from the excimer laser induced water Raman scatter detected at 344 nm yielding $c_{(308)} + c_{(344)}$. Chl a concentration is deduced from its fluorescence at 685 nm; these data show a rather poor signal-to-noise ratio because of the excitation in near UV which is not very appropriate for the purpose of chl a detection.

Concerning Gelbstoff, two profiles are shown, the upper curve being derived from the fluorescence measured at 380 nm, the lower one at 500 nm. According to the statements given above, equ. (4) and Fig. 4, the fluorescence detection channel closer to the excitation and water Raman scattering wavelengths yields more accurate data because of the deteriorating effect of dissolved and particulate matter and their spectrally varying light attenuation. The upper curve will thus reflect the true Gelbstoff concentration with better reliability, and the importance of a suitable choice of the fluorescence detection wavelength is clearly seen.

A synchronized measurement of airborne and shipboard data was done at the beginning of the track, km 0. In situ values of 2.6 µg/l of chl a and 1.4 mg/l

584

of Gelbstoff were found which gives an idea of the scaling of the profiles.

Compared with the data shown in Fig. 9, a correlation of the measured parameters is not found here. The chl a distribution is very irregular. The light
attenuation coefficient shows an increasing tendency in front of the Weser and
Elbe mouth as it has to be expected, however, no distinct signatures are observed. In contrast to this, Gelbstoff data reveal the presence of two water
masses with a virtually constant concentration, separated by a strong gradient
at km 48. The poor correlation with the light attenuation is probably a result
of the competiting influences of suspended minerals and phytoplankton on this
parameter.

The assumption that the hydrographic situation in the surface layer is characterized by two homogeneous water masses separated by a front of the physical
parameters as outlined by the Gelbstoff concentration, has to be confirmed by
in situ data of temperature and salinity.

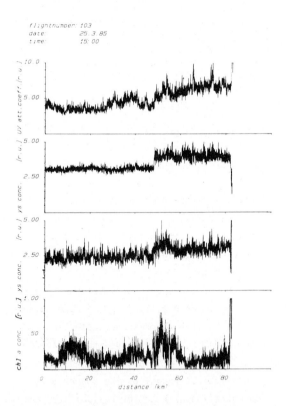

Fig. 9. Results of lidar measurements performed on a west-east track on latitude 54°00' N, starting at 7°34' E and ending on the shore of the Elbe estuary.
The light attenuation at UV wavelengths and the Gelbstoff (ys) and chl a concentrations are shown. For an explanation of the two Gelbstoff profiles see
text. Flight height was 200 m, signal repetition rate was 10 Hz.

Adriatic Sea

In August 1984 the experiment ADRIA 84 was performed in the northern Adriatic Sea under the auspices of the Commission of the European Communities, ISPRA
Establishment. The goal of the experiment in which a large number of institutions have been involved was an investigation of the complicated hydrodynamic
processes present in this area in the summer period, and an examination of the
potential of airborne and satellite remote sensing for their study (Schlittenhardt, 1984, 1985).

The Oceanographic Lidar System was operated in the period of August 29 - 31
on flight tracks which were assumed to cover the areas of highest interest: a
north-south track on longitude 12°40' E passing nearby the Po delta, a second
track on 13°00' E in the open Adriatic where waters with a lower influence of
river run-off were expected. Flight height was 200 m throughout the measurements, and both the excimer and the dye laser were operated at a repetition
rate of 2.5 Hz.

Results of a flight on August 30 are shown in Fig. 10a,b. From the excimer
laser induced signals, the UV light attenuation coefficient $c_{(308)} + c_{(344)}$
and the Gelbstoff concentration (Fig. 10a), from those of the dye laser, the
blue/green attenuation coefficient $c_{(450)} + c_{(533)}$ and the chl a concentration
(Fig. 10b) are derived according to the methods described in the preceding
sections.

The Po river plume with a high variability of the measured data, and the
existence of different water masses with individual light attenuation and substance concentration are obvious from the data. These water masses are separated by fronts with gradients of typically a factor of 10 over distances of 1 -
3 km.

A comparison with in situ findings obtained in the northern part of the
flight track on 13°00 E yields the following data to be characteristic for the
water mass present in that area: attenuation coefficient $(c_{(455)} + c_{(525)})/2 =$
0.2 - 0.3 m^{-1}, Gelbstoff concentration 0.15 - 0.2 mg/l, chl a concentration
1 - 2 µg/l. The lower sensitivity limit of OLS in limpid waters with low sediment load is thus in the order of 0.01 mg/l for Gelbstoff and 0.1 µg/l for
chl a.

The hydrographic conditions on the outer track turn out to be very stable
within the period of investigation. The situation found on the track near the
coastline is considerably different: the Po river plume scarcely identified on
August 29 shows a strong evolution towards the open sea on the following days.
Fig. 11 displays this evolution from the Gelbstoff data, the other measured parameters yielding a very similar tendency.
The further interpretation of these data and a comparison with in situ and satellite observations are now in progress. It is expected that, in view of the

586

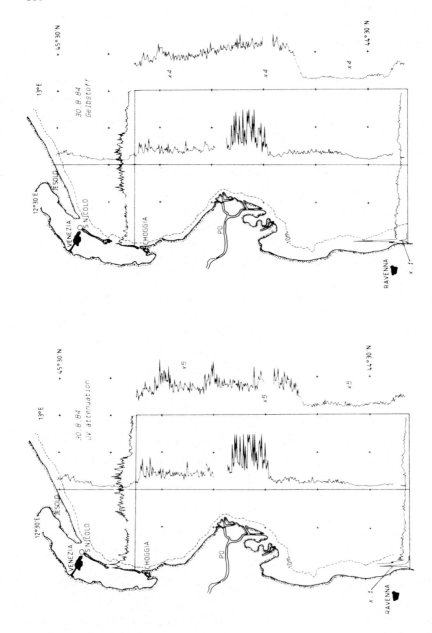

Fig. 10a. Distribution of the UV attenuation coefficient (left) and the Gelbstoff concentration (right) in the northern Adriatic Sea measured with lidar on August 30, 1984. Scaling of the profiles is identical except for the track on longitude 13°00' E which is given with higher resolution. Flight height was 200 m, signal repetition rate was 2.5 Hz.

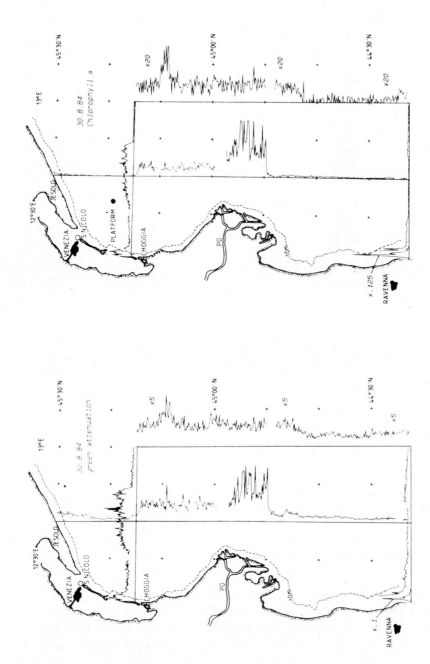

Fig. 10b. Distribution of the blue/green attenuation coefficient (left) and the chl a concentration (right) in the northern Adriatic Sea measured with lidar on August 30, 1984. Scaling of the profiles is identical except for the track on longitude 13°00' E which is given with higher resolution. Flight height was 200 m, signal repetition rate was 2.5 Hz.

588

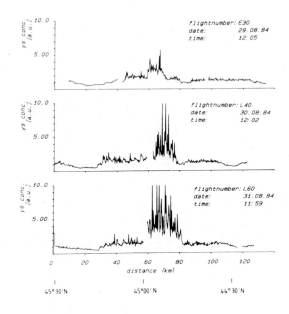

Fig. 11. Results of lidar measurements of Gelbstoff (ys) concentrations performed in the northern Adriatic on longitude 12°40' E in the period of August 29 - 31, 1984. Gelbstoff data show a drastic increase in front of the Po delta over the period of investigation.

enormous number of data obtained, the results of ADRIA 84 will enable a very detailed delineation of the hydrographic and biological processes in the northern Adriatic Sea.

CONCLUSIONS

The optical properties of Gelbstoff and its high concentration found in coastal zones present problems for the investigation of these waters with optical satellite radiometry. The evaluation of chlorophyll concentrations from water leaving radiance data measured with these sensors is hampered by the spectrally selective light absorption of Gelbstoff.

On the other hand, this material can be sensitively measured with lidar due to the specific and intense fluorescence emission. This is demonstrated in experiments performed in the German Bight and the Adriatic Sea. Moreover, lidar allows an evaluation of light turbidity and of chlorophyll concentrations without deteriorating interference of Gelbstoff or other water column constituents. A combination of synoptic lidar measurements with satellite imagery results thus in a calibration of satellite data in terms of these parameters.

Taking into account the good stability, Gelbstoff can be utilized as a tracer substance for the investigation of characteristic water masses. In addition to the investigation of dynamic processes, this characteristic enables a study of transport and mixing of river water in coastal areas.

A quantitative delineation of these processes necessitates some further investigations of Gelbstoff, particularly an evaluation of more reliable data on its fluorescence efficiency and stability with respect to changing chemical conditions in coastal waters.

Efforts will be done in order to establish procedures for an absolute calibration of lidar data without the need of further ground truth information. This will yield an important improvement of oceanographic lidar remote sensing.

ACKNOWLEDGEMENTS

The experiments in the German Bight were performed in cooperation with the Institute for Marine Research, Bremerhaven. The project ADRIA 84 was initiated, realized and supported by the Commission of the European Communities, ISPRA Establishment. We wish to express our thanks to Prof. G. Krause, Bremerhaven, and to Dr. P. Schlittenhardt, Ispra.

We are grateful to DFVLR Oberpfaffenhofen for making available the research aircraft. We are especially indebted to Mr. H. Finkenzeller and Mr. P. Vogel for their comprehensive support and the logistical coordination during the experiments.

The development of the Oceanographic Lidar System is financed by a grant from the Bundesministerium für Forschung und Technologie; research on water masses and frontal systems by a grant from the Deutsche Forschungsgemeinschaft.

REFERENCES

Bricaud, A., Morel, A. and Prieur, L., 1981. Absorption of dissolved organic matter of the sea (yellow substance) in the UV and visible domains. Limnol. Oceanogr., 26: 43-53.
Browell, E.V., 1977. Analysis of laser fluorosensor systems for remote algae detection and quantification. Report NASA TN D-8447. NASA Langley Research Center, Hampton, 39 pp.
Diehl, P. and Haardt, H., 1980. Measurement of the spectral attenuation to support biological research in a plankton tube experiment. Oceanol. Acta, 3: 89-95.
Fischer, J. and Graßl, H., 1984. Radiative transfer in an atmosphere - ocean system: an azimutally dependent matrix - operator approach. Appl. Opt., 23: 1032-1039.
Gordon, H.R., 1982. Interpretation of airborne oceanic lidar: effects of multiple scattering. Appl. Opt., 21: 2996-3000.
Gordon, H.R., Clark, D.K., Brown, J.W., Brown, O.B., Evans, R.H. and Broenkow, W.W., 1983. Phytoplankton pigment concentrations in the Middle Atlantic Bight: comparison of ship determinations and CZCS estimates. Appl. Opt., 22: 20-36.

Günther, K.P., 1985. A quantitative description of the chlorophyll a fluores-
    cence reduction due to global irradiation in the surface layer. Proc. 17.
    International Liège Colloquium on Ocean Hydrodynamics.
Haekel, W., 1982. Untersuchungen zur Schwermetallbindung durch Huminstoffe in
    Ästuarien. PhD thesis, Universität Kiel, 147 pp.
Hoge, F.E. and Swift, R.N., 1982. Delineation of estuarine fronts in the Ger-
    man Bight using airborne laser - induced water Raman backscatter and fluor-
    escence of water column constituents. Int. J. Remote Sensing, 3: 475-495.
Højerslev, N.K., 1980. On the origin of yellow substance in the marine envi-
    ronment. In: Studies in physical oceanography, Københavns Universitet, In-
    stitut for Fysisk Oceanografi, Report Nr. 42: 39-56.
Kalle, K., 1937. Meereskundliche chemische Untersuchungen mit Hilfe des Zeiss-
    schen Pulfrich Photometers. VI: Die Bestimmung des Nitrats und des Gelb-
    stoffs. Ann. Hydr. u. Marit. Meteorol., 276-282.
Kung, R.T.V. and Itzkan, I., 1976. Absolute oil fluorescence conversion effi-
    ciency. Appl. Opt., 15: 409-415.
Lundgren, B., 1979. Spectral transmittance measurements in the Baltic. Køben-
    havns Universitet, Institut for Fysisk Oceanografi, Report Nr. 30, 38 pp.
Nyquist, G., 1979. Investigation of some optical properties of seawater with
    special reference to lignin sulfonates and humic substances. PhD thesis,
    Göteborgs Universitet, 200 pp.
Russo, M.C., 1983. A study of the inherent optical properties of yellow sub-
    stance from the northern part of the Adriatic Sea. In: P.M. Schlittenhardt
    (Editor), Workshop on Remote Sensing of Coastal Transport in the Northern
    Adriatic Sea, Proceedings, Joint Research Centre: 59-74.
Schlittenhardt, P., 1984. ADRIA 84. A joint remote sensing experiment. Tech-
    nical Note No. 1.05.E2.84.91. Commission of the European Communities, Joint
    Research Centre, Ispra, 33 pp.
Schlittenhardt, P., 1985. ADRIA 84. A joint remote sensing experiment. 2. Re-
    port: a short summary of the experiment. Technical Note No. 1.05.E2.85.20.
    Commission of the European Communities, Joint Research Centre, Ispra, 58 pp.
Slusher, R.B. and Derr, V.E., 1975. Temperature dependence and cross sections
    of some Stokes and anti-Stokes Raman lines in ice Ih. Appl. Opt., 14: 2116 -
    2120.
Smith, R.C. and Baker, K.S., 1981. Optical properties of the clearest natural
    waters (200 - 800 nm). Appl. Opt. 20: 177-184.
Walrafen, G.E., 1967. Raman spectral studies of the effects of temperature on
    water structure. J. Chem. Phys., 47: 114-126.

WATER DEPTH RESOLVED DETERMINATION OF HYDROGRAPHIC PARAMETERS FROM AIRBORNE LIDAR MEASUREMENTS

D. DIEBEL-LANGOHR, T. HENGSTERMANN and R. REUTER

Universität Oldenburg, Fachbereich Physik, P.O.Box 2503, 2900 Oldenburg,

(Federal Republic of Germany)

ABSTRACT

The lidar method applied from low flying aircraft over the ocean has a potential for the depth resolved investigation of hydrographic parameters. This
method requires laser pulses in the nanosecond range which are directed towards
the water surface. Fluorescence and Raman scatter of water column constituents
are collected by a telescope system and recorded as time resolved signals. They
contain information on the depth distribution of these substances. The lidar
signals depend on the shape of the laser pulse, the fluorescent and scattering
decay times of the substances under investigation, their distribution with
depth, and the time response of the instrument, and are mathematically described
by a convolution of these parameters. Algorithms for the evaluation of depth
profiles from lidar data are presented and applied to various models of stratified surface layers. Time resolved measurements were performed over the northern
Adriatic Sea in September 1984 with the Oceanographic Lidar System (OLS) of the
University of Oldenburg. Profiles of the attenuation coefficient obtained in
this experiment are presented.

INTRODUCTION

Since 1963 the potential of lidar measurements for the derivation of range
resolved parameters of the atmosphere has been demonstrated. Gaseous constituents are identified with the Differential Absorption Lidar (DIAL) (Zuev, 1983;
Uchino, 1980); the detection of Mie scattered light allows the mapping of
aerosol structures (Allen, 1972); Rayleigh and Raman scattering is used for
temperature profiling (Schwiesow, 1981; Arshinov, 1983). One of the crucial
problems for atmospheric lidar measurements is the large dynamic range of the
signals due to the $1/R^2$ dependence with distance R (Allen, 1972; Shimizu, 1985;
Harms, 1978).

In the past years the possibility of measuring depth resolved hydrographic
parameters with lidar technique has been studied (Hoge, 1983). Time resolved
scattering and fluorescence signals contain information on the vertical distribution of the attenuation coefficient and of the concentration of Gelbstoff and
chlorophyll a. The dynamic of the signals is determined here by the exponential
signal compression due to the light attenuation in the water column whereas for
flight heights of typically 100-500 m the $1/R^2$ dependence is less important. The
measurement of profiles is limited to a few attenuation lengths. To obtain a

high depth resolution within this range laser pulses with a length of only some nanoseconds and a signal detection at time intervals small compared to the total signal duration are required.

A lidar return signal depends on the shape of the exciting laser pulse, the attenuation and the substance concentrations along the light path, the decay function of the scatter or fluorescence process, and the response function of the lidar system. Mathematically the detected signal is a convolution of all these functions. An interpretation of lidar measurements without first performing a deconvolution results in a reduced range resolution. This is not problematic for atmospheric studies where a resolution of some meters is generally sufficient. For hydrographic profiling of the surface layer of the ocean where the penetration depth of blue/green laser light is limited to about 20 meters in most cases, a depth resolution of 1 meter or better should be achieved.

In the following, algorithms are presented to evaluate vertical profiles of the attenuation coefficient and the concentration of Gelbstoff from time resolved water Raman scattering and Gelbstoff fluorescence signals. It is not necessary to consider the decay functions since the decay times of these optical interactions are shorter than the 2 ns time resolution of the Oceanographic Lidar System (OLS) of the University of Oldenburg. The rise and fall times of the photomultipliers used are about 2 ns, the rise time of the amplifier is less than 3 ns, the fall time less than 5 ns per decade. So in a first approximation the lidar signal can be assumed to be a convolution of the laser pulse shape and of a function containing the attenuation coefficient and the concentration of Gelbstoff.

THEORETICAL BACKGROUND
Relation between time resolved lidar signals and depth resolved hydrographic parameters

Fig. 1 illustrates the principle laser pulse shape $P_L(t)$ and the water column return signal $P(\lambda,t)$ measured at the detection wavelength λ. $P(\lambda,t)$ can include fluorescence and scattering contributions. The water column is in general composed of water itself (index w), suspended minerals (index p), Gelbstoff (index y), and chlorophyll a (index c). The signal $P(\lambda,t)$ is determined by their concentrations n_w, n_p, n_y, n_c, by the fluorescence and scattering efficiencies $\eta_w(\lambda)$, $\eta_p(\lambda)$, $\eta_y(\lambda)$, $\eta_c(\lambda)$, and by the attenuation coefficient $c(\lambda_L,\lambda)$ which means here the sum of the attenuation coefficients at the laser and at the detection wavelength. n_p, n_y, n_c, and $c(\lambda_L,\lambda)$ are functions of the water depth z.

A signal $S(\lambda,z)$ from depth z is first dependent on the attenuation of the laser light and of the light excited at depth z in the water column between $0 \leq z' \leq z$, and on the concentrations of the fluorescent and scattering sub-

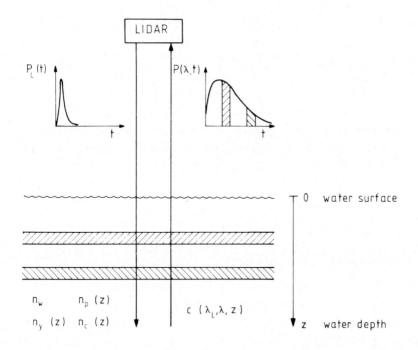

Fig. 1. Illustration of the abbreviations used. $P_L(t)$ laser pulse shape, $P(\lambda,t)$ lidar signal detected at wavelength λ, z water depth, n_w, n_p, n_y, n_c, concentrations of water, suspended minerals, Gelbstoff, chlorophyll <u>a</u>, $c(\lambda_L,\lambda)$ sum of the attenuation coefficients at the laser and at the detection wavelength.

stances at depth z. The depth function $S(\lambda,z)$ can be written as (Browell, 1977):

$$S(\lambda,z) \sim [n_w \cdot n_w(\lambda) + n_p(z) \cdot n_p(\lambda) + n_y(z) \cdot n_y(\lambda) + n_c(z) \cdot n_c(\lambda)] \quad .$$

$$\exp[- \int_0^z c(\lambda_L,\lambda,z') \, dz'] / (z + mH)^2 \qquad z \geq 0$$

(1)

where H is the flight height and m the refractive index of water. The depth z is related to a time t being twice the time the light needs to pass through a water layer of depth z. The transformation with t = 2 zm/v (v light velocity) yields:

$$S(\lambda,t) \sim [n_w \cdot n_w(\lambda) + n_p(t) \cdot n_p(\lambda) + n_y(t) \cdot n_y(\lambda) + n_c(t) \cdot n_c(\lambda)] \quad .$$

$$\exp[- (v/2m) \int_0^t c\ (\lambda_L,\lambda,t')\,dt'] \ / \ ((vt/2m) + mH)^2 \qquad t \geq 0 \tag{2}$$

The signal $P(\lambda,t)$ is composed of signals from all water depths. At a fixed time t the lidar system is not only receiving a signal from a fixed depth z but from a water layer with a thickness corresponding to the duration of the laser pulse. This is mathematically expressed by a convolution of the depth function $S(\lambda,t)$, and the laser pulse $P_L(t)$:

$$P(\lambda,t) \sim \int_{-\infty}^{\infty} P_L(t') \cdot S(\lambda,t-t')\,dt' \qquad S(\lambda,t) = 0 \text{ at } t < 0 \tag{3}$$

Algorithms for the derivation of depth profiles

The algorithms to calculate depth profiles of the attenuation coefficient and the substance concentrations from the measurable lidar signals have to start with the solution of the convolution integral (3) in order to express $S(\lambda,t)$ by the quantities $P_L(t)$ and $P(\lambda,t)$.

One way is to perform a Fourier analysis which yields:

$$S(\lambda,t) \sim \mathfrak{F}^{-1}[\mathfrak{F}(P(\lambda,t))/\mathfrak{F}(P_L(t))] \tag{4}$$

where $\mathfrak{F}$ means the Fourier transformation of the function which follows.

Alternatively the laser pulse shape $P_L(t)$ can be approximated by the δ-function $\delta(0)$ which would be correct for an infinitely short laser pulse at time $t = 0$. Then the convolution integral (3) can be solved directly. The result of this Delta pulse analysis is:

$$S(\lambda,t) \sim P(\lambda,t) \tag{5}$$

The next step is to calculate the depth profiles from the functions $S(\lambda,t)$ which are obtained from lidar signals measured at different detection wavelengths.

As an example the evaluation of the attenuation coefficient and of the concentration of Gelbstoff is discussed.

In the case of water Raman scattering detected at a wavelength λ_R where other signal contributions, e.g. fluorescence of chlorophyll a or Gelbstoff are negligible, (2) simplifies to:

$$S(\lambda_R,t) \sim n_w \cdot n_w(\lambda_R) \cdot$$

$$\exp[-(v/2m) \int_0^t c(\lambda_L,\lambda_R,t')\,dt'] / ((vt/2m) + mH)^2 \quad t \geq 0 \tag{6}$$

With signals detected time resolved at time intervals of length Δt one can form the quotient

$$S(\lambda_R,t) / S(\lambda_R,t+\Delta t) = \exp[(v\cdot\Delta t/2m) \cdot c(\lambda_L,\lambda_R,t)] \tag{7}$$

assuming $c(\lambda_L,\lambda_R,t') \cong c(\lambda_L,\lambda_R,t)$ within the small time interval $t \leq t' \leq t+\Delta t$ and $(vt/2m)+mH \cong (v(t+\Delta t)/2m)+mH$. This yields for the attenuation coefficient:

$$c(\lambda_L,\lambda_R,t) = (2m/v\Delta t) \ln[S(\lambda_R,t) / S(\lambda_R,t+\Delta t)] \tag{8}$$

The depth function (2) is reduced to

$$S(\lambda_Y,t) \sim n_y(t) \cdot n_y(\lambda_Y) \cdot$$

$$\exp[-(v/2m) \int_0^t c(\lambda_L,\lambda_Y,t')\,dt'] / ((vt/2m)+mH)^2 \quad t \geq 0 \tag{9}$$

for a Gelbstoff fluorescence signal if the corresponding efficiency n_y is dominating at the wavelength λ_Y. For a detection wavelength λ_Y chosen close to λ_R the attenuation coefficients $c(\lambda_L,\lambda_R)$ and $c(\lambda_L,\lambda_Y)$ can be assumed to be nearly equal, and (9) can be written as:

$$S(\lambda_Y,t) \sim [(n_y(t) \cdot n_y(\lambda_Y))/(n_w \cdot n_w(\lambda_R))] \cdot S(\lambda_R,t) \tag{10}$$

and it follows:

$$n_y(t) \sim S(\lambda_Y,t) / S(\lambda_R,t) \tag{11}$$

Discussion of the algorithms

Model signals of water Raman scattering and Gelbstoff fluorescence were computed according to (3) based on a realistic laser pulse shape P_L, and on model depth profiles of the Gelbstoff concentration n_y and of the attenuation coefficients $c(\lambda_L,\lambda_R)$ and $c(\lambda_L,\lambda_Y)$ which are set to be identical.

The example given in Figs. 2a and 2b shows that the profiles reconstructed by application of the Fourier analysis and the Delta pulse analysis fit the given model depth profiles well. The disadvantage of the Delta pulse analysis is that it only allows a calculation of smoothed values of the attenuation coefficient and the Gelbstoff concentration. This smoothing effect is a consequence of the

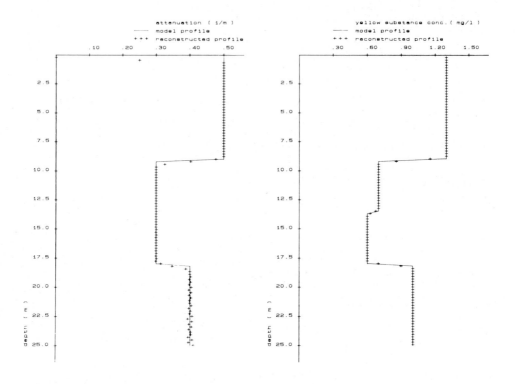

Fig. 2a. Depth profiles of the attenuation coefficient and the Gelbstoff concentration reconstructed with the Fourier analysis.

finite halfwidth of the laser pulse which is approximated by a δ-function. In the case of the Fourier analysis the agreement of the calculated profiles with the model profiles is mainly determined by the frequency range used.

A realistic lidar system is recording signals with a finite amplitude resolution. The high time resolution which is necessary for the derivation of depth profiles has as consequence a low amplitude resolution. For example the OLS detects signals at 2 ns time intervals with an amplitude digitization of only 6 bit. For examination of the limited amplitude and time resolution of the digitization process on the quality of the reconstructed depth profiles, the model signals described above are digitized with different amplitude resolutions before applying the Fourier and the Delta pulse analysis. The time resolution is always 2 ns. In order to enhance the resolution at low signal levels the utilization of a logarithmic amplifier is assumed which converts the exponential decay of $P(\lambda,t)$ into a linear one.

It can be demonstrated that for a given accuracy for the derived profiles the

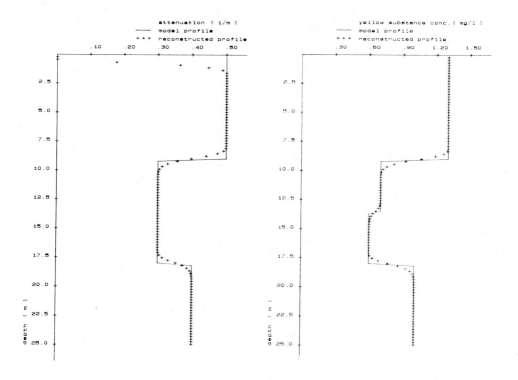

Fig. 2b. Depth profiles of the attenuation coefficient and the Gelbstoff concentration reconstructed with the Delta pulse analysis.

Fourier analysis necessitates a higher amplitude resolution than the Delta pulse analysis. Fig. 3a and Fig. 3b show that the profiles of the attenuation coefficient and of the Gelbstoff concentration can be computed with an error of less than 10% with the Delta pulse analysis with 10 bit signal resolution compared to 14 bit needed for the Fourier analysis.

The maximum error depends on the time and the amplitude resolution, the logarithmic amplification, and the attenuation coefficient, and can easily be calculated for the Delta pulse analysis. In Fig. 4 the relative maximum error for the attenuation coefficient is plotted as function of the number of bits for the digitization with different values of the attenuation coefficient. The time resolution of 2 ns and the amplifier characteristics are fixed. An important result is that the 6 bit amplitude resolution of the OLS is too low for a single shot determination of depth profiles of hydrographic parameters.

598

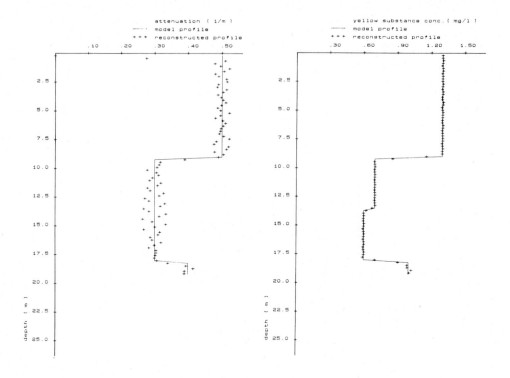

Fig. 3a. Depth profiles of the attenuation coefficient and the Gelbstoff concentration reconstructed with the Fourier analysis for a 14 bit amplitude resolution.

EXPERIMENTAL RESULTS

In 1984 first lidar measurements with the OLS for the derivation of the depth
dependent attenuation coefficient were carried out over the northern Adriatic
Sea. To use the spectral range where water shows the highest light transmission
the OLS was working with an excitation wavelength of 450 nm resulting in a water
Raman scatter wavelength of 533 nm. The signals were logarithmically amplified
and recorded at time intervals of 2 ns with an amplitude resolution of 6 bit.
An averaging over 20 signals which corresponds to an averaging over 0.5 km
flight distance for the 2 Hz repetition rate used was necessary to increase the
amplitude resolution. The laser pulses with a peak power of 1MW had a halfwidth
of 6 ns which allowed to apply the Delta pulse analysis according to (5). For
the chosen flight track on the latitude of the Lagoon of Venice the Gelbstoff
fluorescence contribution at the Raman scattering wavelength was less than 10%
so that the approximation (6) with the following calculations (7) and (8) could

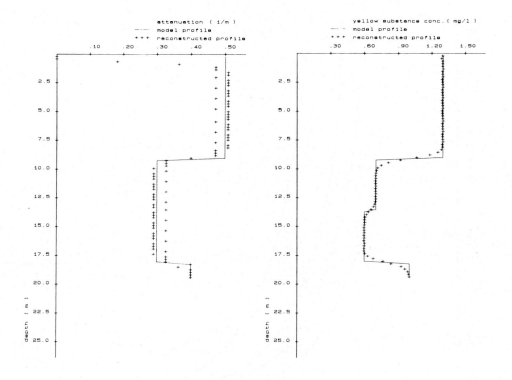

Fig. 3b. Depth profiles of the attenuation coefficient and the Gelbstoff concentration reconstructed with the Delta pulse analysis for a 10 bit amplitude resolution.

be used.

   With the described equipment, profiles of the attenuation coefficient with a depth resolution of about 1 m could be derived from the measured signals for 5-6 attenuation lengths with the Delta pulse analysis. Profiles calculated with this algorithm start at the value 0 and are not realistic for the first meter of the water column as a consequence of the approximation of the real laser pulse shape by a δ-pulse.

   Along the track from 45°19'N, 12°39'E to 45°19'N, 12°57'E two flights were performed only differing by a time delay of 20 minutes and navigation uncertainties of about 200 meters. Therefore the fine structures of the depth profiles cannot be expected to be identical for both flights, but the good correspondence of the main characteristics is obvious from Fig. 5 : except for the profiles obtained in the west the attenuation coefficient has a strong maximum

600

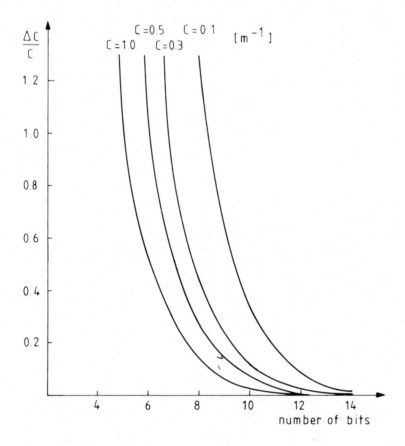

Fig. 4. Relative maximum error for the attenuation coefficient calculated with the Delta pulse analysis as function of the amplitude resolution for different values of the attenuation coefficient. The time resolution is 2 ns.

of 0.6-0.7 m^{-1} at a depth of 2 m, falls down to a broad minimum of 0.2-0.3 m^{-1} centered around 6 m water depth, and increases to a value of about 0.5 m^{-1} for greater depths. In the western part of the track the attenuation coefficient is relatively homogeneous with depth.

CONCLUSIONS

In October 1985 flights with an improved lidar equipment were carried out over the German Bight. The repetition rate was increased to 10 Hz, and a simultaneous measurement of a logarithmically amplified Gelbstoff fluorescence signal was performed at a detection wavelength close to the water Raman detection wavelength. Since Gelbstoff fluorescence at the Raman scatter wavelength cannot be neglected in this area of the North Sea, it is necessary to subtract that back-

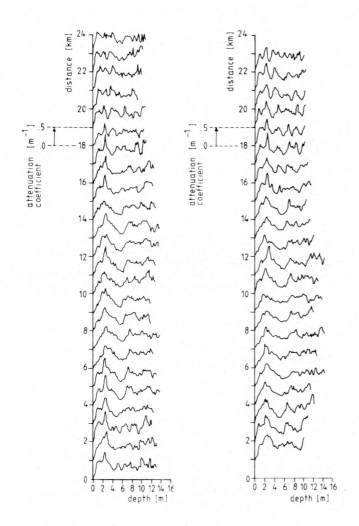

Fig. 5. Profiles of the attenuation coefficient c = c(450nm) + c(533nm) in the Adriatic Sea, derived with the Delta pulse analysis. Flight track from 45°19'N, 12°39'E to 45°19'N, 12°57'E. August 31, 1984.

ground before calculating the depth profile of the attenuation coefficient. This is possible by using the measured Gelbstoff fluorescence signal if the ratio of the Gelbstoff fluorescence efficiencies at the two detection wavelengths is known. Furthermore this signal will allow the evaluation of the depth dependent Gelbstoff concentration.

ACKNOWLEDGEMENTS

The experiment ADRIA '84 was initiated and organized by Dr. P. Schlittenhardt from the European Communities, ISPRA Establishment.

We are indebted to the Flight Department of the DFVLR, Oberpfaffenhofen, for
the availibility of the aircraft and for the support during ADRIA '84.

The development of the Oceanographic Lidar System was financed by a grant
from the Bundesminister für Forschung und Technologie, Bonn.

We are grateful to our colleagues at the University of Oldenburg, Dr. K.P.
Günther, K. Loquay, R. Zimmermann for their support concerning the preparation
and performance of the experiment.

REFERENCES

Allen, R.J., Evans, W.E., 1972. Laser Radar (LIDAR) for Mapping Aerosol Struc-
    tures. The Review of Scientific Instruments, 43 : 1422-1432.
Arshinov, Yu.F., Bobrovnikov, S.M., Zuev, V.E., Mitev, V.M., 1983. Atmospheric
    temperature measurement using a pure rotational Raman lidar. Applied Optics,
    22 : 2984-2990.
Browell, E.V., 1977. Analysis of laser fluoresensor systems for remote algae
    detection and quantification. Report NASA TN D-8447 : 39 pp.
Harms, J., Lahmann, W., Weitkamp, C., 1978. Geometrical compression of lidar
    return signals. Applied Optics, 17 : 1131-1135.
Hoge, F.E., Swift, R.N., 1983. Airborne detection of oceanic turbidity cell
    structure using depth-resolved laser-induced water Raman backscatter. Applied
    Optics, 22 : 3778-3786.
Schwiesow, R.L., Lading, L., 1981.Temperature Profiling by Rayleigh-Scattering
    Lidar. Applied Optics, 20 : 1972-1979
Shimizu, H., Sasano, Y., Nakane, H., Sugimoto, N., Matsui, I., Takeuchi, N.,
    1985. Large scale laser radar for measuring aerosol distribution over a wide
    area. Applied Optics, 24 : 617-626.
Uchini, O., Maeda, N., Shibata, T., Hirino, M., Fujuwara, M., 1980. Measure-
    ment of Stratospheric Vertical Ozone Distribution with a Xe-Cl Lidar : Esti-
    mated Influence of Aerosols. Applied Optics, 19 : 4175-4181.
Zuev, V.V., Zuev, V.E., Makushkin, Yu.S., Marichev, V.N., Mitsel, A.A., 1983.
    Laser sounding of atmospheric humidity : experiment. Applied Optics, 22 :
    3742-3746.

A QUANTITATIVE DESCRIPTION OF THE CHLOROPHYLL A FLUORESCENCE REDUCTION DUE TO
GLOBAL IRRADIATION IN THE SURFACE LAYER

K. P. GÜNTHER

University of Oldenburg, FB 8-Physics,P.O. Box 2503

D - 2900 Oldenburg (Federal Republic of Germany)

ABSTRACT

   The measurement of the chlorophyll a fluorescence from low flying aircraft
by lidar method allows to monitor phytoplankton distribution synoptically as
well as their temporal and spatial evolution.
   The fluorescence signals depend on the water attenuation, the chlorophyll a
concentration and on the in vivo fluorescence efficiency, which varies with environmental parameters. Neglecting the influence of nutrient availability, temperature and phytoplankton composition, the global irradiation in the surface
layer can result in a reduction of the in vivo fluorescence efficiency up to a
factor 4.
   Based on a comprehensive photochemical model of the photosynthetic apparatus,
a quantitative description of the reduction of the in vivo fluorescence efficiency due to photosynthetic active radiation is presented.
   In situ and lidar measurements were performed. Both data sets show a daily
cycle of the in vivo fluorescence efficiency and are compared with the results
of the model.

INTRODUCTION

   The synoptic measurement of chlorophyll a distributions in open sea and coastal waters became of great importance during the last decade due to the development of fluorescence lidar systems operated from low flying aircraft (Kim, 1973;
Mumola et al., 1975; Farmer et al., 1979; Bristow et al., 1981; Hoge and Swift,
1981;Hoge and Swift, 1983;Diebel-Langohr et al.,1983,1985).The fluorescence signals of phytoplankton depend on the chlorophyll a concentration, the in vivo
fluorescence efficiency and on the light attenuation of the water. Large fluctuations in the penetration depth of the laser beam due to variations in the
concentration of other substances such as suspendend sediments and dissolved organic matter (yellow substance) influence the chlorophyll a fluorescence readings
even under constant phytoplankton concentration. Bristow et al. (1981) proposed

successfully to normalize the lidar fluorescence data to the water Raman scattering signal, which depend to a good approximation solely on the optical attenuation depth. The normalized fluorescence data showed a high correlation with ground truth chlorophyll a concentration extracted by wet chemistry methods.

Since the beginning of laser remote sensing of chlorophyll a, laboratory tank tests of single species phytoplankton fluorescence excited by laser systems were performed to investigate the in vivo fluorescence efficiency under controlled conditions. In order to differentiate the color groups of phytoplankton found in the sea a multiwavelength laser excitation system was investigated by Mumola et al. (1975). A linear relationship between the extracted chlorophyll a concentration and the four wavelength lidar fluorosensor data was found during the log phase growth of single species phytoplankton cultures (Brown et al., 1978, 1981). Under nutritional stress, in vivo chlorophyll a fluorescence shows an increase up to a factor of 2-4 (Kiefer, 1973 b,c; Blasco, 1975; Slovacek et al., 1977).

Moreover, the reduction of the in vivo fluorescence efficiency due to high light is a well known phenomenon since the work of Loftus and Seliger (1975), Kiefer (1978 a,b), Heaney (1978), Vincent (1979) and Abbott et al. (1982). This photoinhibition can be observed clearly in homogeneous water masses with low turbulent mixing and nearly constant chlorophyll a concentration profiling the fluorescence and the downwelling irradiance.

Thus, for the interpretation of actively remote sensed fluorescence and in situ data in terms of chlorophyll a concentrations it is necessary to find a concept to correct the fluorescence efficiency for the impact of environmental factors. Due to the fact that the nutritional concentration cannot be monitored by a remote sensor until now, the main interest of this work was to investigate the influence of light in the surface layer on the fluorescence efficiency.

Three different approaches are possible. Laboratory or in situ data normalized to constant chlorophyll a concentration are analyzed using numerical methods e.g. fit procedures giving an empirical understanding of what happens with the molecular fluorescence efficiency with increasing light. Second, time series of light and fluorescence are investigated by calculating the coherence spectrum and its confidence interval using fast Fourier transformation (Abbott et al.,1982) assuming a linear relationship between the two parameters. Third, based on a comprehensive photochemical model of the photosynthetic apparatus, where the fluorescence originates, the reduction of the fluorescence efficiency due to photosynthetic active radiation is described in terms of physical rate constants and other photosynthesis related parameters.

THEORETICAL BACKGROUND

According to present knowledge, photosynthesis and in vivo chlorophyll a fluorescence of algae and higher plants occur in the cellular organelles called chloroplasts of typical dimension of 3 - 10 μm. Within this structure, pigments embedded in membranes, called thylakoids, absorb the visible light energy. Common to all oxygen producing algae and higher plants is chorophyll a , which acts as a light harvesting protein complex and in special chlorophyll a protein complexes as reaction center, where the excitation energy is transformed to photochemical energy. Typical absorption bands of in vivo chlorophyll a are found in the blue at 440 nm and in the red at 670 nm. Other pigments found in the thylakoids, chlorophyll b and c, biliproteins (phycoerythrin, phycocyanin and allophycocyanin) and carotinoids (carotenes and xanthophylls) enhance the spectral absorption in the 480 to 660 nm band.

The pigment composition is controlled by environmental conditions as nutritional supply, temperature and the spectral light intensity. Under high light, the cellular chlorophyll a content is reduced compared to low light in less than a generation time (Riper et al., 1979), whereas the carotinoid synthesis is stimulated to protect the photosynthetic unit by reducing the efficiency of energy transfer to the reaction centers or by photochemical quenching.

The excitation energy from the accessory pigments is fed to two different photochemical reaction centers via exciton transfer, a resonant dipol-dipol interaction described by the Förster-mechanism showing a strong dependence on the distance of the two interacting molecules and on the relative orientation of the dipols (Förster, 1948). Calculations show that for efficient energy transfer the typical distance is less than 10 nm. The typical time constant of exciton transfer to the trapping center is in the picosecond range. At the reaction centers the transformation of excitation energy into photochemical energy occurs by charge separation. At present, one assumes that the reaction center is a molecular complex consisting of the chlorophyll a molecule accepting the excitation energy and of an electron donor and acceptor, respectively. The electron acceptor of reaction center II is termed Q (for quencher) and is probably a specialized plastoquinone or phaeophytin molecule (Klimov et al., 1977). The electron donor of reaction center II is the water splitting complex. The photochemical energy from photosystem II is transported by redox reactions to photosystem I. The electron donor for reaction center I usually is a plastocyanin molecule, while the electron acceptor is possibly a bound iron-sulfur protein or a flavoprotein. The photochemical energy

from photosystem I is brought to the Calvin cycle to drive the biochemical formation of ATP in the cyclic and NADPH in the noncyclic electron transport.

Figure 1 shows a schematic diagram of the photosynthetic unit, divided in the
light harvesting chlorophyll proteins (LHCP), the antenna pigments of photosystem
I and II, the photochemical reaction centers I , II and the simplified electron
transport chain, connecting the reaction centers. Only one substructure of the
electron transport chain is shown, the so-called plastoquinone pool, which stores
the electrons coming from reaction center II. The plastoquinone pool has a regu-

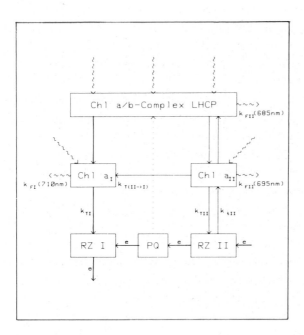

Fig. 1: Schematic diagram of the structure and organization of a photosynthetic
unit.
Chl $\underline{a}$ / $\underline{b}$ - complex : light harvesting chlorophyll proteins with
fluorescence emission at 685 nm
Chl $\underline{a}_I$, Chl $\underline{a}_{II}$: antenna pigments of photosystem I, II
fluorescence emission at 730 nm, 695 nm
RZ I, RZ II : reaction center I, II for charge separation
PQ : plastoquinone pool
k_{FI}, k_{FII} : rate constants of fluorescence from photosystem
I, II, respectively
k_{TI}, k_{TII}, k_{tII} : rate constants of energy transfer from antenna
to reaction center and vice versa
$k_{T(II \to I)}$: rate constant of energy transfer from antenna II
to antenna I, called spillover
The dotted line indicates an interaction of the plastoquinone pool with
the light harvesting chlorophyll proteins by enzymatic reactions (phosphorylation of LHCP).

latory function, concerning the state of the thylakoid membrane (Horton and Black, 1980; Horton et al. 1981). The highly reduced state of plastoquinone induces an enzymatic reaction, the phosphorylation of the light harvesting chlorophyll proteins.This,in turn,changes the structure of the thylakoid membrane (Barber, 1983) and thus the distance between the accessory pigments and the reaction centers. The energy transfer to reaction center II is reduced. Due to the fact that the coupling between reaction center I and the light harvesting chlorophyll proteins is very weak (Andersson and Anderson, 1980), the transfer of excitation energy to the reaction center I is nearly unaffected. In total, the input of electrons from reaction center II to the plastoquinone pool is reduced while the demand of electrons by reaction center I remains nearly constant. This change of energy transfer affects the in vivo chlorophyll a fluorescence as well as the rate of photosynthesis. The time constant for this reversible process is in the minute range.

For the discussion of the in vivo chlorophyll a fluorescence it is important to understand where the observable fluorescence originates. The analysis of fluorescence spectra of chloroplasts of higher plants at 77 K (Goedheer, 1964) and of measurements of the in vivo fluorescence of algae in the pico- to nanosecond range (Moya and Garcia, 1983; Haehnel et al., 1983) lead to the conclusion that at least three different chlorophyll a proteins act in the photosynthetic unit. The fluorescence emission of the light harvesting chlorophyll proteins has its maximum at 685 nm, while those of the antenna chlorophylls of photosystem II and I are at 695 and 730 nm, respectively (Fig.1). At physiological temperatures the in vivo fluorescence spectra of algae show only one maximum at 685 nm with a bandwidth of about 10 nm and a shoulder at 730 nm, corresponding to the weak fluorescence of the antenna of photosystem I.

In general, for the observed chlorophyll a fluorescence F_{II} one can write:

$$F_{II} = I_{abs} * \Phi_{FII}$$

where I_{abs} is the absorbed light intensity and Φ_{FII} is a variable fluorescence efficiency, depending on environmental parameters. According to the bipartite model of Butler and Kitajima (1975) which incorporates the internal structure and organization of a photosynthetic unit described above, the variable fluorescence efficiency Φ_{FII} can be formulated in the following way:

$$\Phi_{FII} = \beta * \psi_{FII} * f(A_{II})$$

where β is the energy distribution parameter describing the relative amount of absorbed light energy going to photosystem II, ψ_{FII} is a constant fluorescence efficiency depending on the desexcitation rate constants of the fluorescent antenna chlorophyll a molecules and $f(A_{II})$ a model function describing the connection of photosynthetic units in the thylakoids. The parameter A_{II} represents the relative number of open reaction centers. In the open state the reaction center can accept excitation energy from the antenna for charge separation. The closed state is termed the oxidized state where excitation energy is transferred back to the antenna. This process enhances the probability of fluorescence, thermal desactivation and the so-called spillover to photosystem I. The model of Butler and Kitajima (1975) can explain the fast increase of the chlorophyll a fluorescence of dark adapted cells after the onset of continuous light due to closed reaction centers.

The reduction of the fluorescence due to high light, called photoinhibition, is not included in the bipartite model. Introducing an intensity dependent energy distribution parameter $\beta(I_{PhaR})$ describing the light dependent state of the thylakoid membrane, and an intensity dependent parameter $A_{II}(I_{PhaR})$, it is possible to model the photoinhibition in a quantitative way. As mentioned above, the membrane state is regulated by the redox state of the plastoquinone pool (Horton and Black, 1980; Allen et al., 1981). In turn the redox state of the plastoquinone is controlled by the amount of light absorbed by the cell. Assuming $\beta(I_{PhaR})$ decreases from a maximum level β_{max} under low light to a minimum level β_{min} with increasing light exponentially, an intensity parameter I_1 determines the state of the membrane. In contrast, the light dependence of $A_{II}(I_{PhaR})$ is modeled by an exponential decrease with increasing light determined by a light parameter I_0, indicating the adaption of the reaction centers according to the growth conditions, e.g. shade or light adapted cells.

With these assumptions, confirmed by biochemical and physiological results as well as by inspecting the transient behaviour of the chlorophyll a fluorescence after the onset of continuous light, the so-called Kautsky effect (Kautsky and Hirsch, 1931), one can calculate the relative variation of the fluorescence efficiency due to the global irradiation. It is important to note that the light influencing phytoplankton is restricted to a wavelength band from 350 nm to 750 nm, the so-called photosynthetic active radiation. In Figure 2, the relative decrease of the fluorescence efficiency with increasing light is shown. The dark value of the fluorescence efficiency is set to 1. Calculating the normalized fluorescence with the expanded bipartite model, a second light parameter, the excitation light, had to be introduced. Assuming a short excitation pulse of typical pulse width of some µs, the energy distribution parameter β is not influenced by the excitation pulse due to the long time constants observed for phosphorylation of the light harvesting chlorophyll proteins. The impact of the excitation pulse is thus restricted to the parameter A_{II}. Moreover, the number of open reaction centers is influenced by the photosynthetic active radiation too.

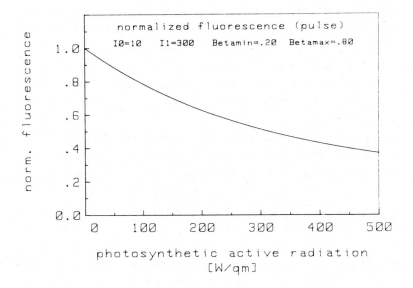

Fig. 2: Diagram shows the normalized chlorophyll a fluorescence F$_n$ depending on the photosynthetic active radiation (PhaR) according to the expanded bi-partite model. Excitation was modeled by short pulses influencing the reaction centers exclusively. The parameter are choosen in accordance with the results presented in Figure 3.

EXPERIMENTAL RESULTS

During the Fluorescence Remote Sensing Experiment, FLUREX '82, continuous ground truth measurements with an in situ fluorometer and an in situ attenuation meter, held at constant water depth, were performed at the research platform NORDSEE to investigate the daily cycle of the fluorescence efficiency due to the impact of photosynthetic active radiation. Additionally, a two-channel fluores-cence lidar system was installed at the top of the research platform to monitor the chlorophyll a fluorescence simultaneously.The system,described by Gehlhaar et al. (1981), was modified for the specific detection of the in vivo chloro-phyll a fluorescence at 685 nm and the water Raman scattering at 650 nm. The excitation wavelength of the flashlamp pumped dye laser was adjusted to 532 nm with pulse energies of about 350 mJ. The lidar signals gave the depth integrated chlorophyll a fluorescence and the effective water turbidity, resulting from the beam attenuation at the laser wavelength and form the diffuse attenuation at the Raman scattering wavelength. Ground truth water samples were collected at regular intervals. Subsamples were used for chlorophyll a extraction according to the method of Whitney and Darley (1979). The rest of the water samples was used to determine the total beam attenuation coefficient over the spectral region from 400 nm to 800 nm with a laboratory photometer. The cuvette length was one meter. The detector solid angle was $0.27°$ with a spectral bandwidth of 5 nm. In a second step, the attenuation due to dissolved organic matter (yellow substance) was determined after filtration with 0.2 μm filters. Bidestilled water was used as reference standard.

A first analysis of the fluorescence data shows a high correlation of the Raman corrected fluorescence lidar signals and the in situ fluorescence data over the whole experimental period with a correlation coefficient of .97. Both fluorescence data vary by a factor of 4 in time while the analysis of the extracted chlorophyll a values reveals a relatively constant chlorophyll a concentration, indicating that the in situ and the lidar signals are both affected by environmental factors.

To analyse the influence of daylight on the fluorescence efficiency in detail, the continuously recorded in situ fluorescence data are normalized to constant chlorophyll a concentration taking into account the result of the analysis of the water samples and of the continuously recorded in situ attenuation data at 670 nm. With a good approximation the influence of yellow substance on the diffuse attenuation coefficient at 670 nm can be neglected due to yellow substance concentrations of less than .74 mg/l. Additionally, the influence of particulate matter on the attenuation was nearly constant in time and less than $.1m^{-1}$. With these findings the continuous in situ attenuation data can be considered to be proportional to the chlorophyll a concentration. Assuming that the in situ fluorescence data are proportional to the chlorophyll a concentration and the fluorescence efficiency, the normalization of the fluorescence readings can be accomplished by dividing the fluorescence data by the water and suspended sediment corrected attenuation values. With this approximation, the normalized fluorescence data in relative units can be regarded as a measurement of the fluorescence efficiency. The mean square error of the normalized fluorescence is approximately 18 %.

For the four days, April 20 to 23, Figure 3 to 6 show the daily cycle of fluorescence efficiency together with the daily cycle of the photosynthetic active radiation given by the dotted lines. The solid line shows the result of the expanded photosynthetic model. The fluorescence efficiency is set to 1 during night. The model parameters β_{min}, β_{max}, I_0 and I_1 are fitted by a computer program. For all days, only the parameter β_{min} had to be changed from .2 to .3 to describe the observed photoinhibition with a high correlation.

On April 20, the weather was calm without clouds. At noon, the maximum value of the photosynthetic active radiation was about 350 Wm^{-2}. The fluorescence efficiency was reduced to about 60% of the fluorescence efficiency at night. With increasing radiation the fluorescence efficiency decreased and vice versa. For April 21, the mean photosynthetic active radiation was 150 Wm^{-2} with short clear ups. For this cloudy day, the model parameter β_{min} had to increase to .25 for optimizing the theoretical results to the measured data. The fluorescence inhibition of typical 25% to 30% was clearly seen for both data sets. The maximum deviation was in the expected error of the experimental data. On April 22, the weather was partly cloudy with short clear ups similar to April 21, showing a maximum irradiance of about 470 Wm^{-2}. The fluorescence efficiency showed a daily cycle due to the influence of the photosynthetic active radiation again. Although the maximum irradiance on April 22 was higher compared to April 20, the fluorescence efficiency was reduced to about 50% compared to 60%, suggesting that the short clear ups at noon had minor influence due to an internal regulation mechanism with time constants greater than the shortest increase of the light during this day. The best approximation of the model to the in situ data could be achieved by increasing the parameter β_{min} to 0.3. For April 23, the same parameter set as for April 22 was found minimizing the deviation of the model and the measured data. A good correlation between photosynthetic active radiation and fluorescence efficiency was observed until the afternoon. The increase of the global irradiation in the late afternoon was not observed in a comparative decrease of the measured fluorescence efficiency. Due to a nonconsistent increase of the night values up to 1.2, it was assumed that a systematic error had been introduced.

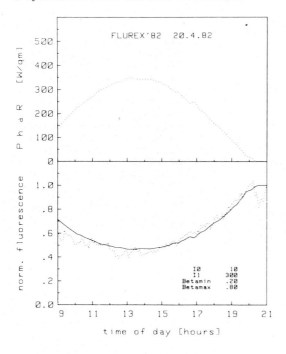

Fig. 3: Daily cycle of the relative chl <u>a</u> fluorescence efficiency and of the photosynthetic active radiation on April 20, 1982.
dotted line: measured data; solid line: results of the proposed model for the fluorescence efficiency taking into account the influence of the photosynthetic active radiation

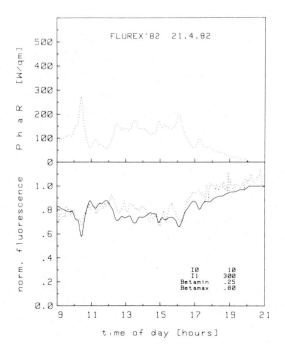

Fig. 4: Daily cycle of the relative chl a fluorescence efficiency and of the photosynthetic active radiation on April 21, 1982. dotted line: measured data; solid line: results of the proposed model for the fluorescence efficiency taking into account the influence of the photosynthetic active radiation

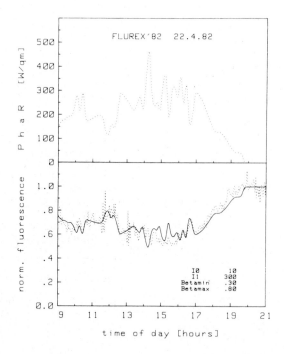

Fig. 5: Daily cycle of the relative chl a fluorescence efficiency and of the photosynthetic active radiation on April 22, 1982. dotted line: measured data; solid line: results of the proposed model for the fluorescence efficiency taking into account the influence of the photosynthetic active radiation

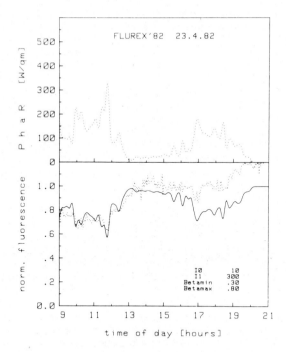

Fig. 6: Daily cycle of the relative chl a fluorescence efficiency and of the photo-synthetic active radiation on April 23, 1982.
dotted line: measured data; solid line: results of the proposed model for the fluorescence efficiency taking into account the influence of the photosynthetic active radiation

DISCUSSION

For the interpretation of the model parameters introduced to expand the bipartite model of Butler and Kitajima (1975) for a quantitative description of the in vivo fluorescence reduction due to high light a sensitivity study was made with respect to the fluorescence and the photosynthesis efficiency. The parameter I_0, describing the light dependence of the state of the reaction center II, can range from 1 Wm^{-2} to 30 Wm^{-2} giving an error of about 2% with respect to the fluorescence reduction. Regarding the influence of I_0 on the efficiency of photosynthesis, which can be deduced by the bipartite model, it can be seen, that I_0 determines the light value for maximum photosynthesis (Fig.7).
A comparison with the empirical formula of Steele (1962) for the description of photosynthesis efficiency shows that the free parameter of Steele is reciprocal related to I_0 and describes the light value of maximum photosynthesis, too. Additionally, investigations of Ryther and Menzel (1959) confirm that for shade adapted algae the maximum of photosynthesis is typical at 10 Wm^{-2} and for

614

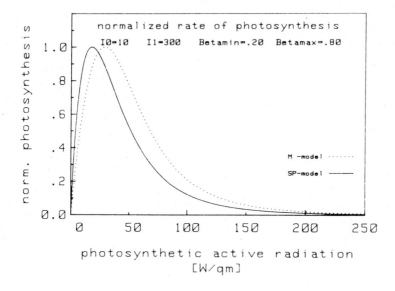

Fig. 7: Diagram shows normalized rate of photosynthesis according the expanded bipartite model. The dotted line represents the result of the "matrix-model", intrduced by Butler and Kitajima (1975). In this model, an exchange of excitation energy between photosynthetic units is assumed, in contrast to the "separate package model". The solid line illustrates the results of the "separate package model" for the rate of photosynthesis. For the reduction of the fluorescence efficiency both models give similar results.

sun adapted algae at 70 Wm^{-2}. With I_0 = 30 Wm^{-2}, the maximum is in the range observed by Ryther and Menzel indicating that the algae population under investigation during FLUREX '82 is assumed to be shade adapted. The parameter I_1 determines the fluorescence reduction due to high light levels. Varying I_1 in the range from 200 Wm^{-2} to 400 Wm^{-2}, the normalized fluorescence efficiency changes by about 12% compared to I_1 = 300 Wm^{-2}. According to the analysis, the parameter I_1 describes the light adapted state of the membrane and can be related to the threshold level for photoinhibition introduced by Kiefer (1973a), Heaney (1978) and Vincent (1979). The parameter I_1 is nearly insensitive to the efficiency of photosynthesis. To analyze the influence of β_{min} and β_{max} on the normalized fluorescence efficiency F_n, an approximation can be given, assuming I_0 and I_1 as constant for the time of investigation:

$$F_n \sim (\beta_{min}/\beta_{max}) + (\beta_{min}/\beta_{max} - 1) * exp(- I_{phar}/I_1)$$

where I_{phar} is the intensity of the photosynthetic active radiation. The approximation shows, that F_n is determined by the ratio of β_{min} to β_{max}. A variation of β_{min} and β_{max} with fixed ratio reveals that the approximation of F_n compared to the exact evaluation is within an error of 2%. Analyzing the influence of β_{min} and β_{max} on the efficiency of photosynthesis, one can show that β_{max} determines the light level of maximum photosynthesis, while the maximum of photosynthesis is to a first approximation independent of β_{min}. Assuming that the ratio β_{min} to β_{max} describes the physiological state of the membrane due to phosphorylation, an increase of this ratio is equivilant to a minor decrease of photoinhibition at high light levels. In the limit of $\beta_{min} \rightarrow \beta_{max}$, no photoinhibition is expected according to the expanded bipartite model. The data of Kiefer (1973a, 1973b) and Blasco (1975) as well as own investigations confirm that under nutritional stress the photoinhibition of algae is reduced compared to algae in the exponential growth phase resulting in a higher fluorescence efficiency. Thus, it can be assumed that the ratio β_{min} to β_{max} depends on the nutrient availability and on environmental stress factors.

In conclusion, the four parameters introduced by the proposed model can be reduced to two parameters determining the photoinhibition of the in vivo fluorescence efficiency due to high light. For actively remote sensed chlorophyll a fluorescence as well as for in situ fluorescence data it is necessary to correct the readings with respect to the impact of the photosynthetic active radiation. The daily cycle of the global irradiation influences the in vivo fluorescence efficiency as shown in this paper. The transformation of in vivo fluorescence data to chlorophyll a values can be optimized introducing the expanded bipartite model for data evaluation. The typical error for data processed by the proposed method is in the range given by standard extraction methods.

ACKNOWLEDGEMENTS

The experiment at the research platform NORDSEE during FLUREX'82 was financed by a grant from the Bundesministerium für Forschung und Technologie.

REFERENCES

Abbott, M.R., Richerson, P.J. and Powell, T.M., 1982. In situ response of phytoplankton fluorescence to rapid variations in light. Limnol. Oceanogr., 27: 218-225.
Allen, J.F., Bennett, J., Steinback, K.E. and Arntzen, Ch.J., 1981. Chloroplast protein phosphorylation couples plastoquinone redox state to distribution of

excitation energy between photosystems. Nature, 291: 25-29.

Andersson, B. and Anderson, J.M., 1980. Lateral heterogeneity in the distribution of chlorophyll-protein complexes of the thylakoid membranes of spinach
chloroplasts. BBA, 593: 427-440.

Barber, J., 1983. Membrane conformational change due to phosphorylation and the
control of energy transfer in photosynthesis. Photobiochem. Photobiophys., 5:
181-190.

Blasco, D., 1975. Variations of the ratio in vivo-fluorescence/chlorophyll and
its application to oceanography. Effects of limiting different nutrients, of
night and day and dependence on the species under investigation. NASA Technical Translation, TTF-16: 317 pp.

Bristow, M., Nielsen, D., Bundy, D. and Furtek, R., 1981. Use of water Raman
emission to correct airborne laser fluorosensor data for effects of water
optical attenuation. Appl. Opt., 20: 2889-2906.

Brown, C.A., Farmer, F.H., Jarrett, O. and Staton, W.L.,1978. Laboratory studies of in vivo fluorescence of phytoplankton. Proc. Fourth Joint Conf. Sens.
Environ. Pollutants. American Chem. Soc. 782-788.

Brown, C.A., Jarrett, O. and Farmer, F.H., 1981. Laboratory tank studies of single species of phytoplankton using a remote sensing fluorosensor. NASA TP-1821.

Butler, W.L. and Kitajima, M., 1975. A tripartite model for chlorophyll fluorescence. In: M. Avron (Editor), Proceedings of the Third International Congress
on Photosynthesis. Elsevier, Amsterdam, 13 pp.

Diebel-Langohr, D., Günther, K.P. and Reuter, R.,1983. Lidar applications in remote sensing of ocean properties. Int. Coll. on Spectral Signatures of Objects
in Remote Sensing, Conf. Proc., Bordeaux.

Diebel-Langohr, D., Günther, K.P., Hengstermann, Th., Loquay, K., Reuter, R. and
Zimmermann, R., 1985. An airborne lidar system for oceanographic measurements.
In: Optoelektronik in der Technik, Tagungsberichte LASER 85 - Optoelektronik,
München 1. - 5. Juli 1985, Springer Verlag (in press).

Farmer, F.H., Brown, C.A., Jarrett, O., Campbell, J.W. and Staton, W.L., 1979.
Remote sensing of phytoplankton density and diversity in Narragansett Bay
using an airborne fluorosensor. In: Proceedings Thirteenth International
Symposium on Remote Sensing of the Environment, 23-27 Apr. 1979.Environmental
Research Institute of Michigan, Ann Arbor. 1793-1805.

Förster, Th., 1948. Zwischenmolekulare Energiewanderung und Fluoreszenz. Annalen
der Physik, 2: 55-75.

Gehlhaar, U., Günther, K.P. and Luther, J., 1981. Compact and highly sensitive
fluorescence lidar for oceanographic measurements. Appl. Opt., 20: 3318-3320.

Goedheer, J.C., 1964. Fluorescence bands and chlorophyll a forms. BBA 88:
304-317.

Haehnel, W., Holtzwarth, A.R. and Wendler, J., 1983. Picosecond fluorescence kinetics and energy transfer in the antenna chlorophylls of green algae.
Photochem. Photobiol. 37: 435-443.

Heaney, S.I., 1978. Some observations on the use of the in vivo fluorescence
technique to determine chlorophyll a in natural populations and cultures
of freshwater phytoplankton. Freshwater Biol. 8: 115-126.

Hoge, F.E. and Swift,R.N., 1981. Airborne simultaneous spectroscopic detection
of laser-induced water Raman backscatter and fluorescence from chlorophyll a
and other natural occuring pigments. Appl. Opt. 20: 3197-3205.

Hoge, F.E. and Swift, R.N., 1983. Airborne dual excitation and mapping of phytoplankton photopigments in a Gulf Stream Warm Core Ring. Appl. Opt. 22:
2272-2281.

Horton, P. and Black, M.T., 1980. Activation of adenosine 5' triphophate-in-
duced quenching of chlorophyll fluorescence by reduced plastoquinone. The
basis of state I - state II transitions in chloroplasts. FEBS Lett. 119:
141-144.

Horton, P., Allen,. J.F., Black, M.T. and Bennett, J., 1981. Regulation of phosphorylation of chloroplast membrane polypeptides by redox state of plastoquinone. FEBS Lett., 125: 193-196.

Kautsky, H. and Hirsch, A., 1931. Neue Versuche zur Kohlensäureassimilation. Naturwissenschaften, 19: 964.

Kiefer, D.A., 1973a. Fluorescence properties of natural phytoplankton population. Mar. Biol. 22: 263-269.

Kiefer, D.A., 1973b. Chlorophyll a fluorescence in marine centric diatoms: responses of chloroplasts to light and nutrient stress. Mar. Biol. 23: 39-46.

Kiefer, D.A., 1973c. The in vivo measurements of chlorophyll by fluorometry. In: L.H. Stevenson and R.R. Colwell (Editors), The Belle W. Baruch Library in Marine Science: Estuarine Microbial Ecology. University of South Carolina Press.

Kim, H.H., 1973. New algae mapping technique by use of an airborne laser fluorosensor. Appl. Opt., 12: 1454-1458.

Klimov, V.V., Klevanik, V.A., Shuvalov, V.A. and Kravsnovsky,A.A., 1977. Reduction of phaeophytin in the primary light reaction of photosystem II. FEBS Lett. 82: 183-186.

Loftus, M.E. and Seliger, H., 1975. Some limitations of the in vivo fluorescence technique. Chesapeake Sci. 16: 79-92.

Moya, I. and Garcia, R.,1983. Phase fluorimetric lifetime spectra. I. In algal cells at 77 K. BBA 722: 480-491.

Mumola, P.B., Jarrett, O. and Brown, C.A., 1975. Multiwavelength lidar for remote sensing of chlorophyll a in algae and phytoplankton. The use of lasers for hydrographic studies. NASA-SP 375: 137-145.

Riper, D.M., Owens, T.G. and Falkowski, P.G.,1979. Chlorophyll turnover in Skeletonema costatum, a marine plankton diatom. Plant. Physiol. 64: 49-54.

Slovacek, R.E. and Hannan, P.T., 1977. In vivo fluorescence determination of phytoplankton chlorophyll a . Limnol. Oceanogr. 22: 919-925.

Steele, J.H., 1962. Environmental control of photosynthesis in the sea. Limnol. Oceanogr., 7: 137-150.

Vincent, W.F., 1979. Mechanisms of rapid photosynthetic adaption in natural phytoplankton communities. I.Redistribution of excitation energy between photosystem I and II. J. Phycol. 15: 429-434.

Whitney, D.E. and Darley, W.M., 1979. A method for the determination of chlorophyll a in samples containing degradation products. Limnol. Oceanogr., 24: 183-186.

NIMBUS-7 COASTAL ZONE COLOR SCANNER PICTURES OF PHYTOPLANKTON GROWTH ON AN UPWELLING FRONT IN SENEGAL

C. DUPOUY[1], J.P. REBERT[2], and D. TOURE[3]

1 ORSTOM, Centre de Météorologie Spatiale, 22302 Lannion (France)

2 ORSTOM, IFREMER Centre de Brest, 29273 Brest (France)

3 Centre de Recherche Océanographique de Dakar-Thiaroye, Dakar B.P. 2241 (Sénégal)

ABSTRACT

Two CZCS pictures (11 January 1981 and 1 March 1981) of phytoplankton blooms during the wind-induced upwelling in Senegal waters are presented. The influence of upwelling on the mesoscale spatial variability of the surface temperature and biomass fields is shown. Strong frontal phytoplankton features dominate in the first period of upwelling and a more homogeneous bloom is found at the end of the upwelling period. We try to relate the instantaneous fields to the frontal dynamical processes.

INTRODUCTION

Ever since the Meteor expeditions in the early 1950's and then the CINECA, A. Von Humbolt and JOINT cruises, the West African upwelling system has been the subject of intensive systematical research (Hagen, 1974, Schemainda et al., 1975). In 1982, this culminated in the publication of an I.C.E.S. report on the Canary Current. Less is known on the West African coast under the 15°N latitude (fig. 1). Specific data over Senegal waters are provided by Berrit (1961 and 1962), Rossignol (1973), Rebert (1974, 1977, 1983), Touré (1982, 1983).

In general, the Senegalese system shows the properties of an upwelling system related to the trade winds on the eastern boundaries of oceans (Mittelstaedt, 1983). The northeastern winds trigger the Ekman transport which results in upwelling of cold and nutrient rich deep water (mixing of South and North Atlantic Central Water). At the latitude of Senegal, the upwelling occurs from December to March, related to the southern position of the Inter Tropical Convergence Zone ("dry season") and forms a characteristic cold plume (fig. 2).

620

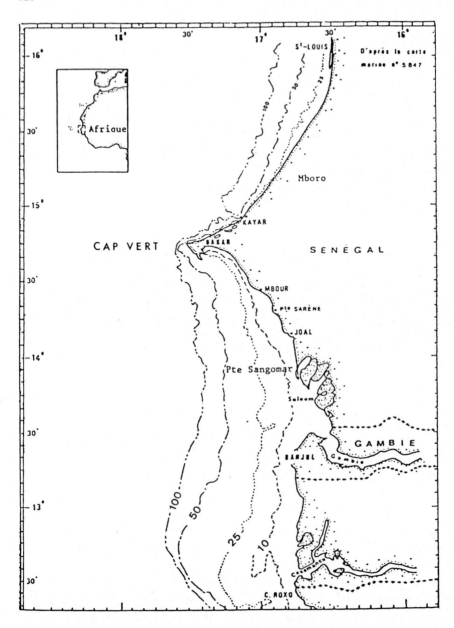

Fig. 1. Situation of the studied area on the West African coast.
Major bathymetric features of the Senegalese continental
shelf. Note the difference between the deep narrow continental shelf to the north of Dakar, and the broad and
shallow one in the south. The width varies from 30 km at
St Louis, to 10 km above the Cap to 100 km in front of
Gambia.

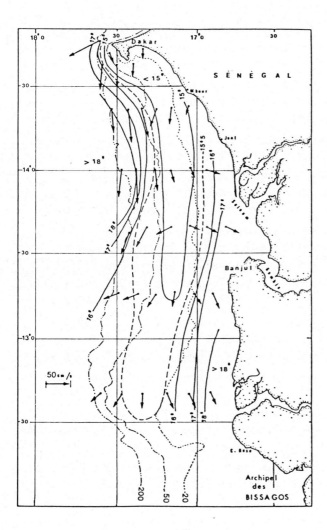

Fig. 2. Currents at 5 m, from the 20th to 31st of March 1974
(Rebert, 1974) on the Senegalese continental shelf. Note
the maximal source of divergence between isobaths 20 m
and 50 m.

Phytoplankton productivity is greatly increased in the upwelling zone of nutrient-rich waters. Annual primary production reaches 250 g.C m-2 (Schemaida et al., 1975). This high productivity places the Senegal shelf as the second most productive region, after the Cap Blanc (325 g.C m-2 year -1).

The usefulness of visible imagery for determining phytoplankton production in the upwelling areas has been well demonstrated on the upwelling off Peru (Feldman, 1985), on the Californian coast (Laurs et al., 1984, Traganza et al., 1982), on the Benguela current region off South Africa (Shannon et al., 1984) and over Senegal (Stürm, 1983 ; Viollier and Stürm, 1984). Since 1982, ORSTOM and the C.R.O.D.T., in cooperation with ESA, have established a continuous satellite coverage of the Gulf of Guinea ("LISTAO" operation, Citeau et al., 1984) in order to study the natural environment of the listao (main tuna species), using the infrared data of the geostationnary satellite METEOSAT. What is missing, however, is the understanding of the evolution of the distribution and abundance of phytoplankton, first link of the food chain. The purpose of this study is to classify the phytoplankton responses to the upwelling event in order to explain the primary production cycle in this region. In a first step, two images of NIMBUS-7 (Coastal Zone Color Scanner) were used to provide maps of phytoplankton biomass during the 1981 upwelling. In addition, the C.R.O.D.T. has conducted sea-going expeditions over the continental shelf break since 1966, providing an oceanographic knowledge of the region.

I - MATERIAL AND METHODS

I.1 - Nimbus-7 C.Z.C.S. experiment

The Coastal Zone Color Scanner experiment was built specifically to provide an estimate of phytoplankton pigment concentrations in the surface layer of the ocean, by measuring the upwelled spectral radiances in four channels in the visible (443, 520, 550, 670 nm), and one channel in the near infrared (750 nm); an infrared channel (11.5 μm) gives simultaneous images of the superficial temperature (first microns) (Hovis et al., 1980). Nimbus-7 has a circular sun-synchronous ascending node orbit with a path width of 1956 km, spatial resolution of 825 m x 825m and a repeat cycle of 26 days.

I.2 - <u>NIMBUS-7 - CZCS data</u>

The two CZCS images presented were recorded on January 11, 1981 at 1250 UT, orbit 11198 and on March 1, 1981 at 1230 UT orbit 11875. They correspond respectively to the beginning of the Senegalese upwelling (January) and to its maximal phase (March), according to Touré (1983) and Rebert (1983). The area studied extends from the Bissagos in the south (12°N, 16°W) to Saint Louis (16°N, 16°W), and out to sea about 20°W (see fig. 1). It covers mainly the continental shelf and open waters on a width of 2° of longitude.

I.3 - <u>Processing of CZCS data and chlorophyll algorithm</u>

In order to determine subsurface upwelling radiances from CZCS data, the atmospheric contribution due to scattering and aerosol effects must be removed. The atmospheric corrections were achieved as follows. The absolute reflectance values were calculated from the raw channel data (Viollier, 1982), taking into account the decrease of sensitivity of the sensor. The Rayleigh scattering contribution was calculated using an algorithm from Viollier et al. (1980). The aerosol component is estimated from the channel 4 reflectance at 670 nm (Gordon and Clark, 1980). For the latter, a negative value was chosen for n, the Angström exponent, expressing the exponential dependence between aerosol optical thickness and wavelength. The influence of residual water reflectance at 670 nm was taken into account using an empirical coefficient, v, of the proportionality between channels 3 and 4. The parameters used in the processing are summarized and discussed in annex I.

Pigment concentration maps (chlorophyll a + phaeopigments) were obtained from satellite-derived true water reflectances in the three channels : 443, 520 and 550 nm, using an algorithm proposed by Gordon et al. (1980). The concentration of "chlorophyll" C (in $mg.m^{-3}$) is related to the ratio of reflectances $R(\lambda)$ as follows :

$$\log C = a + b \log \frac{R\,443}{R\,550} \quad \text{for low values of C, C} < 1.5 \text{ mg.m}^{-3} \quad (1)$$

$$\log C = c + d \log \frac{R\,520}{R\,550} \quad \text{for high values of C, C} > 1.5 \text{ mg.m}^{-3} \quad (2)$$

The coefficients a, b, c, d were determined empirically from combined measurements of pigment and sea reflectances at 443, 520 and 550 nm (Gordon et al., 1980). These values were chosen

because they correspond to a very broad spectrum of water types
and concentrations at sea. The CZCS-derived values represent the
average pigment concentrations to a depth of 1 optical attenuation length, which varies from 17 m to 1.6 m over the pigment
concentration range of 0.1 - 10 mg.m^{-3} (Smith and Baker, 1978).
The expected accuracy on chlorophyll determination from CZCS is
log C $\pm$ 0.5 (Gordon et al., 1980).

The obtained scenes were geometrically remapped on a Mercator
projection (GIPSY software, Belbeoch, 1983). For the calibration
of scenes, see annex I. The infrared images were used, despite
noisy scan lines, due to detecting errors. They were enhanced
so that sea surface temperature ranges from warm (black) to cold
(white). Land, clouds and dust winds were masked in white. The
chlorophyll maps were enhanced so that chlorophyll poor waters
(C < 1mg.m^{-3}) appear in black and rich waters (C > 5 mg.m^{-3}) appear in white.

The dust wind observed on the "chlorophyll" image of the 1st
of March 1981 masks only the blue to-near infrared channels, and
does not affect the infrared image. The aerosols, mainly composed of sand dust of land origin and pushed by trade winds over
the sea are strongly reflectant but have the same temperature
as the sea. On the contrary, the aerosols of maritime origin
were well eliminated by the atmospheric corrections.

II - RESULTS

The CZCS images from the 11th of January 81 and 1st of March
81 provide two dimensional synoptic views of sea surface temperature and subsurface reflectances. They show a number of surfa-
ce-expressed features of the upwelling dynamics.

II.1 - Sea Surface Temperature (SST) fields (CZCS)

II.1.1 - 11 January 1981 temperature map

On the 11th of January (figure 3a), the SST field is mainly
composed of two water masses : the warmer stratified water mass
between the longitudes of 19°W and 18°W (tropical waters according to Rossignol, 1978) and the colder mixed water mass produced by the coastal upwelling. Frontal instabilities are seen at
the interface of the two water masses.

The upwelled waters show three main patterns. The most striking feature is the cold plume under the Cap Vert, which coincides with the maximal divergence of the currents (see fig. 2,

Rebert, 1963) and the main region of vertical motion on the middle of the shelf. The western limit of the plume follows exactly the 200 meters isobath at 17°30W (fig. 1). It forms the main frontal system between two opposing hydrodynamic features : the upwelling divergence and the tropical waters convergence.

The southern limit, at the latitude of the Gambia (13°N), is distorted by a long feature extending to the west and then to the north (13°N, 18°W). A wedge of warmer water appears near the coast in front of M'bour (14°N).

North of Cap Vert (the Great Coast from Dakar to Saint-Louis), cold plumes perpendicular to the coast and gyres of 100 kms wide are visible. One is obviously turning around the cape showing a cyclonic rotation. The cold hammer shaped plume above the peninsula is related to the northern upwelling system.

In offshore waters, 200 km from the coast and around 19°W, 15°N, there appear instabilities and double eddies which are characteristic of the frontal mechanisms between water masses of different density.

In January, the northern part of the continental shelf (Great Coast) is colder than the southern one. The image of the 12th of January, not shown here, reveals the same oceanographic patterns.

The meteorological file of merchant ships SST indicates that temperatures of the southern shelf show a north-south gradient from 21°C under the Cap to 26°C at the Bissagos.

II.1.2 - 1 March 1981 temperature map

The main temperature feature of the 1 March scene (fig. 3b) is the intensification of the upwelling patch under the Cap Vert. The cold plume forms meanders directed to the south and parallel to the coast. A narrow central vein of cold water marks the zone of maximal divergence between the 20 m and 50 m isobaths (compare fig. 1 and 2). The gradient between tropical and upwelling waters is sharper than in January and is indented by three gyres. Similar gyres have been recorded by Meteosat during five successive days in February 82 (Roy, 1982). They are quasi-permanent during the upwelling period.

The eddies of the southern plume are strictly cyclonic and have a diameter of 30 km. The southern gyre is apparently anti-cyclonic and has a greater diameter of 80 km (persistence of January pattern).

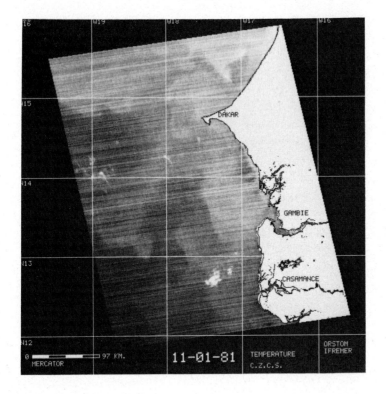

Fig. 3a. CZCS image in the thermal infrared of the upwelling
cold plume on January 11, 1981 over Senegal waters.

Since January, the cold tongue is expending further to the
south, as it is generally observed from cruises measurements.

At the latitude level of the Saloum (14°N), the plume is detached from the coast which causes a secundary coastal front fitting the 20 meters isobath.

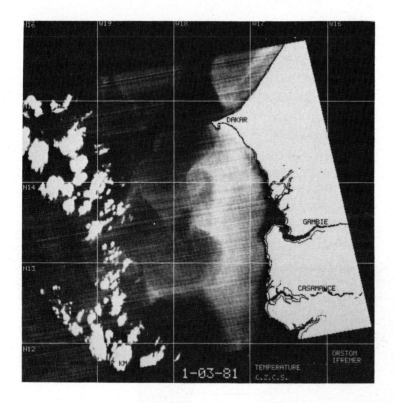

Fig. 3b. CZCS image in the thermal infrared of the upwelling
cold plume on March 1, 1981 over Senegal waters.

Comparing the northern and southern parts of the image, the
upwelling plume is colder than the Cap Vert shelf water.

The merchant ship SST indicates a gradient from 18°C under
the cape to 22°C at the Bissagos latitude.

II.2 - <u>Turbidity map</u>

The coastal turbidity is formed by suspended material related
or not to phytoplankton (organic or inorganic matter). This material increases the reflectances over the whole visible spectrum (Morel and Prieur, 1980), and then in the three CZCS channels. The CZCS reflectance at channel 3 (550 nm) can be used to
measure the total seston (Zbinden, 1981).

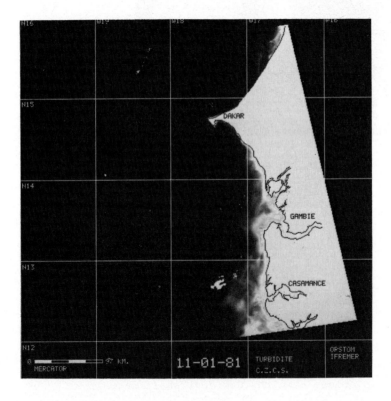

Fig. 4. CZCS map of total suspended matter on January 11, 1981
over the Senegalese continental shelf showing distinctive bathymetric features within the 20 m isobath, especially in front of Casamance and Gambie.

On the picture of Senegal, the high reflectances observed
along the coast (40 km wide bands at the Bissagos latitude) are
caused by a resuspension of bottom sediments due to swell. The
limit of the coastal turbidity follows the 20 m isobath ; a
strong coastal mixing effectively occurs between the coast and
this isobath. In the south, the reflecting band shows features
which give a good mapping of the submarine topography (see fig.
4 in January). In the north, the most reflecting band appears at
St Louis (outlet of the Senegal river). These two very highly
reflecting bands correspond to high mud content in the bottom
sediments (Domain, 1982), rather than to a river run-off of particles (no rain fall during the dry season).

Off this turbid zone, the waters have stricly an oceanic origin and are only influenced by the phytoplankton content (case
1 waters, Morel and Prieur, 1980).

II.3 - Chlorophyll a distributions (CZCS)
II.3.1 - 11 January 1981 chlorophyll map

The most impressive characteristic of this image (see fig.
5a) is its considerable spatial heterogeneity. The patches of
phytoplankton are elongated and form narrow meanders and ribbon-
like features in some very specific places. Under the Cap Vert
the crown shape of the bloom is remarkable. Discrete patches, only 5 miles broad and 15 miles long, appear at the western edge
of the upwelled plume at 17.50°W, which corresponds to the thermal front with tropical waters. The center of the plume inside
the continental shelf is depleted in chlorophyll.

In the north, the enrichment is inhomogeneous and follows the
patterns of the cold gyres. There, the patches are larger and
show their maximal concentration at the end of the eddies.

Some offshore phytoplankton patterns are also visible on the
fronts between warm and cold waters ($2.5 \ mg.m^{-3}$) and follow the
temperature features.

The greater concentrations ($5 \ mg.m^{3}$) are found in the northern part of the upwelling, above the Cap Vert, at the outlet
of the Gambia river and in coastal area off M'bour (14°N).

The turbid waters at the south of the scene are also rich in
chlorophyll ($4 \ mg.m^{3}$).

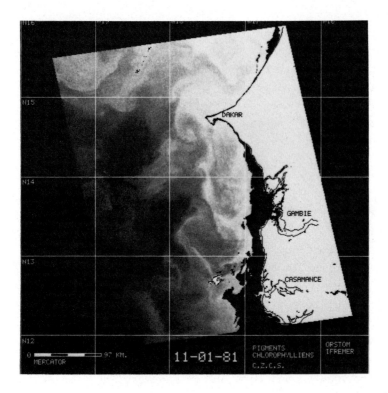

Fig. 5a. CZCS phytoplankton pigment concentration map over Sene-
        gal waters on January 11, 1981. Processed using algo-
        rithm of Gordon et al. (1980), and ratio of channels
        2 (520 nm) and 3 (550 nm). The concentrations range
        from low content water (C_3 < 1 mg.m^{-3}) in black to high
        content water (C > 5 mg.$^{-3}$) in white. The high turbidi-
        ty coastal band appears in black ; land, clouds and
        dust sand in white. Note the crown shape of the chloro-
        phyll enrichment on the plume.

II.3.2 - 1 March 1981 chlorophyll map

    The chlorophyll distribution shows a greater homogeneity
(fig. 5b). Two very large patches of high concentration (8
mg.m^{-3}) are found : the southern bloom (40 miles wide) is rejected on the shelf. The northern one follows the vein of cold
water. They both form a strong gradient with the open waters
(1.5 mg.m^{-3}).

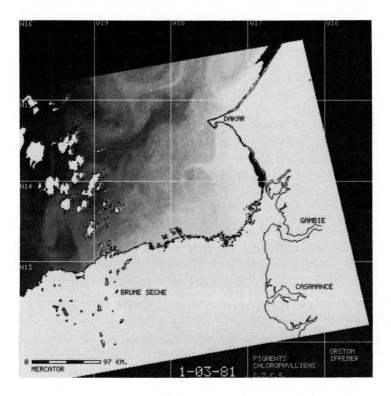

Fig. 5b. CZCS phytoplankton map over Senegal waters on March 1,
1981. Same processing and same chlorophyll scale as
January (legend fig. 5a). Note the uniform aspect of
the phytoplankton bloom.

No color eddies are visible except one curved effect on the
plume which reflects the position of the cold cyclonic gyre.

Offshore, there is an intrusion of relatively poor water forming a "river", extending from the south east towards the Cap
(1.7 mg.m^{-3}), creating offshore color fronts, corresponding with
the thermal fronts.

III - <u>DISCUSSION</u>

III.1 - <u>What we know about the upwelling mechanism</u>

The horizontal circulation over the continental shelf is well shown in figure 2 (Rebert, 1983). The upwelled waters form a cold plume under the peninsula. Inside the shelf, the surface currents are directed southward and tend to be parallel to the wind direction. At the shelf break, a southward surface current of 70 cm.s^{-1} and a bottom northward current of 10 cm.s^{-1} have been measured (Rebert, 1983 ; Teisson, 1981 : see fig. 6). Along-shore, Kelvin-type waves induced by wind variations propagate northward (Crépon et al, 1984). Offshore, the main circulation is directed northward.

Due to the shelf topography (see fig. 2), the upwelling modes differ from north to south (Rebert, 1983). To the south of the Cap Vert, where the shelf is shallow, two cells of vertical cir-culation occur : the first cell will create an offshore front, where cold waters dive under the lighter ones ; the second cell is coastal. To the north of the peninsula, only one cell of up-welling occurs because of the deepness of the shelf.

III.2 - <u>Is the evolution of the thermal field well shown by CZCS?</u>

At the beginning of the cold season, the upwelling upward mo-tion does not disturb the stratification and affects the layer under the thermocline causing only a weak cooling (Teisson, 1983). At its maximal phase, the upwelling causes the intrusion of the thermocline to the surface, which cools the whole mixed layer. This evolution is well shown by the CZCS. Effectively, from January to March, the aspect of the upwelling plume changed from a slightly cool surface field with a weak offshore gradient to a cold patch with a sharp frontal region characterized by cy-clonic gyres. The March gyres may be generated by waves on the continental shelf, or by instabilities between the two water masses. The intensification of the vertical motions is accompa-nied by a change of the general circulation (fig. 7). In Janua-ry, northern advection process is stronger than the southern up-welling. In January 1981, advection could be the consequence of 10 days of easterly winds (ASECNA data), pushing the northern cold gyres to the south around the Cap.

In March, above the Cap, circulation changes as seen on the

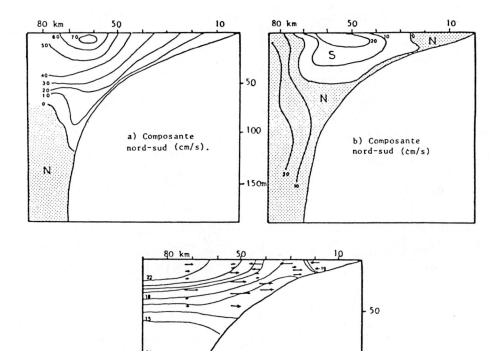

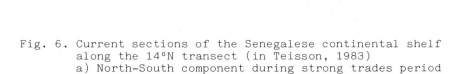

Fig. 6. Current sections of the Senegalese continental shelf
along the 14°N transect (in Teisson, 1983)
a) North-South component during strong trades period
(26 March 1974).
b) North-South component during weak trades period
(11 February 1977).
c) East-West component and isotherms (12 April 1977).

634

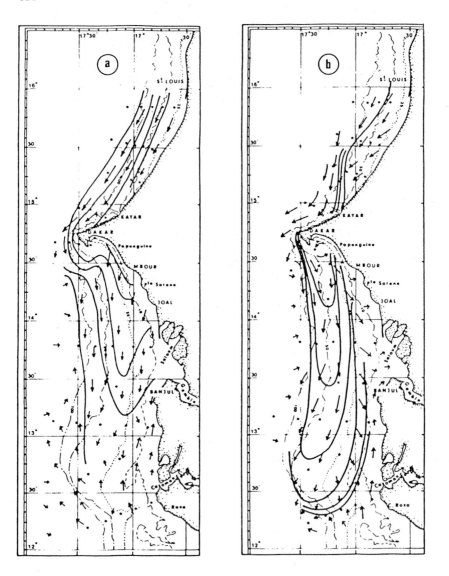

Fig. 7. Superficial circulation and repartition of isotherms
during the cold season on the Senegalese continental
shelf.
a) weak upwelling period (December-January) and advection
b) strong upwelling period (February-April) (from Rebert,
1983).

March 1981 image. For Rebert (1983), there is convergence of warm waters to the coast due to westward rotation of winds. This process may also result from the blocking of the upwelling by a Kelvin front generated at the coastal discontinuity of the cape and propagating northward (Crépon at al., 1984). In fact, northwesterly winds have been recorded at St Louis and Dakar during the 10 days preceding the scene, which could favor the piling-up hypothesis.

The wedge of warm coastal shallow waters (especially seen in March) is caused by solar heating, without influence of upwelling.

III.2.1 – Relation with the coastal wind measurements

The intensification of the upwelling system can be related to the change of the coastal winds direction. The winds on the Senegalese coast are characterized by a moderate intensity (2 to 6 $m.s^{-1}$, compared to 10 $m.s^{-1}$ in Mauritania). Their direction changes progressively from northeast to northwest between December and June (Kirk and Speth, 1985).

The ASECNA coastal measurements show that in January 1981, there were continental thermic winds (0 to 90 degrees) with an intensity of 5 $m.s^{-1}$, and in March predominant maritime trades (330° – 30° and 330° – 60°) with the same intensity. As the temperature response is maximal to this last category (Portolano, 1981), the intensification of the upwelling between January and March 1981 can be related to a change in direction rather than to an increase of the wind stress.

III.3 – Is the evolution of the chlorophyll field well seen by CZCS ?

In January, the CZCS scene shows that the spatial chlorophyll distribution is patchy and associated with the frontal zones.

One day later, (the 12th of January scene could not be processed quantitatively) the eddies patterns in the north are unchanged. Over the shelf break, the width of one of the crown patches has almost doubled. As the surface of the other patches remains unchanged, this indicates that the main zone of growth is well located on the front.

In March, at the contrary, the patches are broad indicating a more global enrichment of the cold water mass.

III.3.1 - Relation with the sea-truths

The sea-truths data (R/V "Laurent Amaro" cruises from January to April 1981) allow an approach of the seasonal evolution of the shelf waters characteristics. Only secchi-disk depths, nutrients were taken (Touré, 1982).

The secchi-disk depth (S.D.D.) remains heterogeneous from January to February 81, showing usually minimal values (14m) on the 200m isobath. In April 81, homogeneously low SDD are measured on the shelf. A sharp frontal zone is formed above the shelf break, with values varying from 9 to 15m in a few kilometers (see fig. 8).

As no river runoff occurs during the dry season, the SDD are directly related to marine particulate matter (no dissolved matter from the rivers). The relation SDD-chl a (Touré, 1983) varies from month to month, indicating a seasonal evolution of the water components from sand and mud (before December) to living cells (until May). Then, the evolution of the SDD is a good index of the phytoplankton evolution during the upwelling season. The SSD values indicate high chlorophyll concentration (maximum of 12 mg.m^{-3}). These high values correspond to the moderate coastal concentrations found at Dakar for the last ten years (Gallardo, 1981), and to usual values found in the Baie de Gorée (Dia, 1982 ; Touré, 1983).

In conclusion, the SSD indicates an evolution between heterogeneous phytoplankton repartition in the first period of the upwelling towards an homogeneous bloom over the shelf at the end of the period. This coincides with the evolution shown by the CZCS chlorophyll maps.

III.4 - Processes of phytoplankton growth

The patches observed in the upwelling area result from the combination of upwelling rate, growth rate, the grazing pressure, the sinking of cells and nutrient input and regeneration.

The upwelling provides the nutrients to the surface waters as the origin of deep waters is a mixing of 80 % SACW and 20 % NACW (Rebert, 1983). The nitrate concentrations measured in the Baie of Gorée are high (15 µ atg.l-1) during the whole period of upwelling.

The residence time of the rich upwelled waters can be estimated. They describe a helical motion which takes one month, with

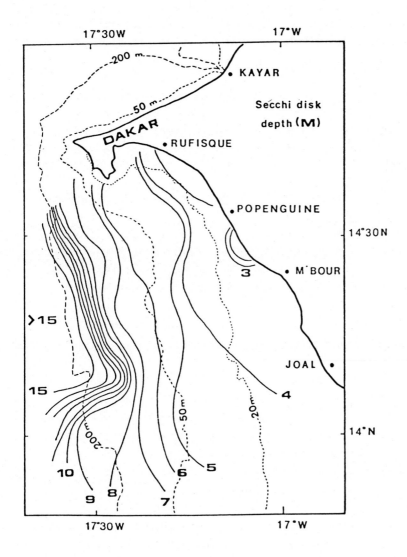

Fig. 8. Secchi-disk depths (in meters), during the CRODT R/V "Laurent Amaro" cruise from the 6th to 10th of April 1981, after the maximal phase of upwelling (from Touré, 1983).

one week in the euphotic zone (Teisson, 1981). This time is sufficient for a bloom developpment since the phytoplankton growth
rate typically varies from one division per day to one division
per 10 days.

The availability of light can be estimated from the SDD, since Z 1 %, depth of the euphotic zone is approximately 2.2 x Z
10 %, Z 10 % being equal to SDD (Hojerslev, 1981). The corresponding euphotic zone depth varies from 8m at the coast to 28m offshore.

III.5 - Contribution of the CZCS to the understanding of biolo-
gical processes

In January 1981, the blooms are not associated strictly to
the cold plume. They occur mainly on the cold side of the shelf
break front. The phytoplankton patches may be associated to nutrient supplies highly connected to vertical motions at the
shelf break. The strict superposition of cold and high nutrient
waters is the rule in the Mauritanian upwelling (Coste and
Minas, 1982). This could not be the case here, at least in
January.

In the plume center, strong mixing may limit the primary production. The newly upwelled waters are often found depleted in
chlorophyll (Morel, 1982) because the algal growth is not achieved yet or rather because the new upwelled waters must first be
"conditioned" by mixing with surface waters (Barber et al. 1971).

In March 1981, the blooms occur over a large part of the cold
plume, on the inner part of the shelf. The CZCS "turbidity" map
shows that the suspended matter content is there low enough to
allow a production. The algae may also avoid the cold waters of
the central vein. The grazing pressure may also be too high on
the extern side of the front. The more homogeneous aspect of the
bloom may indicate a wider distribution and supply of nutrients
over the shelf, which is caused by diffusion process following
two months of vertical motion and continuous input of nutrients.

In conclusion, we could see the manifestation of two productive systems, one young in January with new produced biomass,
and one mature in March, at its stable phase.

The difference between the two systems observed in 1981 could
also be explained by the model of Hill and Johnson (1974). The
two cells of vertical circulation, coastal and offshore are

separated by a convergence zone situated just above the 200 m
isobath at the shelf break. On this cell, downwelling motion com-
pensate the vertical upward motions. In January 1981, the frontal bloom of the cold plume may correspond to this convergence
zone. The bloom of March 1981 may then correspond to an accumulation of biomass due to the coastal upwelling cell.

CONCLUSION

The CZCS (Coastal Zone Color Scanner) provides new information on the phytoplankton variability during the upwelling period over the Senegalese shelf. The two scenes of the 11th of
January and 1st of March 1981 give two images of very distinct
phases of the upwelling (infrared scenes) and two characteristic
responses of phytoplankton (chlorophyll maps) to these events.
The great spatial resolution of the sensor allows a detailed description of the patchiness of the blooms over the shelf, expe-
cially at the beginning of the upwelling when the blooms are
frontal.

The CZCS is the only tool for biologists to understand the
fugitive phenomena related to highly localized or varying processes. The images of ocean color provide an information which is
greatly dependent of the infrared information but which comple-
ments it.

This work, despite its ponctual conclusions lets appear the
possibility of a classification of the phytoplankton responses
to the highly varying processus of upwelling : frontal at the
shelf break or more homogeneous on the coast, with more amplitude on the northern or southern part of the continental shelf,
the peninsula of the Cap Vert marking the limit between two ecosystems. The CZCS two-dimensional fields may bring information
on the mechanisms of phytoplankton development and be useful in
determining the regions of greater sensitivity of algae populations to the combination of physical controlling factors.

AKNOWLEDGEMENTS

This study would not have been possible without the data base
from the Centre de Recherche Océanographique de Dakar Thiaroye.
We thank Dr J. Pagès for his helpful comments. We aknowledge the
advice and help of M. Viollier for atmospheric corrections. We
wish also to thank IFREMER for the data processing facilities,

the Antenne ORSTOM and the Centre de Météorologie Spatiale for
scientific support, photographic work and for assistance in the
preparation of the manuscript.

ANNEXE I

CALIBRATION OF CZCS DATA

The retrieval of the pigment (chlorophyll + pheopigments) concentrations from CZCS data has been a subject of intensive studies using sea truths measurements of sea reflectances and chlorophyll, added to modelling of the atmosphere layers (Gordon
et al., 1980 ; Viollier et al., 1980). Even in the worst case,
the chlorophyll concentration can be estimated to within a factor of 2 (Gordon et al., 1980). A better accuracy was found in
the English Channel (log C $\pm$ 0.19) by Holligan et al. (1983).

The main unknown in the processing is the aerosol contribution
to the total observed radiance in CZCS channels.

The aerosol content of the atmosphere is extremely variable
in space and time, and the scattering of the aerosols is not pre-
dictable a priori.

At the latitude of Senegal, the main source of aerosols added
to normal haze is the sand wind that comes from the land during
the trade periods. When checking series of Meteosat images in
the visible channel over this area from year to year, one can
see that these dust hazes occur regularly from December to· March
(specially during the 1980's).

The optical properties of these particles have been measured
along the African coast during cruises between Dakar and Abidjan
by Cerf (1982). The values of the Angström exponent (expressing
the spectral dependence of the aerosol thickness) are found very
low (even n = 0), compared to those known at midlatitude countries. This indicates a natural origin of the aerosols and specific properties of the mixing of maritime and continental tropical airs.

Despite these uncertainties, we have determined quantitatively the "chlorophyll" concentrations. The parameters used for the
processing of the 2 CZCS scenes were :
   - Angström exponent (aerosol turbidity) : -0.5 ;
   - coefficient of proportionality between channels 3 and 4
     V = 1.5 ;

- coefficients c, d of equation (2) for retrieval of chlorophyll concentrations (Gordon et al., 1980) c = 0,3 ;
  d = -3.73. The ratio of channels 2 and 3 (520 and 550 nm)
  has been used because the scene is in a coastal region,
  where the chl a concentrations are high.

The result of the total processing is shown below in table 1,
where are expressed sea reflectances in the three CZCS channels
for different pixels of the image, representative of the different water masses.

Latitude Longitude pixel	CZCS reflectances (in %) at 440, 520, 550 nm	Chlorophyll mg.m^{-3}
11 January 81 (orbit 11198)		
1 17.40 W, 15 N	0.7 1.1 1.2	5
2 16.40 W, 15.4 N	1.8 3.4 3.8	5
3 19 W, 14 N	2.1 1.5 1.4	2.5
1 March 81 (orbit 11875)		
1 17.10 W, 14.1 N	0.6 1.3 1.4	8
2 16.40 W, 15.4 N	1.5 3.4 3.8	5.4
3 18.50 W, 14.5 N	3.8 2.8 2.3	1.7

Table 1 : Reflectances (in %) for CZCS images of 11 January 81
and 1 March 81 and derived chl a values calculated
from equation (2).

From these reflectance values, two types of waters can be
easily distinguished, the turbid waters with a mixed influence
of resuspended matter and living phytoplanktonic cells (pixels
2), and the pure case 1 waters with the alone influence of phytoplankton biomass (pixels 1, high C ; pixels 3, low C). Clear
water pixels were found offshore out of the scenes since the
growth of phytoplankton occurs within the 200 miles band along
the Senegalese coast.

The CZCS reflectance values are comparable with the measurements made during the GUIDOM - CINECA cruises in the Mauritanian region (Morel, 1982).

REFERENCES

Abbott, M.R. and Zion, M.F., 1985. Satellite observations of phy-
    toplankton variability during an upwelling event. Contin.
    Shelf Res. 4 : 661-680.
Barber, R.T., Dugdale, R.C., Mac Isaac, J.J., and Smith, R.L.,
    1971. Variations in phytoplankton growth associated with the
    source and conditioning of upwelling water. Invest. Pesq. 35,
    171-193.
Berrit, G.R., 1962. Contribution à la connaissance des varia-
    tions saisonnières dans le Golfe de Guinée. Cah. Océanogr.
    14 : 633-643.
Citeau, J., Guillot, B. et Laé, R., 1984. Opération LISTAO :
    Reconnaissance de l'environnement inter-tropical à l'aide des
    satellites Météosat et Goes-E. Télédétection numéro 10.
The Canary Current, 1982. Studies of an upwelling system. Rap-
    port et procès verbaux des réunions du Conseil International
    pour l'Exploration de la mer. Symposium held in Las Palmas,
    April 1978, volume 10.
Cerf, A., 1980. Atmospheric turbidity over West-Africa. Contribu-
    tions to Atmospheric physics, 53 : 414-429.
Coste, B. and Minas, H.J., 1982. Analyse des facteurs régissant
    la distribution des sels nutritifs dans la zone de remontée
    d'eau des côtes mauritaniennes. Oceanol. Acta, 5 : 315-324.
Crépon, M., Richez, C. and Chartier, M., 1984. Effects of coast-
    line geometry on upwellings. J. Phys. Oceanogr. 14 :
    1365-1382.
Dia, A., 1984. Observations océanographiques effectuées en 1982.
    Doc. Arch. 126 CRODT-ISRA.
Domain, F., 1979. Etude des températures de la mer au voisinage
    de Mauritanie et du Sénégal. Télédétection 3, 42p.
Domain, F., 1982. Répartition de la matière organique de la cou-
    verture sédimentaire du plateau continental ouest-africain.
    Rapp. p-v. Réun. Cons. int. Explor. Mer., 180 : 339-341.
Feldman, G.C., 1985. Variability of the productive habitat in
    the Eastern Equatorial Pacific. Proceedings of Symposium on
    Vertical motion in the Equatorial upper ocean and its effects
    upon living resources and the atmosphere, Paris, May 1985.
Gallardo, Y., 1981. On two marine ecosystems of Senegal separa-
    ted by a peninsula, in : J.C.J. Nihoul (Editor), Ecohydrodyna-
    mics, Elsevier, Amsterdam, pp 141-154.
Gordon, H.R., Clark, D.K., Mueller, J.L. and Hovis, Z.A., 1980.
    Phytoplankton pigments from Nimbus-7 Coastal Zone Color Scan-
    ner : comparisons with surface measurements. Science, 210 :
    63-66.
Gordon, H.R. and Clark, D.K., 1981. Clear water radiances for at-
    mospheric correction of Coastal Zone Color Scanner imagery.
    Appl. Opt. 20 : 4175-4180.
Hagen, E., 1974. A simple scheme of the development of cold wa-
    ter upwelling circulation cell along the Northwest African
    coast. Beiträge zur Meereskunde 33 : 115-125.
Hill, R.B. and Jonhson, J.A., 1974. A theory of upwelling over
    the shelf break. J. Phys. Oceanogr. 4 : 19-26.
Hojerslev, N.K., 1981. The colour of the sea and its relation to
    surface chlorophyll and depth of the euphotic zone. Eurasep
    Newsletter n°2.
Holligan, P., Viollier, M., Dupouy, C. and Aiken, J., 1983. Sa-
    tellite studies on the distribution of chlorophyll and dino-
    flagellate blooms in the eastern English channel. Contin.
    Shelf Res. 2 : 81-96.

Hovis, W.A., Clark, D.K., Anderson, F., Austin, R.W., Wilson, W.H., Baker, E.T., Ball, D., Gordon, H.R., Mueller, J.L., El-Sayed, S.Z., Stürm, B., Wrigley, R.C. and Yentsch, C.S., 1980. Nimbus-7 Coastal Zone Color Scanner : system description and initial imagery. Science, 210 : 60-63.

Kirk, A. and Speth, P., 1985. Wind conditions along the coasts of Northwest Africa and Portugal during 1972-79. T. O. A. N. 30 : 15-16.

Laurs, R.M., Fielder, P.C. and Montgomery, D.R., 1984. Albacore tuna catch distributions relative to environmental features observed from satellites. Deep Sea Res. 31 : 1085-1099.

Mittelstaedt, A. and Prieur, L., 1980. Analysis of variations in ocean color. Limnol. Oceanogr. 22 : 709-722.

Morel, A., 1982. Optical properties and radiant energy in the waters of the Guinea Dome and the Mauritanian upwelling area in relation to primary production. Rapp. P. -v. Réun. Cons. Explor. Mer., 180 : 94-107.

Portolano, P., 1981. Contribution à l'étude de l'hydroclimat des côtes sénégalaises. Doc. Sci. CRODT-ORSTOM.

Rebert, J.P., 1983. Hydrologie et dynamique des eaux du plateau continental sénégalais. Doc. Sci. 89. CRODT-ISRA.

Rossignol, M., 1973. Contribution à l'étude du "Complexe Guinéen" Centre ORSTOM de Cayenne-Océanogr. Doc. Arch. 17, 143p.

Schemainda, R., Nehring, D. and Schulz, S., 1975. Ozeanologische Untersuchungen zum Produktionspotential der nordwestafrikanischen Wasserauftriebsregion 1970-2973. Geod. Geoph. Veroff. 4, 16, 85 p.

Shannon, L.V., Mostert, S.A., Walters, N.M. and Anderson, F.P., 1983. Chlorophyll concentrations in the southern Benguela current region as determined by satellite (Nimbus-7 Coastal Zone Color Scanner). J. Plankton Res. 5 : 565-583.

Shannon, L.V., Schlittenhardt, P. and Mostert, S.A., 1984. The Nimbus-7 experiment in the Benguela Current Region off Southern Africa, February 1980. 2. Interpretation of imagery and oceanographic implications. J. Geophys. Res. 89 : 4968-4976.

Smith, R.C. and Baker, K., 1978. The Bio-optical state of ocean waters and remote sensing. Limnol. Oceanogr. 23 : 247-259.

Stürm, B., 1983. Selected topics of Coastal Zone Color Scanner (CZCS) data evaluation. in : A.P. Cracknell (editor), Remote Sensing Applications in Marine Science and Technology, pp 137-167.

Teisson, C., 1983. Le phénomène d'upwelling le long des côtes du Sénégal. Caractéristiques physiques et modélisation. Doc. Arch. 123 CRODT-ISRA.

Touré, D., 1982. Observations océanographiques effectuées en 1981. Doc. Arch. 125 CRODT-ISRA.

Touré, D., 1983. Contribution à l'étude de l'upwelling de la Baie de Gorée et de ses conséquences sur le développement de la biomasse phytoplanctonique. Thèse de 3ème cycle. Paris VI, 151 p.

Viollier, M., Tanré, D. and Deschamps, P.Y., 1980. An algorithm for remote sensing of water color from space. Boundary Layer Meteorol. 16 : 247-267.

Viollier, M., 1982. Radiometric calibration of Coastal Zone Color Scanner on Nimbus-7 : a proposed adjustement. Appl. Opt., 21 : 1142-1145.

Viollier, M. and Stürm, B., 1984. CZCS data analysis in turbid coastal water. J. of Geoph. Res. 89, 4977-4985.

Williams, S.P., Szajna, E.F. and Hovis, W.A., 1985. Nimbus-7
    Coastal Zone Color Scanner (CZCS). Level 2 data product
    user's guide. NASA Technical Memorandum 86202.
Zbinden, R., 1981. Les suspensions de la Baie du Mont Saint-
    Michel ; étude microgranulométrique et radiométrique. Thèse
    de 3ème cycle. Paris VI, 302p.

Abbreviations

ASECNA : Agence pour la Sécurité de la Navigation Aérienne.

CRODT   : Centre de Recherche Océanographique de Dakar-Thiaroye.

CZCS   : Coastal Zone Color Scanner.

IFREMER : Institut Français de Recherche pour l'Exploitation de
          la Mer.

NASA   : National Aeronautics and Space Administration.

ORSTOM : Institut Français de Recherche Scientifique pour le
          Développement en Coopération.

MANGROVE ECOSYSTEM STUDY OF CHAKORIA SUNDERBANS AT CHITTAGONG WITH SPECIAL
EMPHASIS ON SHRIMP PONDS BY REMOTE SENSING TECHNIQUES

O. QUADER, M.A.H. PRAMANIK, F.A. KHAN and F.C. POLCYN
Bangladesh Space Research and Remote Sensing Organization (SPARRSO), Dhaka
(Bangladesh)

ABSTRACT

Mangrove ecosystem of Chakoria Sunderbans have been greatly influenced by
man's activities. In order to allow for a suitable habitat for shrimp culture
much of this ecosystem has been altered. The objective of this study was to
identify shrimp-ponds and monitor temporal changes and their impact on the
Chakoria Sunderbans Mangrove Ecosystem, using aerial photography and Landsat
MSS digital data.

The low altitude aerial photographs which were taken in 1975, 1981 and 1983
have been used for mapping in this study. These maps were used to identify the
shrimp-ponds and to assess the spatial coverage and temporal change in
mangrove ecosystems. Several additional maps showing shrimp-ponds allotted by
the Government to the public and also showing the locations of rivers, coastline and other natural features were also used. Landsat CCTs were collected
for two dates (March 7, 1976 and December 3, 1980). These tapes were analysed
by using the IBM/360 computer and a software program known as LARSYS, which
was developed by the Laboratory for Application of Remote Sensing (LARS),
Purdue University, USA. The Multitemporal Landsat MSS data were very useful in
identifying changes in the Mangrove Ecosystem of Chakoria Sunderbans. But the
boundaries of the shrimp-ponds are not detectable in MSS data. Because of the
improve ground resolution, the use of Thematic Mapper (TM) data for monitoring
coastal mangrove ecosystem offers many advantages over MSS data.

INTRODUCTION

The Bangladesh part of the Sunderbans ranks among the largest mangroves in
the world with a total area of 590,000 ha, of which about 410,000 ha are land
area. Minor areas of mangrove forest occur near Chittagong. Furthermore
mangrove species are planted extensively as coastal afforestation to protect
embankments and new accretions. In the Chittagong district, the mangrove ecosystem has been greatly influenced by man's activities. In order to allow for
a suitable habitat for shrimp culture much of this ecosystem has been altered.
The areas occupied by mangrove vegetation have been recently cut.

The objective of this study was to monitor temporal changes in the Chakoria
Sunderbans mangrove ecosystem of Chittagong district using Landsat MSS (Multispectral Scanner) data collected at orbital altitudes and aerial photographs
taken in 1975, 1981 and 1983.

Study Area

The mangrove forest of Chakoria Sunderbans has demarcated boundaries sur-
rounded by waterways on all sides. Chakoria Sunderbans is a delta at the mouth
of the river Matamuhuri and is a low lying saline swamp. The soil within the
mangrove ecosystem consists of rich alluvium, sand, gravel and other materials,
deposited by running water. In the forests there are many low lying islands
which are submerged at high tides and they are intersected by a system of
canals and creeks. The forest has been divided into two blocks which have
been further subdivided into 21 compartments. In the early 50's these forests
were stocked with 20 types of mangrove genus like *Heriteria Avicinia Rohinias,
Aegiceras, Braguiera, Carpa, Coriops, Soneretia, Exocaria, Phoenix, Tamarine,
Rhizophora,* etc ... Some of the species attained a height of 40 to 50 feet.
Recently, the condition of the forests has been deteriorating and the weeds
like Cauba kanta and Nunia kanta form dense canopies and occupy the areas once
dominated by timber species.

Various species of marine shrimp spawn in the sea waters up to a depth of
100 meters but their larval stages are planktonic and require less saline
waters for survival and further development. The young shrimp move in the
coastal rivers, canals, creeks and mangrove swamps with tidal currents. The
young shrimp remain in these areas up to pre-adult stages of growth. As the
salinity tolerance of the young shrimp increases, they migrate back to the
sea for further growth, development of gonads and spawning. The mangrove forest
of Chakoria Sunderbans provide ideal ecological conditions for early stages of
marine shrimp and thus constitute their nursery ground where they can be
trapped and cultured in impoundments. The species of commercial shrimp which
are available in abundance in their early stages of growth in the canals and
creeks of Chakoria Sunderbans which are cultured on a large scale are :
giant tiger shrimp (Bagda Chingri) and white shrimp (Sada Chingri).

METHODOLOGY
Data Collection

Several types of data have been collected for studying the impact of shrimp
culturing on the Chakoria Sunderbans Mangrove Ecosystem. These include : low
altitude aerial photography, thematic map data (maps displaying the location
of mangrove vegetation) and Landsat MSS digital data.

The low altitude aerial photography was collected in January 1975, December
1981 and December 1983. These black and white photographs and color infrared
photography were collected at a scale of 1:30,000 and 1:50,000 respectively
over the whole country and the coastal areas of Bangladesh. These photographs
were taken to provide insights into the location of natural resources and

distribution of mangrove vegetation in the region. These maps were traced from
aerial photography to help depict the location and distribution of shrimp beds
and magrove ecosystems. The tracing method consisted of enlarging the scale of
the photographs to that of the base maps and using a sheet of mylar to trace
information from the photos to the map surfaces by mapograph. These maps correspond to the dates of the aerial photography and were used to assess the
spatial coverage and temporal change in mangrove ecosystems. Several additional maps (predrawn) were obtained from the Bangladesh Fisheries Directorate to
show the location of shrimp farms allotted to the private entrepreneurs and
location of rivers, coastlines and other natural features.

Landsat Data

Landsat MSS data were collected for two dates (March 2, 1976 and December 3,
1980). The multitemporal Landsat MSS data were very useful in identifying
changes in the mangrove ecosystem of Chakoria Sunderbans. These data were also
effective in categorising or classifying landcover features other than
mangrove vegetation. The multispectral characteristic of the data allowed the
analyst to make detailed decisions regarding the type of landcover found in
the area and the temporal variation in landcover features. The MSS data used
have four separate spectral bands. These bands occupy selective portions of
the electromagnetic spectrum (band 4, 5 to 6 micrometers; band 5, 6 to 7
micrometers; band 6, 7 to 8 micrometers and band 7, 8 to 11 micrometers).

Data Analysis

Before information could be extracted from Landsat MSS data, a number of
computer based algorithms or processors were implemented to process the data.
These digital data processing programs are part of an integrated system of
software known as LARSYS. They include : CDISPLAY, PICTUREPRINT, CLUSTER,
MERGESTATISTICS, SEPARABILITY, CLASSIFYPOINTS, and PRINT RESULTS and are used
to display the data and implement various pattern recognition techniques in
order to extract information for training class purposes and final classification of landcover features.

RESULTS AND DISCUSSIONS

Mangrove ecosystems represent an important part of the world's coastal
ecosystems. They provide important functions in protecting coasts from
extensive erosion, extending and building islands and providing energy input
into fisheries (Odum, 1971).

Prior to 1977 the Chakoria Sunderbans ecosystem remained primarily unaltered
by man's activities. However, recently (after 1977) considerable change has

occurred. These changes are mainly in the form of destruction of mangrove forests for shrimp culturing.

Interpretation of aerial photographs : the three maps made from aerial photographs were used to help determine temporal changes in mangrove ecosystem (Fig. 1,2,3). The first map made from 1975 aerial photographs depicts considerable forest cover encompassing nearly all of Chakoria Sunderbans (Fig. 1). There were no shrimp farms found within the mangrove forest at this date. The second map produced (1981) showed that considerable change had occured in the mangrove forest (Fig. 2). Approximately 42 shrimp ponds were found within the confines of the mangrove forest and portions of the forest had been destroyed. Evaluation of the third map (1983) depicts even greater deforestation of Chakoria Sunderbans than evident in 1981 (Fig. 3). Additional shrimp ponds were also constructed since the 1981 aerial photography was imaged. These maps categorized the shrimp ponds into 4 different classes, Symbol A are shrimp ponds where shrimp culture is presently occuring, Symbol B is where shrimp ponds will be prepared, Symbol C indicates locations where mangrove forest has been destroyed for shrimp pond activities and D represents areas in which mangrove forest still exist in its natural state. It is evident from the maps and table that shrimp ponds are increasing in abundance both in and outside of Chakoria Sunderbans. Interpretation of aerial photography was based upon tonal variations and spatial relationships. Stereocopic viewing was also helpful in interpretation of spatial and temporal changes within the ecosystems.

Classification of Landsat data : classification of 1976 Landsat data revealed extensive coverage of mangrove forest in the Chakoria Sunderbans tidal estuary. Other types of vegetation and natural features were also present. However, the main interest of mangrove vegetation was dominant. Four distinct spectral classes of mangrove were classified in the study area using the March 2, 1976 Landsat data set. These spectral classes are probably related to the density of the mangrove forest in the area, however, it is possible that the differences observed are due to variations in species composition. The location of canals and open waterways was also detectable. No shrimp ponds however could be identified. Classification of 1980 Landsat MSS data showed that considerable mangrove deforestation had occurred throughout along major canals which divide the areas occupied by mangrove forest. The locations of canals and streams were difficult to identify on this classification. This was due to destruction of forest near waterways in the area. Large differences spectrally between the canals and mangrove vegetation allowed waterways to be located easily on classification maps prepared from 1976 Landsat MSS data.

649

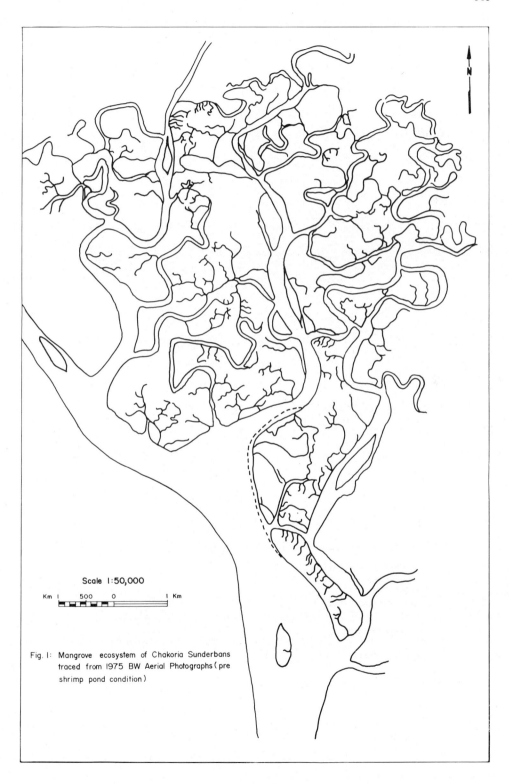

Scale 1:50,000

Km 1   500   0   1 Km

Fig. 1: Mangrove ecosystem of Chakoria Sunderbans
        traced from 1975 BW Aerial Photographs (pre
        shrimp pond condition)

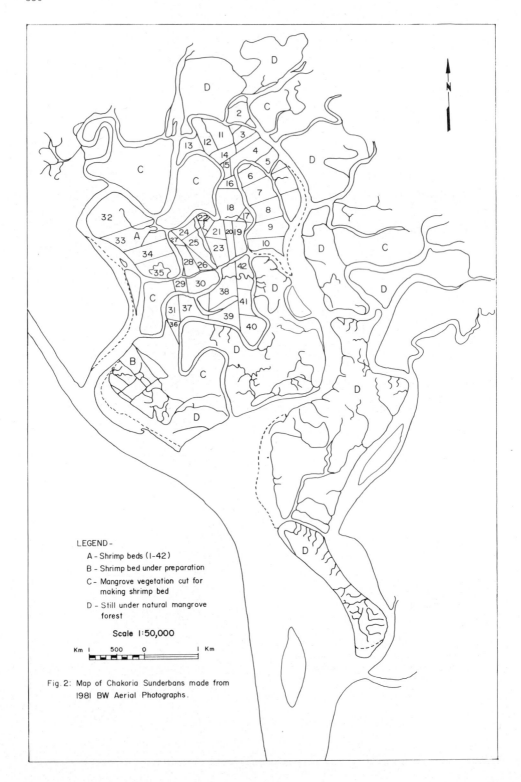

LEGEND-

A - Shrimp beds (1-42)

B - Shrimp bed under preparation

C - Mangrove vegetation cut for making shrimp bed

D - Still under natural mangrove forest

Scale 1:50,000

Km 1   500   0   1 Km

Fig. 2: Map of Chakoria Sunderbans made from 1981 BW Aerial Photographs.

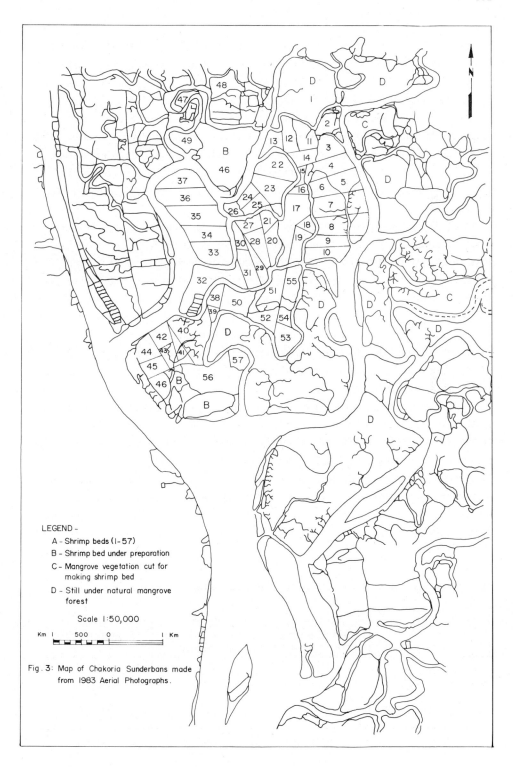

LEGEND -

   A - Shrimp beds (1-57)

   B - Shrimp bed under preparation

   C - Mangrove vegetation cut for
       making shrimp bed

   D - Still under natural mangrove
       forest

       Scale 1:50,000

Km 1   500   0   1 Km

Fig. 3: Map of Chakoria Sunderbans made
       from 1983 Aerial Photographs.

CONCLUSIONS

The result of mangrove forest destruction in this area could have serious impacts such as : (i) decreases in shrimp and fish population, (ii) increased discharge of nutrients in a less decomposed form into open ocean water and (iii) the destruction of nursery ground for marine fauna.

These problems may be overcome by planting mangroves on the embankments and near the shore of the embankments of shrimps farms. This will ensure retention of sufficient mangrove vegetation needed for production of shrimps on a sustained basis, at the same time the trees will act as shelter belt against cyclones and tidal surges that frequently occur in these areas during monsoons.

Remotely sensed data collected by orbital platforms offer a data base in which measurements of spatial and temporal changes in delicate marine ecosystems can be made. It is hoped that the changes occurring within the Chakoria Sunderbans mangrove ecosystems can be effectively monitored in the future using these and more advanced data sources. Analysis of Landsat MSS data has indicated that deforestation of Chakoria Sunderbans is occurring at a rapid rate. The implications of this occurrence are not at this time clear, however additional analysis of Landsat MSS data and other data sources such as SPOT and TM, may provide information on the effects of these activities on the plant and animal communities of the area.

The use of Thematic Mapper (TM) data for monitoring coastal mangrove ecosystems offers many potential advantages over MSS data. The major advantage of TM data is their ability to discriminate very small differences in vegetative cover. If TM data had been used in this analysis it may have been possible to discriminate smaller differences in vegetation present within mangrove ecosystems. Small changes in the Chakoria Sunderbans ecosystems may have been detected. Discrimination of individual shrimp ponds may be also possible using Thematic Mapper data.

TABLE

Summary of the areas of each category for each date in hectare

	1975 (black & white) aerial photographs	1981 (black & white) aerial photo	1983 (IR aerial photo)
Mangrove Forest	2882.86	1780.15	1788.73
Mangrove Forest cut	-	364.63	158.74
Shrimp bed under preparation	-	90.21	113.16
Shrimp ponds	-	674.25	980.20
Total	2882.86	2909.24	3040.83

ACKNOWLEDGEMENTS

Deepest appreciation is made to Dr. Louis Bartlouci, Technical Director, LARS, Purdue University, USA, Mr. Carlos Valenzuela for advising and assisting the work, Mr. James Clinthorne for assistance in analysing and preparing the report materials, Mrs. Maryleen Kleeper for typing the report. Mr. Abul Hossain of SPARRSO typed the manuscript. We gratefully acknowledge the support of FAO and SPARRSO to carry out the research work in the USA.

REFERENCES

ADB, Fisheries, 1980, Utilization of Chakoria Sunderbans Mangrove Forestry for Aquaculture. A feasibility study report of Fisheries Directorate. Bangladesh, Appendix p 5-2, p 1-3

FAO, 1982, Management and Utilisation of Mangrove in Asia and the Pacific FAO Environment paper 3. M-08 ISBN 92-5-101221-0, p 6, 82

KLEMAS, V. and BARTLETT, D.S., 1980, Remote Sensing of Coastal Environments and Resources, Proceedings of the Fourteenth International Symposium on Remote Sensing of Environment, Ann Arbor, Michigan, p 543-562

ODUM, E.P., 1971, Fundamentals of Ecology, W.B. Sanders Company, Philadelphia

SWAIN, P.H. and DAVIS, S.M., editors, 1978, Remote Sensing, The Quantitative Approach, McGraw-Hill, Inc. New York

BIOLOGICAL PROCESSES ASSOCIATED WITH THE PYCNOCLINE AND SURFACE FRONTS IN THE
SOUTHEASTERN BERING SEA

TERRY E. WHITLEDGE[1] and JOHN J. WALSH[2]

[1]Oceanographic Sciences Division, Oceanographic Sciences Division, Brookhaven
National Laboratory, Upton, NY 11973

[2]Department of Marine Science, University of South Florida, St. Petersburg, FL
33701

INTRODUCTION

The factors such as light and nutrient concentration conducive to primary
production processes in the ocean environment are well known in the qualitative
sense but only an approximation of these essential ingredients can be quantified
especially when the absolute amounts are often not very important. The availability of nutrients and light are subject to a wide range of environmental
factors that vary widely in both space and time. It is useful to study an
oceanic area that has high primary production and a small amount of relative
water movement and mixing processes that render biological production measurements difficult. It is also useful to have well defined regions such as
surface fronts or pycnoclines which are areas of increased production and can be
used to focus the observations program.

The Processes and Resources of the Bering Sea (PROBES) program on the southeastern Bering Sea shelf studied the wide (~600 km) and flat shelf which
increases the distances between the active sites of shelf processes to better
enable observations to be made (McRoy et al., 1985). In addition, this portion
of the southeastern Bering Sea shelf has very small net current velocities (<1
cm sec^{-1}) (Kinder and Schmacher, 1981) and internal waves are not present.

This paper will show the persistence of surface fronts and strength of the
pycnocline that is coupled to the primary production and the phytoplankton
biomass.

Physical setting

The southeast Bering Sea was sampled for five years in the region between the
Pribilof Island and the Aleutian chain. A standard across-shelf transect of 19
stations was occupied nearly fifty times between the 1500 m and the 40 m
isobaths (Fig. 1). This standard transect crossed three frontal regions that

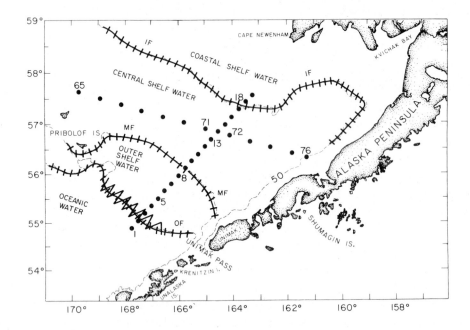

Fig. 1. PROBES sample locations across the southeastern Bering Sea shelf. Approximate locations for the outer front (OF), middle front (MF) and inner front (IF) are shown as hatched lines. The track of the shipboard surface map is shown as zig-zag along the outer front.

occurred on the outer shelf (170 m), middle shelf (100 m), and the inner shelf (50 m). The mean current velocities are estimated to be 5 to 10 (slope), 1 to 5 (outer shelf), <1 (middle shelf), and 1 to 3 (inner shelf) cm sec^{-1} so the central part of the shelf between the 50 and 100 m isobaths exhibit very little net movement and the predominant flow is tidal (Coachman, 1985).

The frontal regions

   The frontal regions can be clearly defined by both physical and biological measurements. The most extensive observation that have been collected which show the lateral extent of a front was obtained by surface underway mapping at the outer front. The zig-zag track of the ship at night (Kelly et al., 1975) was oriented along the frontal region (Fig. 1). The areal distribution of nitrate (Fig. 2) showed a higher concentration at the front but there was a substantial amount of structure in the observed features. The nitrate and chlorophyll (Fig. 3) were lower in the southeast where the gradients of both were also smaller. The non-uniform distributions of nitrate and chlorophyll varied on a spatial scale of about 10 miles across the front. In the northwest

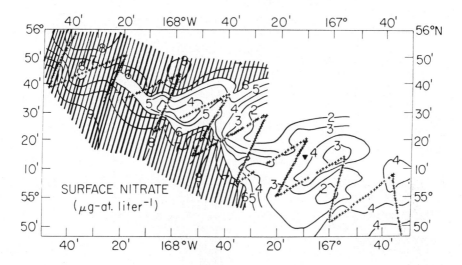

Fig. 2. The distribution of surface nitrate (ug-at 1^{-1}) along the outer front as determined by shipboard underway surface mapping at night. The highlighted area has concentrations greater than 6 ug-at 1^{-1}.

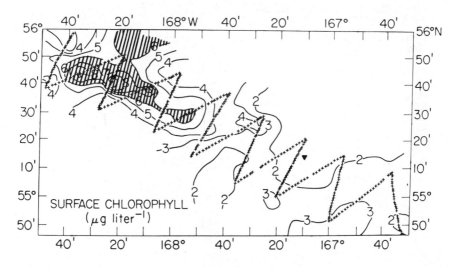

Fig. 3. The distribution of surface chlorophyll (ug 1^{-1}) measured by in vivo fluorescence along the outer front as determined by shipboard underway surface mapping at night. The highlighted area has concentrations greater than 6 ug 1^{-1}.

section of the mapped area the concentration of nitrate increased twofold on either side of the front which coincided with a twofold decrease of chlorophyll. The scales of both nitrate and chlorophyll features corresponded closely

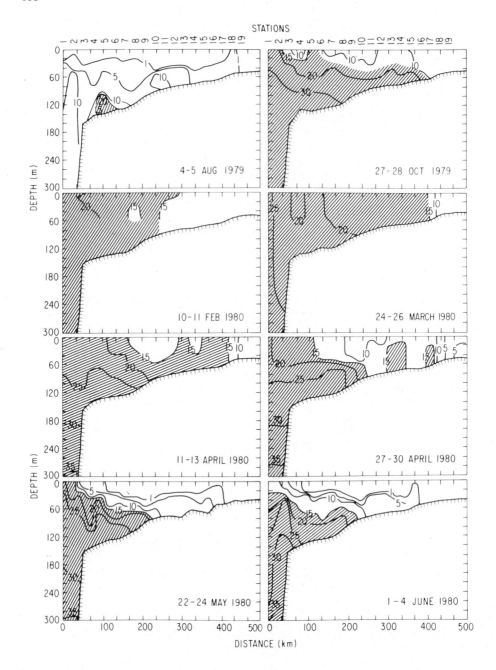

Fig. 4. The across shelf vertical distribution of nitrate (ug-at l^{-1}) for autumn through spring. The highlighted areas have concentration greater than 15 ug-at l^{-1}.

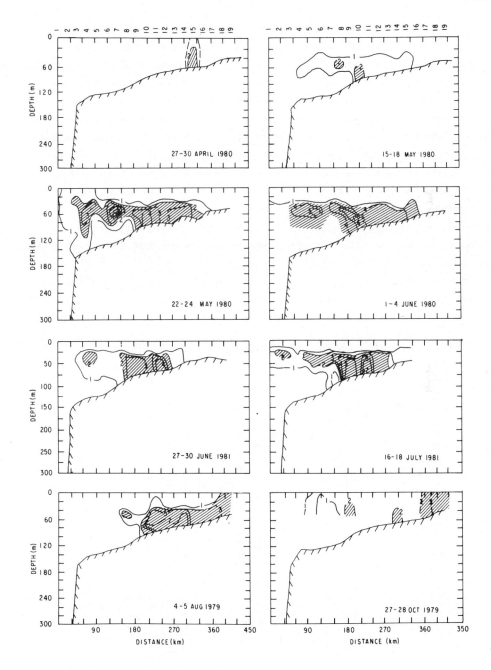

Fig. 5. The across shelf vertical distribution of ammonium (ug-at l^{-1}) for spring through autumn. The highlighted areas have concentrations greater than 2 ug-at l^{-1}.

with only a slight displacement of chlorophyll maximum to the northwest from the nitrate minimum. The seasonal changes in the across shelf distribution of nitrate (Fig. 4) are apparent, however, the frontal features always present with the exception of winter. As the nitrate declines in concentration the same relative distribution patterns persist from spring through autumn.

The ammonium distributions across the shelf has an interesting pattern only during late spring and summer when degradation processes have started recycling the phytoplankton from the spring bloom. The major region of ammonium production occurs in the middle domain between the inner and middle fronts (Fig. 5). As the ammonium concentrations increase in the bottom layer during the spring, the boundaries between inner and middle fronts are clearly defined as an extension of the subpycnocline water to the outer domain. This mid-depth offshore flow is probably the balancing effect of the onshore movement of bottom water on the outer 200 km of the transect.

Primary production processes require light which can only be obtained in the upper layer of the water column so an analysis on the upper 40 m of the water column approximates the conditions that exist March through June during the

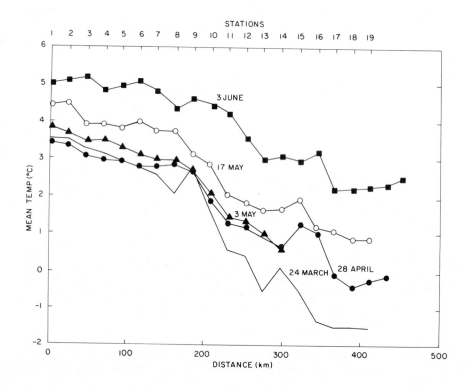

Fig. 6. The mean temperature (°C) in the upper 40 m of the water column on the PROBES across shelf transect of stations from 24 March - 3 June 1980.

maximum primary production. The mean temperature in the upper 40 m of the water column across the shelf (Fig. 6) has the highest gradient across the middle front (Sta 9-13) in March, but in May the inner front (Sta 16-17) and middle front (Sta 8-11) are distinctly apparent in the transect stations. The relative across-shelf distribution of mean temperature remains unchanged through the spring and summer time periods.

The concentration of nitrate integrated for the upper 40 m of the water column begins the spring period with a uniform gradient of decreasing nitrate over the offshore end of the transect of approx. 150 km (Sta 1-8) and a very constant level of nitrate for the next 250 km (Sta 8-17) with a very sharp decline at the inshore end of the transect (Sta 17-19) (Fig. 7). The middle portion of the transect (Sta 6-12) centered over the middle front region has an initial rapid decline in nitrate concentration during March and April. The inner front exhibits a similar decline in nitrate which gives the transect a bimodal distribution of nitrate by the beginning of May. Both areas of low nitrate are associated with frontal areas (middle and inner) where the horizontal temperature temperature gradients are largest. The chlorophyll value shows

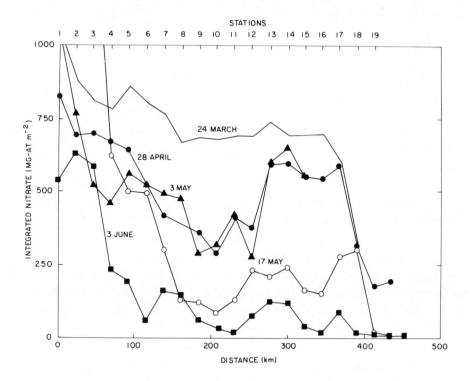

Fig. 7. The integrated nitrate concentrations (mg-at m-2) in the upper 40 m of the water column on the PROBES across shelf transect of stations from 24 March - 3 June 1980.

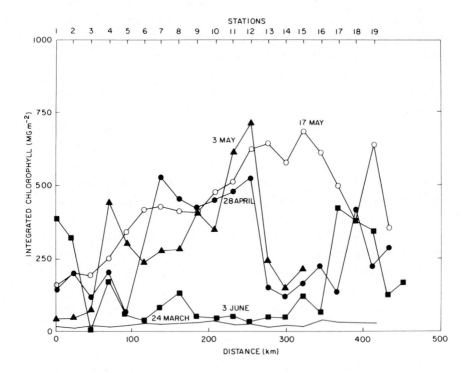

Fig. 8. The integrated extracted chlorophyll concentrations (mg m^{-2}) in the upper 40 m of the water column on the PROBES across shelf transect of stations from 24 March - 3 June 1980.

the expected increase in both frontal regions at this time (Fig. 8).

During May, the middle shelf region (Sta 12-16) between the middle and inner fronts reaches the necessary stability of the water column to undergo an extremely rapid bloom period (Whitledge et al., 1985) which depleted nitrate across the middle and inner shelf to limiting values and produced chlorophyll concentrations as high as 700 mg m^{-3} in the 40 m upper water column. The oceanic and outer shelf locations (Sta 1-6) maintained an extremely large nitrate gradient and no phytoplankton bloom was observed by July. The small temperature increases in the upper water column was only 1.5°C on the offshore end of the transect while a 4.0°C temperature rise occurred near the inner front. The chlorophyll produced at the outer front could have been eaten by zooplankton populations as it was produced and thereby regulated the phytoplankton production rates (Smith et al., 1985; Dagg et al., 1982).

The ammonium concentrations (Fig. 9) are primarily produced in the bottom layer below 40 m by processes consuming phytoplankton production that has settled from the upper layer. The largest ammonium concentrations occur in the middle shelf region (Sta 10-16) in the near bottom environment by shellfish or

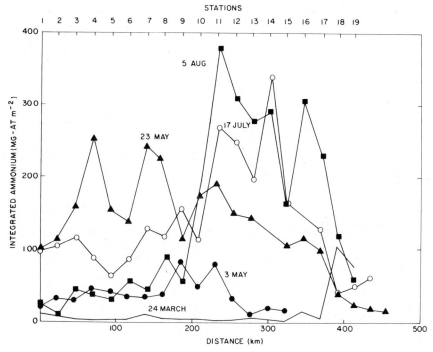

Fig. 9. The integrated ammonium concentrations (mg-at m⁻²) in the uper 100 m of the water or the bottom on the PROBES across shelf transect of stations from 24 March - 5 August 1980.

microbiota while the outer stations (Sta 1-9) have a probable consumption of the phytoplankton by zooplankton throughout the water column (Fig. 10). The ammonium observed on the outer stations at 40-60 m has an origin at the middle shelf and is probably advected offshore in a narrow layer.

The pycnocline

As spring approaches at the end of winter the surface layer warms and the vertical stability of the water column increases (Fig. 6) and enables the phyto- plankton population to secure enought light to initiate the spring bloom (Whitledge et al., 1985). In the shallow shelf region where the depth is 50 m or less, the spring bloom may occur before stratification is initiated because the bottom is acting as the interface instead of the pycnocline. This effect also includes near bottom recycling processes in the mixed layer and would increase the relative contribution of the bottom a great deal over a more typical two-layered system.

As the spring progresses and the pycnocline develops in the middle and outer shelf, the phytoplankton bloom becomes shelfwide until the nitrogen is depleted in the upper layer. At this point in time when nutrients are low, the rate of

664

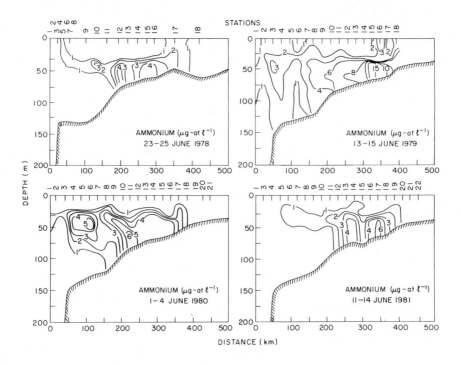

Fig. 10. The distribution of ammonium concentrations (ug-at l⁻¹) across the shelf on the PROBES transect in June of the years 1978-1981.

primary production is dependent on the quantities of nutrients available in the euphotic zone. The bottom layer over the shelf receives a continued flux of nitrate from the shelf edge and the recycled nitrogen regenerated near the bottom. Storm events occur every 5 to 6 days in the Bering Sea and after the storms the upper mixed layer is 5 to 10 m deeper and a net increase of nutrients is observed. Storms of about a 24-hour duration with wind speeds of about 10 m sec⁻¹ or greater are required to have a significant effect. Even early in the spring season while large amounts of nitrate are still present over the shelf, the effects of vertical wind mixing can increase the nitrate concentration by 10% (Fig. 11). One month later, when the nitrate concentrations are very small in the upper 15 m of the water column, a wind mixing event increased the nitrate concentration by 1.5 ug-at l⁻¹ and make the mixed layer 5 m deeper. Coincident increases of 8 and 0.4 ug-at l⁻¹ of silicate and phosphate also occur. A series of storms over the summer could produce 18 mixing events in three months and bring about 700 ug-at l⁻¹ NO_3 into the euphotic zone. The large ammonium production near the bottom of the middle shelf combined with the wind-induced mixing would enhance the vertical flux of recycled nitrogen. The rate of ammonium nitrogen produced was estimated to be a mean of 7.2 mg-at m⁻² d⁻¹ for

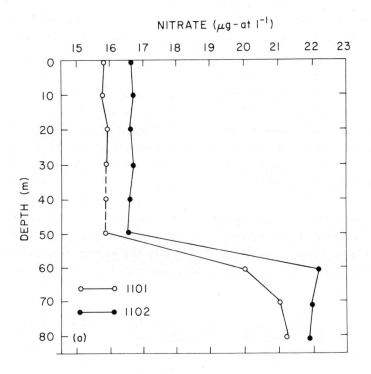

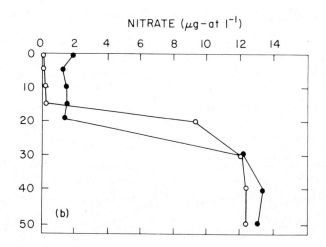

Fig. 11. The concentrations of nitrate (ug-at l⁻¹) in the water column before (o) and after (●) wind mixing on a) 22-24 April and b) 13-16 May 1979.

the years 1979-1981 (Whitledge et al., 1985). Nitrification or eventual mixing/transport must reduce the ambient concentrations of ammonium to undetectable levels in the autumn.

Bering Sea production associated with surface fronts

The question of whether the surface fronts or the pycnocline have a major influence on the primary production in the Bering Sea can be discussed with respect to the relative nitrate losses and chlorophyll production at specified locations. The maximum uptake of nitrate observed during the spring bloom on the middle shelf was a mean of 28.7 mg-at m^{-2} d^{-1} for 1979-1981 (Whitledge et al., 1985).

The nitrate losses between stations 2 and 9, stations 5 and 9, stations 13 and 9, and stations 13 and 18, for given time periods, quantify the difference between the outer, middle, and inner fronts with the domains between the fronts (Table 1). In most instances the nitrate concentrations in the non-frontal

TABLE 1

Maximum and Minimum Integrated Nitrate in Upper 40 m Across the Shelf (mg-at m^{-2}). Maximum values are in bold-face type.

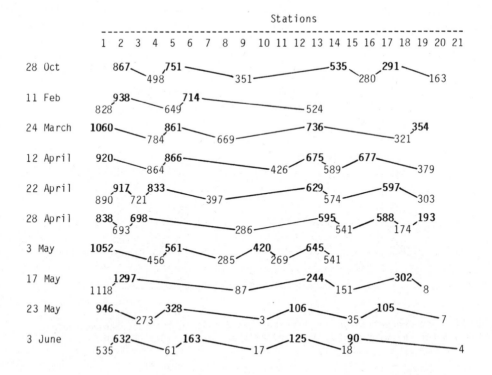

areas did not change as much as the frontal areas.

The locations of maximum and minimum concentrations in the upper layer move 1 or 2 stations positions between consecutive observations during the spring season but these changes probably result from lateral movement in the shelf responses to storm events or other external forcing. Subsequent observations often coincide with the previous distributions. The most notable feature is the pronounced minimum of nitrate between stations 5 and 13 throughout the entire spring. The maximum chlorophyll concentrations (Table 2) coincide with this minimum except near station 15 on 3-17 May. Both the nitrate loss and the chlorophyll appearance were the largest observed. Since they are surface features, outer and inner fronts are apparent in the nitrate and chlorophyll distributions. The middle front which divides the two-layered system from the region with fine scale features between the layers (Coachman et al., 1980) shows most clearly as the outer edge of the large ammonium concentrations in the bottom layer (Table 3). A lag in the production of ammonium in the bottom layer after the intense spring phytoplankton bloom causes the maximum ammonium concentration to appear in June through August.

TABLE 2

Maximum and Minimum Integrated Chlorophyll in Upper 40 m Across the Shelf (mg m^{-2}). Maximum values are in bold-face type.

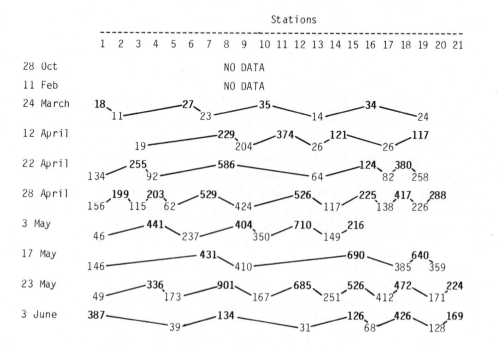

TABLE 3

Maximum and Minimum Integrated Ammonium in Upper 40 m Across the Shelf
(mg-at m^{-2}). Maximum values are in bold-face type.

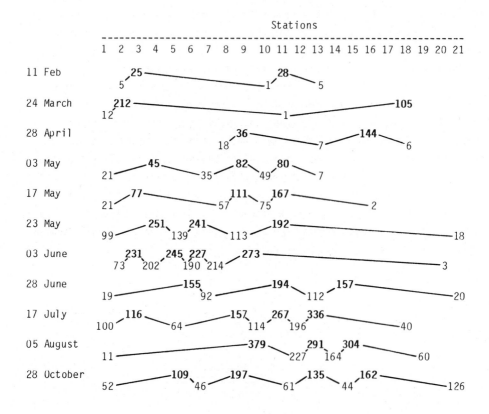

From a comparison of nitrate differences (Table 4), it is apparent that there
is a gradient of nitrate concentration across the shelf in the southeastern
Bering Sea between March and June and a table of differences confirm that the
outer front has a larger nitrate concentration than the middle front (mean
difference = 686 mg-at m^{-2}) and the middle front has a higher nitrate
concentration than the inner front (mean difference = 98 mg-at m^{-2}) (Table 4).
In addition, the central shelf water represented by station 13 has greater
concentrations of nitrate than the middle or inner front even after the large
decrease between 3 and 17 May.

Chlorophyll increases in the upper 40 m of the water column (Table 5) were
typically inversely related to the nitrate concentrations, however, the variance
is even greater than nitrate due to the changes in storm mixing but also by
sinking and/or herbivore grazing processes which change the vertical

TABLE 4

Differences of Maximum and Minimum Nitrate Concentrations Between Frontal and Nonfrontal Locations over a 40 m Surface Layer (mg-at m^{-2}). Values were calculated from maximum and minimum table of integrated nitrate across the shelf.

	Outer front- middle front	Outer shelf- middle front	Central shelf- middle front	Central shelf- inner front
24 Mar	391	192	67	415
12 Apr	494	440	249	296
22 Apr	520	436	232	326
28 Apr	552	412	309	421
03 May	767	276	360	–
17 May	1210	–	157	236
23 May	943	1325	103	99
03 Jun	615	146	108	121

distribution of phytoplankton cells. As the spring progresses, the middle front had more chlorophyll than the central shelf water but this changed markedly during the maximum chlorophyll appearance between 3 and 17 May. By May 17 both the outer shelf and the central shelf had more chlorophyll than the middle front.

DISCUSSION

The intensely sampled southeast Bering Sea shelf during the PROBES project presents a very good set of data that can be used to investigate the significance of frontal regions on nutrient and chlorophyll distributions. A large number of repeated stations over a four-year period allows for a very complete view of the response of nitrate and chlorophyll distributions around the fronts under changing conditions. This analysis was persued to demonstrate that the

TABLE 5

Differences of Maximum and Minimum Chlorophyll Concentrations Between Frontal and Nonfrontal Locations over a 40 m Surface Layer (mg m^{-2}). Values were calculated from maximum and minimum table of integrated chlorophyll across the shelf.

	Outer front- middle front	Outer shelf- middle front	Central shelf- middle front	Central shelf- inner front
24 Mar	17	8	21	20
12 Apr	355	145	348	91
22 Apr	452	494	522	316
28 Apr	327	323	409	300
03 May	664	473	561	–
17 May	264	-21	-280	-50
23 May	852	565	650	221
03 Jun	-253	95	103	395

fronts were always identifiable by either nutrient or chlorophyll concentrations. The mean primary productivity rates have been well delineated using PROBES observations and simple models of phytoplankton growth (Sambrotto et al., 1985; Walsh and McRoy, 1985), however, the physical structure of surface fronts and interaction with the pycnocline is still required before the the observed variability in production can be simulated. Further work on the development of such a simulation model is the next logical step in this work.

ACKNOWLEDGMENTS

The authors would like to thank all of the scientists, technicians, and students associated with the PROBES project. A true joint effort was needed to collect the many field data. This research was supported by the National Science Foundation Grant No. DPP 76-23340 to the University of Alaska and under the auspices of the United States Department of Energy under contract No. DE-AC02-76CH00016 to Brookhaven National Laboratory.

REFERENCES

Coachman, L.K., T.H. Kinder, J.P. Schumacher, and R.B. Tripp. 1980. Frontal systems of the southeastern Bering Sea shelf. In: Stratified Flows, 2nd IAHR Symposium, Trondheim, June 1980, T. Carstens and T. McClimans (eds.), Tapir, Trondheim. pp. 917-933.

Coachman, L.K. 1985. Circulation, water masses, and fluxes and transport on the southeastern Bering Sea shelf. Cont. Shelf Res. 4: (in press).

Dagg, M.J., J. Vidal, T.E. Whitledge, R.L. Iverson, and J.J. Goering. 1982. The feeding, respiration, and excretion of zooplankton in the Bering Sea during a spring bloom. Deep-Sea Res. 29: 45-63.

Kelly, J.C., T.E. Whitledge, and R.C. Dugdale. 1975. Results of sea surface mapping in the Peru Upwelling System. Limnol. Oceanogr. 20: 784-794.

Kinder, T.H. and J.D. Schumacher. 1981. Circulation over the continental shelf of the southeastern Bering Sea. In: The Eastern Bering Sea Shelf: Oceanography and Resources, Vol. 1, D.W. Hood and J.A. Calder (eds.), University of Washington, Seattle, pp. 53-75.

McRoy, C.P., D.W. Hood, L.K. Coachman, J.J. Walsh, and J.J. Goering. 1985. Processes and Resources of the Bering Sea Shelf (PROBES): The development and accomplishments of the project. Cont. Shelf Res. 4: (in press).

Sambrotto, R.N., H.J. Niebauer, J.J. Goering, and R.L. Iverson. 1985. Relationship between vertical mixing, nitrate uptake and phytoplankton growth in the southeast Bering Sea middle shelf. Cont. Shelf Res. 4: (in press).

Smith, S.L. and J. Vidal. 1985. Variations in the distributions, abundance, and development of copepods in the southeastern Bering Sea in 1980 and 1981. Cont. Shelf Res. 4: (in press).

Walsh, J.J. and C.P. McRoy. 1985. Ecosystem analysis in the southeastern Bering Sea. Cont. Shelf Res. 4: (in press).

Whitledge, T.E., W.S. Reeburgh, and J.J. Walsh. 1985. Seasonal inorganic nitrogen distributions and dynamics in the southeastern Bering Sea. Cont. Shelf Res. 4: (in press).

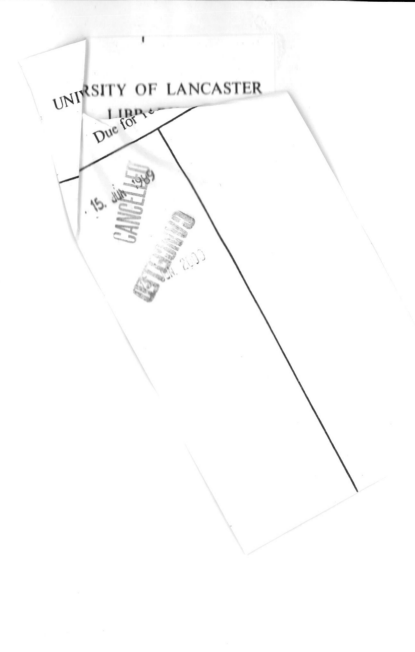